CONTINENTS ADRIFT
AND CONTINENTS AGROUND

Readings from

SCIENTIFIC AMERICAN

CONTINENTS ADRIFT AND CONTINENTS AGROUND

With Introductions by

J. Tuzo Wilson

Ontario Science Centre

W. H. Freeman and Company

San Francisco

Cover illustration: A portion of the San Andreas Fault
a few miles north of Frazier Park, California.
North is toward the top. Photograph is from
"The San Andreas Fault," by Don L. Anderson, pp. 87–102.

Most of the *Scientific American* articles in *Continents Adrift and Continents Aground* are available as separate Offprints. For a complete list of more than 900 articles now available as Offprints, write to W. H. Freeman and Company, 660 Market Street, San Francisco, California 94104.

Library of Congress Cataloging in Publication Data

Main entry under title:

Continents Adrift and Continents Aground.

Includes bibliographies and index.

1. Continental drift—Addresses, essays, lectures.
2. Plate tectonics—Addresses, essays, lectures. I. Wilson, John Tuzo, 1908- II. Scientific American.
QE511.5.C65 551.1'3 76-46564
ISBN 0-7167-0281-9
ISBN 0-7167-0280-0 (pbk)

Printed in the United States of America

9 8 7 6 5 4 3 2

PREFACE

This book brings together seventeen articles from *Scientific American* that describe the latest scientific revolution—the revolution in ideas about the behavior of the earth's surface. Formerly, most scientists thought of the earth as a rigid body with fixed continents and permanent ocean basins, but now most of them consider the brittle surface of the earth to be broken into six large plates and several smaller ones, which very slowly move and jostle one another like blocks of ice on a river that is breaking up in the spring thaw. They believe that the thin plates of this surface layer are "floating" on a deeper layer that is slowly deformable. Each continent does not constitute one plate, but rather each is incorporated with surrounding ocean floor into a plate that is larger than the continent, just as a raft of logs may be frozen into a sheet of ice. These plates have repeatedly collided and joined, broken apart, and rejoined in different patterns. As they have done so ocean floors have been reabsorbed, but the continents have been modified and remain. These new discoveries have united geological and geophysical studies in ways not possible before.

This scientific revolution, like others before it, was long in the making, but it was not until the late 1960s that it achieved wide acceptance. Until that time most earth scientists regarded the theory of continental drift as heresy, but at a meeting of the world's geophysicists held in Moscow in August 1971 it became clear that suddenly a majority of earth scientists throughout the world had accepted this revised version of continental drift as the prevailing orthodoxy. The new form is called global plate tectonics.

Since that time, *Scientific American* has devoted much attention to the revolution with the result that this book contains eleven recent articles and only six reprinted from its predecessor, *Continents Adrift.* These articles have been grouped into four sections, each with a separate introduction.

The first section comprises three articles outlining the development and acceptance of the theory of continental drift and its modern form, plate tectonics, and two articles on the nature of the earth's interior and the bearing of conditions there upon the movements observed on the earth's surface. The second section consists of four articles, three of which were in the first edition, and which describe in detail the four kinds of boundaries between plates. Any two plates may be moving apart by sea-floor spreading, they may be sliding laterally with respect to one another along transform faults, or they may be coming together either by subduction, in which one plate overrides another to form deep ocean trenches and island arcs, or by crumpling of the surface to form high mountains and plateaus. The third section gives the history of the fracturing of the former supercontinent of Pangaea and the separation of the parts to form the present continents. Separate articles give some details of the growth of each of the three major ocean basins. The fourth section illustrates how well the new or revised theory can explain the observed distribution of both fossils and living creatures and how it can suggest much about the distribution of ore deposits and petroleum resources, and hence assist prospectors and drillers in their search for mineral wealth.

In preparing this edition, lack of space has made it necessary to omit all five articles that constituted the first section of *Continents Adrift.* They dealt with observations that preceded the acceptance of the theory of continental drift. Their chief interest is now historical, for they paved the way for recognizing the validity of the theory. Lack of space has also led to the omission of four other excellent articles; some had been made redundant by material in more recent articles, and others dealt with details not pertinent to the main thrust of the argument. In their number and diversity, the recent articles describe fairly completely and in the words of original contributors the history, the nature, and some of the consequences of the modern revolution in the earth sciences.

In calling this reversal of opinion a scientific revolution, I am following those historians who hold that most branches of science developed out of the practical experience and technology of such men as miners, seafarers, farmers, and foundrymen. At first, each branch of science was no more than a codification of the knowledge accumulated over many generations; as such, it offered few surprises and provided little basis for prediction. Historians of science—notably T. S. Kuhn—have pointed out that, as the quantity of knowledge increased, each branch of science reached a stage in which theoreticians reinterpreted the lore of practical men by new and subtler formulations. Many of these new theories seemed at first sight and in the light of older ideas to be contrary to reason, but they were eventually accepted because they were shown to be consistent with the body of accumulated observation and to offer superior interpretations.

In his book *The Structure of Scientific Revolutions* (2nd ed.; University of Chicago Press, 1970), Kuhn cites as the classic example of a scientific revolution the acceptance of Copernican astronomy and the abandonment of the Ptolemaic belief that the earth is the center of the solar system. That the earth is the center of the universe and that it rests on a fixed support was the obvious and early interpretation. To realize that the earth is spinning freely in space and that the sun, not the earth, is at the focus of the solar system required a prodigious feat of imagination. Lest we dismiss the Ptolemaic astrologers too lightly, we should recall that they devised the calendar we use today and invented methods by which to predict eclipses accurately and to navigate the ships whose voyages led to the discovery of the New World and the first circumnavigations of the globe. They had developed and practiced a form of science, but it was an inferior form, limited by its earth-centered premise, which made it incapable of achieving any proper understanding of the true nature of the planets or the stars. Changing the basic point of view created a new form of science with a different frame of reference. It was this change in the manner of interpreting the observations that constituted the scientific revolution. The importance of Copernicus' achievement was demonstrated by the way it provided opportunities for Galileo, Kepler, and Newton to make a burst of great discoveries.

Similarly, the change from phlogistic alchemy to modern chemistry, the change from a belief in many separate creations to a belief in organic evolution, and the change from classical physics to the acceptance of radioactivity, quantum mechanics, and relativity in modern physics were scientific revolutions that led to great developments. All took the form of changes in basic premises and advances in theory that reinterpreted past observations without destroying their validity.

Today many earth scientists believe that within the past decade a scientific revolution has occurred in their own subject. As in other cases, the new belief demands a reinterpretation of the observations of both geologists and geophysicists, which incidentally demonstrates the interdependence of the two disciplines. The acceptance of continental drift has transformed the earth sciences from a group of rather unimaginative studies based upon pedestrian interpretations of natural phenomena into a unified science that holds the promise of great intellectual and practical advances. This has made the present an exciting time to study the earth, for history shows that every scientific revolution has opened the way for fresh discoveries in the succeeding decades.

May 20, 1976 *J. Tuzo Wilson*

Note on cross-references: References to articles included in this book are noted by the title of the article and the page on which it begins; references to articles that are available as Offprints, but are not included here, are noted by the article's title and Offprint number; references to articles published by SCIENTIFIC AMERICAN, but which are not available as Offprints, are noted by the title of the article and the month and year of its publication.

CONTENTS

I

MOBILITY IN THE EARTH

MOBILITY IN THE EARTH

I

INTRODUCTION

The concept that continents have moved is now several centuries old, but ideas about the way in which they move have evolved through three stages; only recently has the last of these been widely accepted. The three stages correspond to three beliefs: first, that early in the earth's history some cataclysm forced the continents apart; second, that each of the continents has moved slowly and is still moving like a ship through the ocean floor; and third, that the whole surface of the earth, including both continents and ocean basins, has been broken into a few huge plates that are in motion relative to one another. These plates are larger than continents and carry them about like moving sheets of ice transporting rafts of logs frozen into their surface.

Traditionally, earth scientists regarded the earth as rigid and the continents as fixed. Once the Americas had been discovered and the shores of the Atlantic mapped, many observers, including such well-known savants as Francis Bacon, Buffon, Placet, and von Humboldt, commented upon the similarity in the shapes of the coasts of the continents on opposite sides of the Atlantic, although generally in vague terms. Not until 1857 and 1858, respectively, did Richard Owen and Antonio Snider clearly express the view that the Atlantic Ocean had opened by the separation of opposing continents. They considered that this had happened in catastrophic fashion early in the earth's history. In 1882 Osmond Fisher proposed that the separation of the moon from the Pacific basin had been the cause of this cataclysmic opening of the Atlantic. The recent analysis and dating of samples from the ocean floors and from the moon has completely demolished this view. The moon is not made of crustal materials and most of its features formed more than three billion years ago; the Atlantic, on the other hand, started to open only within the past 200 million years.

A century ago geologists began to make the observations and draw the conclusions that were eventually to lead them away from the notion of a rigid earth to the second stage, the concept of a mobile one. They observed the great folding and shortening of the strata in alpine mountains and concluded that the two sides of such ranges had moved together, although at first they regarded the motions as small. Geophysicists discovered by indirect instrumental means that this compression had created deep roots of light surface rocks that effectively buoy up the mountains. Earth scientists also observed that Scandinavia and other regions from which great ice sheets have recently melted are rising, and they reasoned that to compensate for such vertical motions some horizontal in-flow of material is required. They concluded that some part of the earth's interior must be mobile, which would allow the observed motions of the surface. This led to the theory of isostasy, that continents and ocean basins float in approximately hydrostatic balance with one another: continents and mountains are high because the crust beneath them is light, while ocean basins are deep because their crust is dense. The rates of motion involved are so slow, a few centimeters a year at most, that it is not necessary to suppose the mobile part of the interior to be liquid, but rather hot solid rock capable of creep.

Meanwhile, geologists and biologists had noted an increasing number of similarities between continents, particularly between those of the Southern Hemisphere, in which they included India. These views were well set out by Eduard Suess, an Austrian geologist, in a four-volume work entitled *Das Antlitz der Erde* (translated into English as *The Face of the Earth* and published between 1885 and 1909). To explain the observed relationships, he proposed that the southern continents had once been joined to form a huge supercontinent, which he called Gondwanaland, parts of which had sunk to form the Indian and South Atlantic oceans, thus separating the continents. It was soon pointed out that there was a conflict between this proposal and the theory of isostasy. In an attempt to resolve the difficulty, some biologists suggested that narrow land bridges or chains of islands had once lain across the oceans to provide migration routes, and that these had been small enough to sink in spite of isostasy.

Between 1908 and 1912 Frank B. Taylor and Alfred L. Wegener first presented and published an alternative proposal that the continents are slowly moving about. This marked the beginning of the second stage in the evolution of ideas about the earth's behavior. Internal evidence suggests that they arrived independently at these first statements of slow displacement, because Taylor's arguments emphasized the need for drift to compress the Alpine and Himalayan chains, whereas Wegener was much more influenced by similarities between the two sides of the Atlantic Ocean and by evidence of climatic changes. Although Taylor published first, he did not pursue the subject, with the result that Wegener became the chief proponent of continental drift. Between 1915 and 1928 he published four editions of the book *Die Entstehung der Kontinente und Ozeane,* which were translated into several languages, including English *(The Origin of Continents and Oceans).* World War I delayed discussion of Wegener's theory, but between 1920 and his death in 1930 the theory provoked great controversy. Several geologists in Europe (notably E. Argand, A. Staub, and A. Holmes) and others in the Southern Hemisphere (led by A. L. du Toit and, later, A. Maach, S. Warren Carey, and F. Ahmad) agreed with him, but most North American geologists did not, with the notable exceptions of W. A. J. M. van Waterschoot van der Gracht and, for a time, R. A. Daly. Most geophysicists agreed with Harold Jeffreys that drift was physically impossible because of the difficulty of moving continents of solid rock through ocean floors of solid rock in a rigid earth.

For the twenty-five years following Wegener's death most earth scientists regarded the theory as an aberration and did not consider or discuss it. Only two small groups of geologists continued to offer strong support. One group working in the Alps needed to explain the hundreds of kilometers of shortening that mapping and structural analysis suggested. The other, working in the southern continents, favored drift as an explanation of the many features of stratigraphy and the many past and present forms of life common to these lands. It was not until the ocean floors and the earth's interior were explored that an answer was obtained to this puzzle. This third stage of the theory of continental drift is based upon the concept that the earth's surface is broken into a few great plates, each including both continents and ocean floor, and that these plates, rather than continents alone, move relative to one another. This is the modern theory of global plate tectonics.

The first paper in this section is devoted to a statement of Wegener's views on continental drift, the next two describe how others developed them into the modern theory of plate tectonics, and the last two papers discuss our present ideas about the nature of the earth's interior and how the surface plates may be moving with respect to it.

"Alfred Wegener and the Hypothesis of Continental Drift," by the Oxford geologist Anthony Hallam, describes the life and contributions of the man who named the hypothesis of continental drift and did more than any other to establish it. He was a successful German astronomer and meteorologist with a passion for exploring Greenland, where in 1930, at the age of fifty, he died of exhaustion on his third expedition to that country. In 1912, when he first published his theory, the accepted opinion was that the earth had cooled from a molten state, becoming uniformly layered in the process. Continued contraction had generated mountains and caused some areas of the surface to collapse, forming ocean basins. These were considered to be permanently fixed in position, but temporary land bridges were supposed to have risen between them, enabling plants and animals to migrate from one continent to another.

Wegener pointed out several contradictions and inadequacies in the theory, including its failure to explain why some coasts matched those opposite to them so well, and why mountains form linear belts instead of being widely and uniformly distributed. He showed that two levels, those of the ocean floor and of the continental plains, dominate the world's topography, suggesting that there are two kinds of crust, which remain at different levels. He also pointed out that the rise and fall of land bridges was contrary to the theory of isostasy.

He proposed instead that all the world's land areas, including those of the Northern Hemisphere, had initially been united into a single supercontinent, which he called Pangaea. He held that at the close of Paleozoic time tidal and rotational forces caused Pangaea to split and separate into the present continents. A few geologists, notably the leading South African geologist, Alex L. du Toit, supported Wegener. Du Toit made important contributions to the theory. He supported the view that radioactive heating causes currents in the earth's interior to move the surface about, that in Jurassic time the single supercontinent of Pangaea had divided into two, and that subsequently both parts, Gondwanaland and Laurasia, had broken again to form the present continents.

However, most geologists and geophysicists condemned the theory. Some of their reasons now appear to have been wrong. These included the opponents' suggestions that the fit in bathymetry and geology of the two sides of the Atlantic Ocean was not good, that all the ancient animals could have migrated across land bridges, and that if horizontal movement had occurred it would have made vertical movement impossible.

Opponents also raised some objections that were true in themselves but have since been shown not to be valid arguments against continental drift. They showed, for example, that the forces which Wegener had proposed to be the cause of drift were quite insufficient, but they ignored other forces that Holmes and du Toit had suggested to be adequate. They rightly pointed out errors and uncertainties in the astronomical observations Wegener held to prove that Greenland is moving rapidly relative to Europe, but the existence of these errors did not mean that no movement occurred. Again, they argued that because there is no sign of any disturbances on the east coast of North America and some other coasts, the continents cannot be in motion, but this is an argument against the motion of separate continents, not an argument against plate tectonics, where motion occurs along the mid-ocean ridges as well as along some coasts.

Hallam concludes that Wegener's hypothesis was not accepted for half a century because it was premature, because it required the rejection of orthodox geological theories established and accepted for over 70 years, and because Wegener, not having been trained as a geologist, was not regarded as a reliable authority. Even more important was everyone's lack of knowledge of the ocean floors and the earth's interior. As soon as more information about those regions came to light, it established the truth of Wegener's basic ideas, but led to modification of his original concept of continental drift into the currently accepted hypothesis of plate tectonics.

The year 1956 marks the beginning of the turn to acceptance, for at that time a group of English geophysicists first produced convincing evidence from measurements of paleomagnetism in rocks that continents had moved relative to one another, and Maurice Ewing and Bruce Heezen of Columbia University first suggested the continuity of the mid-ocean ridges. As soon as earth scientists realized that this system of ridges constituted the world's largest mountain system they recognized its importance. They also saw that the older tectonic theories had failed to predict the existence of such a vast feature and were incapable of explaining its origin. They soon appreciated that the old theories were completely inadequate.

The second article, entitled "Continental Drift," was written at the time that these ideas were beginning to lead to an acceptance of drift. It describes the theory advanced by H. H. Hess of Princeton University that the mid-ocean ridges are loci of sea-floor spreading. Hess considered the cause of these motions to be currents, heated by natural radioactivity, rising from the deep interior, as Holmes and du Toit had proposed. These break the crust and carry it towards deep ocean trenches and young mountains, where it is carried down by subduction into the interior. F. A. Vening Meinesz had already found evidence for large downfolds beneath oceanic trenches. This paper also discusses ocean islands, showing that many appear to form over hot spots, which are regarded as local heat sources fixed in the interior. The movement of the crust over a hot spot may generate a chain of volcanic islands, like the Hawaiian Islands, which steadily increase in age from a young and active volcanic island at one end of the chain to older and more deeply eroded islands towards the other end.

Between 1965 and 1968 these ideas led to the development of the original theory of continental drift into a new version, called "plate tectonics," which is the subject of the third article, by John F. Dewey. According to this revised theory, the earth's surface and lithosphere is divided into six major plates and about a dozen minor ones. The fact that these plates are rigid, brittle, and entirely above their melting point enables them to move coherently according to a geometry of plate motion, which Dewey describes. These plates move apart and grow at mid-ocean ridges, slide past one another on transform faults, and come together beneath deep ocean trenches and in young mountains. All earthquakes occur in the lithosphere. A few occur within plates but most are confined to narrow belts under the mid-ocean ridges and somewhat wider belts under continental mountains and island arcs and are therefore concentrated along plate boundaries. They occur at depth only where the overlapping of two plates of the lithosphere has forced one of them down into the interior, a process called subduction.

The fourth article, "The Earth's Mantle," by Peter J. Wyllie, describes our knowledge of the earth's interior. At the start, Wyllie distinguishes the crust and mantle from the lithosphere and asthenosphere. All are shells of the earth, but the first pair are distinguished by different chemistries, the second pair by different strengths. The crust is thin and formed of silicious rocks resting upon the more basic rocks of the mantle. Its thickness can be measured because earthquake waves have different velocities in the rocks of the crust and those of the mantle, and can be reflected and refracted at the boundary between them, called the Mohorovičić discontinuity after its Yugoslav discoverer. The lithosphere comprises the crust and the uppermost part of the mantle. This is the strong layer of which the plates are made. It rests upon the asthenosphere, a deeper layer of the mantle so hot that it has lost its brittleness and is capable of slow creep.

As if these distinctions were not complicated enough, there is a further source of confusion discussed in "Hot Spots on the Earth's Surface," by Kevin C. Burke and J. Tuzo Wilson, the final article of this section. It involves the identification of the boundary between the lithosphere and the asthenosphere. This has long been thought to be the top of the low-velocity layer, a zone below the Mohorovičić discontinuity so called because in it, at least under the ocean basins, the velocity of earthquake waves is reduced. This is attributed to temperatures great enough to cause the grains of rock to melt partially, which slows the waves. A complication has been that the low-velocity layer is difficult to detect beneath continents, yet continents move.

Recently it has been suggested that the volume of rock melted is small, perhaps as little as 0.1 percent of the rock, and that this, while sufficient to slow the waves, is not enought to unlock the grains, which still interfere with one another and prevent motion. According to this view, the asthenosphere does not begin at the low-velocity layer, but lies at a deeper level where temperature and pressure combine to allow the solid mineral grains to be sheared, enabling motion to take place. Its level can be calculated on the basis of laboratory experiments but, unlike the Mohorovičić discontinuity and the low-velocity layer, it cannot be detected. This can explain why continents move, even in the absence of an underlying low-velocity layer.

The oceanic lithosphere thickens by cooling, so that its thickness increases approximately in proportion to the square root of its age, being only a few kilometers near the crest of a ridge and 90 kilometers where it is 100 million years old. Beneath a thin and variable thickness of sediment, which tends to increase with time, the lithosphere consists of an upper layer of 5 kilometers of basalt formed as the plates grow at the crest of the mid-ocean ridges and a second layer of 15 kilometers of olivine and pyroxene minerals, the residual material from which the basalt arose. These layers have densities of from 3.0 to 3.3 grams per cubic centimeter. D. L. Anderson proposed a third layer, of eclogite rock, which has a density of about 3.5 grams per cubic centimeter. As the lithosphere cools, underplating causes this third layer to grow, so that both the thickness of the lithospheric plates and their average density increase. The result may be that, whereas young and thin oceanic lithosphere may be lighter than the asthenosphere and can only be forced down during subduction, older lithosphere, which has become denser, may tend to sink by its own weight to form subduction zones.

The situation under continents is even less well known, beyond the fact that the continental lithosphere is probably about 200 kilometers thick and overlies a thin asthenosphere.

The last article goes on to discuss the significance of hot spots. These are isolated volcanic uplifts, like the Island of Hawaii, some of which appear to have produced trails of older islands, like the Hawaiian chain, and thus seem to provide markers showing the movement of the lithospheric plates over the deeper interior. If that is so, then it becomes possible to distinguish between moving and fixed plates. This in turn leads to two kinds of continents, two kinds of ocean basins, and two types of continental coast lines.

Little is yet known of the roots of hot spots, although they are distinctive enough on the surface. It has been observed that the plume of slowly rising hot solid rock, or whatever it is that gives rise to these volcanic centers, uplifts and fractures them, usually forming three or four radial faults. Evidence in the Red Sea and East African rift valleys suggests that some of these faults may join and cause a break from one hot spot to another, and that this composite fracture system may become an axis of spreading. Where this has happened, some of the faults at hot spots are left as blind valleys leading to the spreading ocean, for example, the Jordan Valley and the north end of the African Rift Valley, both of which lead into the Red Sea. Later the sea may close again and then these blind valleys lead to younger mountain systems.

The last stage of this sequence was the first to be recognized, about 1942, by the Soviet geologist Nicholas Shatsky, who called these rifts "aulacogens." It took some time to piece together the rest of the sequence. The name is now used to refer to all stages of the formation of these blind valleys.

1

Alfred Wegener and the Hypothesis of Continental Drift

A. Hallam
February 1975

The jigsaw-puzzle fit of the coastlines on each side of the Atlantic Ocean must have been noticed almost as soon as the first reliable maps of the New World were prepared. The complementary shapes of the continents soon provoked speculation about their origin and history, and a number of early theories suggested that the shapes were not the product of mere coincidence. In 1620 Francis Bacon called attention to the similarities of the continental outlines, although he did not go on to suggest that they might once have formed a unified land mass. In the succeeding centuries several other proposals attempted to account for the correspondence, usually as a result of some postulated catastrophe, such as the sinking of the mythical Atlantis. The first to suggest that the continents had actually moved across the surface of the earth was Antonio Snider-Pellegrini in 1858, but he too attributed the event to a supernatural agency: the Great Flood. Today, of course, the migration of the continents is an essential feature of the theory of the earth's structure that is all but universally accepted by geologists.

Of the various hypotheses that preceded the modern theory of plate tectonics, one version stands out: the one propounded by Alfred Wegener in the early years of the 20th century. Wegener had access to only a small part of the information available today, yet his theory anticipates much that is now fundamental to our conception of the earth, including not only the movement of the continents but also the wandering of the poles, with consequent changes in climate, and the significance of the distribution of ancient plants and animals. There are also parts of his work that have turned out to be wrong, but the most important points in his arguments have been substantiated. His hypothesis stands not merely as a forerunner of the concept that now prevails but as its true ancestor.

Wegener first presented his ideas to the scientific community in 1912, but it was not until 50 years later that they gained general currency. When his view of the earth did replace the older model (in the 1960's), the change represented a radical revision of a well-established doctrine, and it took place only because new evidence, derived from discoveries in geophysics and oceanography, compelled it. In the interim Wegener's theory had at best been neglected, and it had often been scorned. At the nadir proponents of continental drift were dismissed contemptuously as cranks. To understand that reaction we must examine both Wegener's work and the attitudes of his contemporaries. Did Wegener derive his conclusions from reliable evidence, and did he support them with coherent argument? Or did he merely guess, and happen to guess correctly? If his reasoning was plausible, why was his work opposed with such determination and persistence? I shall attempt to answer questions such as these, and in addition to gauge what kind of man Wegener was and to estimate his rank as a scientist.

Wegener had no credentials as a geologist. Born in Berlin in 1880, the son of an evangelical minister, he studied at the universities of Heidelberg, Innsbruck and Berlin and took his doctorate in astronomy. (His most useful accomplishment in the field of his degree was a paper on the Alphonsine tables of planetary motion.) From his early days as a student he had cherished an ambition to explore in Greenland, and he had also become fascinated by the comparatively new science of meteorology. In preparation for expeditions to the Arctic he undertook a program of long walks and learned to skate and ski; in the pursuit of his other avocation he mastered the use of kites and balloons for making weather observations. He was so successful as a balloonist that in 1906, with his brother Kurt, he established a world record with an uninterrupted flight of 52 hours.

Wegener's preparations were rewarded when he was selected as meteorologist to a Danish expedition to northeastern Greenland. On his return to Germany he accepted a junior teaching position in meteorology at the University of Marburg and within a few years had written an important text on the thermodynamics of the atmosphere. A second expedition to Greenland, with J. P. Koch

ANCIENT SUPERCONTINENT incorporated all the earth's large land masses. Wegener's reconstruction of the supercontinent, which he called Pangaea, is shown in the middle illustration on the opposite page. A more recent version, shown at the bottom, differs in details of placement and orientation but preserves the major features of Wegener's proposal. Both maps are based not on the coastlines of the continents but on the edges of the continental shelves. For comparison a map of the world as it appears today is shown at top.

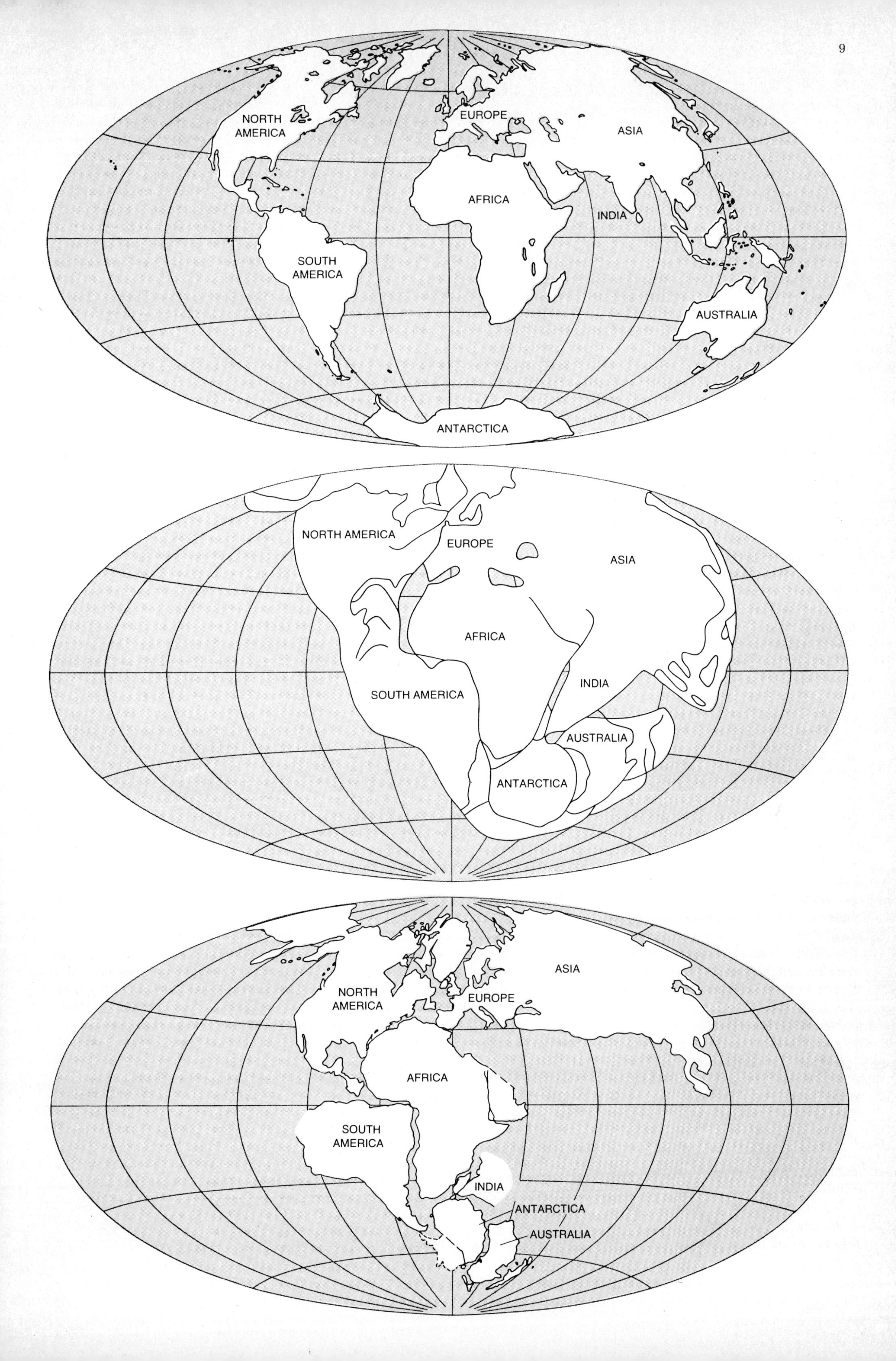
NORTH AMERICA
EUROPE
ASIA
AFRICA
INDIA
SOUTH AMERICA
AUSTRALIA
ANTARCTICA
NORTH AMERICA
EUROPE
ASIA
AFRICA
INDIA
SOUTH AMERICA
AUSTRALIA
ANTARCTICA
ASIA
NORTH AMERICA
EUROPE
AFRICA
SOUTH AMERICA
INDIA
ANTARCTICA
AUSTRALIA

ALFRED WEGENER was by profession a meteorologist; he also participated in three exploratory expeditions to Greenland. In geology and geophysics he was virtually without credentials; nevertheless, it was through his work in these fields that he made his most significant contribution. This drawing of him is based on a photograph made in the 1920's.

of Denmark, followed in 1912. It included the longest crossing of the ice cap ever undertaken; the published glaciological and meteorological findings fill many volumes.

In 1913 Wegener married Else Köppen, the daughter of the meteorologist Wladimir Peter Köppen. After World War I (in which Wegener served as a junior officer) he succeeded his father-in-law as director of the Meteorological Research Department of the Marine Observatory at Hamburg. In 1924 he accepted a chair of meteorology and geophysics at the University of Graz in Austria, where he found that his colleagues were more sympathetic to his research interests than his colleagues in Hamburg had been.

Wegener died while leading a third expedition to Greenland in 1930, probably as a result of a heart attack induced by overexertion. His laudatory obituaries suggest that he had achieved considerable distinction both as a meteorologist and as an Arctic explorer; other sources suggest that in addition he had been a capable organizer and administrator and a lucid and stimulating teacher. His work on continental drift, which will surely be his permanent legacy, had remained a peripheral interest, albeit one that had absorbed him deeply.

Just how Wegener first conceived the idea that the continents could move is not certain. One unauthenticated account has it that he was inspired during a trip to Greenland while watching the calving of glacier ice (the process by which icebergs are born). From his own writings and those of his contemporaries, however, it seems more likely that he came to the theory in the same way that his predecessors had: by noting on a map the complementarity of the Atlantic coastlines. By Wegener's own account the notion first occurred to him in 1910, but a contemporary who knew Wegener as a student maintains that he had shown interest in the matter as early as 1903. Whether the idea had been maturing for a decade or for only two years, it was first presented publicly in January, 1912, in a lecture before the German Geological Association in Frankfurt am Main. The first published reports appeared later that year in two German journals.

The prevailing theory of the structure and evolution of the earth in 1912 could not accommodate drifting continents. Geologists and geophysicists then believed that the earth had been formed in a molten state and that it was still solidifying and contracting. During the process the heavy elements, such as iron, had sunk to the core and lighter ones, such as silicon and aluminum, had risen to the surface to form a rigid crust.

To most geologists of the time the model seemed quite successful in accounting for the more prominent features of the earth's surface. Mountain ranges were produced by compression of the surface during contraction, much as wrinkles develop in the skin of a drying, shrinking apple. On a larger scale the pressure generated by contraction, applied through great arches, caused some regions of the surface to collapse and subside, creating the ocean basins, while other areas remained emergent as continents. Vertical movements of the crust were considered entirely plausible, although movements parallel to the surface were excluded. Thus the continents and the ocean basins were in the long run interchangeable; some continental areas sank faster than the adjacent land and were inundated by the sea; at the same time parts of the ocean floor emerged to form dry land.

The similarity or identity of numerous fossil plants and animals on distant continents was explained by postulating land bridges that had once connected the land masses but had since sunk to the level of the ocean floor. The stratification of sedimentary deposits suggested successive marine transgressions onto the continents and regressions from them. The regressions could be attributed to the subsidence of the ocean basins and the transgressions to the partial filling of the basins with sediment eroded from the continents. At about the time Wegener was devising his hypothesis of continental drift a refinement of the traditional view was proposed in which the vertical movements of the crust are governed by isostasy: the concept that all elements of the system are in hydrodynamic equilibrium. Thus the continents, being less dense than the layer under them, float above the ocean floor.

Wegener detected a number of flaws and contradictions in the contracting-earth model. Moreover, many distinctive features of the earth's surface could not be explained at all by that model, unless they were to be considered the result of coincidence. The most obvious of these features is the correspondence between the Atlantic coasts of Africa and South America. (In plotting the correspondence Wegener employed not the coastline itself but the edge of the continental shelf, which is a more meaningful boundary. The same practice is followed in modern reconstructions.) Another anomaly is the distribution of mountain ranges, which are mainly confined to narrow, curvilinear belts; if they had been produced by the contraction of the globe, they should have been spread uniformly over the surface, as the wrinkles on a dried apple are.

Still another peculiarity was discovered in a statistical analysis of the earth's topography. From calculations of the total area of the earth's surface at each of many land elevations and ocean depths, Wegener found that a large fraction of the earth's crust is at two distinct levels. One corresponds to the surface of the continents, the other to the abyssal sea floor [*see illustrations on pages 14 and 15*]. Such a distribution would be expected in a crust made up of two layers, the upper one consisting of lighter rock, such as granite, and the substratum consisting of basalt, gabbro or peridotite, which would also form the ocean floor. This interpretation is supported by measurements of local variations in the earth's gravity. It is not consistent with a model of the crust in which variations in elevation are the result of random uplift and subsidence; in that case one would expect a Gaussian, or bell-shaped, distribution of elevations around a single median level.

Wegener also found support for his arguments in fossils and distinctive geological features that seemed to cross continental boundaries. In the fossil record an excellent example is provided by the reptiles. Fossils of *Mesosaurus,* a small reptile that lived late in the Paleozoic era, about 270 million years ago, are found in Brazil and in South Africa and nowhere else in the world [see the article "Continental Drift and the Fossil Record," by A. Hallam, beginning on page 186]. The peculiar distribution was traditionally explained by the sinking of a land bridge that was assumed to have connected the continents. On geophysical grounds Wegener rejected this explanation; it violated the principle of isostasy, since the material of the bridge would be less dense than that of the sea floor and could not sink into it. The only reasonable alternative was that the continents had once been joined and had since drifted apart.

The geological evidence is of a similar

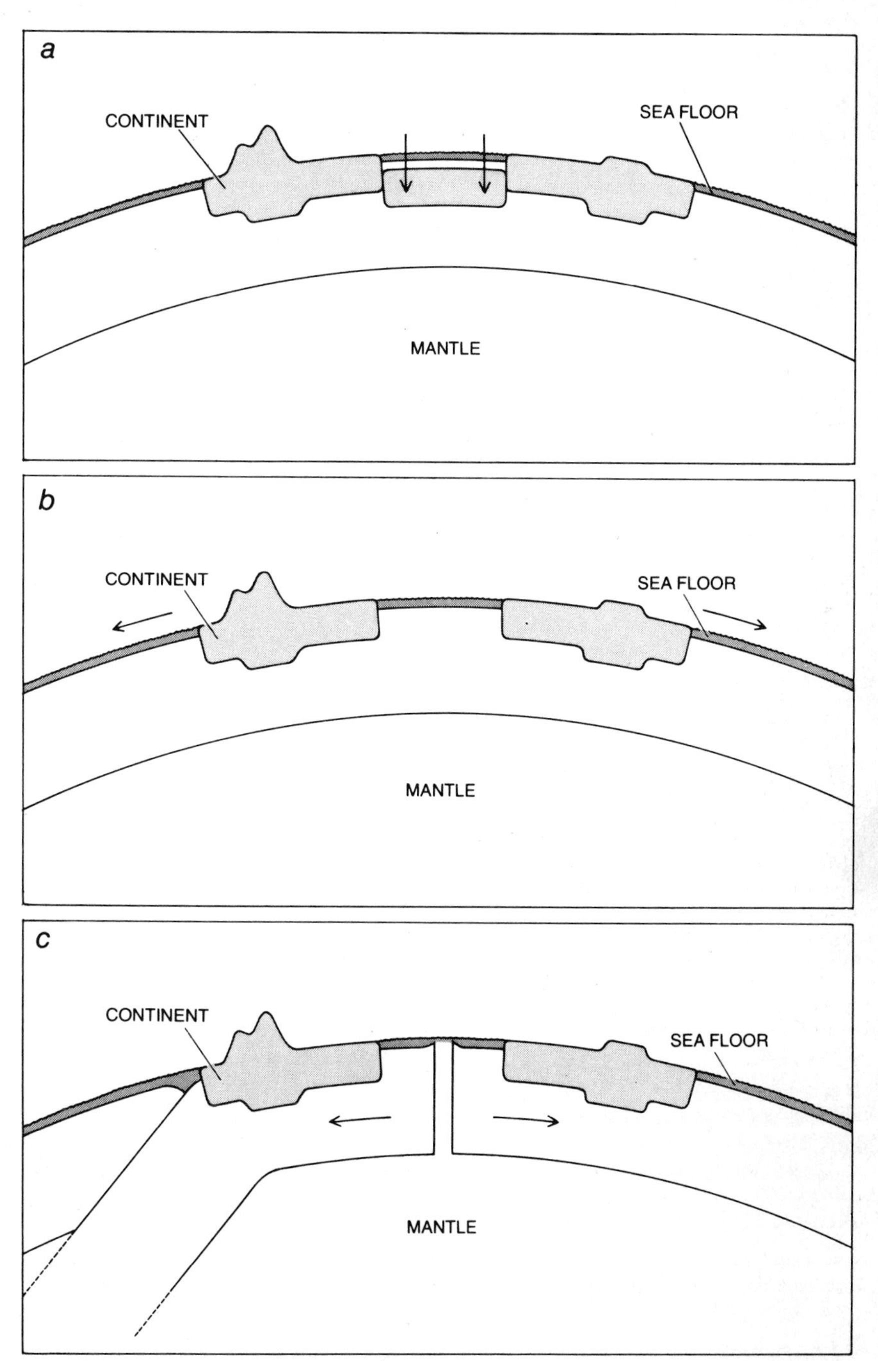

THEORY OF THE EARTH'S STRUCTURE held by most geologists at the beginning of the 20th century was challenged by Wegener. In the traditional view (*a*) the continents were fixed in place and lateral movement was impossible. Vertical motion of the crust, as in the sinking of land bridges, was allowed in some versions of the theory but objected to in others as a violation of the principle of isostasy, which holds that the continents float in hydrodynamic equilibrium on a substratum of denser material. In Wegener's hypothesis (*b*) the continents migrate through the substratum, which acts as a viscous fluid. He suggested that they were propelled by forces related to the earth's rotation, but that idea was soon discredited. In the modern theory (*c*) the continents are carried as passengers on large rigid plates. The plates are forced apart where material wells up to form new sea floor.

nature. For example, large blocks of particularly ancient rock are found both in Africa and across the Atlantic in South America; if the continents are brought together in the proper orientation, the blocks line up precisely [*see illustration on opposite page*]. Wegener himself recognized and described the power of this discovery: "It is just as if we were to refit the torn pieces of a newspaper by matching their edges, then check whether or not the lines of print run smoothly across. If they do, there is nothing left but to conclude that the pieces were in fact joined this way. If only one line was available for the test, we would still have found a high probability for the accuracy of fit, but if we have *n* lines, this probability is raised to the *n*th power."

One further line of evidence on which Wegener relied should be mentioned. Geodetic observations made early in the 20th century seemed to indicate that Greenland was moving westward, separating from Europe at a measurable rate. Such a movement might constitute a direct validation of continental drift, but it has not been confirmed in recent measurements employing more accurate techniques.

In order to resolve these contradictions Wegener formulated a comprehensive theory of the origin of the conti-

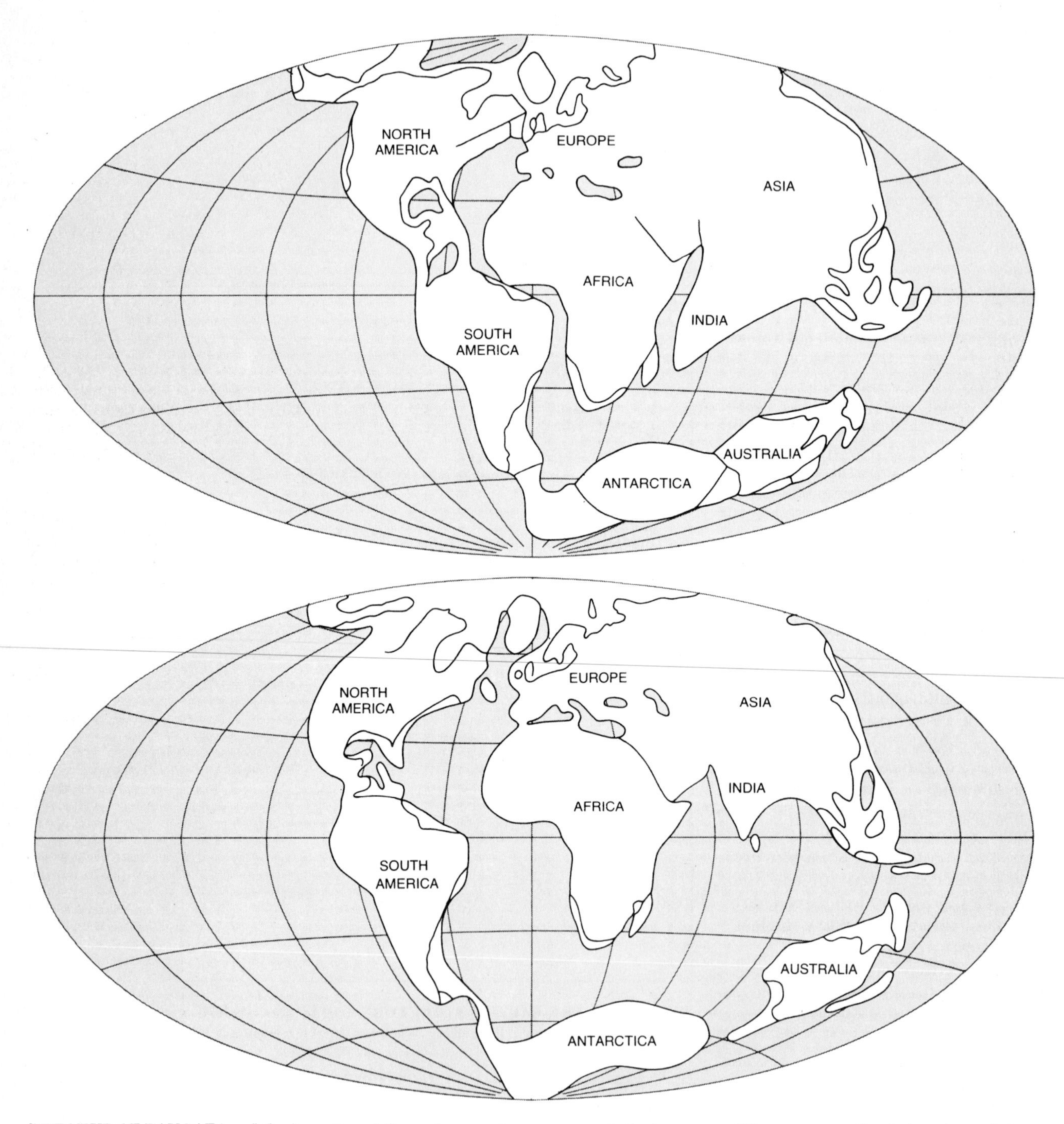

BREAKUP OF PANGAEA and the formation of the modern continents were described by Wegener in two stages. In his account all the continents remained contiguous as late as the Eocene epoch, about 50 million years ago (*upper map*). Even in the early Quaternary period, about two million years ago, South America and Antarctica were connected by an isthmus (*lower map*). Modern workers have considerably revised Wegener's sequence and his dating of events. The breakup actually began about 200 million years ago.

nents. In his reconstruction of earth history all the world's land area was originally united in a single primordial supercontinent, which he named Pangaea from the Greek meaning "all land" [*see illustration on page 9*]. He published his conclusions and evidence in 1915 in a book titled *Die Entstehung der Kontinente und Ozeane* (*The Origin of Continents and Oceans*).

The geophysical basis of Wegener's theory was closely related to the principle of isostasy. Both assume that the substratum underlying the continents acts as a highly viscous fluid; Wegener assumed further that if a land mass could move vertically through this fluid, it should also be able to move horizontally, provided only that a sufficiently powerful motive force was supplied. As evidence that such forces exist he cited the horizontal compression of folded strata in mountain ranges. There is also elegant evidence of the fluid nature of the underlying material: the earth is an oblate sphere, bulging slightly at the Equator, and the size of the bulge is exactly what would be expected for a sphere of a perfect fluid spinning at the same rate. It is a fluid of a special nature. Under short-term stresses, such as those of an earthquake, it acts as an elastic solid; only over the much longer periods of geologic time can its fluid characteristics be observed. Its behavior is analogous to that of pitch, a material that shatters under a hammer blow but flows plastically under its own weight, that is, under the milder but persistently imposed force of gravity.

After World War I Wegener merged his two principal research interests by investigating, with Köppen, changes in world climate through geologic time. By mapping the distribution of certain kinds of sedimentary rock he was able to infer the position of the poles and the Equator in ancient times. His most impressive results were obtained for the Carboniferous and Permian periods, about 300 million years ago [*see illustration on page 16*]. The position of the South Pole was determined from the disposition of boulder beds called tillites, which are formed during the movements of glaciers. In Wegener's reconstruction of Pangaea the pole was just east of what is today South Africa and within ancient Antarctica.

Ninety degrees from the pole Wegener found abundant evidence of a humid equatorial zone. The evidence consists of the vast deposits of coal that stretch from the eastern U.S. to China; fossil plants identifiable within the coals are of a tropical type. Other climatic indicators in sedimentary rock are salt and wind-deposited sand, which suggest the presence of ancient deserts. In the past, as today, deserts were formed mainly in the trade-wind belts on each side of the Equator. In the present Northern Hemisphere the Carboniferous coals are replaced in more recent strata by salt and sand-dune deposits that Wegener interpreted as signifying a southeastward shift in the position of the Equator. The movement was confirmed by a corresponding southeastward shift in the main center of tillite deposits, implying that the pole also shifted in that direction.

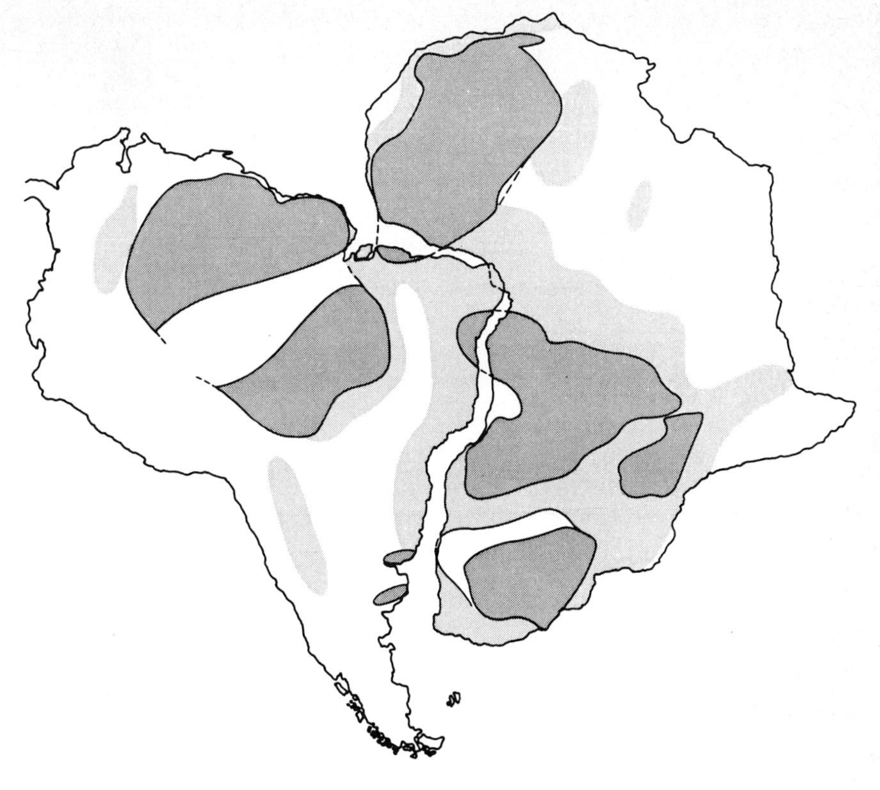

GEOLOGICAL EVIDENCE that the continents once formed a single land mass is provided by distinctive rock formations that can be assembled into continuous belts when Africa and South America are lined up next to each other. Areas shown in dark gray are ancient blocks called cratons; light gray areas represent regions of somewhat younger rock. Wegener considered continuities such as these good evidence for the continental-drift hypothesis.

The last edition of *The Origin of Continents and Oceans* devoted a chapter to ancient climate and contained an extensive discussion of the wandering of the poles. Polar movement is a phenomenon quite distinct from continental drift, but unless the continents are reassembled more or less as Wegener proposed, the distribution of tillites and coal and salt deposits cannot be interpreted coherently.

The last edition also contains more extensive documentation than the earlier ones of similarities in the geology of the southern continents, the outcome of a productive exchange of ideas between Wegener and the South African geologist Alexander L. du Toit. The essential geophysical arguments, however, remain remarkably similar to those proposed in Wegener's first paper, written almost two decades earlier.

The initial reaction of the scientific community to Wegener's hypothesis was not uniformly hostile, but at best it was mixed. At the first lecture in Frankfurt am Main some geologists were provoked to indignation; at Marburg a few days later, however, the audience seems to have been more sympathetic. Following the early publications several prominent German geologists announced their opposition to "continental displacement" (a more accurate rendering of Wegener's term, *Verschiebung*, than "drift"). A number of geophysicists, on the other hand, expressed approval of the concept. Indeed, in 1921 Wegener was able to say that he knew of no geophysicist who opposed the drift hypothesis.

The early publications, including the first edition of *The Origin of Continents and Oceans*, do not seem to have been read much outside Germany; it was not until the third edition was published in 1922 and translated into several other

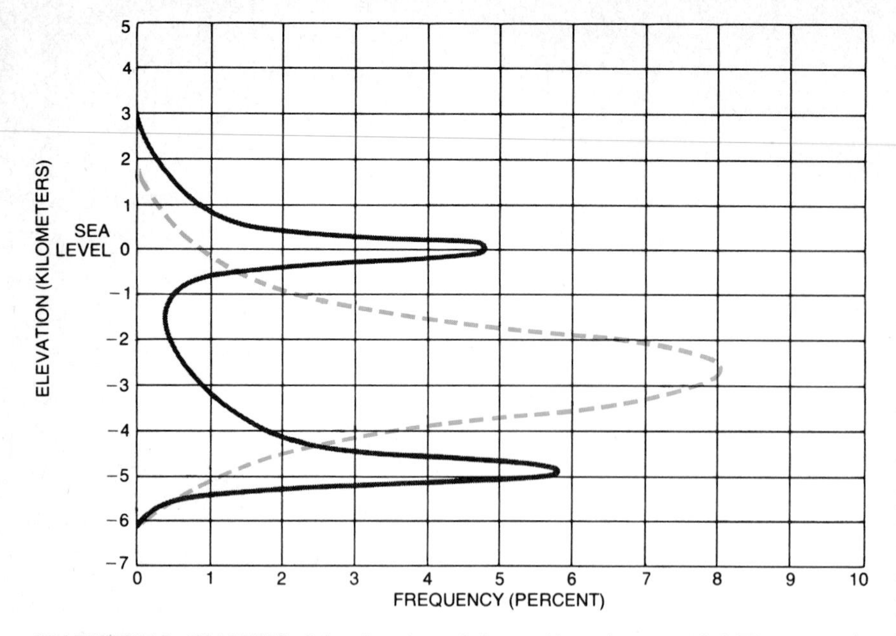

STATISTICAL ANALYSIS of the elevations of the earth's surface provided Wegener with an argument in support of his hypothesis. If topographical features were formed by random uplift and subsidence, a random distribution centered on the mean elevation (*colored line*) would be expected. Actually two levels predominate (*gray line*). One, near sea level, represents average elevation of the continental platforms, the other that of the abyssal sea floor.

languages (including English) two years later that Wegener's hypothesis attracted an international audience. As in Germany, his work was initially given a fair hearing, or at the least it was not dismissed out of hand. At a meeting of the British Association for the Advancement of Science in 1922 discussion of continental drift was "lively but inconclusive," according to the published report. As one would expect, many were skeptical, but the general reception accorded the hypothesis was sympathetic. At about that time several leading geologists on each side of the Atlantic declared themselves in favor of the theory.

The stubborn antagonism to Wegener's ideas that was to be the orthodoxy of geophysics until the 1960's began to develop in the mid-1920's. Two events were instrumental in this hardening of resistance. One was the publication of a treatise titled *The Earth* by Harold Jeffreys of the University of Cambridge; the other was a symposium held in 1928 by the American Association of Petroleum Geologists.

Jeffreys attacked Wegener's theory at what was perhaps its weakest point: the nature of the forces to which Wegener attributed the movement of the continents. Wegener proposed that the westward drift of the Americas could be explained as a consequence of tides in the earth's crust; the postulated force responsible for the northward migration of India and the compression of the crust in the Alpine and Himalayan mountain belts he called *Polflucht,* or "flight from the poles." Jeffreys was able to demonstrate by simple calculations that the earth is far too strong to be even slightly deformed by such forces. If it were not, mountain ranges would collapse under their own weight and the sea floor would be perfectly flat. If the tidal force were strong enough to shift the continents westward, it would also be strong enough to halt the earth's rotation within a year.

These objections are cogent ones, and the mechanism Wegener proposed has long since been abandoned. (The modern theory attributes the movement to the spreading of the sea floor along a system of mid-ocean ridges, where molten rock from the earth's interior wells up.) We can now see, however, that Jeffreys had not refuted the theory merely by demonstrating the inadequacies of the hypothetical motive force. For the most part he simply ignored the empirical evidence, which was the most substantial part of Wegener's argument. Jeffreys dismissed the wandering of the poles as being geophysically impossible.

Of the participants in the American Association of Petroleum Geologists symposium most were hostile, in varying degrees, to Wegener's theory; only one was strongly sympathetic. The proceedings of the symposium are in the main a chorus of criticism. The supposed jigsaw-puzzle fit of the Atlantic continents was inaccurate, the contributors contended, and it did not allow for vertical movements of the crust. The similar rock formations on opposite sides of oceans were not that closely related after all; in any case present similarity did not necessarily imply former contiguity. Ancient animals could have migrated across land bridges. The Carboniferous and Permian tillites of South Africa and other areas were probably not glacial and the Northern Hemisphere coals were probably not tropical. The evidence for the movement of Greenland was inconclusive.

Some of the contributors also demanded that the hypothesis resolve paradoxes that have been approached only in the more recent versions of the theory. For example, they asked why, if the American continents could move laterally by displacing the ocean floor, they crumpled on their western edge, forming the Cordilleran mountain ranges. Did not the compressive force that formed these mountains suggest considerable resistance from the supposedly fluid sea floor? Moreover, why did Pangaea remain intact for most of the earth's history, then abruptly split apart in a few tens of millions of years?

Finally, in addition to questioning Wegener's interpretations and conclusions, some participants in the symposium assailed his credentials and his methods. He was a mere advocate, they protested, selecting for presentation only those facts that would favor his hypothesis. He "took liberties with our globe" and "played a game in which there are no restrictive rules and no sharply drawn code of conduct."

Wegener did not campaign to defend his theory from these criticisms. A middle-aged man with only limited time for research, he felt that he was unable to keep up with the swelling volume of literature and was content to leave the field to younger workers. He did, however, gently rebuke his critics for their partiality. "We are like a judge confronted by a defendant who declines to answer," he said of scientists in relation to the earth, "and we must determine the truth from circumstantial evidence. All the proof we can muster has the deceptive character of this type of evidence. How would we assess a judge who based his decision on only part of the available data?"

After Wegener's death geologists and geophysicists became even more hostile to his account of earth history. In the U.S. the reaction was particularly strong; for an American geologist to express sympathy for the idea of continental drift was for him to risk his career.

Ironically, just as the condemnatory verdict was reached, the theory was significantly strengthened by the contributions of du Toit and of Arthur Holmes of the University of Edinburgh. They eliminated some of Wegener's weaker arguments, marshaled more evidence in support of the hypothesis and provided a more plausible motive mechanism. Holmes suggested that the continents are moved by convection currents in the earth's mantle. His hypothesis was not entirely satisfactory, but it successfully circumvented the criticisms of Jeffreys and his followers. Moreover, it anticipated the modern explanation of why continents move.

In hindsight, and with the knowledge that many aspects of Wegener's theory have been confirmed in the past two decades, we can readily appreciate his accomplishments. He examined critically a model of the earth that was then almost universally accepted. When he discovered weaknesses and inconsistencies, he was bold enough and independent enough to embrace a radical alternative. Furthermore, he had sufficient breadth of knowledge to seek out and perceptively evaluate supporting evidence from a variety of disciplines. The same qualities of mind were applied to the explication of ancient climate and the wandering of the poles.

The intellectual rigor that Wegener brought to his work is illustrated in his own writings and attested to in the words of those who knew him well. In a letter to Köppen written in 1911 Wegener defended his views on continental drift. The letter is reproduced in the biography of Wegener written by his widow. The following passage, which I have translated, has not to my knowledge been published previously in English.

"You consider my primordial continent to be a figment of my imagination, but it is only a question of the interpretation of observations. I came to the idea on the grounds of the matching coastlines, but the proof must come from geological observations. These compel us to infer, for example, a land connection between South America and Africa. This can be explained in two ways: the sinking of a connecting continent or separation. Previously, because of the unproved concept of permanence, people have considered only the former and have ignored the latter possibility. But the modern teaching of isostasy and more generally our current geophysical ideas oppose the sinking of a continent because it is lighter than the material on which it rests. Thus we are forced to consider the alternative interpretation. And if we now find many surprising simplifications and can begin at last to make real sense of an entire mass of geological data, why should we delay in throwing

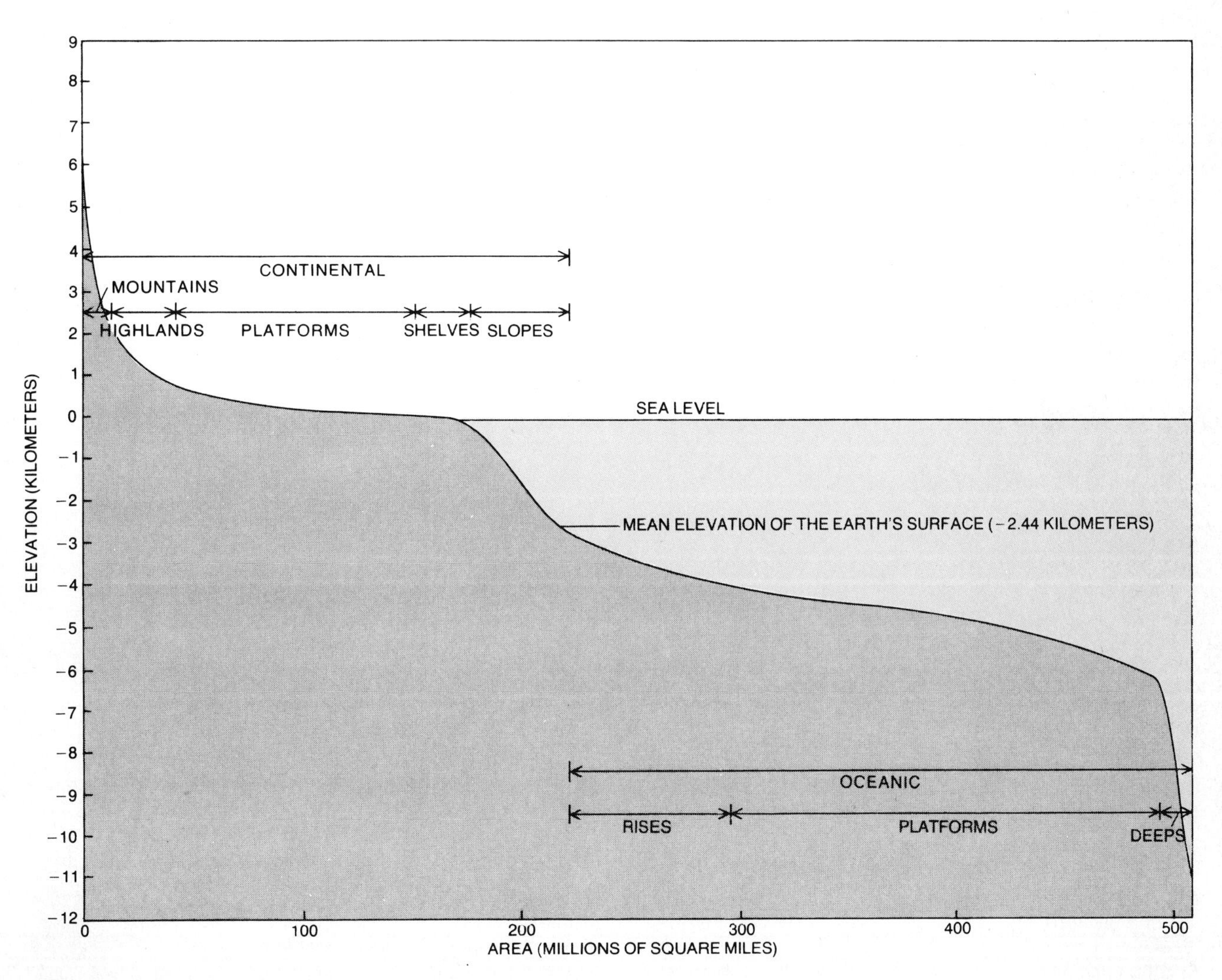

DISTRIBUTION OF ELEVATIONS, when correlated with the area of the earth's surface, confirms that the crust has two fundamental levels. Approximately 300 million square miles of the 510 million square miles of the earth's surface is continental platform or oceanic platform. The distribution conforms to Wegener's model of the earth, in which the continents float in a dense substratum.

the old concept overboard? Is this revolutionary? I don't believe that the old ideas have more than a decade to live. At present the notion of isostasy is not yet thoroughly worked out; when it is, the contradictions involved in the old ideas will be fully exposed." It evidently seemed quite obvious to Wegener, but it is clear that he underestimated the ardor of those committed to the "old ideas."

Wegener's success in constructing from scattered and seemingly unrelated observations a systematic theory of earth history could be attributed to his broad attack on the problem, and perhaps even to his status as a nonspecialist. A glimpse of his approach to scientific problems is provided by an obituary written by Hans Benndorf, a professor of physics and a colleague of Wegener's at Graz. This passage too was translated by me and has not appeared before in English.

"Wegener acquired his knowledge mainly by intuitive means, never or only quite rarely by deduction from a formula, and when that was the case, it needed

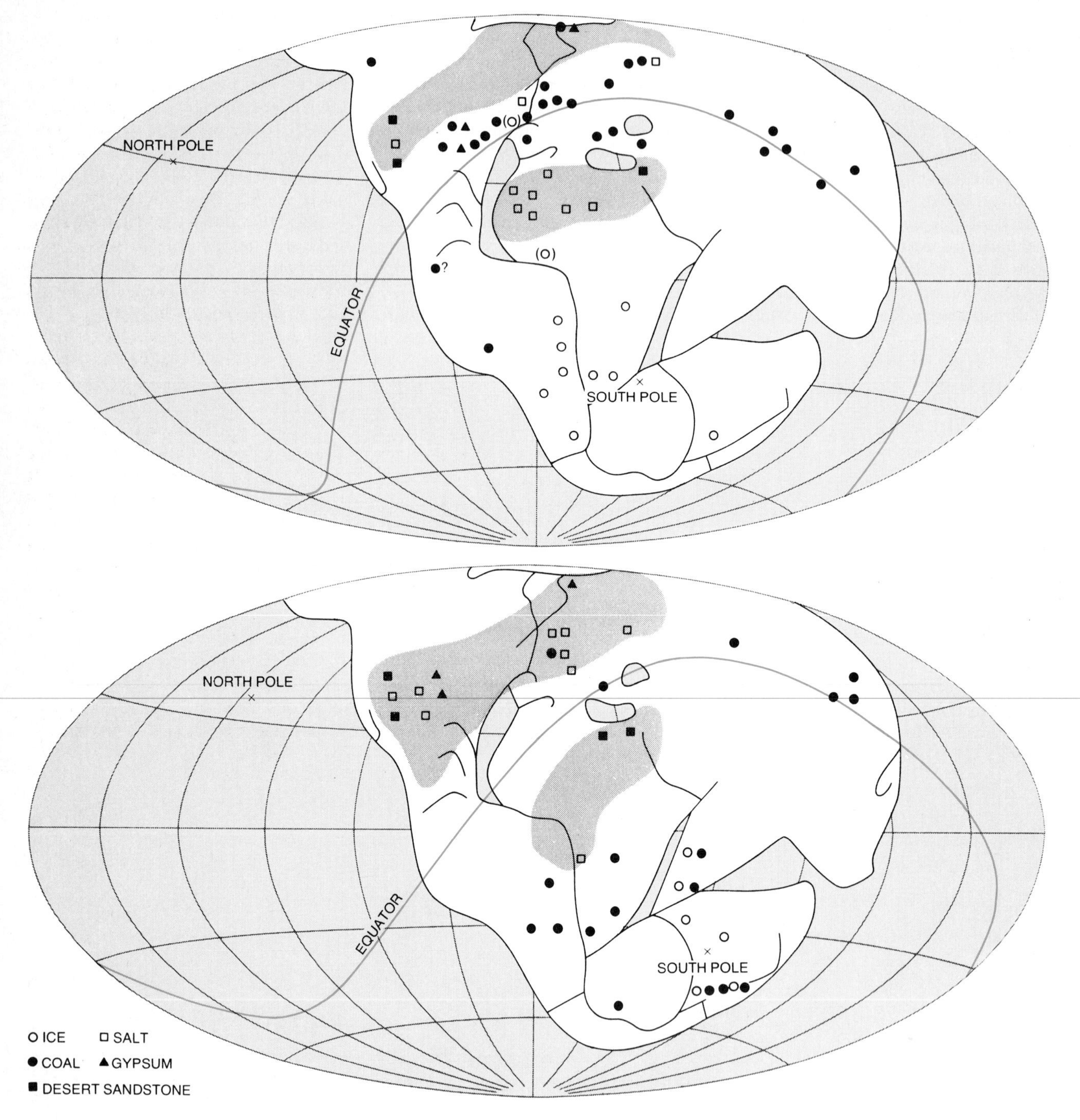

WANDERING OF THE POLES was proposed by Wegener to account for the distribution of ancient climates. Evidence of tropical climate was provided by certain kinds of coal, of polar climate by the tillites that signal glaciation, and of arid climate by deposits of salt, gypsum and desert sandstone. The symbols employed in the maps are identified in the key at left; in addition arid zones, which are characteristic of trade-wind latitudes, are indicated by gray areas. The upper map shows Wegener's reconstruction for the Carboniferous period, about 300 million years ago; the lower map is for the Permian period, about 230 million years ago. (The distortion of the Equator is caused by the projection employed.) The movement of the continents and the movement of the poles are unrelated, but a coherent account of ancient climate can be devised only by rearranging the continents approximately as Wegener did.

only to be quite simple. Also, if matters concerning physics were involved, that is, in a field distant from his own field of expertise, I was often astonished by the soundness of his judgment. With what ease he found his way through the most complicated work of the theoreticians, with what feeling for the important point! He would often, after a long pause for reflection, say, 'I believe such and such,' and most times he was right, as we would establish several days later after rigorous analysis. Wegener possessed a sense for the significant that seldom erred."

Benndorf's assessment of Wegener's method is supported by the remarks of Wilhelm Max Wundt, who knew Wegener as a student in Berlin: "Alfred Wegener started out to tackle his scientific problems with only quite ordinary gifts in mathematics, physics and the other natural sciences. He was never, throughout his life, in any way reluctant to admit that fact. He had, however, the ability to apply those gifts with great purpose and conscious aim. He had an extraordinary talent for observation and for knowing what is at the same time simple and important, and what can be expected to give a result. Added to this was a rigorous logic, which enabled him to assemble rightly everything relevant to his ideas."

If Wegener was in fact the talented and perceptive scientist his contemporaries describe, and if his conclusions were well rooted in evidence and argument, an obvious question arises: Why was the opposition to his ideas so strong, widespread and persistent?

One possible explanation is that Wegener's theory was "premature" at the time he presented it. Gunther S. Stent of the University of California at Berkeley has argued that an idea must be considered premature if it cannot be connected by a series of simple, logical steps to the canonical, or generally accepted, knowledge of the time [see "Prematurity and Uniqueness in Scientific Discovery," by Gunther S. Stent; SCIENTIFIC AMERICAN Offprint 1261.] A related principle, formulated by Michael Polanyi, a British writer on the philosophy and sociology of science, holds that in science there must always be a prevailing opinion of the nature of things, against which the truth of all assertions is tested. Any observation that seems to contradict the established view of the world must be presumed to be invalid and set aside in the hope that it will eventually turn out to be false or irrelevant. This interpretation of how science works suggests that geologists and geophysicists had to be overwhelmed by evidence, as they were in the 1960's, before they could abandon the established doctrine of stationary continents. Wegener's innovations, precisely because they were innovative, had to be held in abeyance until a new orthodoxy could be created from them.

There is almost certainly truth in this analysis, and reassurance that the rejection of Wegener's work was a necessary circumstance in the orderly progress of science; nevertheless, the analysis is not entirely convincing. It could account for indifference to the hypothesis of continental drift but not for the attitude of the many scientists who relegated the theory to the realm of fantasy. It also fails to explain why the traditional model of the earth was retained even after Wegener demonstrated that there were contradictions in it and even though those contradictions were never resolved. Paleontologists continued to rely on vanishing land bridges, for example, at the same time that geophysicists, who had adopted the principle of isostasy, insisted that the sinking of such bridges was impossible.

It has been suggested that the principal impediment to the acceptance of continental drift was the lack of a plausible motive force after Jeffreys had refuted Wegener's initial proposals. If that is the case, however, why was Holmes's convection-current theory given so little consideration? Furthermore, even today the nature of the motor that moves the continents remains uncertain, yet plate tectonics is so well established that those who reject its basic tenets are commonly dismissed as reactionaries.

Perhaps the long travail of Wegener's hypothesis can be best explained as a consequence of inertia. A geologist at the 1928 symposium of the American Association of Petroleum Geologists is reported to have said: "If we are to believe Wegener's hypothesis, we must forget everything that has been learned in the past 70 years and start all over again." It should also be remembered that to the geologists of the time Wegener was an outsider; they must have regarded him as an amateur. Today, of course, we can see that his position was an advantage because he had no stake in preserving the conventional viewpoint. Moreover, we can see that he was not an amateur after all but an interdisciplinary investigator of talent and vision who surely qualifies for a niche in the pantheon of great scientists.

Continental Drift

2

J. Tuzo Wilson

April 1963

Geology has reconstructed with great success the events that lie behind the present appearance of much of the earth's landscape. It has explained many of the observed features, such as folded mountains, fractures in the crust and marine deposits high on the surface of continents. Unfortunately, when it comes to fundamental processes—those that formed the continents and ocean basins, that set the major periods of mountain-building in motion, that began and ended the ice ages—geology has been less successful. On these questions there is no agreement, in spite of much speculation. The range of opinion divides most sharply between the position that the earth has been rigid throughout its history, with fixed ocean basins and continents, and the idea that the earth is slightly plastic, with the continents slowly drifting over its surface, fracturing and reuniting and perhaps growing in the process. Whereas the first of these ideas has been more widely accepted, interest in continental drift is currently on the rise. In this article I shall explore the reasons why.

The subject is large and full of pitfalls. The reader should be warned that I am not presenting an accepted or even a complete theory but one man's view of fragments of a subject to which many are contributing and about which ideas are rapidly changing and developing. If it is conceded that much of this is speculation, then it should also be added that many of the accepted ideas have in fact been speculations also.

In the past several different theories of continental drift have been advanced and each has been shown to be wrong in some respects. Until it is indisputably established that such movements in the earth's crust are impossible, however, a multitude of theories of continental drift remain to be considered. Although there is only one pattern for fixed continents and a rigid earth, many patterns of continental migration are conceivable.

The traditional rigid-earth theory holds that the earth, once hot, is now cooling, that it became rigid at an early date and that the contraction attendant on the cooling process creates compressive forces that, at intervals, squeeze up mountains along the weak margins of continents or in deep basins filled with soft sediments. This view, first suggested by Isaac Newton, was quantitatively established during the 19th century to suit ideas then prevailing. It was found that an initially hot, molten earth would cool to its present temperature in about 100 million years and that, in so doing, its circumference would contract by at least tens and perhaps hundreds of miles. The irregular shape and distribution of continents presented a puzzle but, setting this aside, it was thought that the granitic blocks of the continents had differentiated from the rest of the crustal rock and had frozen in place at the close of the first, fluid chapter of the earth's history. Since then they had been modified *in situ*, without migrating.

This hypothesis, in its essentials, still has many adherents. They include most geologists, with notable exceptions among those who work around the margins of the southern continents. The validity of the underlying physical theory is defended by some physicists. On the other hand, a number of formidable objections have been raised by those who have studied radioactivity, ancient climates, terrestrial magnetism and, most recently, submarine geology. Many biologists have also thought that, although the evolution and migration of later forms of life—particularly since the advent of mammals—could be satisfactorily traced on the existing pattern of continents, the distribution of earlier forms required either land bridges across the oceans—the origin and disappearance of which are difficult to explain—or a different arrangement of the continents.

The discovery of radioactivity altered the original concept of the contraction theory without absolutely invalidating it. In the first place, the age of the earth could be reliably determined from knowledge of the rate at which the unstable isotopes of various elements decay and by measurement of the ratios of daughter to parent isotopes present in the rocks. These studies showed the earth to be much older than had been imagined, perhaps 4.5 billion years old. Dating of the rocks indicated that the continents are zoned and have apparently grown by accretion over the ages. Finally, it was found that the decay of uranium, thorium and one isotope of potassium generates a large but unknown supply of heat that must have slowed, although it did not necessarily stop, the cooling of the earth.

The rigid earth now appeared to be less rigid. It became possible to explain the knowledge, already a century old, that great continental ice sheets had depressed the earth's crust, just as the loads of ice that cover Greenland and Antarctica depress the crust in those regions today. Observation showed that central Scandinavia and northern Canada, which had been covered with glacial ice until it melted 11,000 years ago, were still rising at the rate of about a centimeter a year. Calculations of the viscosity of the interior based on these studies led to the realization that the earth as a whole behaves as though a cool and brittle upper layer, perhaps 100 kilometers thick, rests on a hot and plastic interior. All the large topographical features—continents, ocean basins, mountain ranges and even individual volcanoes—slowly seek a rough hydro-

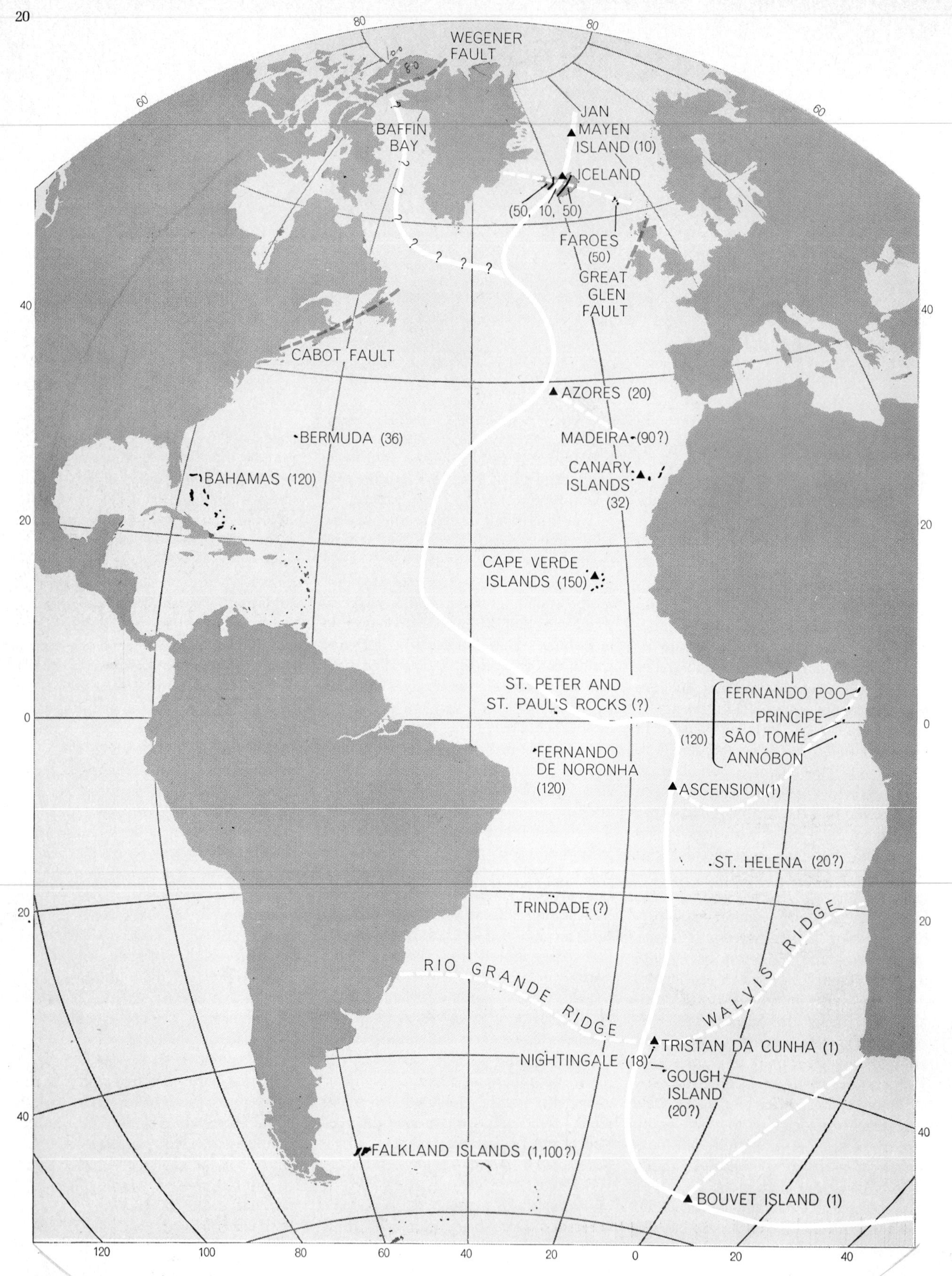

AGE OF ATLANTIC ISLANDS, as indicated by the age of the oldest rocks found on them, apparently tends to increase with increasing distance from the Mid-Atlantic Ridge. The numbers associated with the islands give these ages in millions of years. Geologists divide Iceland into three areas of different ages, the central one being the youngest. The Rio Grande and Walvis ridges are lateral ridges that may have formed as a result of the drifting apart of Africa and South America. Other lateral ridges along the Mid-Atlantic Ridge are also represented. Islands that have active volcanoes are represented by black triangles; most of these islands lie on or near the Mid-Atlantic Ridge. The extension of the ridge into Baffin Bay is postulated. Broken colored lines are faults.

static equilibrium with one another on the exterior. Precise local measurements of gravity showed that the reason some features remain higher than others is that they have deeper, lighter roots than those that are low. The continents were seen to float like great tabular icebergs on a frozen sea.

Everyone could agree that in response to vertical forces the outer crustal layer moved up and down, causing flow in the interior. The crux of the argument between the proponents of fixed and of drifting continents became the question of whether the outer crust must remain rigid under horizontal forces or whether it could respond to such forces by slow lateral movements.

Gondwanaland and "Pangaea"

Suggestions that the continents might have moved had been advanced on various grounds for centuries. The remarkable jigsaw-puzzle fit of the Atlantic coasts of Africa and South America provoked the imagination of explorers almost as soon as the continental outlines appeared opposite each other on the world map. In the late 19th century geologists of the Southern Hemisphere were moved to push the continents of that hemisphere together in one or another combination in order to explain the parallel formations they found, and by the turn of the century the Austrian geologist Eduard Suess had reassembled them all in a single giant land mass that he called Gondwanaland (after Gondwana, a key geological province in east central India).

The first comprehensive theory of continental drift was put forward by the German meteorologist Alfred Wegener in 1912. He argued that if the earth could flow vertically in response to vertical forces, it could also flow laterally. In support of a different primeval arrangement of land masses he was able to point to an astonishing number of close affinities of fossils, rocks and structures on opposite sides of the Atlantic that, he suggested, ran evenly across, like lines of print when the ragged edges of two pieces of a torn newspaper are fitted together again. According to Wegener all the continents had been joined in a single supercontinent about 200 million years ago, with the Western Hemisphere continents moved eastward and butted against the western shores of Europe and Africa and with the Southern Hemisphere continents nestled together on the southern flank of this "Pangaea." Under the action of forces

GREAT GLEN FAULT in Scotland is named for a valley resulting from erosion along the line of the fault. About 350 million years ago the northern part of Scotland was slowly moved some 60 miles to the southwest along this line (*see illustration on opposite page*).

ASPY FAULT in northern Nova Scotia is marked by several cliffs like the one seen here. The fault is part of the Cabot Fault system extending from Boston to Newfoundland (*see illustration on opposite page*) and may represent an extension of the Great Glen Fault.

associated with the rotation of the earth, the continents had broken apart, opening up the Atlantic and Indian oceans.

Between 1920 and 1930 Wegener's hypothesis excited great controversy. Physicists found the mechanism he had proposed inadequate and expressed doubt that the continents could move laterally in any case. Geologists showed that some of Wegener's suggestions for reassembling the continents into a single continent were certainly wrong and that drift was unnecessary to explain the coincidences of geology in many areas. They could not, however, dispute the validity of most of the transatlantic connections. Indeed, more such connections have been steadily added.

It was the discovery of one of these connections that prompted my own recent inquiries into the subject of continental drift. A huge fault of great age bisects Scotland along the Great Glen in the Caledonian Mountains. On the western side of the Atlantic, I was able to show, a string of well-known faults of the same great age connect up into another huge fault, the "Cabot Fault" extending from Boston to northern Newfoundland. These two great faults are much older than the submarine ridge and rift recently discovered on the floor of the mid-Atlantic and shown to be a

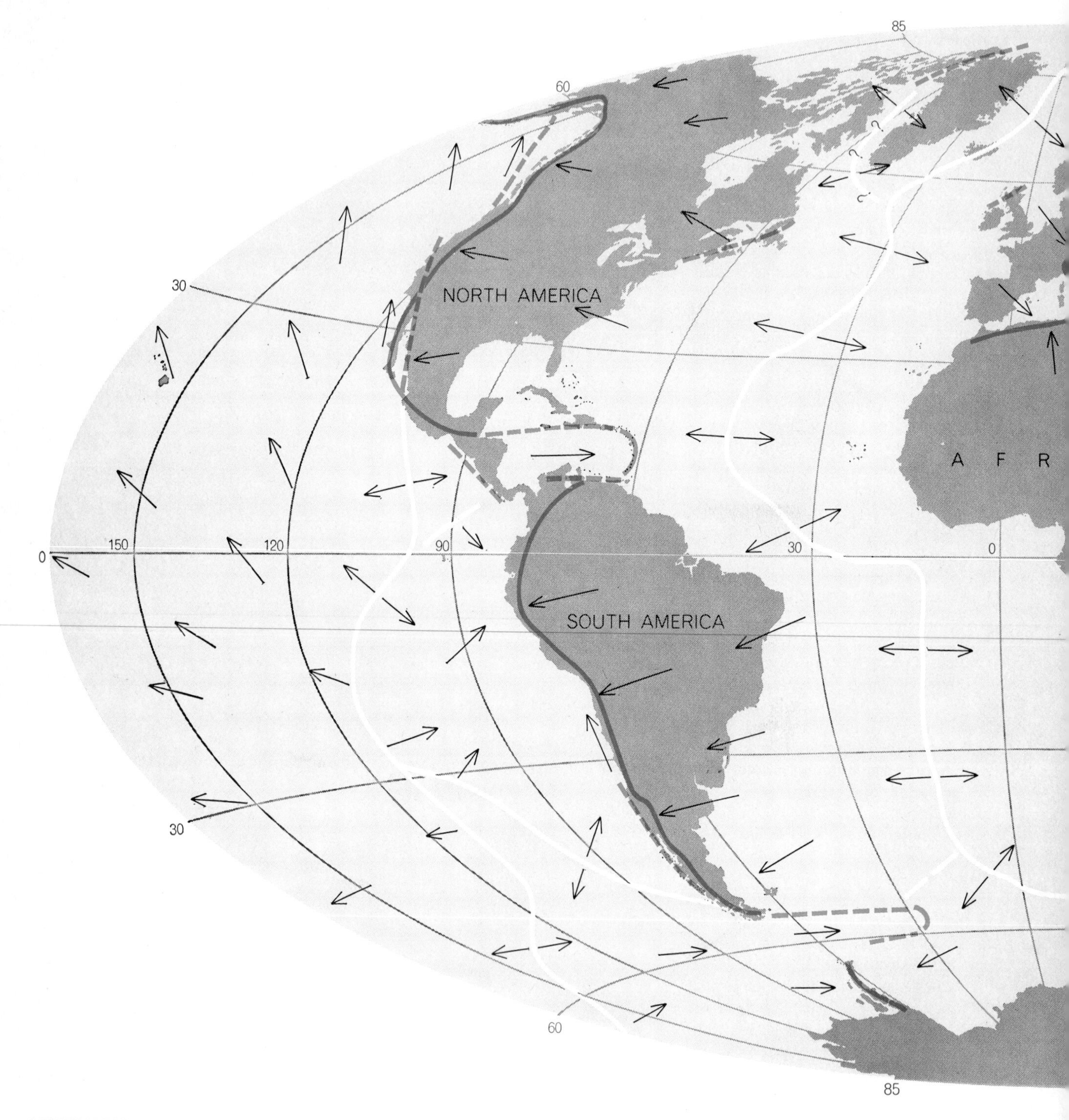

CONVECTION CURRENTS in the earth's mantle may move blocks of crustal material with different effects. Continental mountain chains and island arcs could form where currents sink and blocks meet; mid-ocean ridges, where currents rise and blocks are torn

young formation. The two faults would be one if Wegener's reconstruction or something like it were correct. Wegener also thought that Greenland (where he died in 1930) and Ellesmere Island in the Canadian Arctic had been torn apart by a great lateral displacement along the Robeson Channel. The Geological Survey of Canada has since discovered that the Canadian coast is faulted there.

Many geologists of the Southern Hemisphere, led by Alex. L. Du Toit of South Africa, welcomed Wegener's views. They sought to explain the mounting evidence that an ice age of 200 million years ago had spread a glacier over the now scattered continents of the Southern Hemisphere. At the same time, according to the geological record, the great coal deposits of the Northern Hemisphere were being formed in tropical forests as far north as Spitsbergen. To resolve this climatic paradox Du Toit proposed a different reconstruction of the continent. He brought the southern continents together at the South Pole and the northern coal forests toward the Equator. Later, he thought, the southern continent had broken up and its component subcontinents had drifted northward.

The compelling evidence for the existence of a Gondwanaland during the

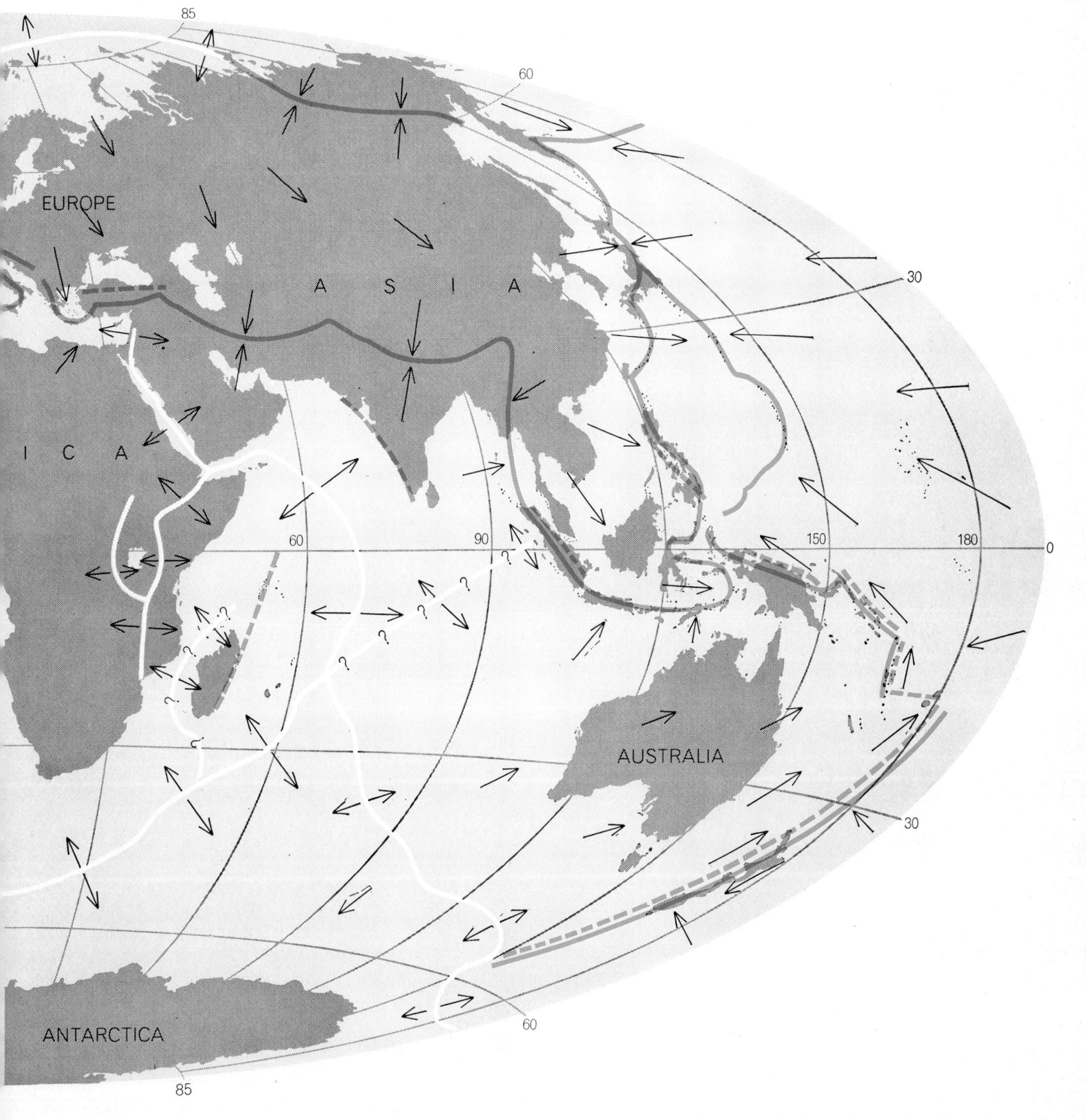

apart. On this assumption arrows indicate directions of horizontal flow of currents at the present time. Solid colored lines represent mountain chains and island arcs; heavy white lines, the worldwide system of mid-ocean ridges; and broken colored lines, faults.

Mesozoic era—the "Age of Reptiles"—has been reinforced by the findings made in Antarctica since the intensive study of that continent began in 1955. The ice-free outcrops on the continent, although few, not only show the record of the earlier ice age that gripped the rest of the land masses in the Southern Hemisphere but also bear deposits of a low-grade coal laid down in a still earlier age of verdure that covered all the same land masses with the peculiar big-leafed *Glossopteris* flora found in their coal beds as well.

Many suggestions have been made as to how to create and destroy the land bridges needed to explain the biological evidence without moving the continents. Some involve isthmuses and some involve whole continents that have subsided below the surface of the ocean. But the chemistry and density of continents and ocean floors are now known to be so different that it seems even more difficult today to raise and lower ocean floors than it is to cause continents to migrate.

Convection in the Mantle

One of the first leads to a mechanism that would move continents came more than 30 years ago from the extension to the ocean floor of the sensitive techniques of gravimetry that had established the rule of hydrostatic equilibrium, or isostasy, ashore. The Dutch geophysicist Felix A. Vening Meinesz demonstrated that a submerged submarine would provide a sufficiently stable platform to allow the use of a gravimeter at sea. Over the abyssal trenches in the sea floor that are associated with the island arcs of Indonesia and the western side of the Pacific he found some of the largest deficiencies in gravity ever recorded. It was clear that isostasy does not hold in the trenches. Some force at work there pulls the crust into the depths of the trenches more strongly than the pull of gravity does.

Arthur Holmes of the University of Edinburgh and D. T. Griggs, now at the University of California at Los Angeles, were stimulated by these observations to re-examine and restate in modern terms an old idea of geophysics: that the interior of the earth is in a state of extremely sluggish thermal convection, turning over the way water does when it is heated in a pan. They showed that convection currents were necessary to account in full for the transfer of heat flowing from the earth's interior through the poorly conductive material of the mantle: the region that lies between the core and the crust. The trenches, they said, mark the places where currents in

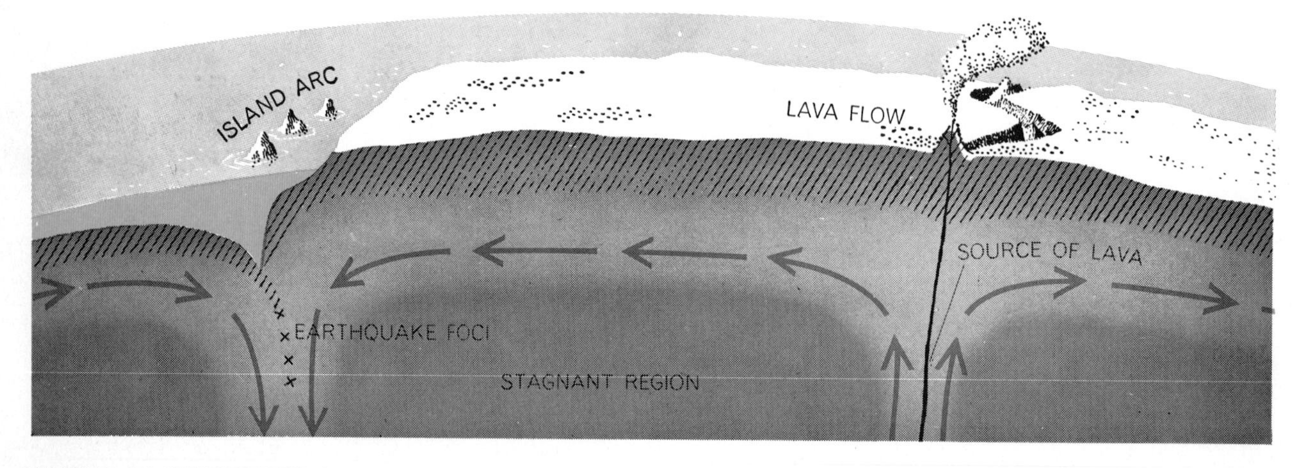

EFFECTS OF CONVECTION CURRENTS, schematized in the two illustrations on this page, provide one possible means of accounting for the formation of median ridges, lateral ridges, mountain ranges and earthquake belts. Rising and separating currents (*arrows at right*) could break the crustal rock and pull it apart; the rift would be filled by altered mantle material (as suggested by H. H. Hess of Princeton University) and lava flows, forming a median ridge. Sinking currents (*left*) could pull the ocean floor down.

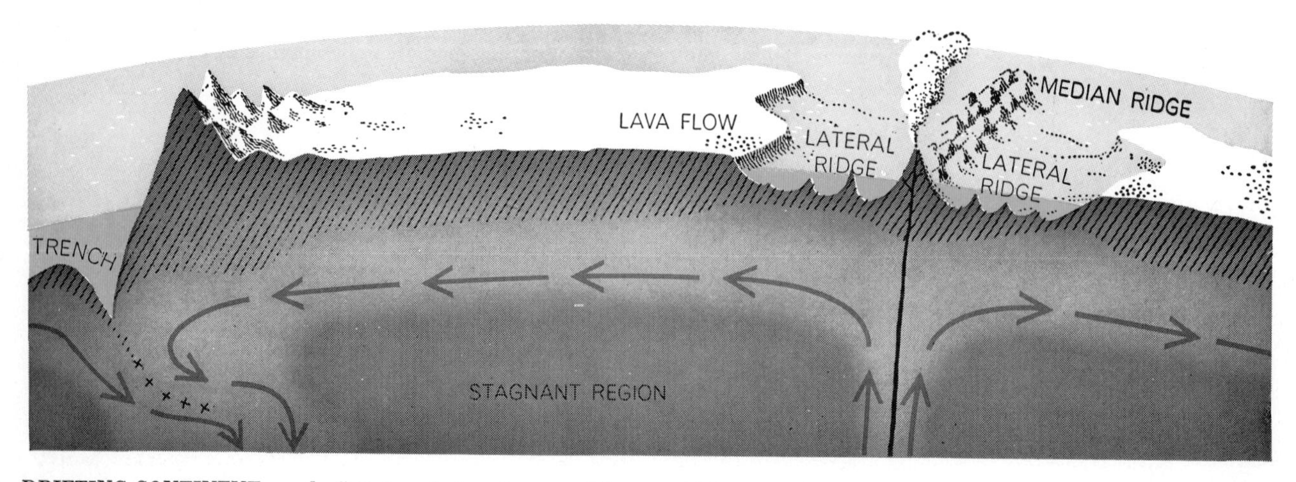

DRIFTING CONTINENT may be "piled up," where it meets sinking currents, to form mountains like those of the Andes (*left*). Since continents are lighter than the mantle material of the ocean floor, they cannot sink but tend to be pushed over sinking currents, which are marked by deep earthquakes. Active volcanoes continue to form over rising currents (*right*); but drift may carry these volcanic piles away to either side of the median ridge. Separated from their source, the inactive cones form one or two lateral ridges.

the mantle descend again into the interior of the earth, pulling down the ocean floor.

Convection currents in the mantle now play the leading role in every discussion of the large-scale and long-term processes that go on in the earth. It is true that the evidence for their existence is indirect; they flow too deep in the earth and too slowly—a few centimeters a year—for direct observation. Nonetheless their presence is supported by an increasing body of independently established evidence and by a more rigorous statement of the theory of their behavior. Recently, for example, S. K. Runcorn of Durham University has shown that to stop convection the mantle material would have to be 10,000 times more viscous than the rate of postglacial recoil indicates. It is, therefore, highly probable that convection currents are flowing in the earth.

Perhaps the strongest confirmation has come with the discovery of the regions where these currents appear to ascend toward the earth's surface. This is the major discovery of the recent period of extraordinary progress in the exploration of the ocean bottom, and it involves a feature of the earth's topography as grand in scale as the continents themselves. Across the floors of all the oceans, for a distance of 40,000 miles, there runs a continuous system of ridges. Over long stretches, as in the mid-Atlantic, the ridge is faulted and rifted under the tension of forces acting at right angles to the axis of the ridge. Measurements first undertaken by Sir Edward Bullard of the University of Cambridge show that the flow of heat is unusually great along these ridges, exceeding by two to eight times the average flow of a millionth of a calorie per square centimeter per second observed on the continents and elsewhere on the ocean floor. Such measurements also show that the flow of heat in the trenches, as in the Acapulco Trench off the Pacific coast of Central America, falls to as little as a tenth of the average.

Most oceanographers now agree that the ridges form where convection currents rise in the earth's mantle and that the trenches are pulled down by the descent of these currents into the mantle. The possibility of lateral movement of the currents in between is supported by evidence for a slightly plastic layer—called the asthenosphere—below the brittle shell of the earth. Seismic observations show that the speed of sound in this layer suddenly becomes slower, indicating that the rock is less dense, hotter and more plastic. These observations have also yielded evidence that the asthenosphere is a few hundred kilometers thick, somewhat thicker than

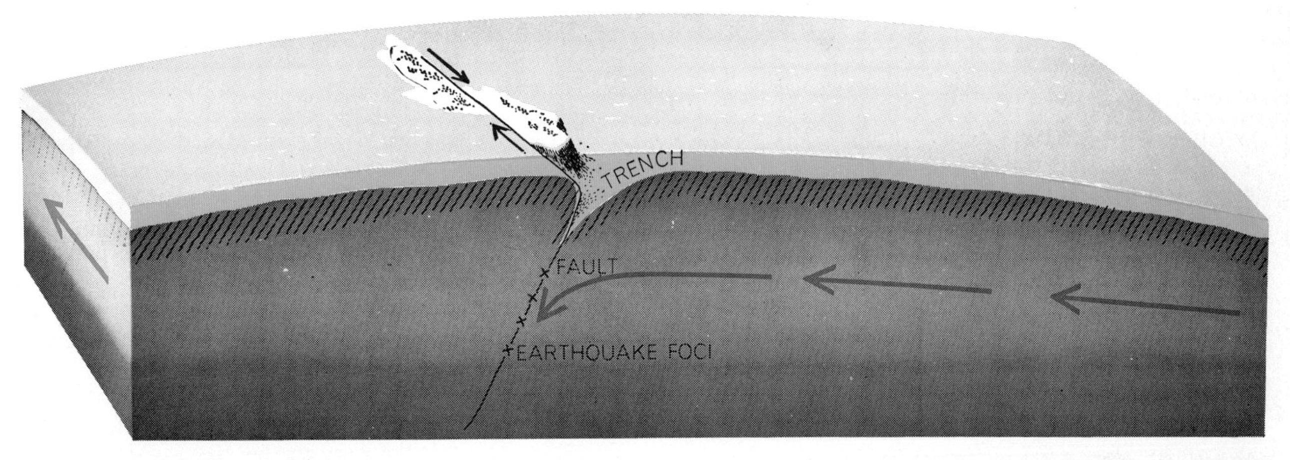

TWO CONVECTION CURRENTS perpendicular to each other suggest a mechanism for producing large horizontal faults such as the one that has offset western New Zealand 300 miles northward. The two convection currents (*arrows indicate direction*) would produce a fault. One current would be forced downward, producing a trench and earthquakes along the sloping surface. Continued flow of the second current would result in a sliding motion, or lateral displacement, along the plane of the fault, shearing the island in two.

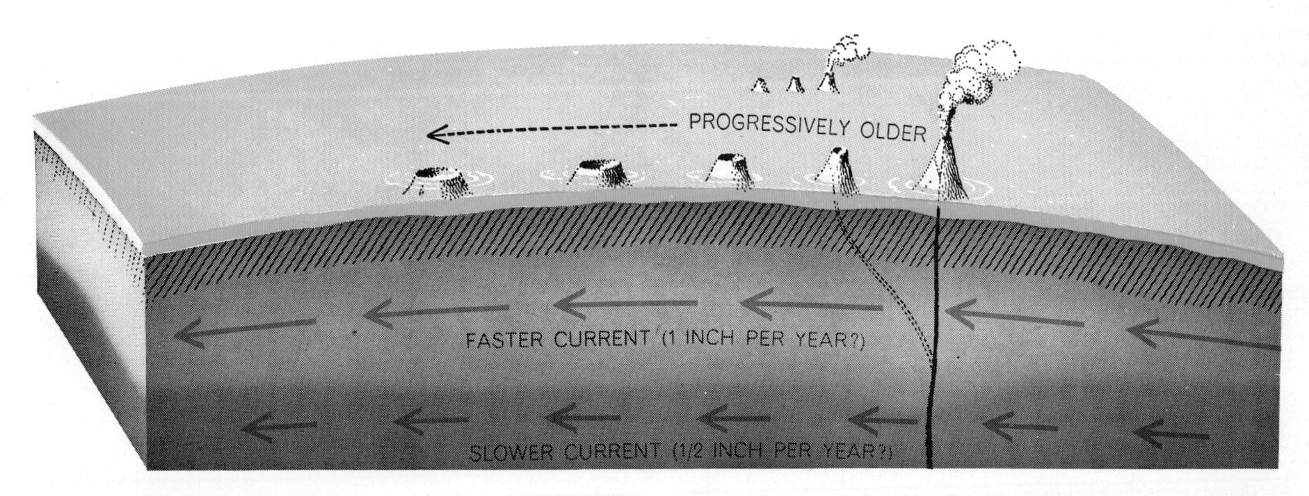

VOLCANIC-ISLAND CHAINS like the Hawaiian Islands must have originated in a process slightly different from that which formed pairs of lateral ridges. The source of lava flow does not lie on a mid-ocean ridge; it is considered that the source may be deep (100 miles or more) in the slower moving part of convection currents. The differential motion carries old volcanoes away from the source, while new volcanoes form over the source. The length of the island chain depends on how long the source has been active.

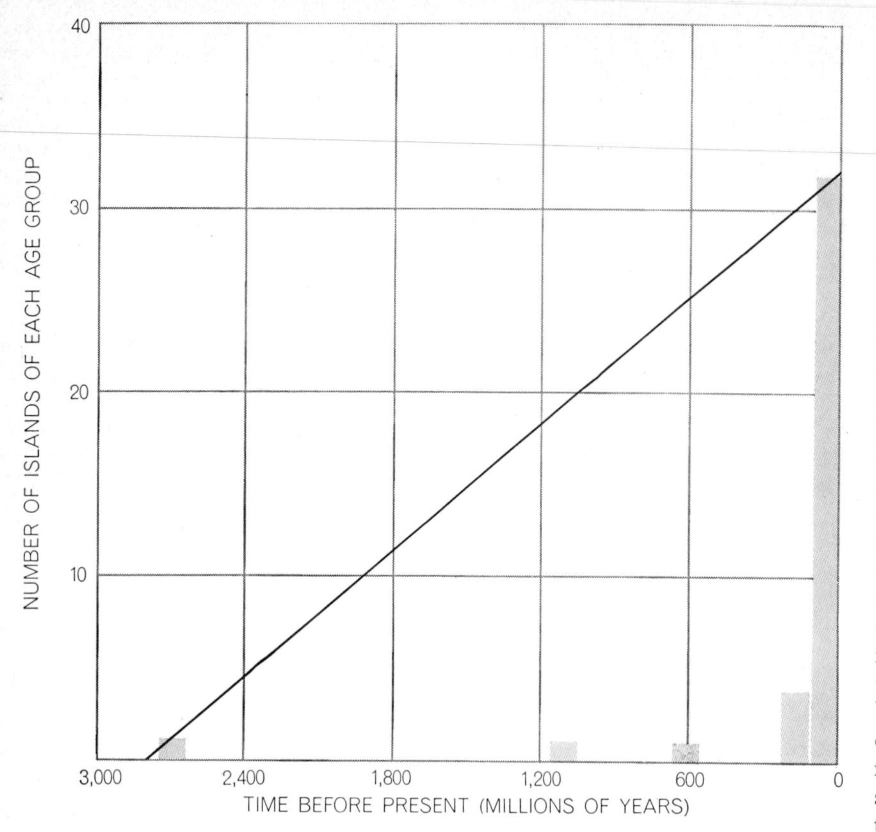

FREQUENCY DIAGRAM shows the age distribution of about 40 islands (in main ocean basins) dated older than "recent" (the number of very young islands is vastly greater). The diagonal line shows the corresponding curve for continental rock ages over equivalent areas.

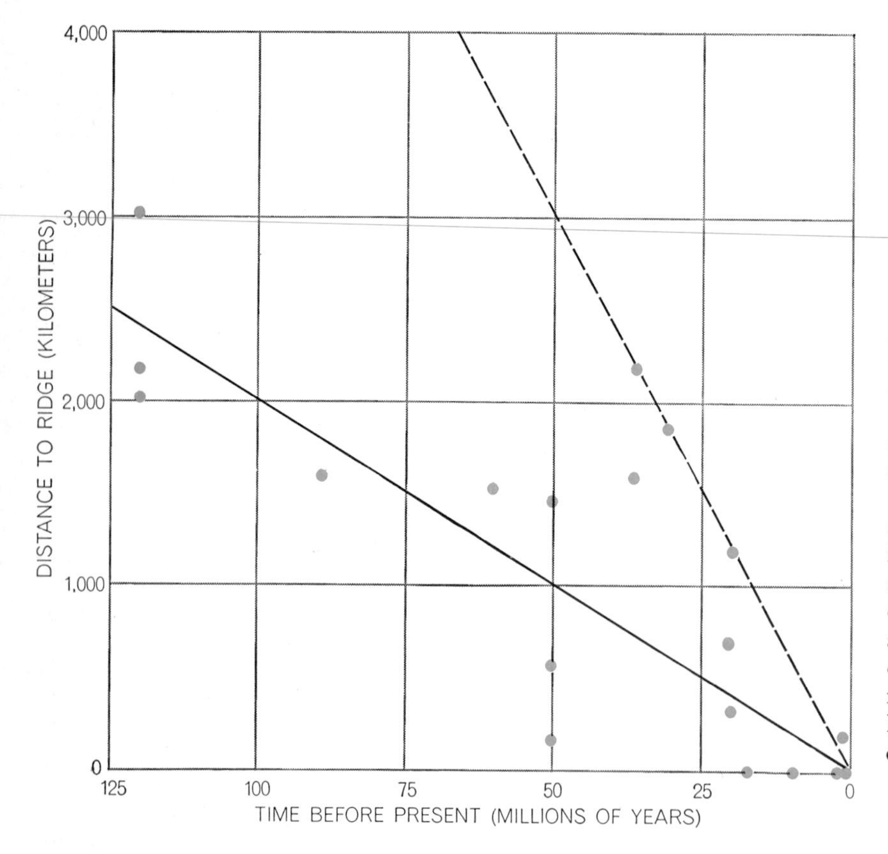

DISTANCE FROM MID-OCEAN RIDGE of some islands in Atlantic and Indian oceans is plotted against age. If all originated over the ridge, their average rate of motion has been two centimeters a year (*solid line*); maximum rate, six centimeters a year (*broken line*).

the crust, and that below it the viscosity increases again.

Here, then, is a mechanism, in harmony with physical theory and much geological and geophysical observation, that provides a means for disrupting and moving continents. It is easy to believe that where the convection currents rise and separate, the surface rocks are broken by tension and pulled apart, the rift being filled by the altered top of the mantle and by the flow of basalt lavas. In contrast to earlier theories of continental drift that required the continents to be driven through the crust like ships through a frozen sea, this mechanism conveys them passively by the lateral movement of the crust from the source of a convection current to its sink. The continents, having been built up by the accumulation of lighter and more siliceous materials brought up from below, are not dragged down at the trenches where the currents descend but pile up there in mountains. The ocean floor, being essentially altered mantle, can be carried downward; such sediments as have accumulated in the trenches descend also and, by complicated processes, may add new mountains to the continents. Since the material near the surface is chilled and brittle, it fractures, causing earthquakes until it is heated by its descent.

From the physical point of view, the convection cells in the mantle that drive these currents can assume a variety of sizes and configurations, starting up and slowing down from time to time, expanding and contracting. The flow of the currents on the world map may therefore follow a single pattern for a time, but the pattern should also change occasionally owing to changes in the output and transfer of heat from within. It is thus possible to explain the periodicity of mountain-building, the random and asymmetrical distribution of the continents and the abrupt breakup of an ancient continent.

Some geophysicists consider that isostatic processes set up by gravitational forces may suffice to cause the outer shell to fracture and to slip laterally over the plastic layer of the asthenosphere. This mechanism would not require the intervention of convection currents. Both mechanisms could explain large horizontal displacements of the crust.

Evidence from Terrestrial Magnetism

Fresh evidence that such great movements have indeed been taking place has been provided by two lines of study in

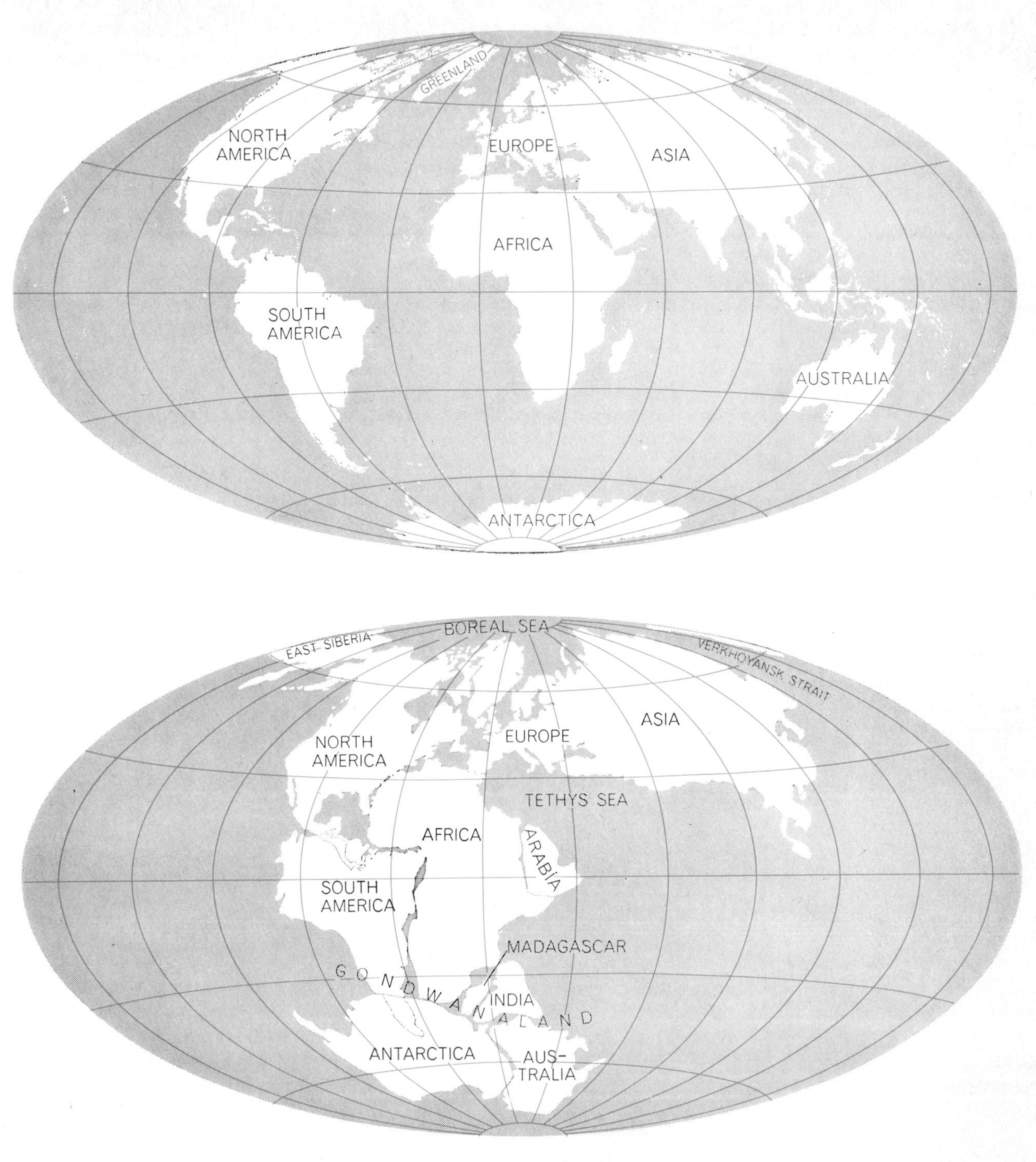

SINGLE SUPERCONTINENT, presumed to have existed some 150 million years ago, would have resembled that depicted in the map at bottom. A present-day map appears at top. In both maps the distortion of the continents is a result of the projection employed.

the field of terrestrial magnetism. On the one hand, surveys of the earth's magnetic field off the coast of California show a pattern of local anomalies in the ocean floor running parallel to the axis of a now inactive oceanic ridge that underlies the edge of the continent. The pattern bears a persuasive resemblance to the "photoelastic" strain patterns revealed by polarized light in plastics placed under stress. More important, the pattern shows that the ocean floor is faulted at right angles to the axis of the ridge, with great slabs of the crust displaced laterally to the west by as much as 750 miles. These are apparently ancient and inactive fractures; now the active faults run northwesterly, as is indicated by the earthquakes along California's San Andreas Fault.

Evidence of a more general nature in favor of continental drift comes from the studies of the "remanent" magnetism of the rocks, to which Runcorn, P. M. S. Blackett of the University of London and Emil Thellier of the University of Paris have made significant contributions. Their investigations have shown that rocks can be weakly magnetized at the time of formation—during cooling in the case of lavas and during deposition in the case of sediments—and that their polarity is aligned with the direction of the earth's magnetic field at the place and time of their formation. The present orientation of the rocks of various ages on the continents indicates that they must have been formed in different latitudes. The rocks of any one continent show consistent trends in change of orientation with age; those from other continents show different shifts. Continental drift offers the only explanation of these findings that has withstood analysis.

Some physicists and biologists are now prepared to accept continental drift, but many geologists still have no use for the hypothesis. This is to be expected. Continents are so large that much geology would be the same whether drift had occurred or not. It is the geology of the ocean floors that promises to settle the question, but the real study of that two-thirds of the earth's surface has just begun.

The Oceanic Islands

One decisive test turns on the age of the ocean floor. If the continents have been fixed, the ocean basins should all be as old as the continents. If drift has occurred, some regions of the ocean floor should be younger than the time of drift.

A survey of the scattered and by no means complete literature on the oceanic islands conducted by our group at the University of Toronto shows that of all the islands in the main ocean basins only about 40 have rocks that have been dated older than the Recent epoch. Only three of these—Madagascar and the Seychelles of the Indian Ocean and the Falklands of the South Atlantic—have very old rocks; all the others are less than 150 million years old. If one regards the exceptions as fragments of the nearby continents, the youth of the others suggests that either the ocean basins are young or that islands are not representative samples of the rock of the ocean floor.

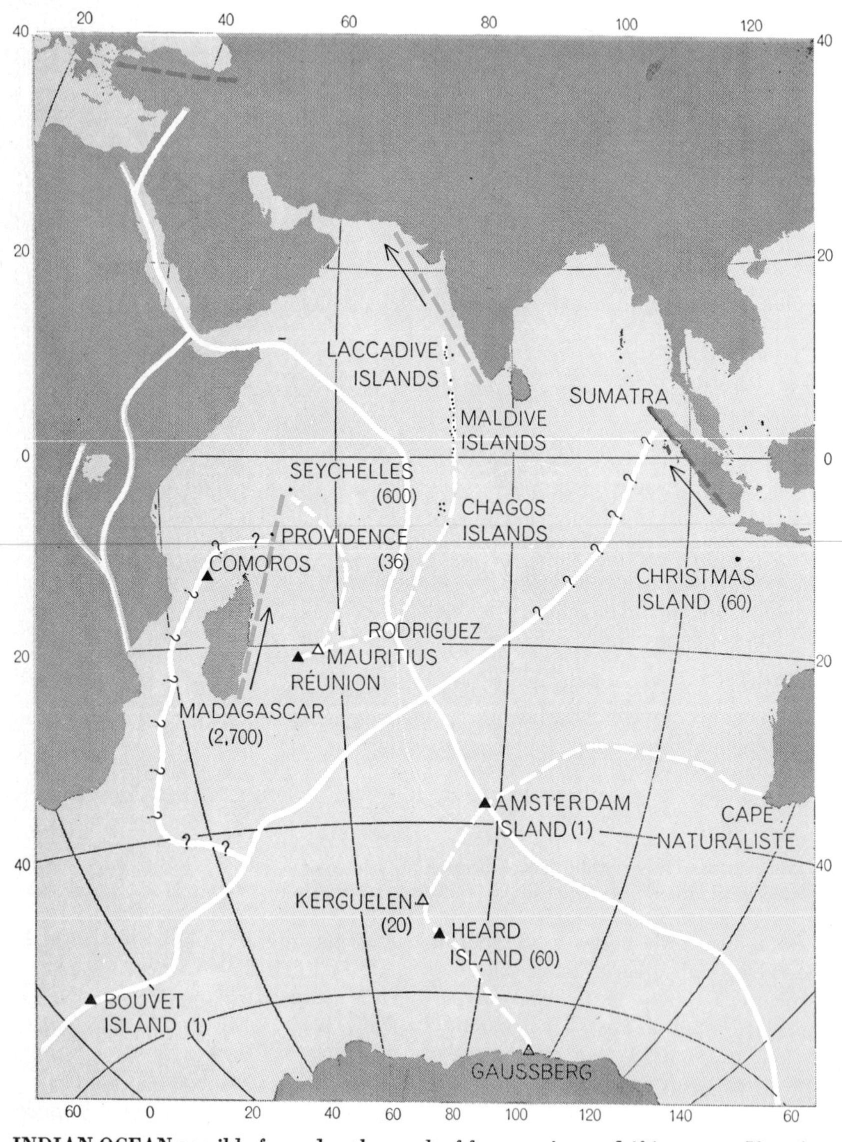

INDIAN OCEAN possibly formed as the result of four continents drifting apart. If so, four median ridges would have formed midway between continents, with pairs of lateral ridges connecting them. Heavy white lines show three known median ridges; there is evidence for one running to Sumatra. Broken white lines are lateral ridges; broken colored lines, faults; open triangles, inactive volcanoes. Numbers give ages in millions of years.

Significantly, it turns out that the age of the islands in the Atlantic Ocean tends to increase with their distance from the mid-ocean ridge. In this reckoning one need not count the island arcs of the West Indies or the South Sandwich Islands, which belong to the Cordilleran system—that is, the spine of mountains running the entire length of North and South America—and so have a continental origin. At least six of the islands on the ridge or very close to it have on them active volcanoes that have had recent eruptions; the most recent was the eruption of Tristan da Cunha, which is located squarely on the ridge in the South Atlantic. Only two of the islands far from the ridge have active volcanoes. If the hot convection currents of the mantle rise under the mid-ocean ridge, it is easy to understand why the ridge is the locus of active volcanoes and earthquakes. The increase in age with distance from the ridge suggests that if the more distant islands had a volcanic origin on the ridge, lateral movement of the ocean floor has carried them away from the ridge. Their ages and distances from the ridge indicate movement at the rate of two to six centimeters a year on the average, in keeping with the estimated velocity of the convection currents.

Of great significance in connection with the mechanism postulated here are the two lateral ridges that run east and west from Tristan da Cunha to Africa on the one hand and to South America on the other. It is reasonable to suppose that these ridges had their origin in a succession of volcanoes that erupted and grew into mountains on the site of the present volcano and were carried off east and west to form a row of progressively older, extinct and drowned volcanoes [*see illustration on page 20*] There are no earthquakes along the la eral ridges and so they are distinctly different in character from the mid-ocean ridge. These ridges meet the continental margins at places that would fit together on the quite independent criterion of the match of their shore lines. One explanation of this coincidence is that the continents were indeed joined together and have moved apart, with the lateral ridges forming trails that record the motion. The two ridges are roughly mirror images of each other, showing that the motion was uniform on each side. Another similar pair of ridges connects Iceland—where the mid-ocean ridge comes to the surface and where the great tension rift is visible in the Icelandic Graben—to Greenland and the shelf of the European continent.

We have therefore advanced two related hypotheses: first, that where adjacent continents were once joined a median ridge should now lie between them; second, that where such continents are connected by lateral ridges they were once butted together in such a manner that points marked by the shoreward ends of these ridges coincided. If this is correct, it provides a unique method for reassembling continents that have drifted apart. One of the major troubles with theories of drift has been that the possibilities are so numerous no such precise criterion existed for putting the poorly fitting jigsaw puzzle together.

Without doubt the most severe test of this double hypothesis is presented by the Indian Ocean. Here four continents—Africa, India, Australia and Antarctica—may be assumed on geological and paleomagnetic evidence to have drifted apart. The collision of India with the Asian land mass could have thrown up the Himalaya mountains at their junction. These continents should accordingly be separated by four mid-ocean ridges. Three such ridges have

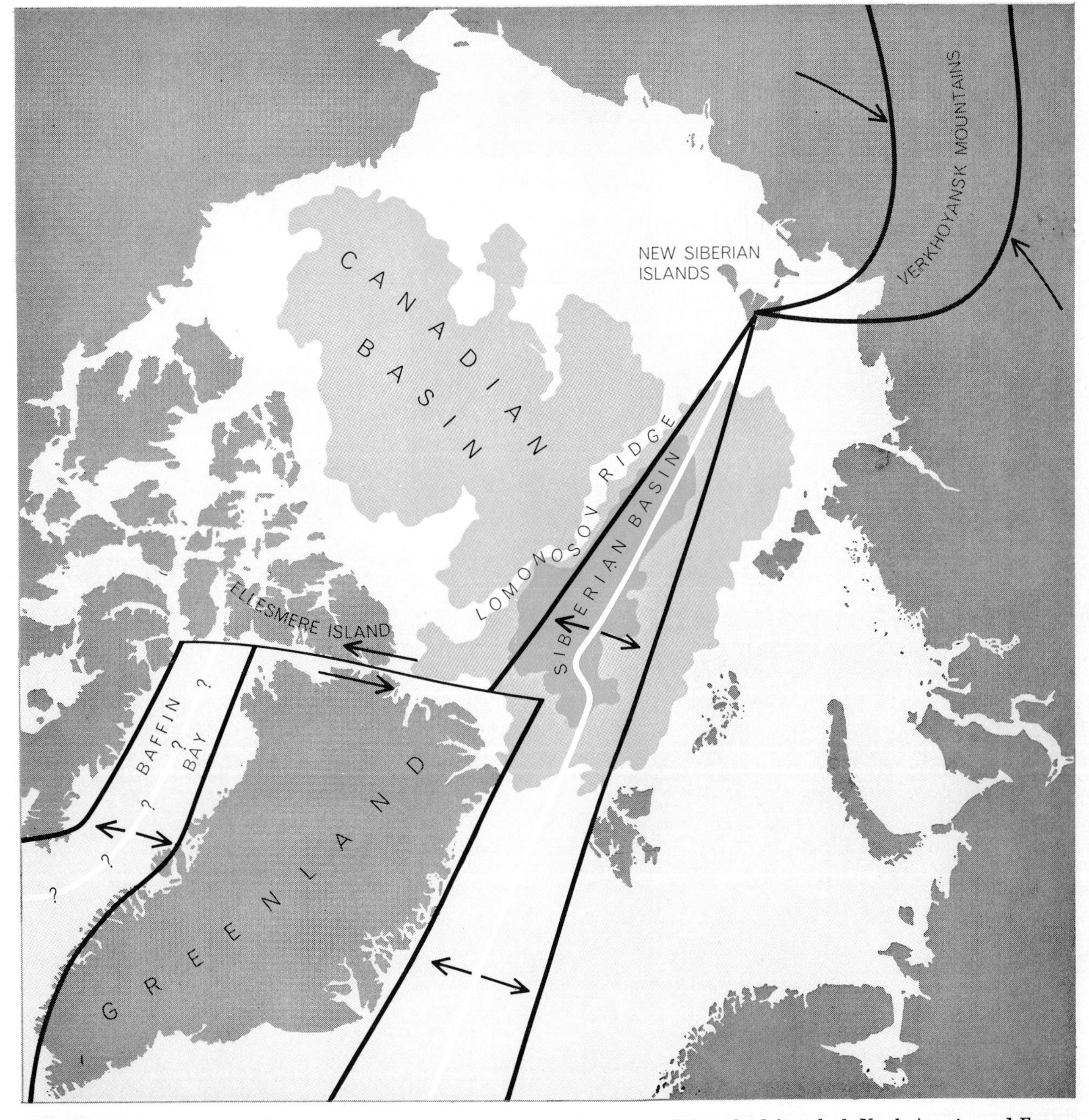

RIFTING OF SUPERCONTINENT to form the Atlantic Ocean could have produced the Verkhoyansk Mountains in eastern Siberia. As shown on this map of the Arctic, the rift spread more widely to the south. The opening of the Atlantic Ocean and Baffin Bay separated Greenland from both North America and Europe. The continents were rotated slightly about a fulcrum near the New Siberian Islands. The resulting compression and uplift would create a mountain range. Opposing arrows mark the Wegener Fault.

already been well established by surveys of the Indian Ocean, and there is evidence for the existence of the fourth. In each quadrant marked off by the ridges there is also, it happens, a lateral ridge! These submarine trails may be presumed to be records of the motion of the continents as they receded from one another. From Amsterdam Island one of these lateral ridges runs through Kerguelen Island to Gaussberg Mountain on the coast of Antarctica; a mirror image of this ridge runs from Amsterdam Island to Cape Naturaliste on Australia. The corresponding ridges connecting Africa and India are distorted by lateral faults running along the coasts of Madagascar and India. Thus in each quadrant there exists a lateral ridge to show how points on Madagascar, India, Australia and Antarctica once lay close together. What is remarkable is not that there is some irregularity in the present configuration of these ridges but that the floor of the Indian Ocean should show such a symmetrical pattern.

The mid-ocean ridge separating Australia from Antarctica has been traced by Henry W. Menard of the Scripps Institution of Oceanography across the

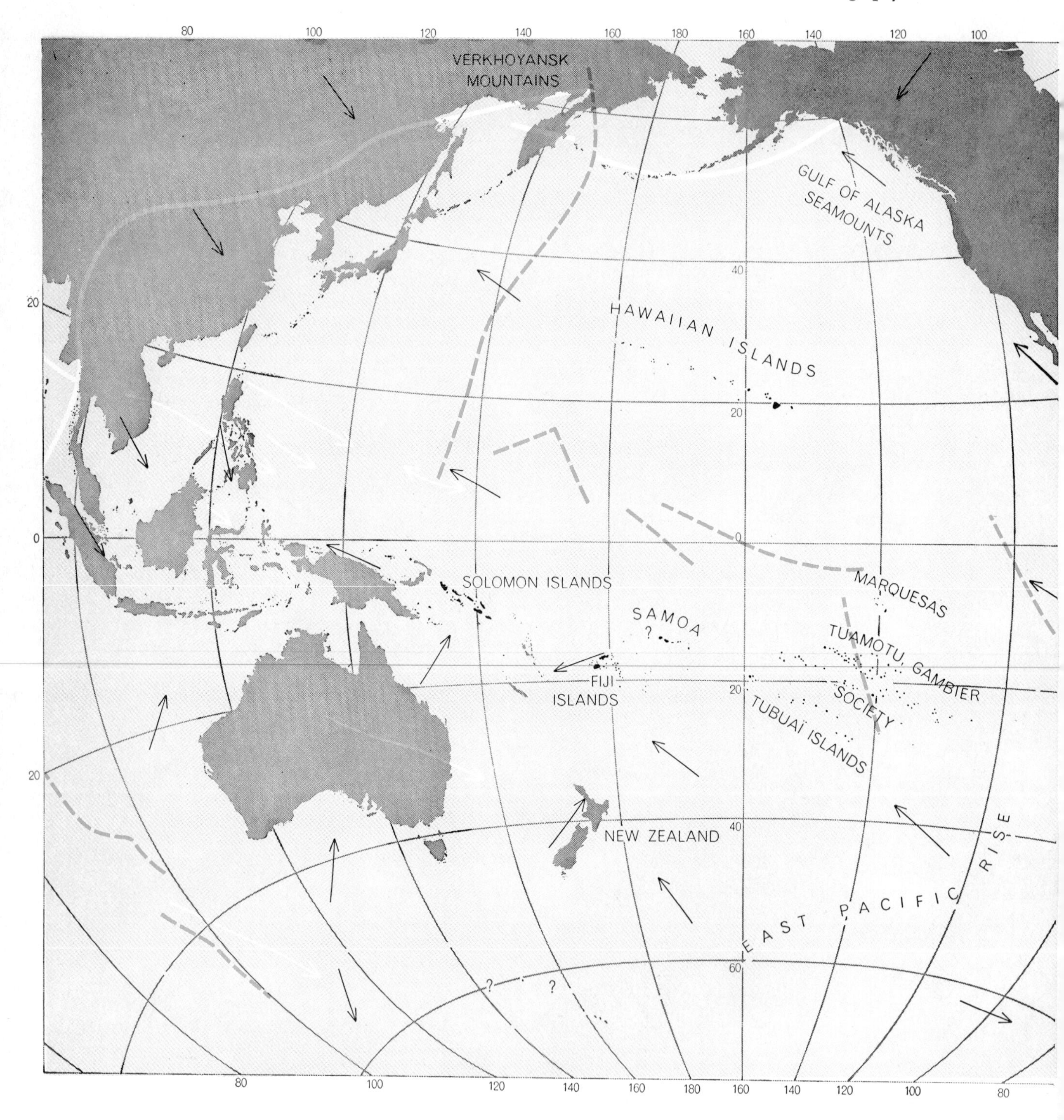

AGE OF PACIFIC ISLANDS appears to increase with increasing distance from the mid-ocean ridge. This is compatible with the idea that the eastern half of the Pacific Ocean has been spreading from the East Pacific Rise (as has been suggested by Robert S. Dietz of the U.S. Navy Electronics Laboratory). Broken colored lines represent faults; the associated arrows indicate the direction of horizontal motion, where known, along the fault. Other arrows show the probable directions of convection flow. Island arcs of the kind

eastern Pacific to connect with the great East Pacific Rise. From the topography of the Pacific floor it can be deduced that this ridge once extended through the rise marked by Cocos Island off Central America and formed the rifted ridge that moved North and South America apart. Another branch of this ridge, running across the southern latitudes, suggests the cause of the separation of South America from Antarctica. The oceanic islands in this broad region of the Pacific form lines that extend at right angles down the flanks of the East Pacific Rise; geologists long ago established that these islands grow progressively older with distance from the top of the rise. Unlike the rest of the continuous belt of mid-ocean ridges to which it is connected, the East Pacific Rise tends to run along the margins of the Pacific Ocean; it has rifted an older ocean apart rather than a continent. The floor of the western Pacific is believed to be a remnant of that older floor.

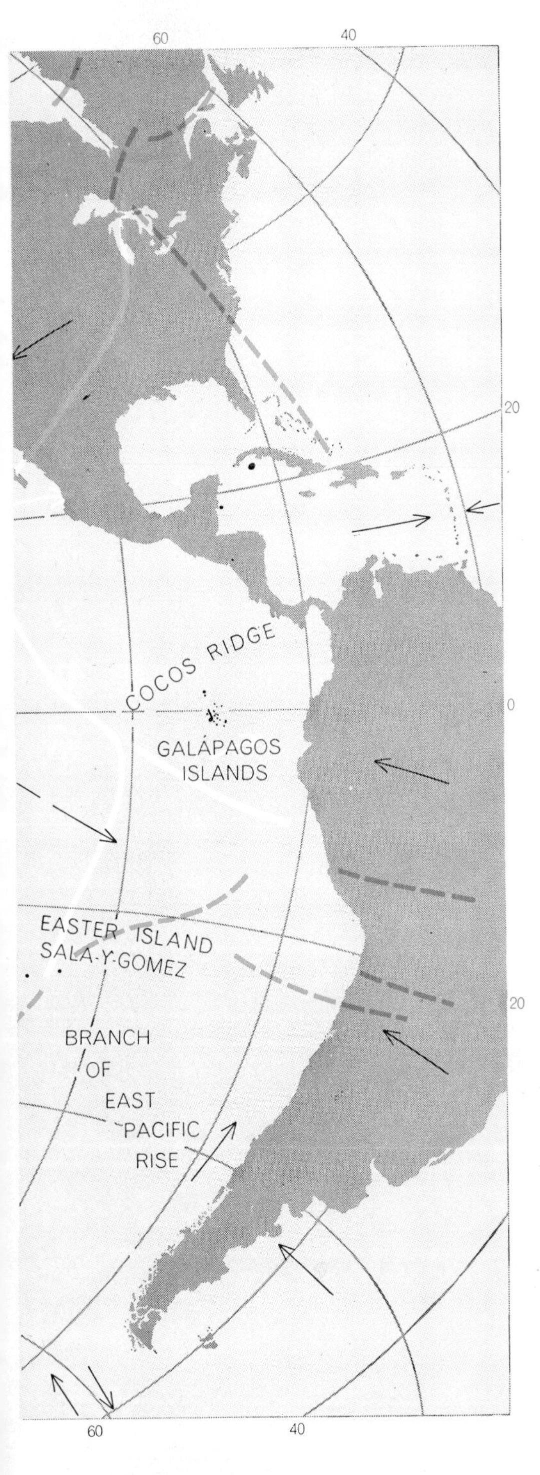

represented by Japan develop where the forces associated with such flow are directly opposed; great horizontal faults, where these forces meet at right angles.

There are therefore enough connections to draw all the continents together, reversing the trends of motion indicated by the mid-ocean ridges and using the continental ends of pairs of lateral ridges as the means of matching the coast lines together. The ages of the islands and of the coastal formations suggest that about 150 million years ago, in mid-Mesozoic time, all the continents were joined in one land mass and that there was only one great ocean [*see illustration on page 27*]. The supercontinent that emerges from this reconstruction is not the same as those proposed by Wegener, Du Toit and other geologists, although all have features in common. The widespread desert conditions of the mid-Mesozoic may have been a consequence of the unusual circumstance that produced a single continent and a single ocean at that time. Since its approximate location with respect to latitude is known, along with the location of its major mountain systems, the climate in various regions might be reconstructed and compared with geological evidence.

It is not suggested that this continent was primeval. That it was in fact assembled from still older fragments is suggested by two junction lines: the ancient mountain chain of the Urals and the chain formed by the union of the Appalachian, Caledonian and Scandinavian mountains may have been thrown up in the collisions of older continental blocks. Before that there had presumably been a long history of periodic assembly and disassembly of continents and fracturing and spreading of ocean floors, as convection cells in the mantle proceeded to turn over in different configurations. At present it is impossible even to speculate about the details.

Breakup of the Supercontinent

If it can be assumed that the proposed Mesozoic continent did exist and spread apart, geology provides some guide to the history of its fragmentation. The present system of convection currents has apparently been constant in general configuration ever since the Mesozoic, but not all parts of it have been equally active all of that time. Shortly before the start of the Cretaceous period, about 120 million years ago, the continent developed a rift that opened up to form the Atlantic Ocean. The rift spread more widely in the south, with the result that the continents must have rotated slightly about a fulcrum near the New Siberian Islands [*see illustration on page 29*]. Soviet geologist have found that the compression and uplift that raised the Verkhoyansk Mountains across eastern Siberia began at about that time. To the south a continuation of the rifting separated Africa from Antarctica and spread diagonally across the Indian Ocean, opening the northeasterly rift. Africa and India were thus moved northward, away from the still intact Australian-Antarctic land mass.

It seems reasonable to suggest, particularly from the geology of the Verkhoyansk Mountains and of Iceland, that at the start of Tertiary time, about 60 million years ago, this convection system became less active and that rifting started up elsewhere. A new rift opened up along the other, northwesterly, diagonal of the Indian Ocean, separating Africa from India and Australia and separating Australia from Antarctica. With the collision of the Indian subcontinent against the southern shelf of the Asiatic land mass, the uplift of the Himalaya mountains began. The proposed succession of activity in the two main ridges of the Indian Ocean would explain why India has moved twice as far north with relation to Antarctica as Australia or Africa has and why the older northeast ridge is now a somewhat indistinct feature of the ocean floor. The younger rift in the Indian Ocean seems to have extended along the East Pacific Rise and Cocos Ridge to cross the Caribbean. A branch also passed south of South America. As these median ridges have continued to widen they have been forced by this growth to migrate northward, forming great shears or faults off the coast of Chile and through California. Indeed, a case can be made out for the idea that every mid-ocean ridge normally ends at a great fault or at a pivot point, as in the New Siberian Islands.

A few million years ago activity in this system decreased, allowing the North and South American continents to be joined by the Isthmus of Panama. The Atlantic rift now became more active again, producing renewed uplift in the

ROBESON CHANNEL separating northwestern Greenland (*upper right*) from Ellesmere Island (*foreground*) marks the Wegener Fault. The latter was named by the author for the German meteorologist who 50 years ago predicted the existence of such a fault and of a great lateral displacement along the length of the channel. Not yet fully mapped, it probably joins a known fault farther southwest.

Verkhoyansk Mountains and active volcanoes in Iceland and the five other still active volcanic islands down the Atlantic. Again the pattern of rifting in the Indian Ocean was altered. The distribution of recent earthquakes shows that the greatest activity extends along the western half of each diagonal ridge from the South Atlantic to the entrance of the Red Sea and thence by two arms along the rift valley of the Jordan River and through the African rift valleys, where the breakup of a continent has apparently begun.

The presently expanding rifts run mostly north and south or northeasterly so that dominant easterly and westerly compression of the outer crust is absorbed by overthrusting and sinking of the crust along the eastern and western sides of the "ring of fire" around the Pacific. For this reason East Asia, Oceania and the Andes are the most active regions of the world. The westward-driving pressure of the South Atlantic portion of the Mid-Atlantic Ridge has forced the continental block of South America against and over the downward-plunging oceanic trench along its Pacific coast. The northwest-trending currents below the Pacific floor have pulled down trenches under the eight island arcs around the western and northern Pacific from the Philippines north to the Aleutians. Even at the surface of the Pacific, the direction of the subcrustal movement is indicated by the strike of several parallel chains of volcanic islands, such as the Hawaiians, which may be thought to have risen like bubbles in a stream from the slower moving deep interior [*see lower illustration on page 25*]. These chains run parallel with the seismically active shearing faults that border each side of the Pacific, along the coast of North America and from Samoa to the Philippines. The compression exerted by the mid-ocean ridge through the southern seas is absorbed, with less seismic activity, along a line from New Zealand, through Indonesia and the Himalaya highlands to the European Alps. In all cases, the angle at which the loci of deep-focus earthquakes dip into the earth seems to follow the direction of subsurface flow—eastward and downward, for example under the Pacific coast of South America; westward and downward under the island arcs on the opposite side of the Pacific.

The theory I have outlined may be highly speculative, but it is indicative of current trends in thought about the earth's behavior. The older theories of the earth's history and behavior have proved inadequate to meet the new findings, particularly those from studies of terrestrial magnetism and oceanography. In favor of the specific details suggested here is the fact that they fit observations and are precise enough to be tested.

3 Plate Tectonics

John F. Dewey
May 1972

It has long been observed that mountains, volcanoes and earthquakes are not randomly distributed over the earth's surface but are found in distinct and usually narrow zones. To account for these evidences of instability in the earth's crust many hypotheses have been put forward. They have included such diverse notions as global expansion, global contraction, the effect of lunar tidal forces and wholesale uplift or foundering of large segments of the earth's crust. One other explanation—continental drift—was advanced from time to time but was unpalatable to most geophysicists because it seemed to violate what was known about the mechanical properties of the earth's crust. Nevertheless, continental drift seemed to explain geological similarities between continents thousands of miles apart. It also explained why some continental margins, for example those of South America and Africa, match each other so precisely.

Within the past 10 years continental drift has been placed on a firm foundation by the development of the concept of sea-floor spreading, originally proposed by the late Harry H. Hess of Princeton University. Sea-floor spreading involves the notion that the floor of the ocean is continuously being pulled apart along a narrow crack that is centered on a ridge that can be traced through the major ocean basins. Volcanic material (liquid basalt) rises from the earth's mantle to fill the crack and continuously create new oceanic crust.

The concept might have been difficult to confirm except for the fortunate fact that the polarity of the earth's magnetic field periodically reverses. It had been observed from magnetometer surveys that rocks of the ocean floor exhibit a zebra-stripe pattern in which the intensity of magnetization changes abruptly in linear ribbons parallel to the nearest oceanic ridge. In 1963 F. J. Vine and D. H. Matthews of the University of Cambridge proposed that the magnetic pattern was evidence of both sea-floor spreading and reversals of the earth's magnetic field. To many geologists the one seemed almost as improbable as the other. Vine and Matthews argued that as the basaltic liquid rose into the axial crack of the oceanic ridges and solidified it would become magnetized in the then prevailing direction of the earth's magnetic field. If new oceanic crust was continuously generated, as Vine and Matthews believed, one should find that each ridge axis is bordered symmetrically by pairs of parallel strips whose direction of magnetization is the same, that is, both normal or both antinormal. The hypothesis was strikingly confirmed in many traverses across oceanic ridges. Furthermore, a time scale of magnetic reversals has been developed showing that the rate of sea-floor spreading is between two and 18 centimeters per year.

It is now clear that virtually all of the present area of the oceans has been created by sea-floor spreading during the past 200 million years, or during the last 5 percent or so of the recorded geologic history of the earth. The creation of new surface area means either that the earth has expanded dramatically or that surface area is somewhere being destroyed at the same rate at which it is being created. There is good evidence that the earth has not expanded more than about 2 percent in the past 200 million years. Thus there must be, in general terms, a global conveyor-belt system or surface motion that links zones of surface creation and surface destruction.

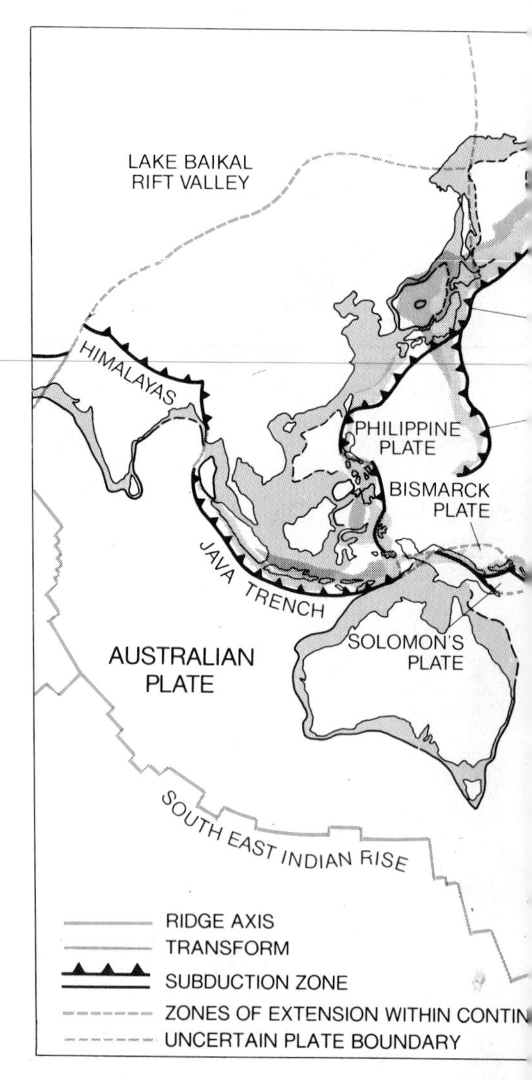

MOSAIC OF PLATES forms the earth's lithosphere, or outer shell. According to the recently developed theory of plate tectonics, the plates are not only rigid but also in constant relative motion. The boundaries of plates are of three types: ridge axes, where plates are diverging and new oceanic floor is generated; transforms, where plates slide past each other, and subduction zones, where plates converge and one plate dives under the leading edge of its neighbor. Triangles indicate the leading edge of a plate.

The concept of sea-floor spreading has now been joined to the earlier idea of continental drift in a single unifying theme called the theory of plate tectonics. The geometric part of the theory visualizes the lithosphere, or outer shell, of the earth as consisting of a number of rigid plates. The kinematic part of the theory holds that the plates are in constant relative motion: they can slide past each other, they can move apart on opposite sides of an oceanic ridge or they can converge, in which case one of the plates must be consumed. Let us now see how various instabilities in the earth's crust can be visualized in terms of plate tectonics.

Earthquakes

Most earthquakes occur in narrow zones that join to form a continuous network bounding regions that are seismically less active. The seismic network is associated with a variety of characteristic features such as rift valleys, oceanic ridges, mountain belts, volcanic chains and deep oceanic trenches [*see illustration below*]. The seismic areas mark the boundaries between plates, which are largely free of earthquakes. There appear to be four types of seismic zone, which can be distinguished by their characteristic morphology and geology.

The first type is represented by narrow zones of high surface heat flow and basaltic volcanic activity along the axes of mid-oceanic ridges where earthquakes are shallow (less than 70 kilometers deep). The axes of the ridges, of course, are the active sites of sea-floor spreading. In Iceland, where the Mid-Atlantic Ridge rises above sea level, the spreading rate has been measured at about two centimeters per year.

The second type of seismic zone is marked by shallow earthquakes in the absence of volcanoes. A good illustration is the region around the San Andreas fault in California and around the Anatolian fault in northern Turkey, along both of which large surface displacements parallel to the fault have been measured [see "The San Andreas Fault," by Don L. Anderson; SCIENTIFIC AMERICAN Offprint 896].

The third type of seismic zone is intimately related to deep oceanic trenches associated with volcanic island-arc systems, such as those around the Pacific Ocean. Earthquakes there can be shallow, intermediate (70 to 300 kilometers) or deep (300 to 700 kilometers), according to where they take place in the steeply plunging lithospheric plate that borders the trench. Thus the earthquake epicenters (the points on the surface above the initial break) define a geologic structure dipping down into the earth away from the trench. These inclined earthquake zones, called Benioff zones, underlie active volcanic chains and have a variety of complex shapes.

The fourth type of seismic zone is typified by the earthquake belt that extends from Burma to the Mediterranean Sea. It consists of a wide, diffuse continental zone within which generally shallow earthquakes are associated with high mountain ranges that clearly owe their existence to large compressive forces. Earthquakes of intermediate depth occur in some areas such as the Hindu Kush and Romania. Although deep-focus

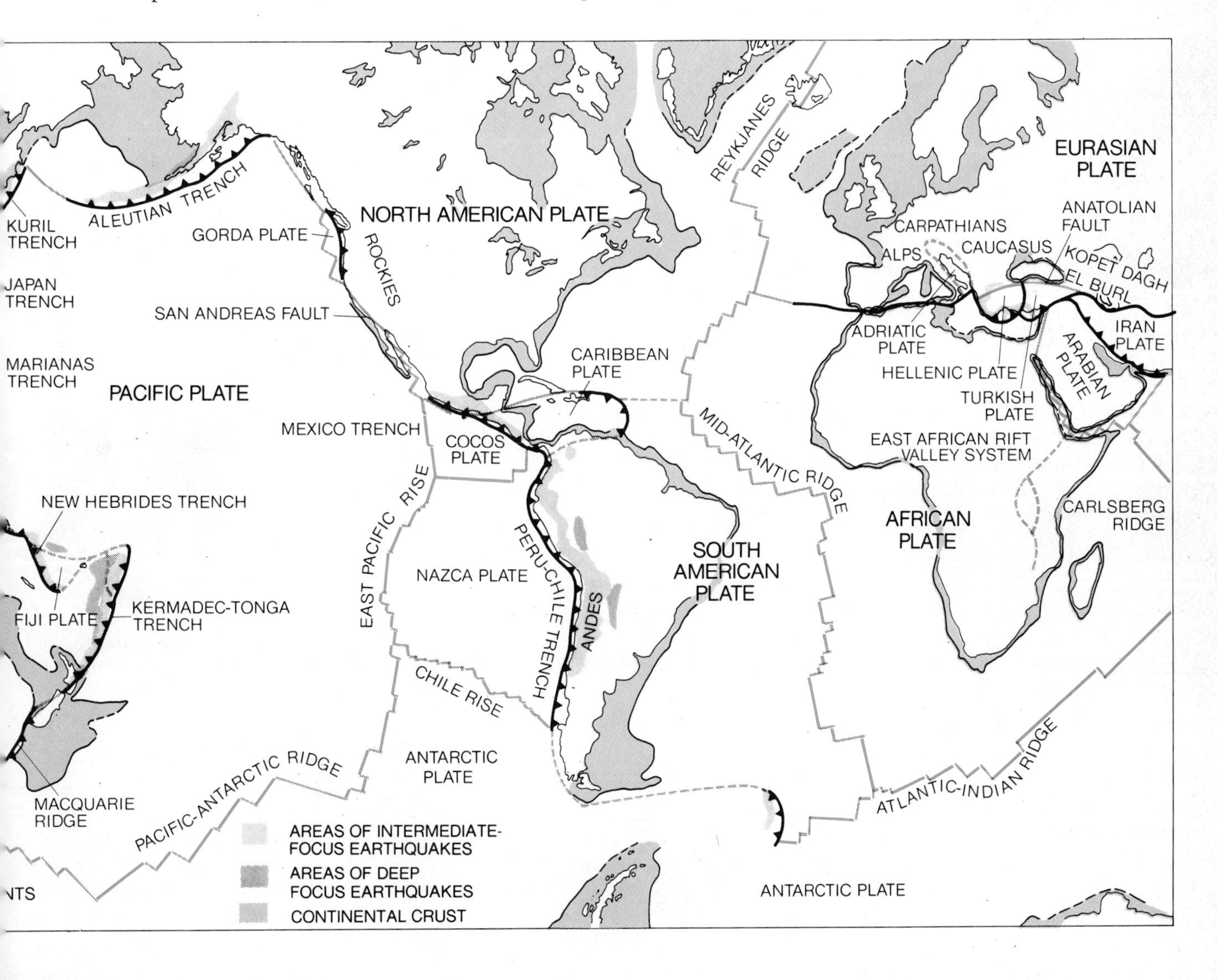

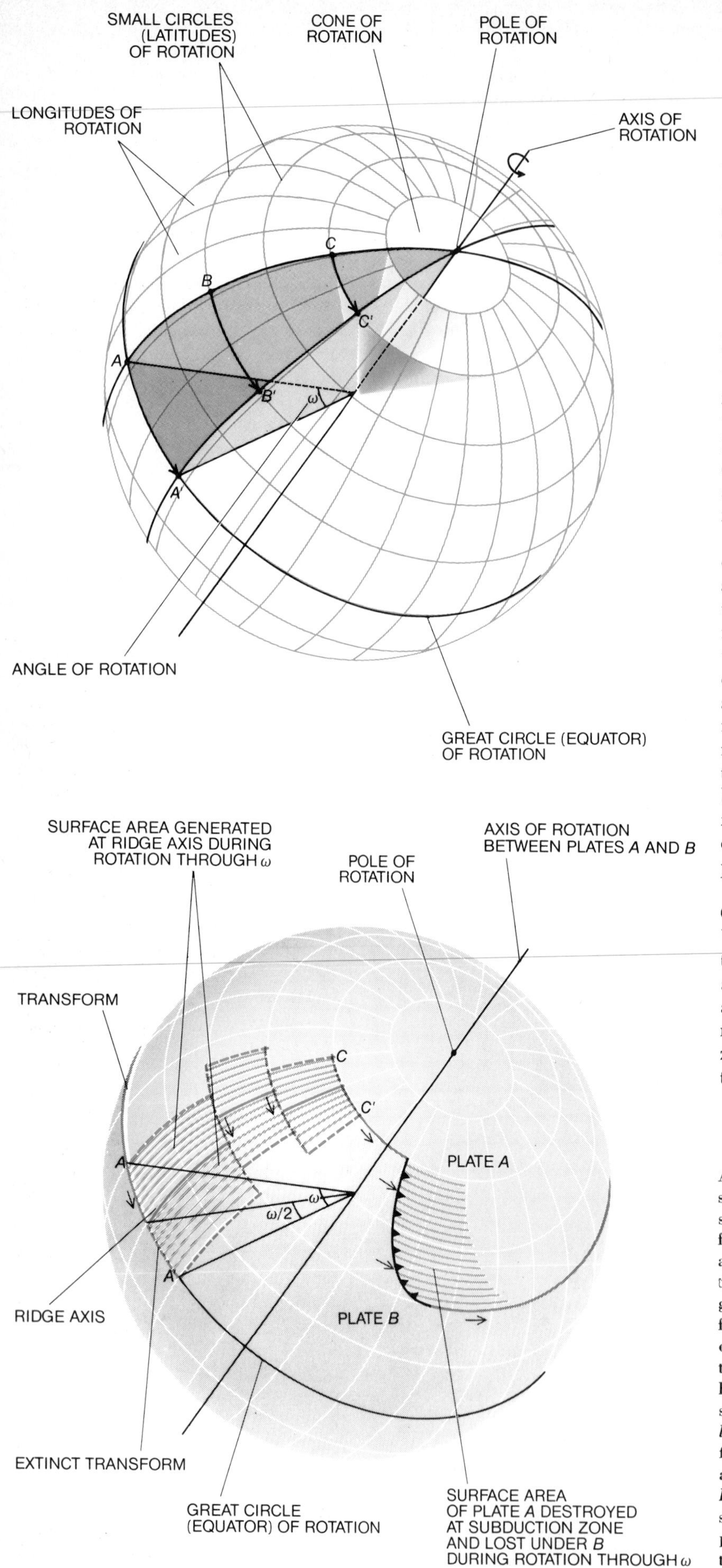

earthquakes are rare, they have been recorded in a few places, for example just north of Sicily under the Tyrrhenian volcanoes.

An earthquake results when stresses accumulate to the point that rocks in the earth's crust break. The breakage is the consequence of brittle failure of the rock (in contrast to plastic deformation, which can relieve stresses slowly). The first seismic waves to leave the region of the break (the hypocenter) are waves of alternate compression and rarefaction generated by the sudden release of elastic energy. After an earthquake one finds that the seismological stations that have received the first waves can be assigned to one of four geographic quadrants. In two of the quadrants, lying opposite each other, the first waves are compressional; in the other two quadrants the first waves are rarefactional.

The quadrants define the orientation of two nodal planes on one of which a sudden slip has presumably produced the earthquake. The intersection of the nodal planes is the null direction, or intermediate stress axis, parallel to which effectively no strain occurs. The line bisecting the quadrant in which the first motion is compressional defines the direction of least principal stress, parallel to which there is extensional strain. The bisector of the rarefaction quadrant defines the direction of the maximum principal stress along which there is compressional strain.

Lynn R. Sykes of the Lamont-Doherty Geological Observatory of Columbia University has applied this analysis to the seismic belts of the world and has shown systematically that the ridge axes are in tension, that there is lateral movement along the second type of seismic zone and that compression dominates the third and fourth types. Thus seis-

AXIS OF ROTATION can be selected in such a way (*top illustration at left*) that a set of two or more points lying on the surface of a sphere (*A*, *B*, *C*) can be moved by a rigid rotation around that axis to new positions (*A'*, *B'*, *C'*), preserving the original geometry of the set. A unique axis can be found only if the initial and final positions of two or more points are known. Similarly, the relative motion of two rigid plates can be described as a rigid rotation around a suitably selected axis of rotation (*bottom illustration at left*). Plate *A* is designated as fixed while plate *B* is rotated anticlockwise, as viewed down the axis of rotation. As plate *B* rotates through angle omega (ω), new surface area is added symmetrically to both plates at the ridge axis, which itself travels through an angle equal to one-half omega.

mology emphasizes that there are three kinds of plate boundary: boundaries across which plates are pulled apart, boundaries along which plates slide past each other and boundaries across which plates converge. Since rock material does not pile up indefinitely in the compressional zones, it follows that somewhere there must be zones in which plates are consumed.

The Plate Mosaic

One can therefore construct a model of global surface displacement involving a mosaic of plates each of which exhibits one or more of the three types of boundary. At ridge axes plates separate and new surface area is generated by the continuous accretion of new oceanic crust at the trailing edges of the plates. At transform faults plates slide past each other and surface area is neither created nor destroyed. At subduction zones one plate is consumed and slides down into the mantle under the leading edge of another plate.

Plates vary greatly in size from the six major plates, such as the plate that carries virtually the entire Pacific Ocean, down to very small plates, such as the plate that is essentially coextensive with Turkey. Moreover, plate boundaries do not always coincide with the margins of continents; many continental margins are peaceful earthquake-free nonvolcanic regions. Hence plates can be partly oceanic and partly continental or they can be entirely one or the other. This fact overcomes one of the traditional objections to continental drift, namely the mechanical difficulty of having a geologically weak continent plow its way across a strong ocean floor. According to the plate-tectonic view, continents and oceans are rafted along by the same crustal conveyor belt.

A look at the boundary around the African plate reveals two important consequences of plate motion. The greater part of the boundary is a ridge axis extending from the North Atlantic into the Indian Ocean and the Red Sea; thus the entire African plate must be growing in area. This behavior in turn means that plates elsewhere on the globe must be getting smaller. The second consequence of the growth of the African plate is that the Carlsberg Ridge in the Indian Ocean is moving away from the Mid-Atlantic Ridge, illustrating one of the essential corollaries of plate kinematics: plate motion is relative. There is no coordinate system within which absolute plate motion can be defined except a system defined in relation to a particular plate or

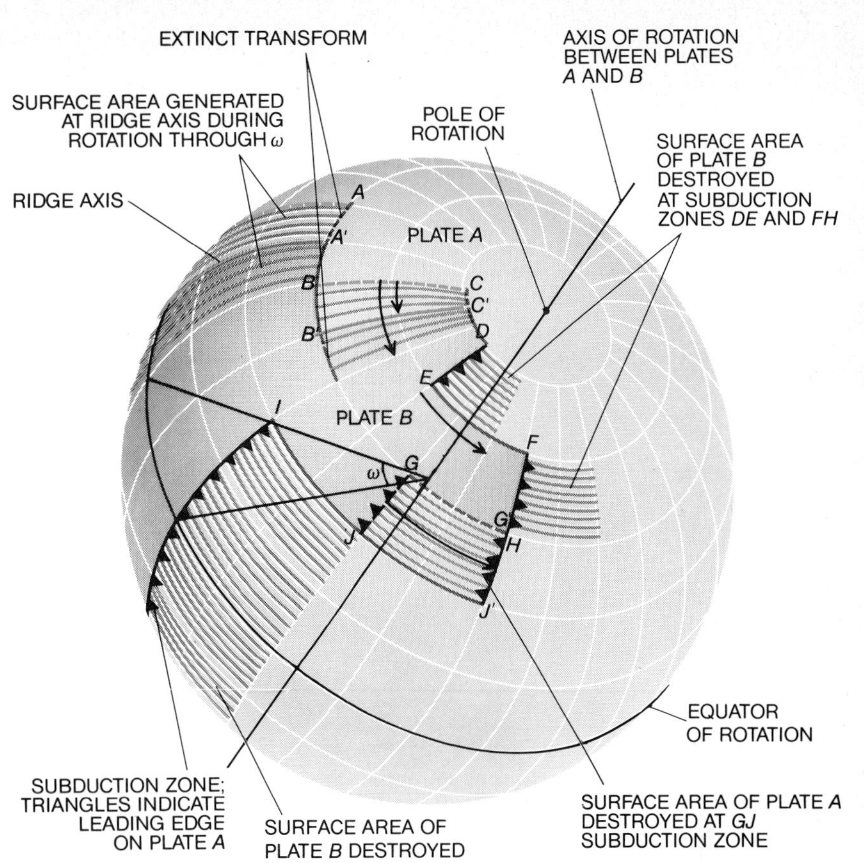

THREE KINDS OF TRANSFORM can exist as segments of a single plate boundary: ridge axis to ridge axis (*AB*), ridge axis to subduction zone (*CD*) and subduction zone to subduction zone (*EF*, *GH*, *IJ*). Plate *A* is again assumed to be fixed while plate *B* rotates anticlockwise. Ridge-axis-to-ridge-axis transforms (*AB*, *A'B'*) maintain a constant length because new surface area is generated symmetrically at ridge axes. Transforms joining ridge axes to subduction zones decrease or increase in length at half the transform slip rate. In the example depicted here *CD* shortens to *C'D*, but if the leading edge at subduction zone *DE* were on plate *A* (as in the case of *GJ*), *CD* would have lengthened. Transform *EF* maintains a constant length, whereas *GH* shortens to zero length and *IJ* lengthens to *IJ'*.

plate boundary that is arbitrarily chosen as being "fixed."

The basic assumption that plates are rigid is essential to plate tectonics and appears to be justified by the fact that excellent restorative fits can be made between many pairs of continental margins. (In making such a fit the margin is typically defined as the 1,000-fathom isobath on the continental shelf adjacent to the continent.) Similar fits can be made with even greater precision between pairs of magnetic anomalies symmetrically disposed on each side of a ridge axis. If plates had been distorted as they evolved, these fits could not be made. As further confirmation of rigidity, profiles produced by seismic reflection have shown that sediments laid down on the oceanic crust as it moves away from a ridge axis form undistorted flat layers.

The fact that rigid plates are in relative motion on a spherical earth means that a displacement between any two plates can be described by a rotation around an axis passing through the center of the earth. The intersection of this axis with the earth's surface is termed the pole of rotation [*see illustration on opposite page*]. This concept was first applied by Sir Edward Bullard, J. E. Everett and A. G. Smith of the University of Cambridge to demonstrate the fit of the continental margins around the Atlantic Ocean. Relative surface motion between two plates proceeds along circles of rotation around the axis of rotation. The circles can be considered as latitudes of rotation from zero radius at the pole of rotation to a maximum at the equator of rotation. Relative plate motion is best described, however, as an angular velocity, since the velocity along rotation circles increases from zero at the pole of rotation to a maximum at the equator of rotation. The nature of displacement across a plate boundary is therefore entirely dependent on its orientation with respect to the circles of rotation.

Of particular interest are boundaries parallel to rotation circles. At these boundaries are faults where surface area is conserved; such faults are called trans-

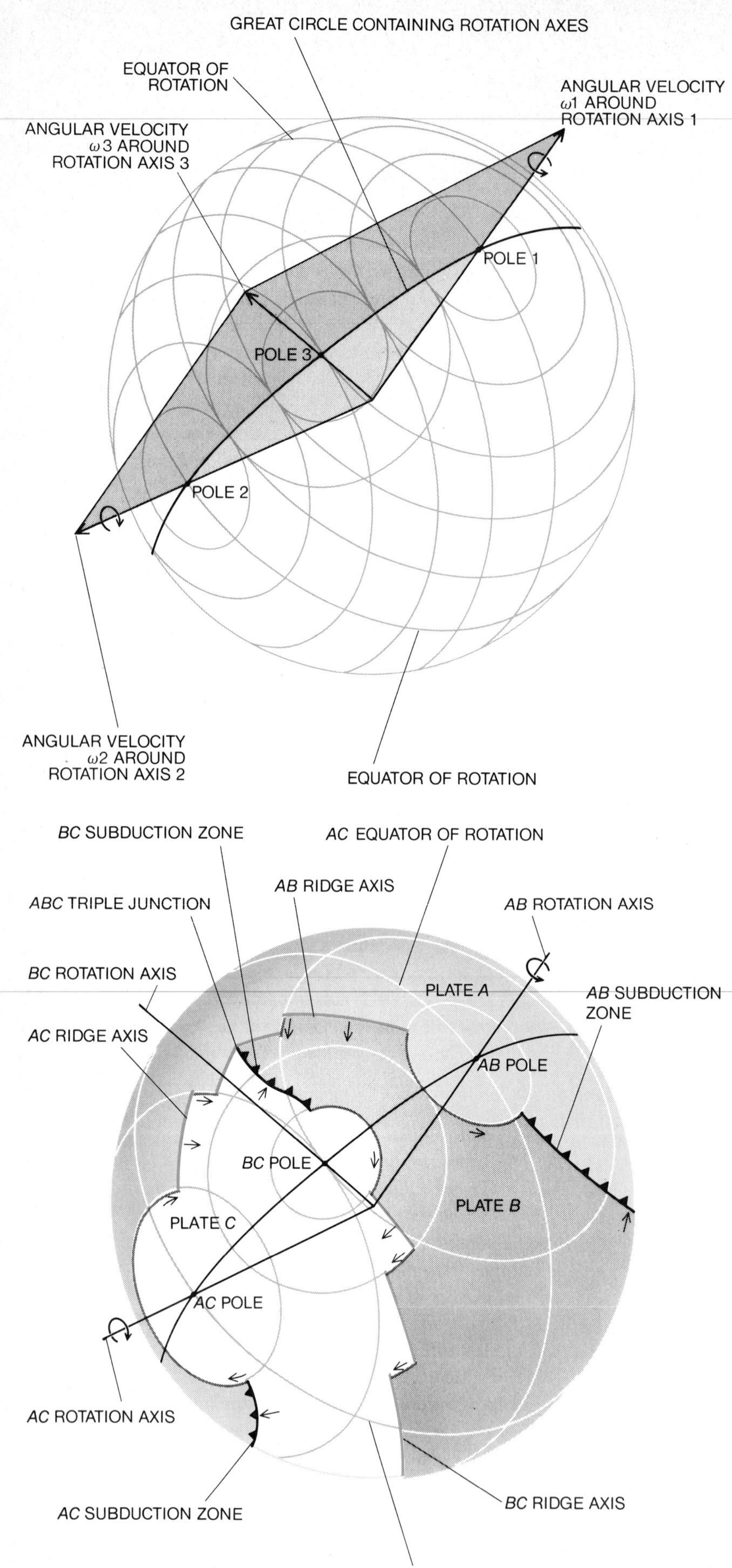

form faults. Great circles drawn perpendicularly to transform faults that are segments of one plate boundary will intersect to define the rotation pole. Plate boundaries oblique to circles of rotation are either ridges or subduction zones, depending on whether plates are separating or converging across them. The increasing rate of plate separation across ridge axes with increasing distance from the pole of rotation is reflected by a progressively increasing distance between particular magnetic anomalies and the ridge axis. Similarly, the rate of plate convergence at subduction zones increases away from the pole of rotation. A particularly good illustration is afforded by the New Zealand–Tonga seismic zone. That part of the zone south of New Zealand has only shallow earthquakes; intermediate-focus earthquakes start in New Zealand and deep-focus earthquakes north of New Zealand [*see illustration on pages 34 and 35*]. This suggests a progessive northward increase in convergence rate across the subduction zone, so that the downgoing plate reaches progressively deeper levels.

Separation rates across ridges where plates are moving apart can be directly deduced from magnetic-anomaly patterns, but there is no direct method for deducing convergence rates across subduction zones where one plate is diving under another and creating a trench. Close attention has therefore been given to plate boundaries where individual segments are a ridge, a transform fault and a subduction zone, because the angular velocity of relative motion deduced for a ridge segment also applies to a trench segment. The angular velocity can be directly translated into a circle-of-rotation velocity for any circle of rotation that crosses the subduction zone,

THIRD AXIS and angular velocity of rotation are defined (*top illustration at left*) as the vector sum of two others. If one knows the angular velocities of two rigid rotations around two axes (*1 and 2*) passing through the center of a sphere, one obtains their vector sum to determine the angular velocity around a third axis (*3*) lying in the same plane as the first two. In the example illustrated the poles of rotation axes *1* and *2* are 90 degrees apart and the angular velocities around them (*ω1 and ω2*) are equal, so that the pole of the third axis lies midway along a great circle between poles *1* and *2*. Similarly (*bottom illustration at left*), if one knows the axes of rotation and the angular velocities that describe the relative motion of plates *A* and *B* and plates *B* and *C*, one can ascertain the rotation axis (*BC*) that describes the relative motion of plates *A* and *C*.

depending on its "rotational latitude." Although there is no apparent geometric reason why ridges should lie parallel to longitudes of rotation, they appear in most cases to approximate this relation. Furthermore, the symmetrical distribution of paired matching magnetic anomalies with respect to the ridge axis where they were generated indicates that new crustal material is added symmetrically to the trailing edges of the diverging plates.

For some reason, perhaps one related to the driving mechanism for plate motion, straight ridge-transform boundaries are a mechanically stable configuration. Subduction zones are generally curved, perhaps also for mechanical reasons, and convergence directions can be right-angled or oblique to them, according to whether they are right-angled or oblique to rotation circles. One may pass in gradations from pure subduction to pure transform motion along a single plate boundary; therefore the orientation of a subduction zone, unlike the orientation of a ridge or a transform fault, is a poor guide to the relative direction of displacement.

The axial spreading zone of oceanic ridges is not a continuous feature. It is interrupted and offset by transform faults that in some places create high submarine cliffs. Transform faults were formerly thought to be lines along which the ridge axis had been displaced from a once continuous zone, since they continue as bathymetric features beyond the offset ends of the ridge axis. J. Tuzo Wilson of the University of Toronto argued, however, that they are simply offsets of the spreading axis and form an integral part of a single boundary. He coined the term transform fault to describe them because they merely transform relative motion between the two ridge segments. Seismic first-motion studies confirmed Wilson's prediction of transform motion; further support is provided by the observation that earthquakes are restricted to that portion between the offset ends of the ridge axis.

The active portion of a transform fault defines part of a circle of rotation. Similarly, the inactive continuation of a transform fault beyond the offset ridge axis defines circles of rotation for the previous history of plate divergence across the ridge axis; it represents earlier circles of rotation "frozen" into adjacent oceanic crust generated earlier. This is of fundamental importance for two reasons. First, the excellent circle-of-rotation lines described by inactive transform faults justify the assumption that relative motion between two plates can be described in

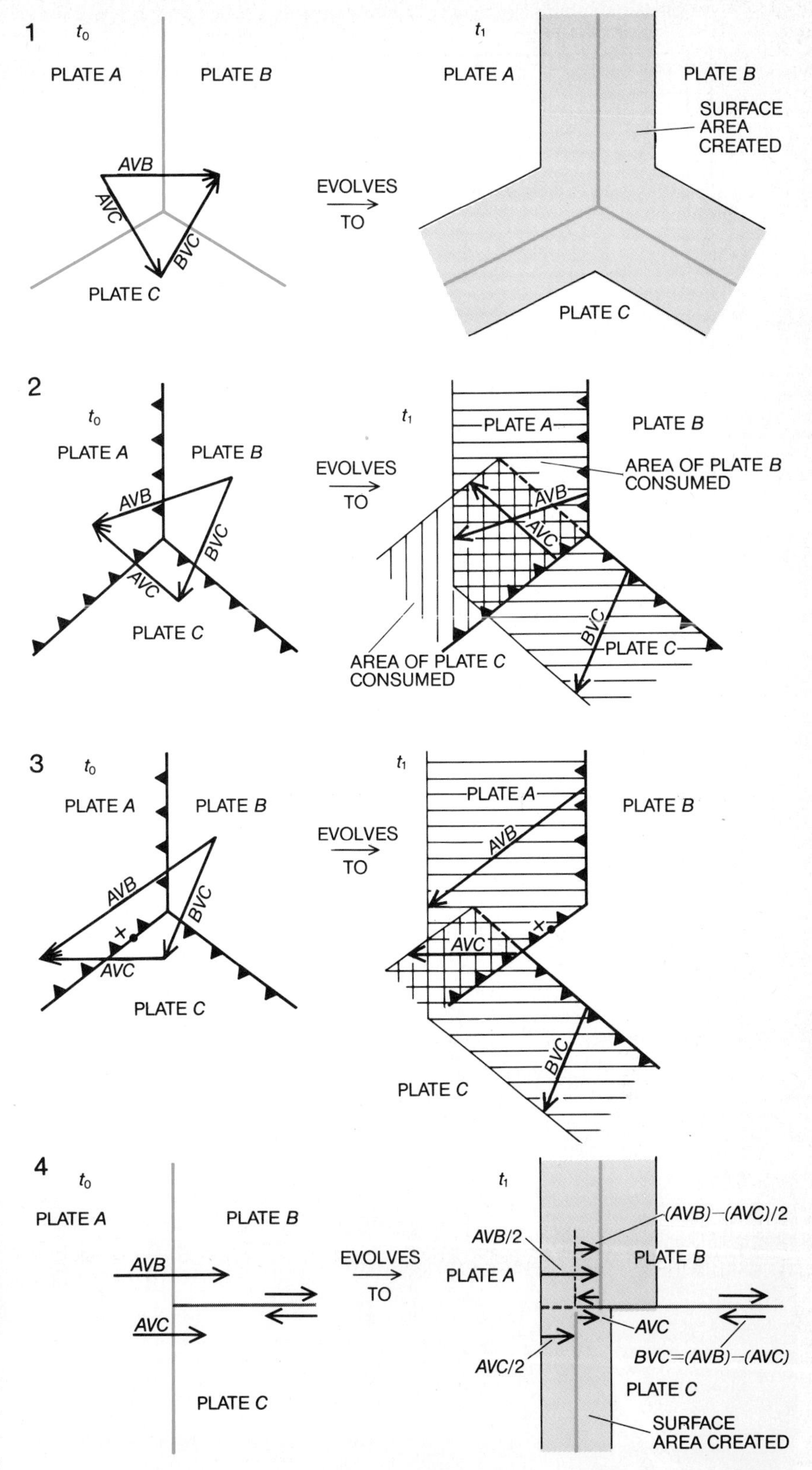

FOUR TRIPLE JUNCTIONS are depicted with their velocity triangles at time t_0 and their configuration at time t_1. Ridge axes are solid color; subduction zones are solid black; transforms are gray. A triple junction involving three ridges (*1*) is always stable. When three subduction zones meet (*2 and 3*) and two leading edges lying on plate *A* do not form a straight line, the triple junction is stable only when the *A* v. *C* velocity vector is parallel to the leading edge of plate *C* (*2*). Otherwise (*3*) the triple junction moves. Thus at some time between t_0 and t_1 the triple junction moves past point *X*. Before the time of movement the relative motion at *X* is *A* v. *C* and thereafter *A* v. *B*. The last diagram (*4*) depicts how it comes to pass that a triple junction involving two ridges and a transform can have the configuration at t_0 only instantaneously since the triple junction evolves immediately to t_1, in which plates *B* and *C* are separated by the transforms and the ridge is thus offset.

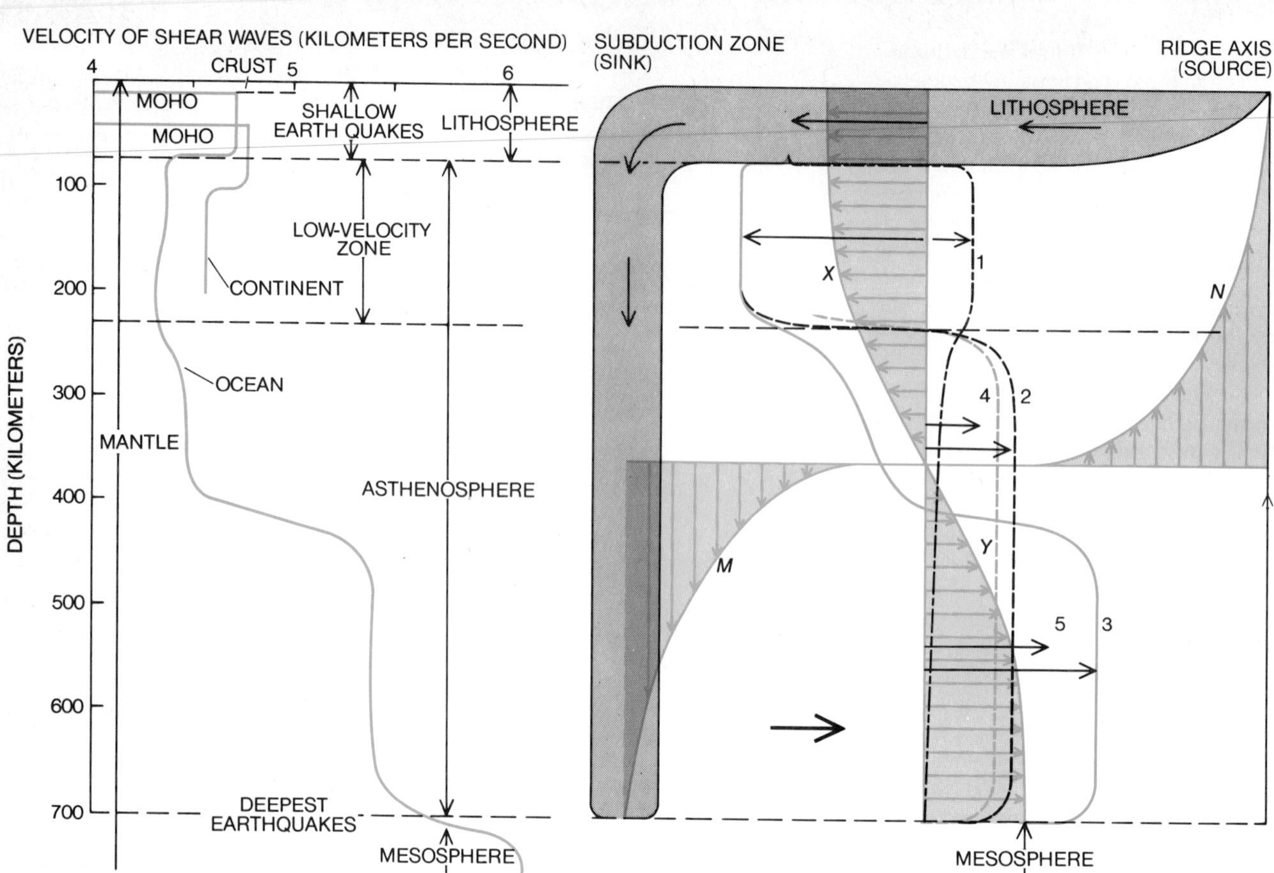

VARYING VELOCITY OF SHEAR WAVES in lithosphere and asthenosphere (*left*) suggests models showing how mass-transfer circuits (*right*) introduce new plate material at ridge axes and consume it at subduction zones. The velocity of shear waves decreases at 70 and 150 kilometers under oceans and continents respectively. Evidently mass-transfer circuits extend to a depth of at least 700 kilometers. Material descending at sinks (under curve *M*) must be balanced by material rising at sources (under *N*), and the upper lateral transfer (under *X*) must be balanced by return flow (under *Y*). Five shapes are suggested for lateral-transfer gradients.

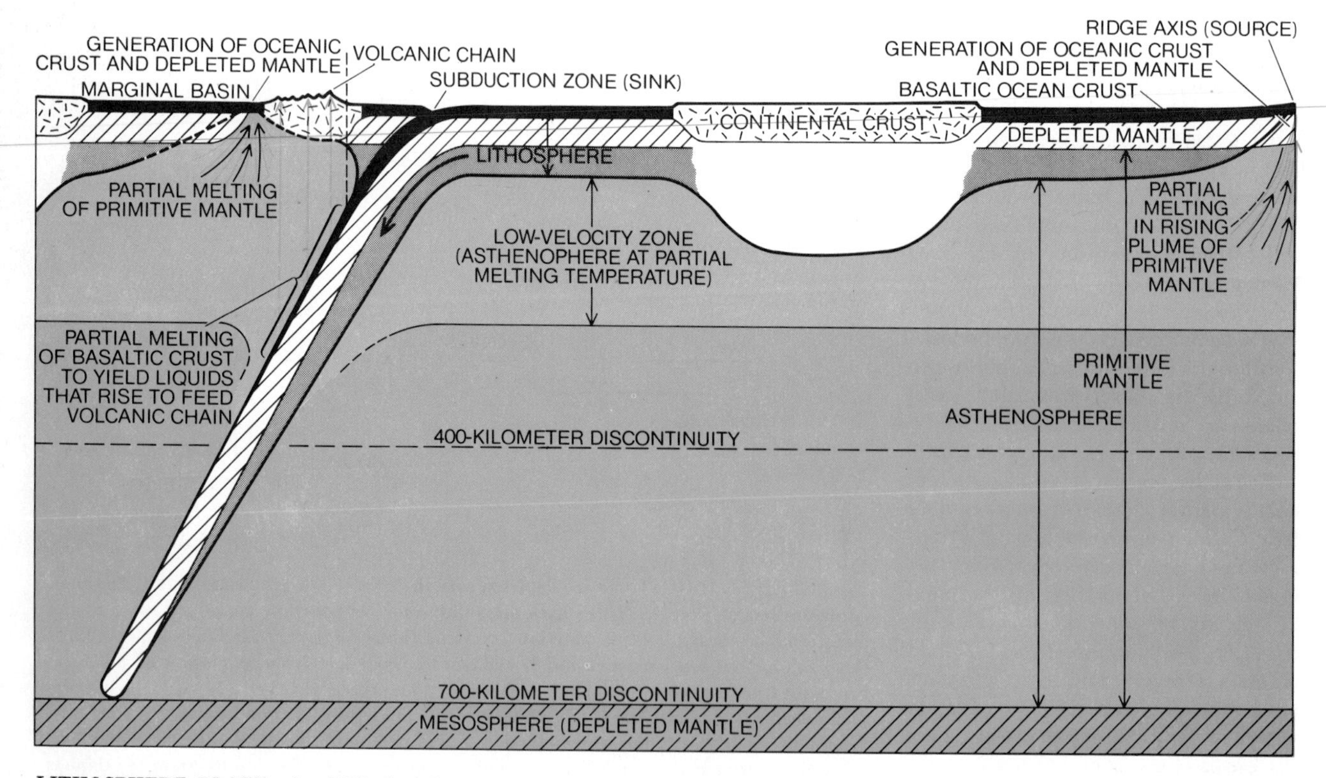

LITHOSPHERE PLATE of solidified rock serves as a thermal boundary conduction layer "floating" on molten or semimolten rock of the asthenosphere. In this schematic view the lithosphere is thicker under the continent it is rafting toward a subduction zone, where the plate descends into the mantle under leading edge of another plate. New oceanic crust is steadily being added at ridge axis.

terms of a rigid rotation around a fixed pole. Second, it provides the key to working out past plate motions. The inactive transform faults give the direction but not the rate of old displacements. Rates, however, can be deduced from the spacing of magnetic anomalies of known ages.

Walter C. Pitman III and Manik Talwani of the Lamont-Doherty Observatory have devised a simple but elegant technique using well-defined pairs of magnetic anomalies of known ages and fracture-zone orientations for computing plate displacements in the central Atlantic Ocean for the past 180 million years. First they assumed that successive pairs of magnetic anomalies were generated at the ridge axis and moved apart on rigid plates. They then found a series of rotation poles around which they could fit the pairs of progressively older magnetic anomalies; the series ended with a rotation that fitted the continental margins of Africa and North America. Reversing this sequence gives the kinematics of plate divergence that records the history of the opening of the central Atlantic Ocean.

A consequence of the symmetrical generation and asymmetric destruction of surface area, at ridge areas and subduction zones respectively, is that transform faults can retain a constant length or can change their length [*see illustration on page 37*]. Transform faults that join ridge axes, or subduction zones with leading edges on the same plate, stay the same length. Where transform faults join subduction zones with leading edges on different plates, they lengthen or shorten, depending on whether the leading edges face away from or toward one another. A transform fault that joins a ridge axis to a subduction zone increases or decreases in length depending on which plate the leading edge lies on.

If the axes and angular velocities of rotation between two pairs of plates (A and B and A and C) are known, the axis and angular velocity of rotation between the third pair (B and C) can be calculated [*see illustration on page 128*]. This means that if ridge-axis segments fall along the boundaries between A and B and A and C, the relative motion of B and C can be found. Xavier le Pichon of the Center for the Study of Oceanography and Marine Biology in Brittany developed this technique for computing the relative motions between the six largest plates and was thus able to work out convergence directions and rates across all the major subduction zones.

With the same technique Pitman has computed the relative motion of Africa and Europe for the past 80 million years. During this time North America and Africa have been parts of separate plates moving apart around a sequence of rotation axes as the central Atlantic has widened. North America and Europe have been similarly moving apart but around a different sequence of rotation axes. Therefore there has been relative motion between Africa and Europe. This motion has been complex but the net effect has been to nearly eliminate a formerly wide oceanic region between the two continents.

Since relative displacements are along circles of rotation, the relative motion between three plates cannot be described by the customary velocity vector triangle except instantaneously at a point. If, however, we are interested in relative motions in an area of the earth's surface so small that it may be regarded as a flat surface, with the result that circle-of-rotation segments are virtually straight lines, the velocity vector triangle is a convenient device for illustrating relative motion. A small area of great interest is one where three plate boundaries join to form a triple junction. Triple junctions are demanded by plate rigidity; this is the only way a boundary between two rigid plates can end. D. P. McKenzie of the University of Cambridge and W. Jason Morgan of Princeton University have analyzed all possible forms of triple junctions with velocity vector triangles and have shown that they can be stable or unstable depending on whether they are able or unable to maintain their geometry as they evolve [*see illustration on page 39*].

Plate Thickness and Composition

So far we have considered those essential aspects of plate geometry and evolution that can be treated from a surface point of view. We have not yet inquired into the thickness and composition of lithosphere plates. It has been known for many years from gravity and seismic-refraction measurements and from general considerations of mass balance that the continents are underlain by a relatively light "granitic" crust about 40 kilometers thick and the oceans by a denser "basaltic" crust only about seven kilometers thick. Both continental and oceanic crust are underlain by a mantle of denser material. The junction between crust and mantle is the "Moho," or Mohorovicic discontinuity. It was the goal of the now abandoned Mohole project to drill through the relatively thin oceanic crust into the mantle.

Plates must be at least as thick as the oceanic and continental crust because some plates have oceanic and continental portions between which there is no differential motion. It was thought for many years that the Moho might be an important physical discontinuity of mechanical decoupling on which large crustal displacements proceed. It is now clear that if there is a zone of decoupling between an outer rigid shell and a less viscous layer below, it is considerably deeper than the Moho.

The best evidence for the thickness of plates comes from seismology. The velocity of seismic waves is dependent on the density and flow properties of the rock through which they pass; it is high in rigid, dense rocks and low in less rigid, lighter rocks. Moreover, an increase in confining pressure increases the velocity of waves and an increase in temperature decreases it. Although confining pressure must increase with depth, recent studies indicate that the velocities of shear waves suddenly decrease below a surface about 70 kilometers under the oceans and about 150 kilometers under the continents [*see top illustration on opposite page*]. Shear-wave velocities then increase with depth, with marked increases between 350 and 450 kilometers and just above 700 kilometers.

These data suggest that an outer rigid layer 70 to 150 kilometers thick (the lithosphere) lies above a weaker and hotter layer (the asthenosphere) that becomes increasingly viscous with depth. The thickness of the lithosphere therefore probably constitutes the thickness of the rigid plates, and the lithosphere is discontinuous at plate boundaries. Earthquakes present a test of this hypothesis, since the cold, rigid lithosphere is probably their source. The distribution of earthquakes should thus provide a guide to the thickness of the lithosphere and to its distribution where it descends in subduction zones into the interior of the earth.

Ridges and transform faults are characterized by earthquakes whose depth extends to about 70 kilometers. The inclined zone of intermediate and deep-focus earthquakes indicates the descent of the lithosphere into the asthenosphere, where it is consumed at subduction zones.

Bryan L. Isacks and Peter Molnar of the Lamont-Doherty Observatory, working with seismic first-motion records, have analyzed the stresses in descending lithospheric plates. They find that the stresses are consistent with those that would be expected if a cold slab of bending lithosphere were to meet increasing resistance as it descended into

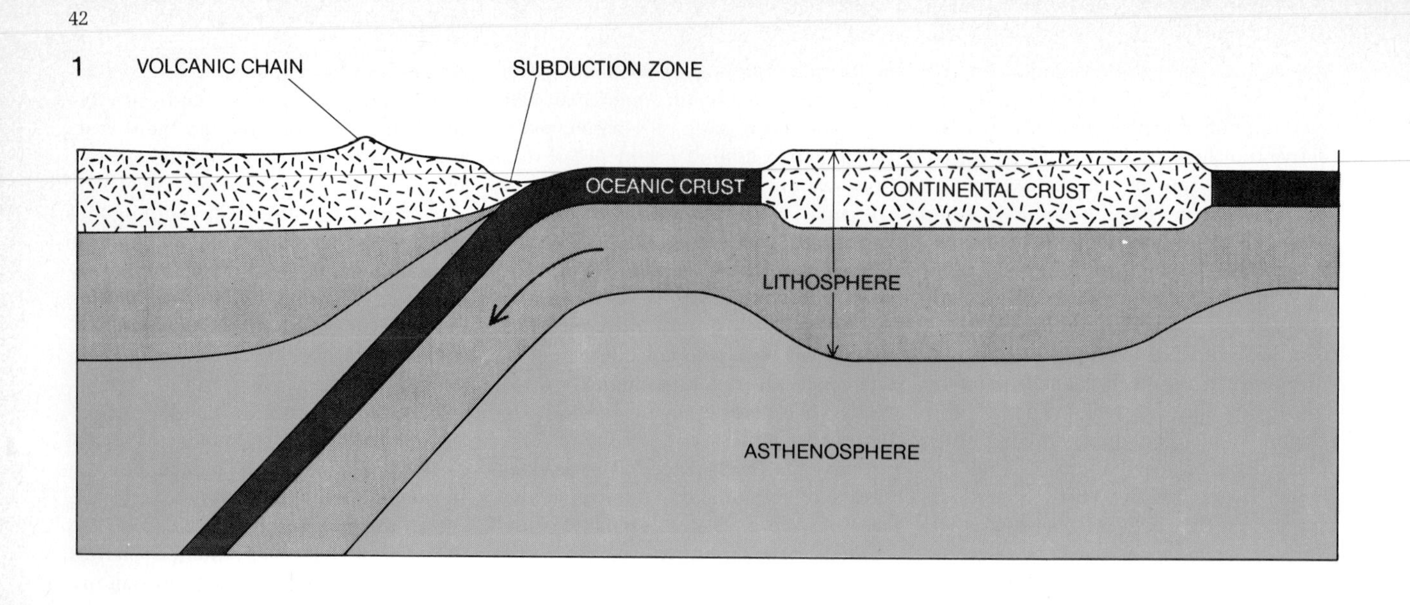

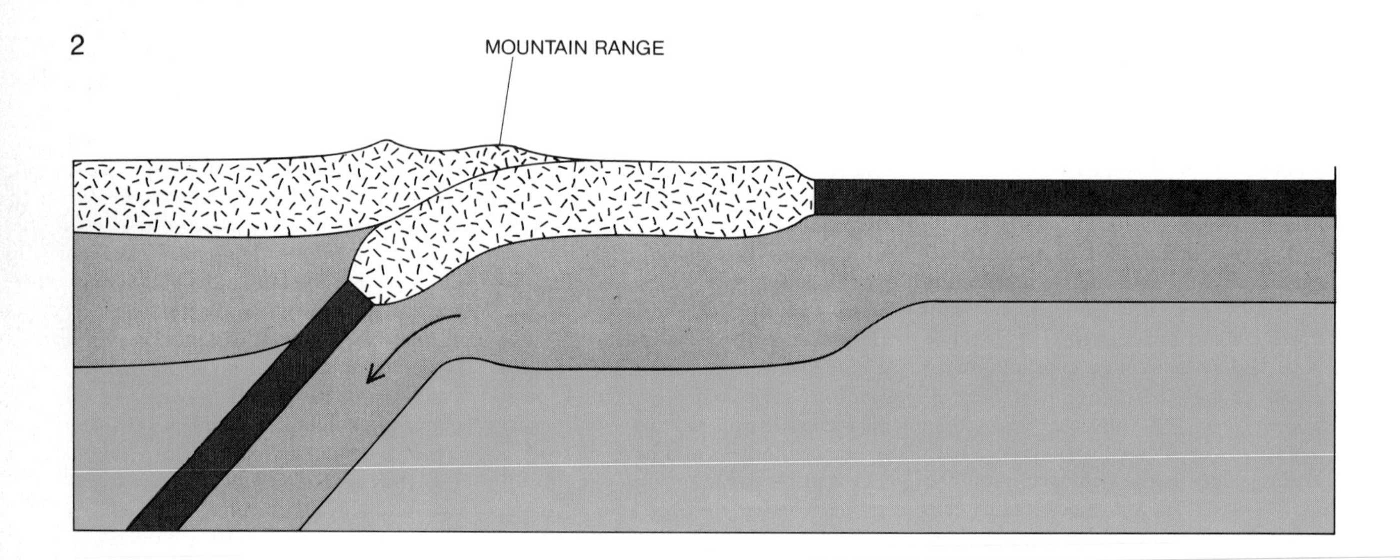

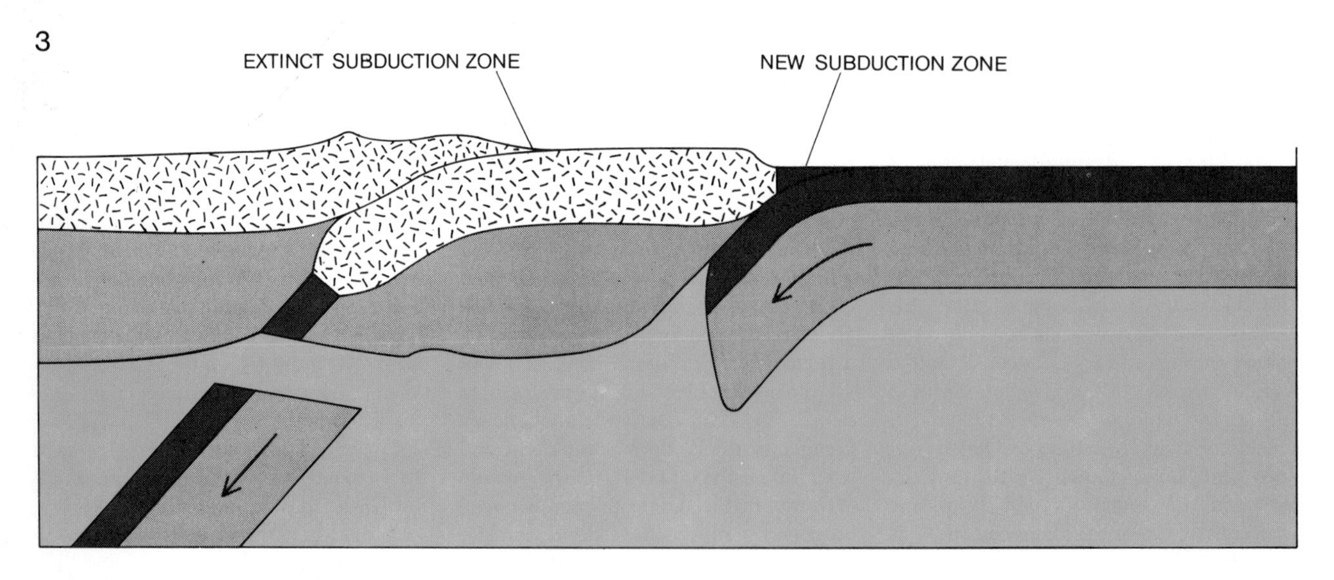

COLLISION OF CONTINENTS occurs when a plate carrying a continent is subducted at the leading edge of a plate carrying another continent (*1*). Since the continental crust is too buoyant to be carried down into the asthenosphere, the collision produces mountain ranges (*2*). The Himalayas evidently formed when a plate carrying India collided with the ancient Asian plate some 40 million years ago. The descending plate may break off, sink into asthenosphere and a new subduction zone may be started elsewhere (*3*).

an increasingly viscous asthenosphere. Where the cross section of the lithosphere is curved (at subduction zones) its upper part is under tension, suggesting elastic bending. Where the lithosphere has descended only a short distance into the asthenosphere it is under tension along its length, indicating low resistance to its descent. Continuously inclined seismic zones, representing slabs of lithosphere that have descended to lower parts of the asthenosphere, are characterized by compression along the slab. This fact suggests that the downgoing lithosphere is put into compression as it meets increasing resistance to its descent. An instructive case is where there is a gap in the inclined seismic zone, suggesting a discontinuity in the downgoing slab. Earthquakes above the gap indicate tension; earthquakes below it indicate compression. Evidently part of the slab has broken off and is descending at a faster rate than the remainder.

The general kinematics of plates—their growth and consumption—requires some form of mass convection, or mass-transfer circuits, in the mantle. Heat flow to the earth's surface is highest along ridge axes; it declines rapidly to a low plateau value across the plates and falls to a minimum at subduction zones. Therefore the lithosphere may represent a cold, rigid boundary conduction layer that is created at the hot ridge "sources" and destroyed at the cold subduction-zone "sinks." Any acceptable model for the geometry of the mass circuits in the earth's mantle must satisfy a number of conditions.

Conditions to Be Met

First, there must be a gross balance between vertical mass transfer at sources and sinks and lateral mass transfer by plate motion and motions in the asthenosphere. Second, the 700-kilometer limit on the depth of earthquakes and the abrupt increase in shear-wave velocity that marks the bottom of the asthenosphere imply that the mass-transfer circuits involve only the lithosphere and the asthenosphere. Third, the boundaries along which new crust is being generated on the earth exceed in length the boundaries where crust is being destroyed, so that plates are generally consumed at individual sinks faster than they are created at individual sources. Fourth, a simple geometric consequence of the fact that plates can change their surface area is that plate boundaries can move in relation to other plate boundaries. This fact implies that mass-transfer circuits must also change their geometry as the plates evolve. Fifth, mass-transfer circuits cannot take the form of simple, regular convection cells linking sources and sinks with an upper lateral transfer and a lower return flow because there is no simple one-to-one relation between sources and sinks.

There are several great circles around the earth along which one can cross two ridge axes without an intervening subduction zone or can cross two subduction zones without an intervening ridge axis. Circuits involving mass-transfer rates of up to 10 centimeters per year must be accompanied by convective heat transfer, because thermal inertia prevents the elimination by conduction of temperature differences between different parts of the circuit. This condition is reflected in the persistence of earthquakes in a plunging slab down to depths approaching 700 kilometers. It is not clear, however, whether convective mass transfer and heat transfer is the cause or a consequence of plate motion. The models for relative lithosphere-asthenosphere motion at the top of page 130 illustrate this difficulty. They are all certainly far too simplistic. Indeed, the relative surface motions of plates may not be a guide to motions in the asthenosphere.

Let us consider a model in which the crust, lithosphere and asthenosphere are linked in a simple source-to-sink system [*see bottom illustration on page 40*]. The lithosphere in this model acts as a cold boundary conduction layer to a hotter asthenosphere, the upper part of which (the low-velocity zone) is probably near its melting temperature. Tension induced by plate separation at the ridge axis reduces the confining pressure in the low-velocity zone under the ridges. Reduction of confining pressure causes partial melting of primitive mantle in the low-velocity zone and the rising of a mush of crystals and liquid, with the result that the ridges are broadly uplifted. As the column of partially molten material rises it undergoes further partial melting; eventually the basaltic liquid rises to fill the crack continuously generated by plate separation. The liquid cools and crystallizes to form the basaltic oceanic crust and leaves under it a depleted layer of mantle.

Where plates descend into the asthenosphere, their leading edge carries a chain of volcanoes; thus one infers that the volcanic rocks are linked in some way with the descending plate. Because the volcanic rocks are less dense than the basalts of the oceanic crust, it is likely that they are formed by partial fusion of the oceanic basalts with other material as the basalts are carried down into the hot asthenosphere. The depleted mantle in the plunging plate is denser than the primitive asthenosphere through which it sinks, both because it has had a lighter basaltic fraction removed from it under the ridge axis and because it is cooler. Therefore once a plate has begun to descend in a particular subduction zone it is likely to continue until the plunging plate meets increasing resistance deep in the asthenosphere.

Since continental crust is only about 40 kilometers thick, whereas plates are 70 kilometers or more thick, the continents ride as passengers on the plates. In the framework of plate tectonics continental drift is no more significant than "ocean-floor drift." Nevertheless, continents, unlike oceans, impose certain important restraints on plate motion. The narrow, sharply defined trenches and the regularly inclined earthquake zones sloping away from trenches indicate that oceanic lithosphere is easily consumed by subduction, probably because it has a thin, dense crust. Intracontinental seismic zones associated with mountain ranges exhibit compressional deformation over a wide area, which implies that continental lithosphere is hard to consume because it has a thick, relatively buoyant crust.

Within the Alpine-Himalayan mountain belt are narrow zones characterized by a distinctive assemblage of rocks, known as the ophiolite suite, whose composition and structure suggest that they are slices of oceanic crust and mantle. If they are, ophiolite zones mark the lines along which continents collided following the contraction of an ocean by plate consumption [*see illustration on opposite page*]. The small oceanic areas within the Alpine belt, such as the Mediterranean Sea and the Black Sea, may be remnants of larger oceans that once lay between Africa and Europe. Evidently lithosphere carrying light continental crust is difficult to consume, as is indicated by the marked scarcity of intermediate-focus and deep-focus earthquakes in zones where continents have collided. Thus it seems that continental collision terminates subduction along the collision zone. This implies that mass-transfer circuits must be drastically rearranged after the collision of continents, since a major sink is eliminated. As a result new sinks may form in oceans elsewhere.

As we have seen, any hypothetical driving mechanism for plate motion must meet a number of conditions. At present some form of thermal convection in the upper mantle seems to hold the most promise, but other mechanisms have been suggested that may play some role in plate dynamics. These mechanisms include the retarding effect of earth tides raised by the gravitational attraction of the moon, the possible pull exerted by a plate "dangling" in the asthenosphere and forces created by plates sliding down the slight grade between sources and sinks. It is also possible that some small plates are mechanically driven by the effects of relative motion between adjacent larger plates. For example, the westward motion of the wedge-shaped Turkish plate with respect to the Eurasian plate may be caused by its being squeezed like an orange seed between the Arabian and the Eurasian plate.

Extinct Plates

It is now certain that plate tectonics has operated for at least the past 200 million years of earth history. During this time virtually all the present oceans were created and others were destroyed. Two hundred million years ago the major continental masses were assembled into the single supercontinent Pangaea [*see illustration below*]. It is therefore legitimate to ask if the breakup of Pangaea some 180 million years ago marked the beginning of plate tectonics. Geologic studies of mountain belts older than 200 million years strongly indicate that they owe their origin to processes operating at plate boundaries that are now extinct. The Ural and Appalachian-Caledonian mountain belts, which lie within ancient Pangaea, have narrow zones where ophiolites are found. These old ophiolite zones, like those in the Alpine-Himalayan mountain belt, thus mark the sites of vanished oceans. This implies that the Urals, for example, were created by the collision of two conti-

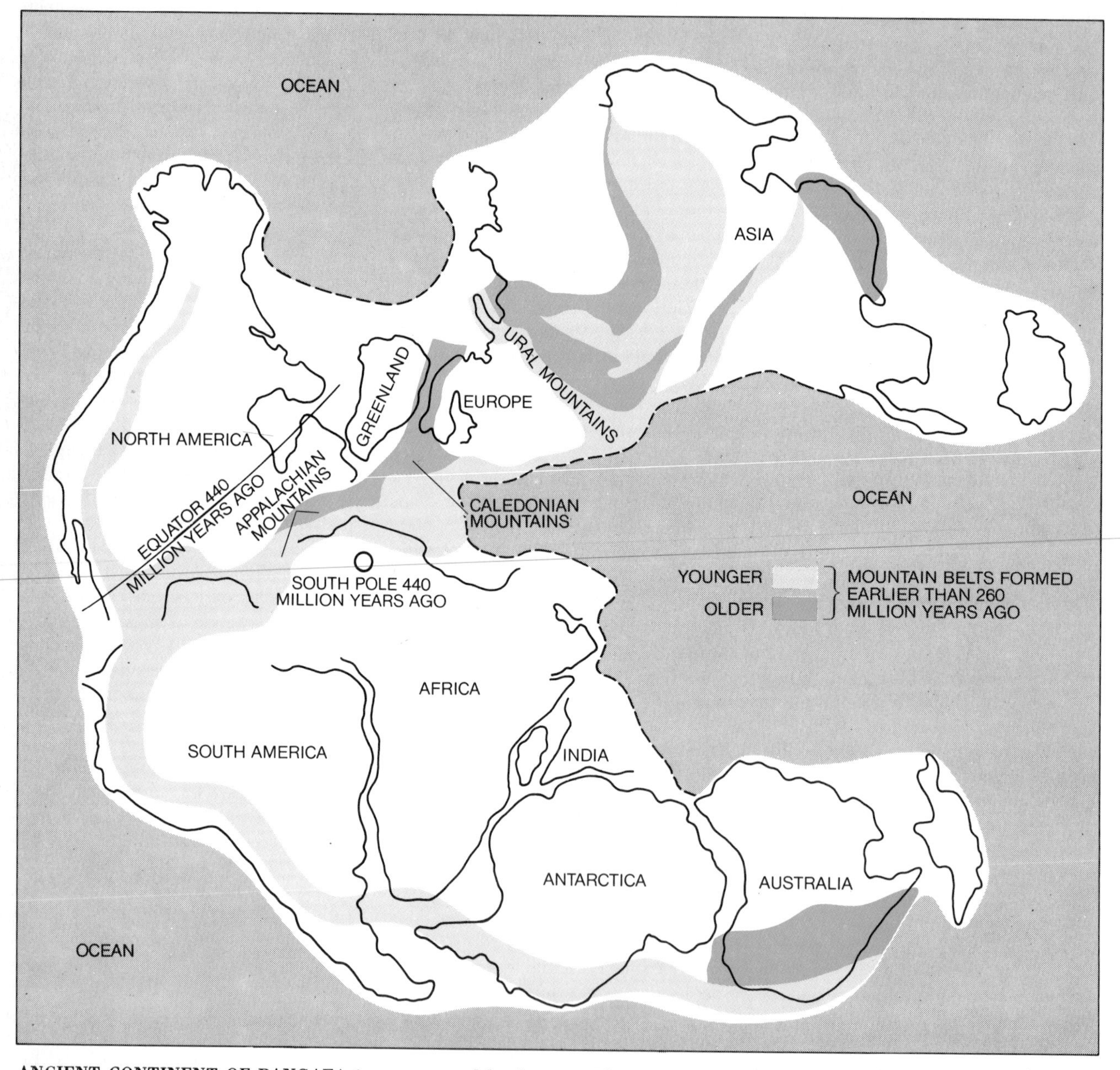

ANCIENT CONTINENT OF PANGAEA is reconstructed by fitting together the major continental land masses. Pangaea started to break up about 200 million years ago with a rift between Africa and Antarctica. Other rifts allowed South America, Australia and India to drift into their present positions. Mountain belts formed more than 260 million years ago are shown in shadings that distinguish younger and older belts. These mountain belts indicate lines of collision between continental fragments antedating Pangaea. Thus a prior collision of North America and Africa formed the younger part of the Appalachians 260 million years ago. Such a collision would explain how Equator and South Pole of 440 million years ago were brought close together after formation of Pangaea.

nental masses and that the ophiolites were generated by sea-floor spreading at a ridge axis before the continents were brought together.

Large-scale horizontal motions of continents before 200 millon years ago are supported by other lines of reasoning. Glacial deposits and other data indicate that about 400 million years ago a south polar ice cap covered the Sahara. At the same time eastern North America lay near the Equator. On the Pangaea reconstruction these close polar and equatorial positions are incompatible: they indicate that Africa and North America were separated by an ocean some 10,000 kilometers wide. The contraction of this ocean and the resulting collision of North America and Africa were probably largely responsible for the growth of the Appalachian mountain belt [see "Geosynclines, Mountains and Continent-Building," by Robert S. Dietz; SCIENTIFIC AMERICAN Offprint 899]. It seems a reasonable assumption that long, narrow, well-defined zones of mountain-building were established along zones of plate convergence. If this is the case, plate tectonics has been operating for the past two billion years.

The absence of well-defined zones of mountain-building older than two billion years suggests, however, that some mechanism other than plate tectonics, at least as we know it at present, was responsible for the evolution of the earth's crust in earlier epochs. The ancient "shield" regions of the continents, which contain rocks older than 2.4 billion years, are characterized by rocks distributed in swirling patterns over areas so wide they can hardly be explained by processes arising at the boundaries of rigid plates. Evidently the shield areas were stabilized about 2.4 billion years ago, and by some 400 million years later a lithosphere with sufficient rigidity to crack into a plate mosaic had developed.

This does not necessarily mean that plate tectonics as we know it today began two billion years ago. Mountain belts older than 600 million years do not have ophiolite complexes like those of the younger mountain belts, indicating that sea-floor spreading before 600 million years ago generated a different type of oceanic crust and mantle. Geological data suggest that plates may have been getting thicker and that plate boundaries may have become more narrowly localized with time.

An exciting corollary of plate tectonics is that it provides a means whereby the total volume of continental crust can increase with time. We have seen that the primitive mantle of the asthenosphere undergoes partial melting to liberate a basaltic liquid that rises and cools to form the oceanic crust on the ridge axis, and that partial melting of the oceanic crust on a descending plate may yield the liquids that erupt to build the volcanic chains on the leading edge of plates. Volcanic rocks, with their deep-level intrusions of liquids that crystallized before they reached the surface, have the same bulk composition as that of the continental crust. The volcanic chains may therefore be sites where strips of embryonic continental crust are generated. Since they lie on the leading edge of plates, their destiny is to collide with other volcanic chains or with various kinds of continental margin. In this way new strips of light continental crust will be added to continental margins.

As we have seen, the arrival of a continental margin at a subduction zone blocks further plate consumption at that site. Thus the oceanic ridge provides an effective means of growing continental crust, but there is no means of destroying such crust. This implies that the total volume of continental crust has been increasing for the past two billion years. One should not conclude, however, that strips of new crust have been added to the continents as a succession of regular concentric rings. Rather, discontinuous strips have been added at different times, reflecting the complex interaction of continental margins with the plate-boundary mosaic.

Although there are geologic phenomena that plate tectonics does not yet obviously explain, and although the driving mechanism is obscure, these deficiencies do not constitute rational objections to the theory of plate tectonics. One of the serious mistakes that many earth scientists made in the past was to reject continental drift because it was not clear how or why it occurred. The remarkable success of plate tectonics has been not only that it provides a consistent logical framework that draws together such diverse phenomena as sea-floor spreading, continental drift, earthquakes, volcanoes and the evolution of mountain chains but also that it has been successfully quantified and tested to the point where its essential core can no longer be questioned.

The essential core of plate tectonics is the geometric evolution of plates and the kinematics of their relative motion. It is of paramount importance to fully explore all the geometric and kinematic aspects of plate evolution if we are ever to understand the dynamics of plate motion and the geologic corollaries of plate tectonics.

4 The Earth's Mantle

Peter J. Wyllie
March 1975

The mantle of the earth is a thick shell of red-hot rock separating the earth's metallic and partly melted core from the cooler rock of the thin crust. Starting at an average depth of from 35 to 45 kilometers (22 to 28 miles) below the surface and continuing to a depth of some 2,900 kilometers, it accounts for nearly half of the earth's radius, 83 percent of its volume and 67 percent of its mass. Its influence on the crust is profound; indeed, the crust and its thin film of ocean and atmosphere are distillates of the mantle, and the driving forces that move the continents slowly about on the earth's surface arise within the mantle. Knowledge of the mantle is therefore crucial to an understanding of the structure and dynamic behavior of the earth. Notwithstanding the inaccessibility of the mantle, a considerable amount of information about it has been assembled by more or less indirect means.

The role of the mantle in influencing conditions at the surface is manifold. For example, during the 4.6 billion years since the origin of the earth in the solar nebula, melting of the more fusible constituents of the mantle has produced lavas that rise to the surface and solidify, adding new rocks to the crust and giving off the water vapor and other gases that go to the atmosphere and the oceans. On another scale gaseous carbon compounds that came from the mantle began the story of life on the earth when they provided the raw material for organic molecules.

Similarly the mantle as a driving force has multiple effects. The surface of the earth is shaped by the action of the mantle, moving very slowly below the crust. Mountains rise and persist because of this movement; without it erosion would wear them down to sea level within 100 million years or so. Movements of the mantle also cause volcanic eruptions, earthquakes and continental drift.

A cross section through the earth shows the concentric layers of the core, the mantle and the crust [*see top illustration on opposite page*]. They differ from one another in composition or physical state or both. The mantle is composed of silicate minerals rich in magnesium and iron, with an average composition corresponding to that of the rock peridotite. (The name comes from the fact that the most abundant mineral in peridotite is olivine, which is more familiar as the transparent green gemstone called peridot.) The mantle is solid but within the relatively thin zone ranging from about 100 kilometers below the surface to about 250 kilometers the rock may be partially melted, with thin films of liquid distributed between the mineral grains. This zone is called the low-velocity layer for reasons that will become apparent below.

The density of the mantle increases with depth from about 3.5 grams per cubic centimeter near the surface to about 5.5 grams near the core. It does not increase smoothly; the curve of density displays distinct steps [*see illustration on page 48*]. They indicate significant changes in the mantle rocks at depths near 400 and 650 kilometers. The distribution of density provides the basic for the calculation that the mantle makes up 67 percent of the mass of the earth.

Plate Tectonics

This static picture of a concentric, layered earth is modified by the theory of plate tectonics, which deals with the movement of lithospheric plates. The lithosphere includes the crust and part of the upper mantle and is distinguished from the asthenosphere below it by the fact that it is cooler and therefore rigid. The theory of plate tectonics provides a dynamic picture of a mobile mantle, with plates of lithosphere about 100 kilometers thick moving laterally over the asthenosphere. The surface of the earth is covered by a few large lithospheric plates and several smaller ones. These shell-like plates move with respect to one another, and the geological activity represented by earthquakes and volcanoes is concentrated along the plate boundaries.

Plate boundaries include divergent and convergent types. Below the crest of an ocean ridge material from the asthenosphere rises, melting as it moves and thus producing lava that is erupted in the central rift valley of the ridge to produce new crust. Convective movements in the mantle cause the plates to diverge from one another as new lithosphere is generated.

At convergent boundaries plates may collide, pushing up crust to form mountains, or one plate may move under another, carrying lithospheric material back into the mantle. The growth of new lithosphere is therefore balanced by the destruction of lithosphere elsewhere. Such boundaries are associated with ocean trenches and with lines of volcanoes, including arcs of volcanic islands and the volcanoes in active mountain ranges.

Many of the physical properties of the earth as a whole have been determined. The planet's size, shape and mass have been measured with precision. Its known volume and mass give a mean density of 5.5 grams per cubic centimeter, which is much higher than the density of the rocks making up the accessible crust. Therefore a significant part of the interior must be composed of material

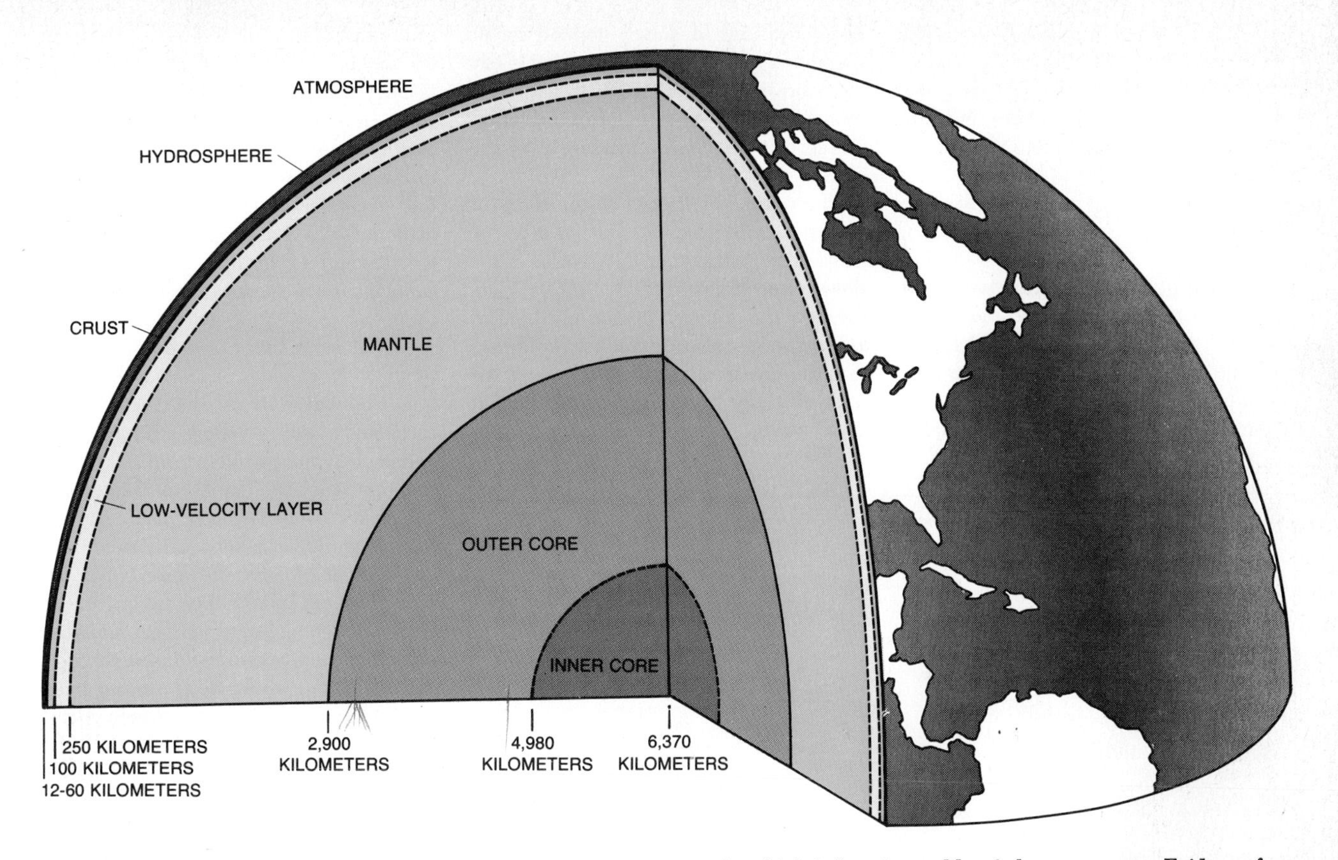

CONFIGURATION OF THE EARTH is portrayed by means of layers alone, without regard to the active processes that go on in the interior. The rocks of the thin crust are cool and rigid. Mantle rock, which is hot, is capable of slow movement. Evidence from earthquake waves indicates that the outer core consists of molten metal. Hydrosphere consists of surface and atmospheric waters.

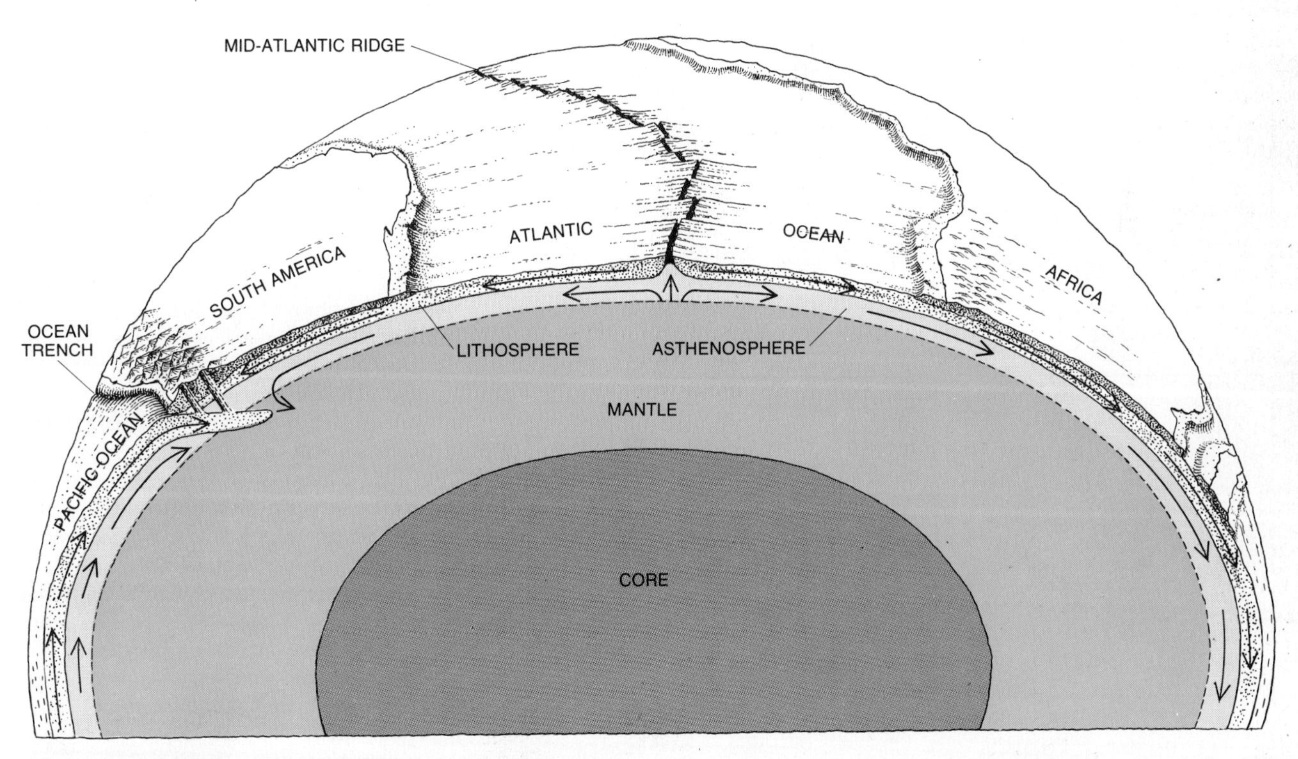

DYNAMIC EARTH is depicted in cross section as it is envisioned in the theory of plate tectonics. Plates of the lithosphere, which includes the crust and part of the upper mantle, migrate laterally over the asthenosphere, which is a hot and perhaps partly molten layer of the mantle. Material from the asthenosphere rises below the crest of an ocean ridge, melting to produce lava that is erupted to form new crust in the ocean floor. The lithospheric plates diverge as new lithosphere is generated from the rising material. The growth of new lithosphere is balanced by the destruction of an equivalent amount of lithosphere at convergent plate boundaries, where the lithospheric layer moves down into the mantle. These boundaries are associated with ocean trenches and lines of volcanoes.

with a density higher than 5.5 grams per cubic centimeter. Studies of the earth's gravitational field and the physical properties of the rotating sphere indicate that the mass is concentrated toward the center.

Evidence on the Mantle

The main body of information about the physics of the earth comes from studies of earthquake waves, which provide the equivalent of X-ray pictures of the interior. The earth rings like a bell when it is shaken by a major earthquake. The vibrations of a bell depend on its shape and physical properties; similarly the vibrations of the earth as recorded by sensitive instruments can be interpreted in terms of the properties of the earth.

The release of energy at the focus of an earthquake produces several types of wave. The primary, or *P*, waves and the slower secondary, or *S*, waves pass through the interior of the earth. The abbreviations also serve as reminders that the energy is transmitted along the path of a ray by different phenomena: the *P* waves are compressional, or push-pull, waves and the *S* waves are shear, or "shake," waves [*see illustration on opposite page*]. *P* waves can be transmitted through both solids and liquids; *S* waves can be transmitted only through materials that can support shear stresses, that is, materials that can be deformed or bent. An *S* wave cannot be transmitted through a liquid, because liquids cannot sustain shear; they flow too easily.

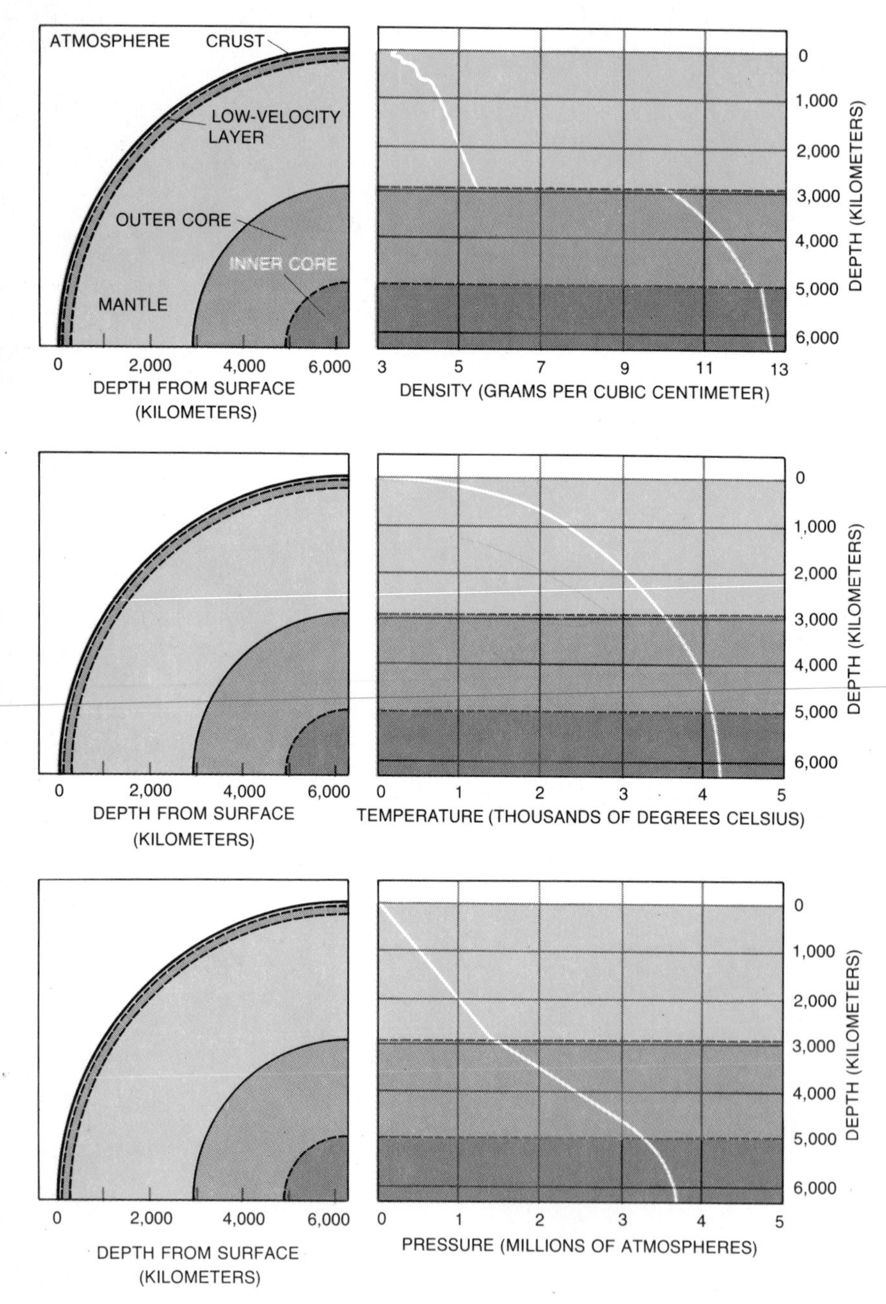

PHYSICAL PROPERTIES of the earth vary with depth from the surface. The depths on the graphs of density, temperature and pressure correspond to the depths on the cross-section diagrams at left. One atmosphere is the air pressure at sea level, 14.7 pounds per square inch.

If the earth were composed of material with uniform properties throughout, the waves from the focus of an earthquake would follow straight lines, and each type of wave would travel at a constant velocity. The times taken for *P* and *S* waves to reach a particular recording station at a known distance from the focus would give the velocity of each type of wave. In actuality, however, the results obtained from many earthquakes show that the waves travel faster through the earth than would be predicted from their known velocities in surface rocks. The results show in addition that the waves that travel the greatest distance have also traveled faster. These findings mean that the velocity of earthquake waves is greater at depth than it is near the surface and also that it increases progressively with depth.

From these observations and others it is known that the waves from earthquakes are refracted and reflected within the earth. Refraction causes them to follow paths that are concave upward. Rays are reflected at levels where there is a distinct change in physical properties between layers. Reflection therefore occurs at boundaries: between the crust and the mantle, the mantle and the core and the inner and the outer core. This wave pattern demonstrates the concentric structure of the earth. *S* waves follow paths that reach as deep as the core-mantle boundary, but they do not pass through the core. This finding is evidence that at least the outer part of the core is liquid.

Measurement of the times of travel of the waves over various paths in the mantle provides a means of calculating the velocity at which the material at each depth within the mantle transmits waves. The wave-velocity profiles show a progressive increase with depth, but the increases occur in a series of steps down to about 1,000 kilometers. These observations indicate that the upper mantle has a layered structure.

The velocity of both *P* and *S* waves decreases in the upper mantle within a layer lying from approximately 100 to 250 kilometers below the surface. This

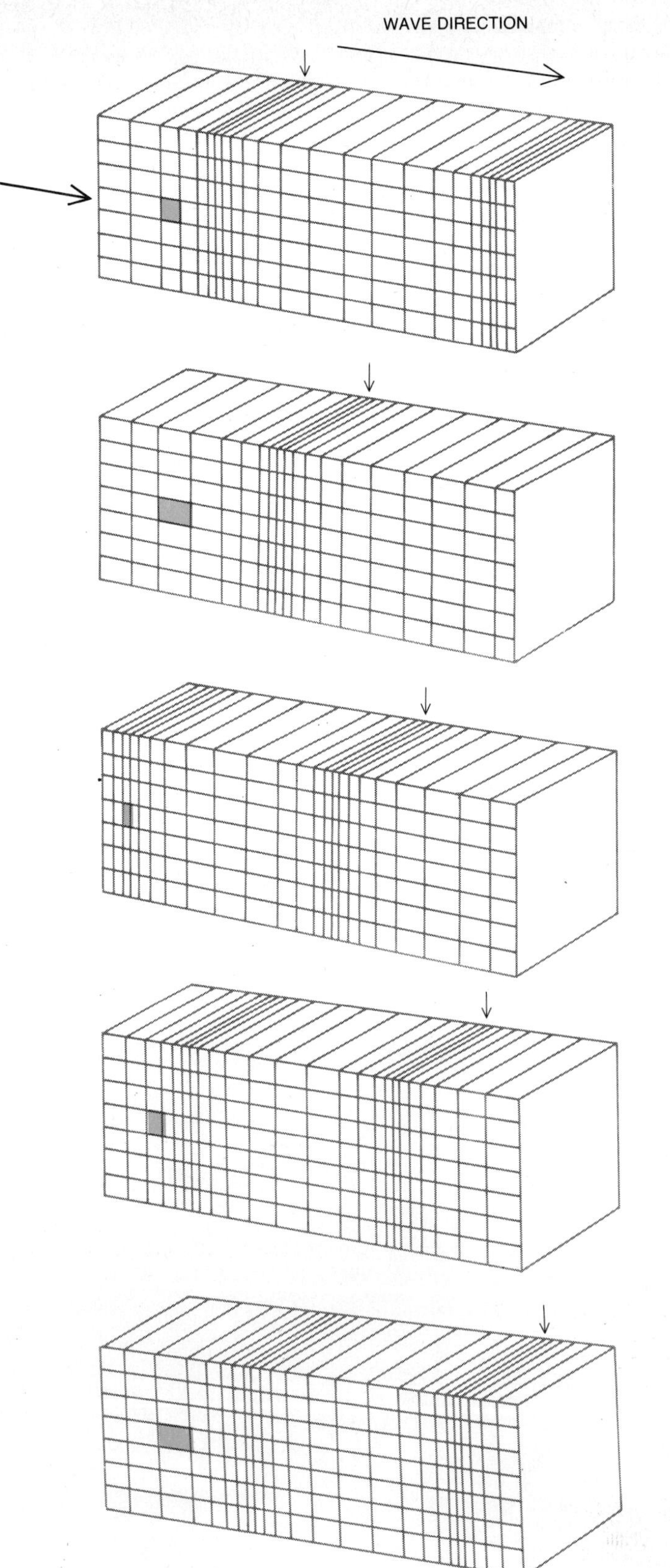

EARTHQUAKE WAVES of two types pass through the interior of the earth from the focus of an earthquake, where energy is released. They are portrayed here passing through a block of rock. At left is a compressional, or *P*, wave, which is started by a sudden push or pull in the direction of the wave path. The action compresses the rock, and the nearby particles move forward. They then rebound to their former positions and beyond, continuing to vibrate in this way for some time. Each small volume of matter (*color*) contracts and expands as the crest of compression moves through the rock. A shear, or *S*, wave (*right*) is started by pressure at a right angle to the wave path. Particles of rock vibrate up and down. A small piece of matter (*color*) undergoes shear deformation. The wave crest transmits energy. The illustration is adapted from *The Heart of the Earth*, by O. M. Phillips (Freeman, Cooper & Company).

is the low-velocity layer that I have mentioned. It is considered to be equivalent to the asthenosphere of the plate-tectonic model. The layer's physical properties, deduced from the wave-velocity profile and other geophysical evidence, are consistent with the presence of an interstitial liquid, which is generally thought to be a molten fraction of the rock in the layer.

Plate Boundaries

Earthquakes provide not only this static picture of the concentrically layered earth but also some aspects of the dynamic picture of plate tectonics. Since earthquakes occur only in rocks that are cool and rigid enough to fracture, the distribution of earthquake foci delineates the boundaries of the stable plates. The depths of the foci along the boundaries associated with ocean ridges are less than 100 kilometers, indicating that the mantle below the lithosphere is hot.

At the convergent boundaries where one plate sinks below another and moves into the mantle the foci of earthquakes are at depths as great as 700 kilometers. The distribution of these deep foci provides a means of mapping the layers of sinking lithosphere. Detailed studies of wave velocities in these regions are consistent with the existence of slabs of cooler lithosphere extending to considerable depths within the mantle.

The information provided by earthquake waves about the structure and physical properties of the mantle is fairly direct. In contrast, the effort to obtain information about the chemical properties of the deep interior requires resorting to mostly indirect evidence. The only chemical samples available are a few rocks in the crust that have been carried up from the topmost layers of the mantle. To form an estimate of the composition of the earth as a whole and of the core and mantle one must rely on the chemistry of extraterrestrial bodies, including stars, the sun and meteorites. The presumption is that such bodies have a composition essentially similar to the earth's. The approach involves the formulation of physical and chemical models for the origin of the solar system and the earth.

It is generally held that the solar system formed about 4.6 billion years ago through the gravitational collapse of matter that previously had been dispersed in interstellar space. A large precursor of the sun, a protosun, was surrounded by a thin, disk-shaped nebula of dust particles and gas. Local aggregation of particles and condensation of gas within the rotating nebula formed small objects that combined to produce planetary bodies. The material of meteorites was also formed during this period. Since the sun accounts for more than 99.6 percent of the mass of the solar system, its composition is effectively the same as that of the system as a whole. The abundances of the elements in the sun and other stars have been measured by spectroscopic methods, whereby each element is identified by its characteristic electromagnetic radiation.

What Meteorites Reveal

Meteorites now travel through the solar system in elliptical orbits that occasionally intersect the earth. There is evidence that they come from the asteroid belt: the swarm of small planetlike bodies orbiting at distances of 2.2 to 3.2 astronomical units from the sun. The meteorites vary widely in chemistry, mineralogy and structure, but for the purposes of this discussion it is sufficient to note the distinctions between two main groups: the iron meteorites and the stony meteorites. The iron meteorites consist essentially of iron-nickel

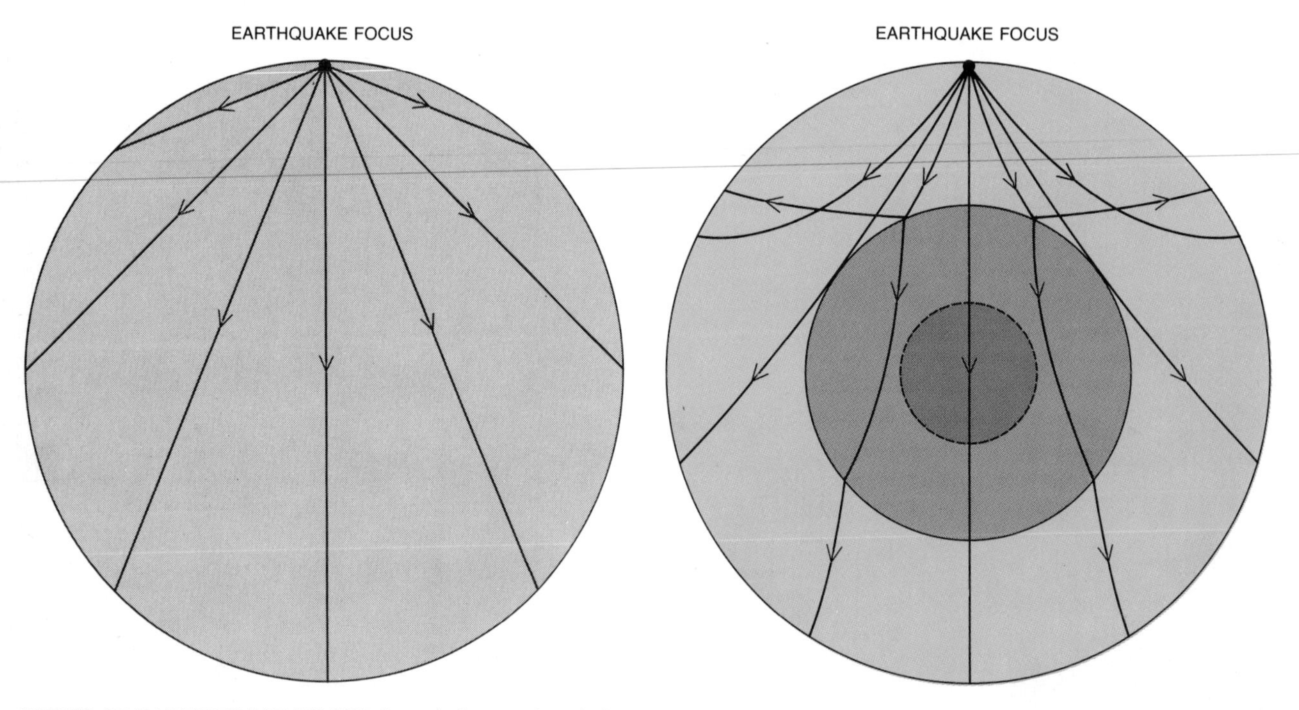

PATHS OF EARTHQUAKE WAVES through the interior of the earth provide information about its structure and the physical properties of its concentric layers. Energy from the focus of an earthquake is transmitted in all directions. If the earth had uniform properties throughout, the wave paths would follow straight lines (*left*) and the wave velocities would be constant. Measurements of the time of travel show, however, that the wave velocities, and therefore the physical properties, change abruptly at certain levels (*right*), thus revealing the concentric layers. *P* waves can be transmitted through both solids and liquids. *S* waves, which cannot pass through liquids, do not pass through the core of the earth, showing that at least the outer core is in the liquid state.

alloy, with the nickel content ranging from 4 to 20 percent; they also contain a small amount of iron sulfide.

Stony meteorites are composed mainly of silicate minerals, together with various proportions of metal alloy and iron sulfide. The relative abundances of nonvolatile elements such as magnesium, silicon, aluminum, calcium and iron are about the same in many types of stony meteorite as they are in the sun and other stars. It is therefore argued that these abundances provide a good basis for estimating the overall abundances of elements in the earth and other planets.

In complex models of the origin and evolution of the solar system the composition of the earth is derived by starting with a volatile-rich stony meteorite and formulating a series of processes and chemical changes that could account for the other types of meteorite and for the present structure of the earth. The calculations yield estimates of the composition of the core and the mantle.

A simpler approach, which gives a similar result, is to select a specific group of stony meteorites and to assume that the average composition of the silicate portion is equivalent to the composition of the earth's mantle. The composition of the earth's core is estimated from the iron sulfide and an appropriate portion of the iron-nickel alloy in the same meteorites. The iron meteorites serve as a check on this calculation.

Whatever procedure is followed, the estimates of the composition of the mantle agree in the following respects: (1) More than 90 percent by weight of the mantle is represented by oxides of silicon, magnesium and iron (SiO_2, MgO and FeO), and no other oxide exceeds 4 percent. (2) The oxides of aluminum (Al_2O_3), calcium (CaO) and sodium (Na_2O) total between 5 and 8 percent. (3) More than 98 percent of the mantle is represented by these six oxides, and no other oxide reaches a concentration of as much as .6 percent. The concentrations of other elements, which are present in trace amounts, are not defined. The oxides are combined in various minerals within the mantle rock.

Of all the rocks found in the crust, only peridotites correspond to the estimates of the composition of mantle rocks made by studying extraterrestrial bodies. It is therefore to such peridotites that geologists interested in the mantle turn for details, including the concentration of trace elements. The problem, of course, is to make sure that the specimen of peridotite one is examining originated

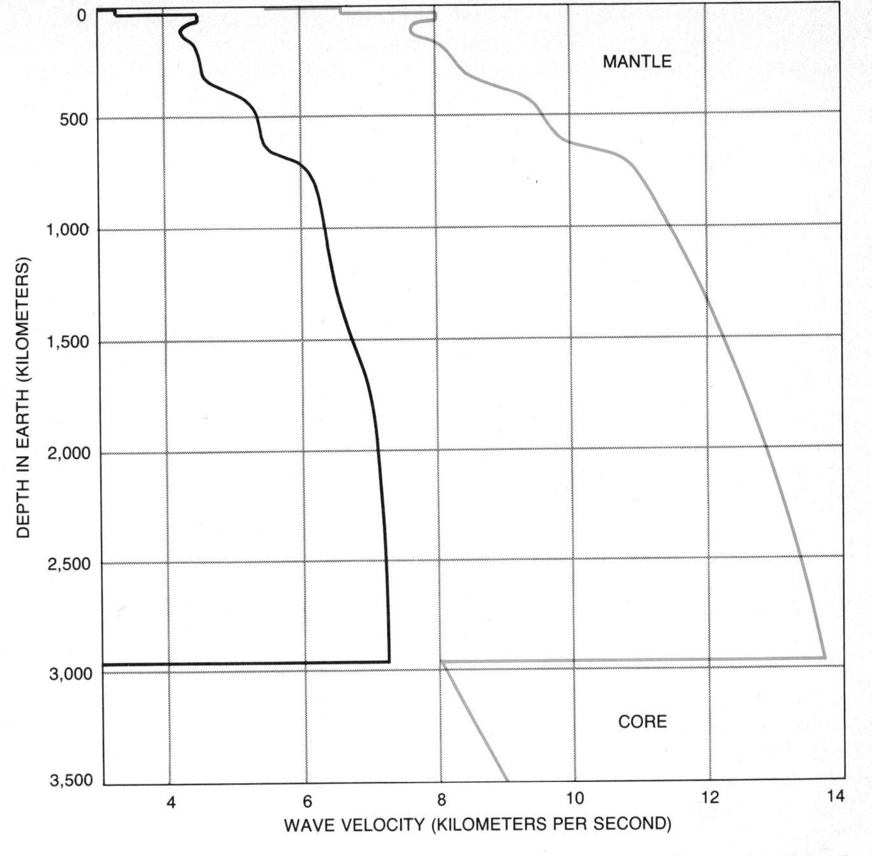

VELOCITY PROFILES of earthquake waves passing through the mantle are plotted for *S* waves (*black*) and *P* waves (*color*). The velocities are affected by the increase in pressure and temperature with depth, but the steps appearing at intervals down to a depth of about 1,000 kilometers correspond to changes in physical properties of the mantle. The decreased velocity between 100 and 250 kilometers is probably due to the presence of partly melted rock, which is an appropriate property for mobile asthenosphere of plate-tectonic model.

in the mantle rather than in the crust. For this reason particular interest attaches to the rounded boulders or nodules that are found in kimberlite "pipes" (cylindrical intrusions, with a diameter of a few hundred meters, that puncture the crust from the mantle in certain regions).

Kimberlites and Lavas

Kimberlites are famous as the rocks that bring diamonds to the surface of the earth. Diamond is a form of carbon that is stable only at very high pressures. One can therefore be confident that kimberlites originate in the depth interval from 150 to 300 kilometers below the surface—well within the upper mantle.

Evidence indicates that a kimberlite pipe originally rose rapidly through the crust as a fluidized system of solids, molten rock and gases, breaking through to the surface with a tremendous blast in a brief volcanic explosion. Fragments of rock ripped from the walls of the pipe and carried upward include nodules of peridotite and subordinate nodules of eclogite, another mantle rock. They are rounded and polished by repeated impact with other gas-driven fragments in the explosive pipe.

Suites of nodules similar to those in kimberlites are also found in certain volcanic lavas. In general they are derived from shallower levels of the mantle than the nodules in kimberlites. Many estimates of the composition of the mantle have been based on the mineralogy and chemistry of nodule suites in kimberlites and lavas. Peridotite rocks found in other geological environments have also served for this purpose, although with them it is particularly important to take into account the nature of the geological association with other rocks and the effect of geological processes in order to be sure that the judgments made on the basis of the evidence in the rocks truly relate to the mantle rather than to events in the crust.

A third approach, in addition to the study of extraterrestrial bodies and mantle rocks, is to calculate a hypothetical peridotite having the chemistry appropriate to yield the volcanic lavas that are derived by partial melting of the mantle. The melting gives rise to the basaltic magma erupted abundantly onto the surface from volcanoes and leaves behind in the mantle a residual peridotite, which can be presumed to lack the more fusible elements that melted and were carried off in the magma. Assigning an appropriate chemistry to the residual peridotite, one arrives at the hypothetical composition of the upper mantle. Pyrolite (pyroxene-olivine rock) is the name given to one of these hypothetical peridotites.

The Heterogeneous Mantle

It was once assumed on the basis of physical models that the upper mantle was homogeneous in composition. As we have seen, however, the detailed geophysical studies of recent years indicate a layered structure. Indeed, the nodules that have been studied confirm the notion that the upper mantle is quite heterogeneous, both chemically and mineralogically. The specimens sample the mantle from the mantle-crust boundary to the sources of kimberlite pipes, a depth interval that could reach 250 kilometers. One could hope to find among the nodules specimens of original mantle peridotite that has never been melted; specimens of depleted residual peridotite from which a molten fraction of the more fusible components has been removed; specimens of the molten fraction that failed to escape to the surface in magma, instead crystallizing at high pressure in the mantle to form eclogite; specimens intermediate between these types, and specimens involving the same mineral assemblage but formed by other processes too complex for consideration here.

Estimates of whole-mantle composition based on the examination of extraterrestrial bodies yield values similar to those obtained from the study of rocks originating in the upper mantle. Together the estimates support the idea that the chemistry of the mantle does not change much from top to bottom. In comparing estimates, however, one finds that the potassium content of the upper mantle according to the hypothetical pyrolite is considerably higher than the amount of potassium found in peridotite rocks derived from the mantle, and both are much lower in potassium than the estimate derived from studying extraterrestrial bodies.

Large uncertainties also remain about the concentration and distribution of other trace elements and of volatile components such as water and carbon dioxide. These components are of fundamental significance for such factors as the generation of heat by radioactive decay (of such elements as uranium and thorium and the radioactive isotope of potassium, potassium 40), melting temperature (in the presence of small amounts of water at high pressure rocks begin to melt at temperatures lower than the melting point of dry rock) and the physical strength of the mantle (where the strength of rock would be significantly reduced by the presence of small amounts of molten material between mineral grains and by interstitial bubbles of gas).

Water and Carbon Dioxide

In certain nodules of peridotite one finds the minerals phlogopite and amphibole, which are hydrous, that is, incorporate water. This finding is accepted as evidence for the existence of water in at least some parts of the upper mantle. It is doubtful that the amount could exceed .1 percent by weight, and it is probable that the distribution of the water is not uniform.

Examination under the microscope has shown that the crystals of olivine and pyroxene in some peridotite nodules from kimberlites and lavas are crowded with tiny cavities up to five micrometers in diameter. Many of them are filled with dense, liquid carbon dioxide trapped at high pressure. This finding indicates the presence of carbon dioxide in some form in the upper mantle.

High-voltage electron microscopy has recently yielded remarkably detailed pictures of crystal defects: the discontinuities in crystal structure that occur within individual grains of minerals. The defects cannot be seen with light microscopes. In the electron micrographs minute bubbles of carbon dioxide are abundant along the discontinuities in the olivine and pyroxene of some peridotite nodules. This evidence suggests that the carbon dioxide was originally dissolved in the solid minerals and was exsolved and precipitated as gas bubbles because of elastic strain near the crystal defects.

The independent determinations of the physical properties and chemistry of

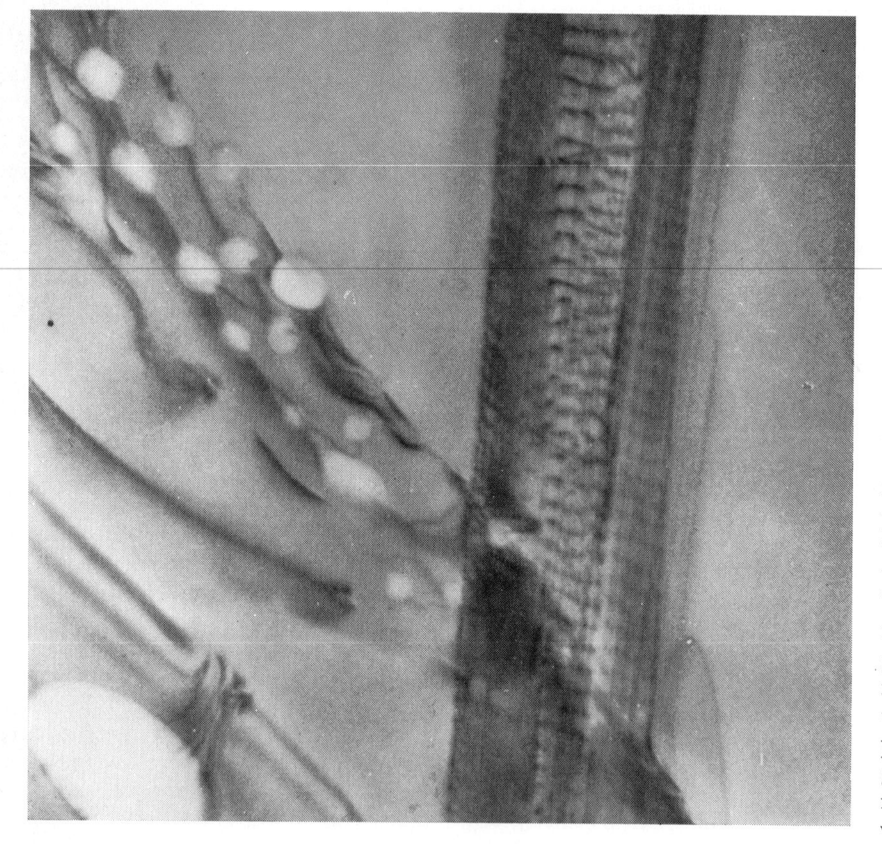

CARBON DIOXIDE BUBBLES appear in this electron micrograph of a specimen of peridotite. The enlargement is 20,000 diameters. The bubbles, which appear in discontinuities in the crystal structure of the rock, indicate the presence of carbon dioxide in the mantle. Micrograph was provided by Harry W. Green II of the University of California at Davis.

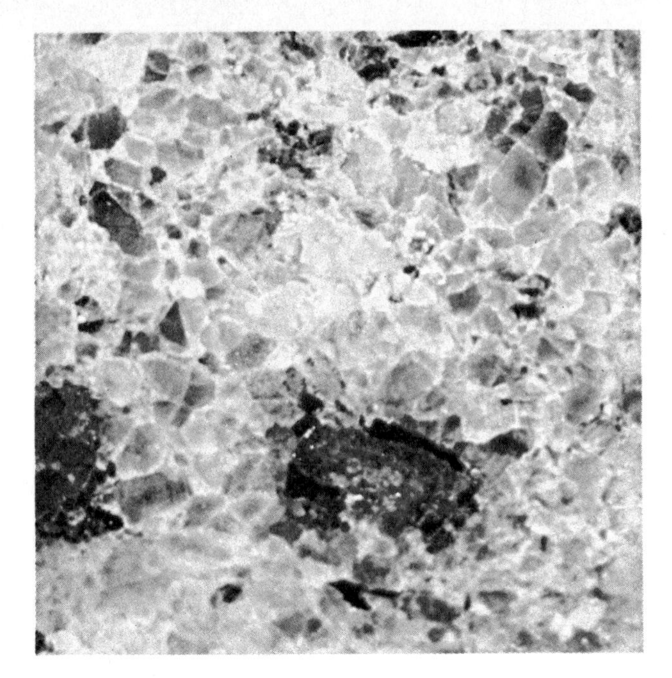

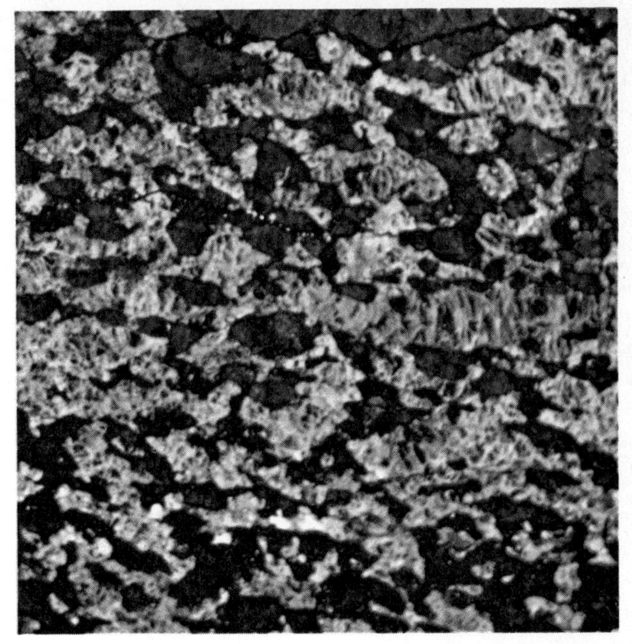

ROCKS FROM MANTLE are a mica-garnet peridotite (*left*) and an eclogite (*right*). The peridotite consists mainly of olivine (*yellowish green*) but also contains garnet (*red*) and orthopyroxene and clinopyroxene (*bright green*). A sample including mica, a mineral with water bound in its structure, is rare. This sample is from Tanzania. The eclogite, which is from a mine in South Africa, is composed of garnet (*red*) and a clinopyroxene (*green*). The minerals are arranged in layers. Mantle rocks are brought to the surface in kimberlite "pipes" and in certain volcanic lava. The photographs were provided by J. B. Dawson of the University of St. Andrews.

the mantle must be consistent with one another. In order to test their consistency one needs to know the physical properties of the estimated composition of the mantle through the range of pressure and temperature existing in the mantle. If one can determine the way the mineralogy varies as a function of pressure and temperature, one has a means of estimating the depth from which a rock sample came and the temperature at that depth when the rock was formed or reached equilibrium with its surroundings.

Pressure and Temperature

The structure of silicate minerals is dominated by the packing of oxygen atoms. The other atoms, which are much smaller, occupy spaces between the oxygens. At low pressure each silicon atom is surrounded by four oxygens whose centers form a tetrahedron; the silicon is said to be in a fourfold coordination. At much higher pressure the oxygen atoms are squeezed closer together. They adjust into a densely packed arrangement with silicon atoms in a sixfold coordination. This readjustment of mineral structure is a phase transition. The steplike changes in value of the physical properties in the mantle are caused by successive phase transitions.

Starting with a material of fixed composition, such as the peridotite thought to exist in the mantle, the phase transitions depend on pressure and temperature. A diagram of the transitions for peridotite has been determined by experiments in laboratory apparatus that carried the pressure up to 200 kilobars (200,000 atmospheres, equivalent to the pressure 600 kilometers below the surface of the earth). The diagram has been extended to higher pressures by indirect methods.

It is known from the nodules in kimberlites and lavas that peridotite in the mantle can crystallize in at least three mineralogical assemblages: plagioclase peridotite, spinel peridotite and garnet peridotite. The high-pressure experiments demonstrate that the assemblages are related through phase transitions. With increasing pressure plagioclase peridotite is transformed first into spinel peridotite and then into garnet peridotite [*see illustration on next two pages*].

Experimental studies show that at still higher pressure garnet peridotite undergoes a phase transition involving an increase in density of almost 10 percent; the dominant olivine of the upper mantle is transformed into a spinel-like material, and the aluminous pyroxene is transformed into a garnet structure that combines in solid solution with the garnet already present. At pressures near 200 kilobars the minerals are further compressed into structures with all the silicon atoms in the sixfold coordination, giving rise to minerals that are unknown at the surface of the earth. This compression results in another increase in density of about 10 percent. The actual pressure at which a phase transition occurs increases with higher temperatures.

At any fixed depth an increase in temperature eventually brings a rock to a point where it begins to melt. This temperature increases with pressure, as shown by the boundary labeled solidus in the phase diagram [*next two pages*]. A rock composed of several minerals melts progressively through an interval of temperature in which solid crystals coexist with liquid. Complete melting is marked by the boundary labeled liquidus.

The effect of temperature can be studied by means of a geotherm, which is a line giving the temperature at each depth in the earth. If the line is drawn on the phase diagram for peridotite, each point on the line occupies one of the phase fields and thus also defines the mineral assemblage for the peridotite at each depth. A cross section through a hypothetical mantle composed of peridotite is constructed by following a geotherm through the phase diagram. Each layer consists of a particular mineral assemblage.

The boundaries between layers of the mantle are presumably at depths where the geotherm crosses phase boundaries. It turns out that these boundaries correspond closely to the depths where the velocity of seismic waves changes. This finding is regarded as being good evidence that the upper mantle does have a composition close to that of the hypothetical peridotite and that the layered structure of the upper mantle is caused by transitions of phase rather than by changes in composition.

The decrease in the velocity of earthquake waves in the low-velocity zone can be explained by the presence of water or carbon dioxide in the upper mantle. Either one could cause a trace of melting in the peridotite of the upper mantle. The result is a change in the physical properties of the rock. If no water were present, a similar effect could perhaps be produced by intergranular carbon dioxide.

During the first half of this century it was widely believed among geophysicists that convection, the upward flow of hotter material and the downward flow of cooler material, could not occur in the solid, rigid mantle. That was one of the reasons the theory of continental drift failed for so long to gain many adherents. Recently, however, a number of models have proposed convection in the mantle as the driving mechanism for the migration of lithospheric plates. Details of the movements in the mantle and the scale of convection remain uncertain, but there is little doubt that the rates are exceedingly slow—so slow that the mantle is effectively motionless within the normal human framework of time.

The lithospheric plates and the continents that ride on them drift at the rate of a few centimeters per year. Suppose the driving material in the mantle moves at a rate of five centimeters per year, which is equivalent to about .005 millimeter per hour. The tip of the hour hand on a typical household clock moves five centimeters per hour, and the movement is not directly apparent to the human eye. Yet the speed is 10,000 times faster than the proposed movement in the mantle. Even so, a movement of five centimeters per year adds up significantly over geologic time; a parcel of rock could move from the bottom of the mantle to the top in 58 million years, which is only a small fraction of the earth's age of 4.6 billion years.

Movements of the Mantle

How is it possible that solid rock can flow, even so slowly? When a blacksmith picks up a bar of cold steel, he is unable to bend it, but when he heats it to a red glow, he can bend it easily, even though it is still a solid bar. Similarly the rocks of the mantle, which are at

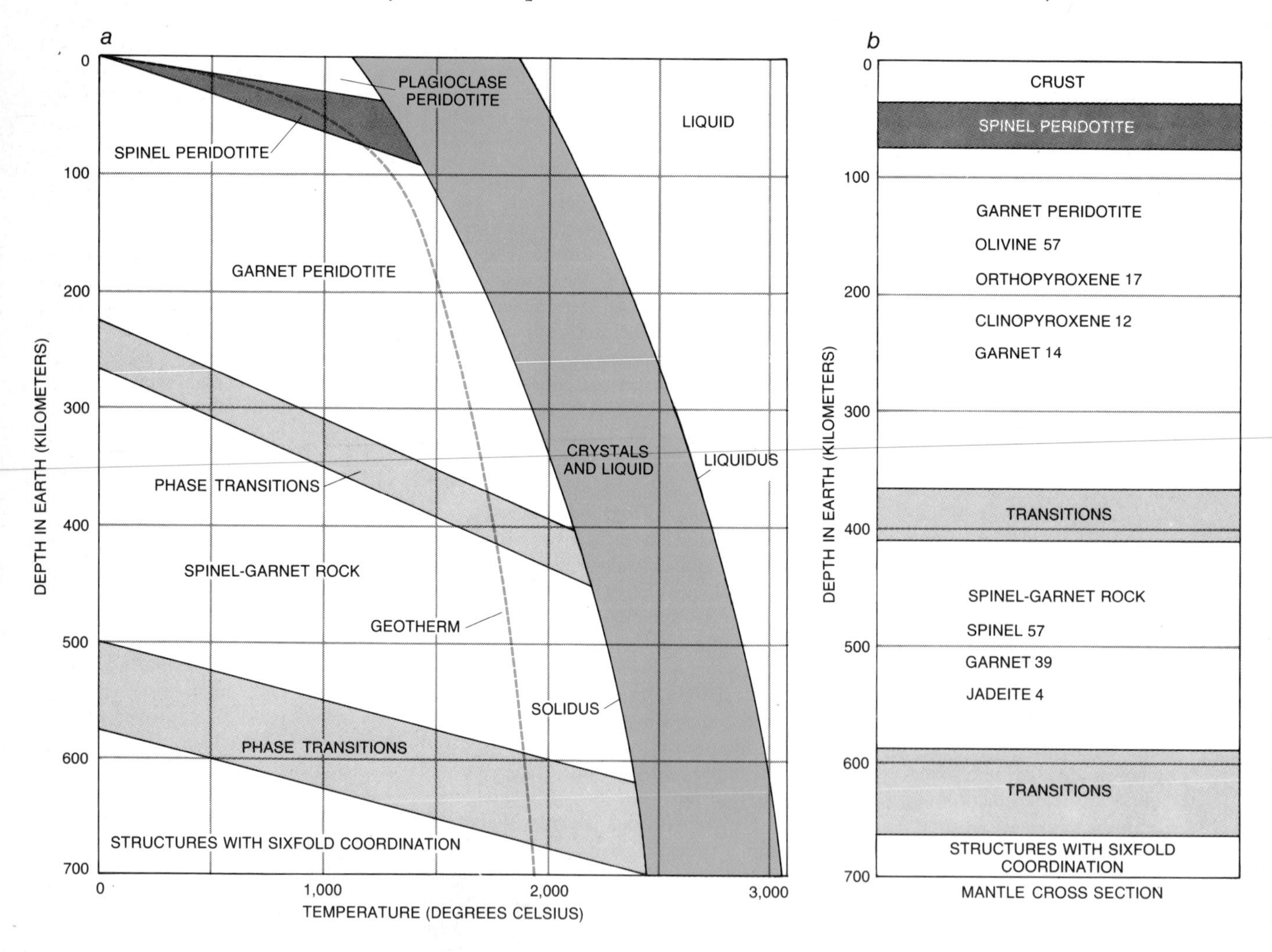

PHASE DIAGRAM for peridotite is stated in terms of pressure, or depth, and temperature. At a fixed depth an increase in temperature eventually brings a rock to a point where it begins to melt. The beginning and end of melting are bounded by the solidus and liquidus curves. Reading downward in the two parts of the diagram at left, following the geotherm (*a*), one sees that the upper mantle has a layered structure owing to successive phase transitions from spinel peridotite in the uppermost region to a rock composed of minerals with elements in a sixfold coordination, that is, having closely packed groups of oxygen atoms enclosing atoms of magnesium, iron and silicon, at a depth of about 600 kilometers. The numerals in the cross-section diagram (*b*) represent the percentages

high temperatures, can be deformed even though they are still in the solid state.

Olivine, pyroxene and peridotite have been subjected to strains at high pressures and temperatures in laboratory experiments. They deform. The deformed products have recently been studied in the high-voltage electron microscope in attempts to establish the mechanisms of plastic flow. A suggestion emerging from this work is that deformation occurs first in individual crystals, which then recrystallize to form a mosaic of new grains. The mechanism of flow varies as a function of pressure and temperature.

Schemes have been proposed for large convection cells extending through the entire thickness of the mantle [*see illustration on page 56*]. An alternative model, based on the argument that the absorption of heat in phase transitions would reduce the driving force, confines

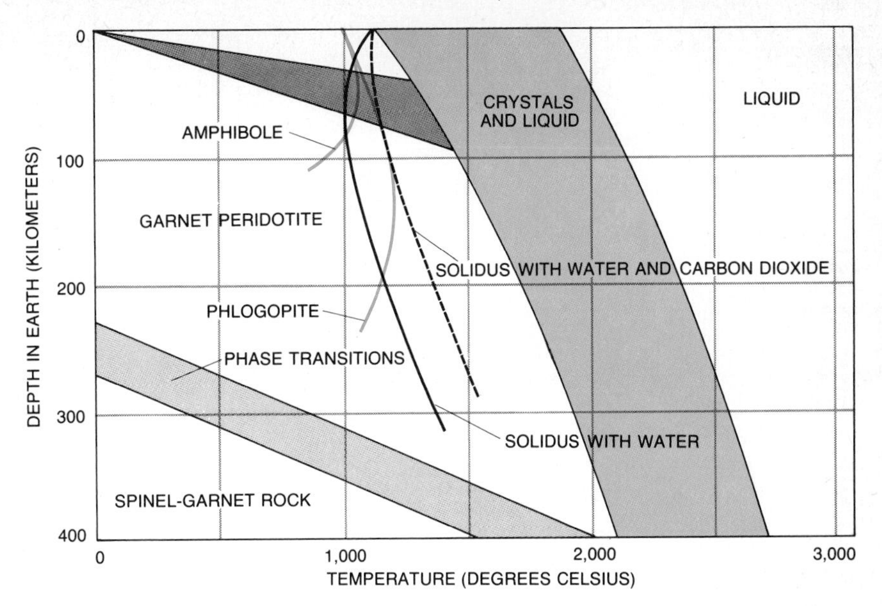

LOW-VELOCITY LAYER in the uppermost mantle appears to be explained by the presence of water and carbon dioxide, which lower the solidus curve, so that rock melts at lower temperature and pressure. The hydrous, or water-bearing, minerals amphibole and phlogopite are stable through a limited temperature range (*color*). Comparing this diagram with the one for dry mantle rock, one sees that the geotherm curve passes from a phase of solid peridotite to a phase of partly melted peridotite at a depth of about 100 kilometers, which corresponds closely with the top of the low-velocity zone observed in the uppermost mantle.

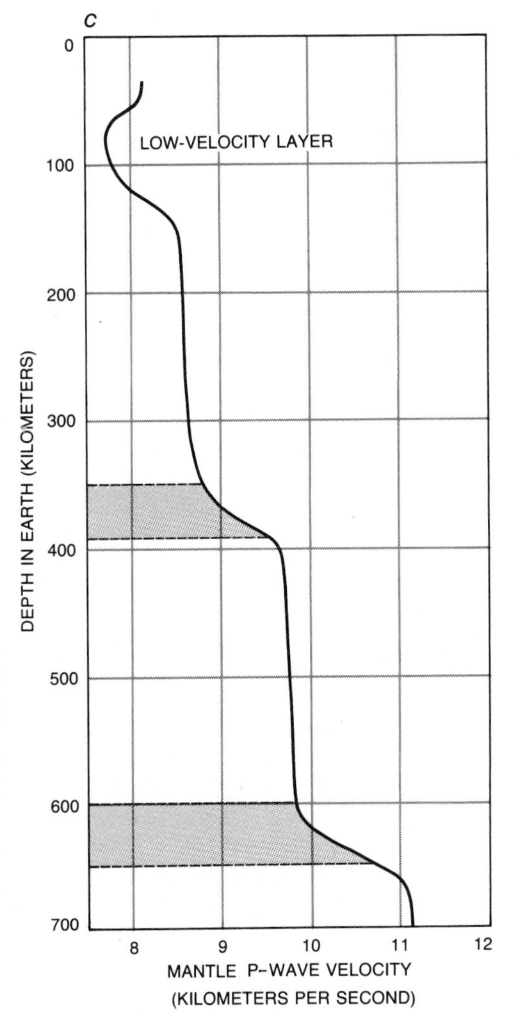

of minerals in the rocks. The profile of *P*-wave velocity (*c*) shows that the depths of actual changes in the mantle indicated by earthquake waves coincide closely with the depths of phase transitions as inferred from the phase diagram and the geotherm.

convection to the upper mantle, above the olivine-spinel transition. A third model limits convection to the asthenosphere, which as a layer of the mantle lies at depths of from 100 to 300 kilometers. Subduction zones, however, where lithospheric material returns to the mantle and thus is involved in convective forces, appear to run as deep as 700 kilometers. Another model for the plate-driving mechanism involves thermal "plumes" in the mantle. According to this argument, all upward movement of mantle material is confined to about 20 plumes, each plume a few hundred kilometers in diameter, rising from the core-mantle boundary. The return flow is accomplished by a slow downward movement of the rest of the mantle.

Where a plume reaches the lithosphere the flow becomes horizontal, spreading radially in all directions. A plume creates a hot spot with volcanic activity at the surface, and it may cause an upward doming of the lithosphere. In such ways the plumes would cause the movement of the lithospheric plates.

The hot-spot hypothesis is currently a hot idea. Many geologists are exploring its implications for various phenomena, including the features of volcanic island chains such as Hawaii. The hypothesis is also under challenge, however, by earth scientists who express doubt that a plume could retain its coherence as it rises through 2,800 kilometers of mantle.

The vertical movement of mantle material in any kind of convective system causes changes in the temperature distribution; at a given depth the temperature is increased where hotter material is rising and decreased where cooler material is sinking. The movement changes the shape of the geotherm curve from time to time and from place to place as the convection proceeds. The compositions of the minerals in mantle peridotite vary as a function of pressure and temperature, as laboratory experiments dealing with the phase diagrams of peridotite and its constituent minerals have established.

A parcel of mantle peridotite is subjected to changes in temperature and pressure as a result of convective motions in the mantle, and its mineralogy will adjust by recrystallization as it strives to attain equilibrium in the changing environment. The movements are slow enough so that an equilibrium mineralogy is normally attained. If the rock is suddenly transported to the surface, however, as in a kimberlite eruption, the time for readjustment is not enough and the sample reaches the surface with the mineralogical signature corresponding to its position and temperature where it last reached equilibrium in the mantle.

The compositions of coexisting minerals in peridotite at any particular pressure and temperature have been measured directly in the laboratory experiments. Using these data for the purpose

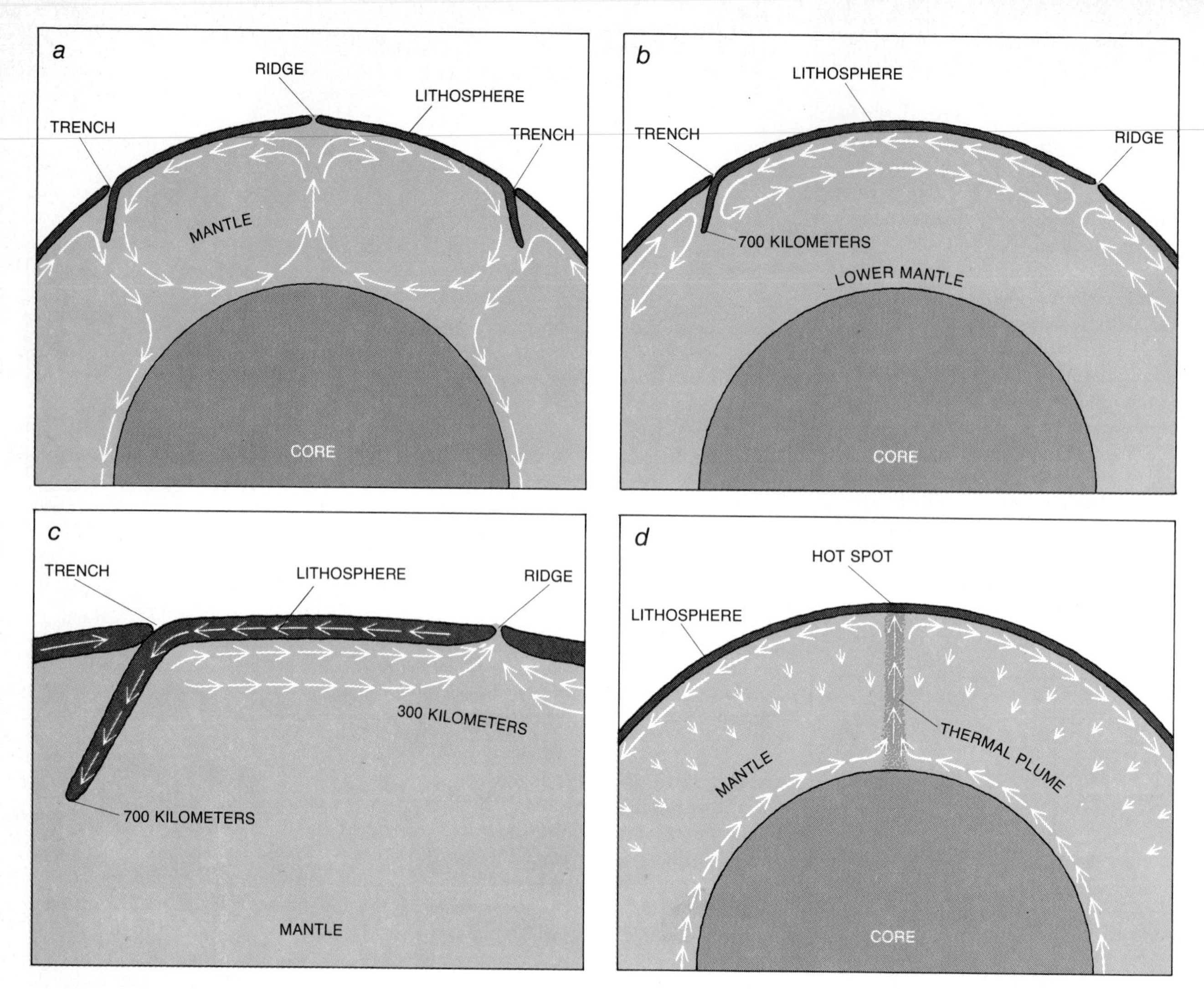

MODELS OF CONVECTION have been proposed to explain how activity in the mantle drives the lithospheric plates. In convection warmer material moves upward and colder material moves downward. One model (*a*) holds that convection cells extend through the entire mantle. According to a second model (*b*), they are confined to depths above the phase transition from spinel to olivine. A third model (*c*) confines movements of the mantle to the asthenosphere. In the thermal-plume model (*d*) all upward movement is confined to a few thermal plumes, and the downward flow is accomplished by slow movements of the remainder of the mantle.

of calibration, it is now possible to take samples of mantle peridotite, such as the nodules from kimberlites, to measure the composition of the minerals in each sample and thus to estimate the pressure (or depth) and temperature at their source in the mantle.

Ancient Geotherm

The application of these methods to nodules from kimberlites has produced intriguing results during the past year or two. A suite of nodules collected from a single kimberlite pipe gives a series of points, each specified for a given nodule by the estimated pressure and temperature of equilibrium before eruption. The locus of these points on a diagram of depth (pressure) and temperature corresponds to the geotherm that existed at the time of eruption of the kimberlite. In other words, each kimberlite nodule contains stored within its mineralogy the record of its equilibrium pressure and temperature in the mantle before it was abruptly carried upward. The experimental mineralogists engaged in this work have found that the results from kimberlite pipes in South Africa give fossil geotherms with normal gradients down to depths of 150 kilometers or so but with steeper gradients at deeper levels; apparently the temperature was higher than normal at those depths for some interval of time before the kimberlite pipes erupted nearly 100 million years ago.

These results are of great interest to geophysicists trying to establish the thermal history of the earth, the dynamics of the mantle and the driving forces of plate tectonics. One interpretation of the data is that the inflection point in a fossil geotherm corresponds to the top of the asthenosphere at a time about 120 million years ago when the African lithospheric plate began to move rapidly as the Atlantic Ocean opened up. According to this interpretation, frictional heating in the asthenosphere as a result of movement shifted the geotherm to higher temperatures, producing an inflection point at the lithosphere-asthenosphere boundary, whereas the conduction rate of heat through rocks is so slow that the original geotherm in the lithosphere remained effectively unchanged. Another interpretation is proposed by geophysicists who conclude that friction in the asthenosphere is not capable of producing such a large thermal effect. They argue that the steepened portion of the fossil geotherm below the inflection point could be caused by the upward convection of a local thermal plume, which initiated the eruption of the kimberlite.

This example illustrates the way plate tectonics, with its focus on the mantle, has brought together into the same symposium rooms at scientific meetings research workers in areas that once were considered quite distinct from one another. Field geologists, mineralogists, geophysicists and experimental chemists and physicists have together discovered in the minerals of kimberlite nodules information that is grist for the mills of the theoreticians who are trying to work out how the mantle moves now and has moved through the 4.6 billion years of the earth's existence.

A view of the entire earth from a spacecraft shows a large, spinning globe as smooth as a billiard ball. The deepest hole drilled in the surface reaches a depth of only nine kilometers; it is a mere pinprick, penetrating less than .15 percent of the earth's radius. It is therefore quite remarkable that so much is known about the inaccessible mantle.

Uncertainty of Models

Nonetheless, the amount of information now in hand is insufficient for a full understanding of the dynamic behavior of the mantle, which is the key to many geophysical and geological phenomena. The extent of the uncertainty about what is really happening in the mantle can be illustrated by comparing two hypotheses about the Hawaiian volcanic island chain. Each island formed as a result of eruption above a melting region fixed in the asthenosphere. Then the moving lithospheric plate carried the island away, creating over a long period of time the island chain.

According to one interpretation of these events, the melting is localized in a hot spot above a thermal plume. According to another interpretation, the melting is localized by friction in rocks flowing from the asthenosphere into a column moving downward through the mantle. The column is said to form a gravitational anchor, maintaining the region of downward flow in a position more or less directly above it.

These diametrically opposed hypotheses are put forward by respected earth scientists. Is Hawaii to be explained by a rising thermal plume or by a sinking gravitational anchor? I am confident that before too many years have passed the accumulation of additional evidence and the refinement of hypotheses will have placed much closer constraints on the picture that can be drawn of the structure and dynamics of the mantle.

5

Hot Spots on the Earth's Surface

Kevin C. Burke and J. Tuzo Wilson
August 1976

Scattered around the globe are more than 100 small regions of isolated volcanic activity known to geologists as hot spots. Unlike most of the world's volcanoes, they are not always found at the boundaries of the great drifting plates that make up the earth's surface; on the contrary, many of them lie deep in the interior of a plate. Most of the hot spots move only slowly, and in some cases the movement of the plates past them has left trails of extinct volcanoes. The hot spots and their volcanic trails are milestones that mark the passage of the plates.

That the plates are moving is now beyond dispute. Africa and South America, for example, are receding from each other as new material is injected into the sea floor between them. The complementary coastlines and certain geological features that seem to span the ocean are reminders of where the two continents were once joined. The relative motion of the plates carrying these continents has been reconstructed in detail, but the motion of one plate with respect to another cannot readily be translated into motion with respect to the earth's interior. It is not possible to determine whether both continents are moving (in opposite directions) or whether one continent is stationary and the other is drifting away from it. Hot spots, anchored in the deeper layers of the earth, provide the measuring instruments needed to resolve the question. From an analysis of the hot-spot population it appears that the African plate is stationary and that it has not moved during the past 30 million years.

The significance of hot spots is not confined to their role as a frame of reference. It now appears that they also have an important influence on the geophysical processes that propel the plates across the globe. When a continental plate comes to rest over a hot spot, the material welling up from deeper layers creates a broad dome. As the dome grows it develops deep fissures; in at least a few cases the continent may rupture entirely along some of these fissures, so that the hot spot initiates the formation of a new ocean. Thus just as earlier theories have explained the mobility of the continents, so hot spots may explain their mutability.

Plate Tectonics

The modern theory of plate tectonics divides the superficial regions of the earth into two layers. The lithosphere, the outermost layer and the only one directly accessible to us, is cold and rigid. Below it is the asthenosphere, which is white hot and capable of being slowly deformed. The asthenosphere is not liquid, although there is a small amount of melted rock in the earth's interior. The asthenosphere is a solid, but one that flows under stress. It is not unlike ice, which seems brittle in the form of an ice cube but is quite plastic in a glacier flowing down a mountain valley.

The distinction between lithosphere and asthenosphere is based on rigidity and to a large extent reflects differences in temperature. An older distinction, based on chemical composition, divides the upper earth into the crust and the mantle. The boundary between these layers does not correspond to that between the lithosphere and the asthenosphere. The crust is the upper portion of the lithosphere, and the lithosphere also contains the topmost part of the mantle. The asthenosphere usually lies entirely within the mantle.

Under the oceans the crust is composed primarily of basalt; the continents, on the other hand, are made largely of granitic rock. Granite is lighter than basalt, and the continents are considerably thicker than the oceanic crust, with the result that the continents float well above the ocean floor. It was once proposed that the continents move through the ocean floor like ships, but that hypothesis had to be abandoned. Actually the continents are carried by the lithosphere like rafts locked in the ice of a frozen river.

The lithosphere is broken into about a dozen plates, in which the continents are firmly anchored. The plates separate from one another at the crests of the mid-ocean ridges, where new lithosphere is created. The ridges wind through all the world's oceans and constitute the largest mountain system on the earth. At the crests of the ridges undersea volcanism adds new material to the plates, pushing them apart. The opposite process—the consumption of lithospheric plates—is observed where the plates converge and overlap. In those regions, called subduction zones, one plate plunges under another and is reabsorbed into the mantle.

The movement of the lithospheric plates is thought to be associated with large-scale convection currents in the mantle. The currents may actually drive the plate movements, but too little is known about convection in the mantle to warrant firm conclusions.

Hot Spots and Plumes

Almost all volcanic activity is confined to the margins of the plates. Along the full length of the mid-ocean ridges there is undersea volcanism in which the lava erupted is predominantly basalt. At convergent plate boundaries lavas are formed by the melting of lighter constituents of the subducted plate. The upwelling lava can create an island arc, such as the arcs of the Philippines, Japan and the Aleutians, or a volcanic mountain system, such as the Andes and the Cascade Range of the Americas. The lavas associated with convergent plates differ from the basalts of the mid-ocean ridges. They are called andesite lavas and they contain more silicon, calcium, sodium and potassium than basalt and less iron and magnesium.

Volcanism that is not associated with plate margins accounts for a small proportion of the world's volcanic activity, probably much less than 1 percent. It is these few isolated volcanoes that have been named hot spots. They are distin-

HOT SPOT in North Africa is an isolated group of volcanic mountains surrounded by the Sahara. The group is called Tibesti, and it lies in northeast Chad, near the Libyan border. The photograph was made from an altitude of about 920 kilometers by the LANDSAT earth-resources satellite. Dark blotches on the landscape are relatively recent lava flows. Two large, recent volcanic craters are visible; the one at the lower right is Emi Koussi, which has an elevation of 3,415 meters and is the highest peak in the region. Some older craters are recognizable, but they have been severely eroded. At Tibesti and at other African hot spots lavas of different ages are piled on top of one another, suggesting that the continent is stationary with respect to the hot spots and probably has been for 30 million years.

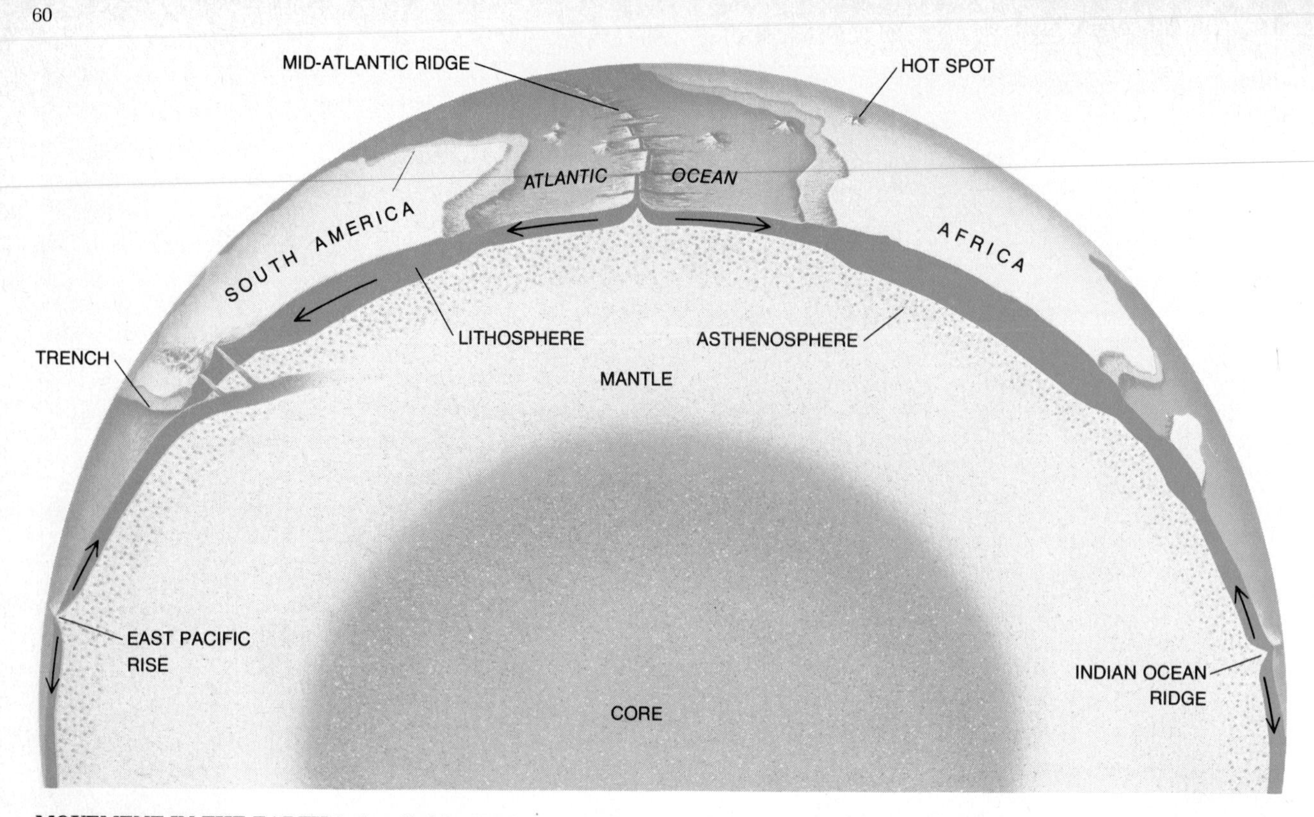

MOVEMENT IN THE EARTH is described by the theory of plate tectonics. The lithosphere, the cool and rigid layer that includes the crust of the earth, is broken into about a dozen large plates. These plates move over the asthenosphere, a layer that is hotter and capable of slow deformation. The crust is the top of the lithosphere; the rest of the lithosphere and all of the asthenosphere are parts of the mantle. Lithospheric plates move apart as new material is added to them along mid-ocean ridges. Where two plates come together one dives under the other and is reabsorbed; this process, called subduction, results in extensive volcanic activity. Hot spots are small volcanic regions typical of neither mid-ocean ridges nor subduction zones. Unlike most other volcanoes, they are often found far from plate margins, and even when they are near a plate margin, they can be distinguished by the volume of lava they eject and by its composition.

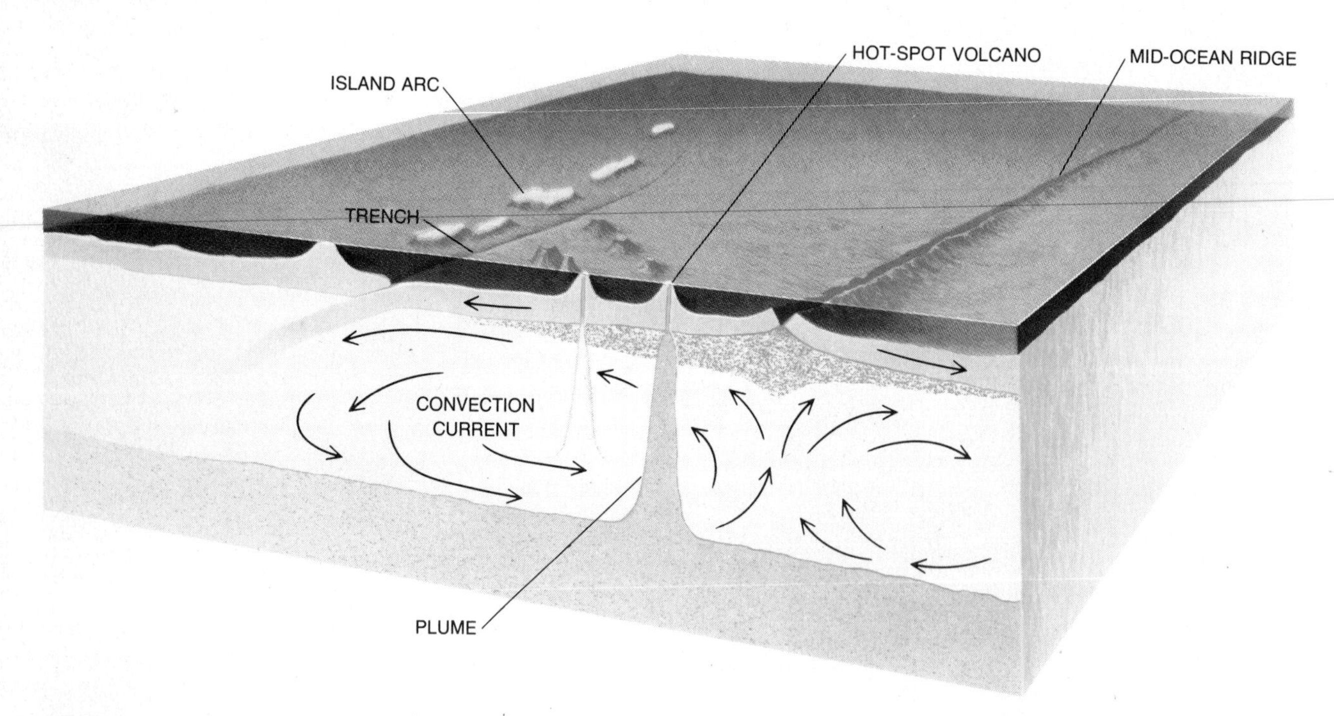

SOURCE OF A HOT-SPOT VOLCANO is thought to be a "plume" rising from deep within the mantle. Differences in composition between the lavas ejected at hot spots and those characteristic of plate-margin volcanism suggest that the two kinds of lava come from different parts of the mantle; indeed, the source of the hot-spot lavas may have been isolated for as long as two billion years. Much of the mantle is probably stirred by convection currents, so that the plumes must originate in some region isolated from this circulation. For example, they might come from a stagnant zone in the middle of a convection cell, or from a layer below the reach of the mantle currents. A lithospheric plate moving over a plume leaves a trail of volcanoes that grow older with distance from present site of volcanic activity.

guished by their very isolation: in the middle of a rigid lithospheric plate, far from centers of seismic activity, a hot spot may be the only distinctive feature in an otherwise monotonous landscape. Almost all hot spots are regions of broad crustal uplift, and this swelling is distinct from the smaller-scale mountain-forming or island-forming activity characteristic of all volcanoes. Finally, the lavas associated with hot spots differ from those found both at the mid-ocean ridges and at subduction zones. The hot-spot lavas are basalts, like those of the ocean ridges, but they contain larger amounts of the alkali metals (lithium, sodium, potassium and so on). Alkali-rich lavas are rare at plate margins.

The mechanism that generates hot spots must be sought in the mantle. They may be surface manifestations of "plumes": rising, columnar currents of hot but solid material. The plumes might well up from below the asthenosphere, at a phase-change boundary a few hundred miles inside the mantle. The distinctive composition of the hot-spot lavas argues that their source is isolated from the general circulation pattern of the mantle. For example, plumes might be generated in stagnant regions in the center of a circular convection current, or they might come from a very deep layer of the mantle, below the region that is effectively stirred by convection. The circulation of the mantle is still poorly understood, however, and for the moment any attempt to explain the origin of hot spots must remain speculative. Here we are concerned mainly with surface manifestations, which are not strongly dependent on the exact source of the magma. It is possible to formulate a consistent interpretation of hot spots even without a detailed model of the earth's interior.

Island Chains

Perhaps the most prominent and most easily recognized hot spot is the one that has formed the Hawaiian Islands. On an expedition to the South Seas in 1838, James Dwight Dana, an American geologist, noted that these islands become progressively older as one proceeds northwest from Kilauea and Mauna Loa, the active volcanoes on Hawaii itself. (Dana estimated the ages of the islands from the extent to which they had been eroded.)

It is now apparent that all the islands in the Hawaiian chain were created by a single source of lava, over which the Pacific plate has passed on a course proceeding roughly toward the northwest. The plate has carried off a trail of volcanoes of increasing age, in much the same way that wind passing over a chimney carries off puffs of smoke.

Dana also called attention to two other chains of Pacific islands whose trend is parallel to that of the Hawaiian chain. These are the islands of the Austral Ridge and the Tuamotu Ridge; the latter group includes Pitcairn Island. Like the Hawaiian Islands, these chains become older toward the northwest, and in each of them the most recent volcanic activity is near the eastern terminus. It would be difficult to ignore the inference that all three chains were generated as a result of the same plate motion. Indeed, from the configuration of the islands the apparent course of the plate can be mapped.

Leonhard Euler, the 18th-century Swiss mathematician, proved that on the surface of a sphere the only possible motions are rotations. It is therefore always possible to describe the movement of a lithospheric plate as a rotation around a pole. (The pole does not have to pass through the plate itself.) W. Jason Morgan of Princeton University has been able to show that the Hawaiian, Austral and Tuamotu chains could all have been generated by the rotation of the Pacific plate around the same pole. Employing a somewhat different approach, Jean-Bernard Minster and his colleagues at the California Institute of Technology have deduced from observed rates of sea-floor spreading that if the African plate has been stationary, then the Pacific plate must have moved along the trajectory defined by the Hawaiian chain.

At the western end of the Hawaiian Islands a string of submerged mountains, the Emperor Seamounts, strikes to the north. It is appealing to consider the entire system of islands and seamounts as a single chain that has changed direction, and age determinations support that interpretation. The oldest of the Hawaiian Islands, near the bend, are about 40 million years old. The Emperor Seamounts continue the age sequence without interruption, beginning near the bend with an age of 40 million years and continuing to an age of about 80 million years where the chain ends off the Kamchatka Peninsula. Morgan has found he can account for the formation of the seamounts by the rotation of the Pacific plate around a different pole, suggesting a remarkably simple sequence of events: about 40 million years ago the motion of the Pacific plate shifted to a new pole of rotation and thereby changed its direction of migration, causing an abrupt kink in the Hawaiian chain.

Another tantalizing inference from geography suggests a possible confirmation of this theory. The Austral and the Tuamotu island chains also seem to bend sharply at an age of about 40 million years, and each is continued in a line of seamounts. These seamounts are parallel to the Emperor system, and they could have been formed by the rotation of the plate around the same pole. For this conclusion to be accepted, however, it must be shown that the seamounts become progressively older to the north. As yet there are few dates established for the seamount series; those that have been obtained suggest a more complex interpretation.

The reconstruction of plate motions from the tracks of hot-spot volcanoes depends ultimately on the belief that the hot spots themselves are immobile or nearly so. This assumption appears to be justified. Minster and his colleagues have made accurate maps of relative plate motions by methods that do not rely on the hot-spot positions. Their work shows that prominent hot spots throughout the world have not moved in relation to one another during the past 10 million years. Other investigators have compared the positions of hot spots over much longer periods. Those determinations seem to show that groups of hot spots in one ocean have moved with respect to groups in other oceans over the past 120 million years—since the supercontinent Pangaea broke apart. This wandering of groups of hot spots, however, is slow compared with the shifting of the lithospheric plates.

The Population of Hot Spots

A census of the world's hot spots suggests that at least 122 have been active in the past 10 million years. Most of them meet all the particulars of the definition and can be classified without ambiguity. They are centers of volcanism that are not associated with plate boundaries and that form elevated domes with a diameter of up to about 200 kilometers. Also counted in the census, however, are several regions that lie on mid-ocean ridges or close to them; prominent among these are Iceland, the Azores and Tristan da Cunha, a small group of islands in the South Atlantic. The reason for the inclusion of these areas is that they seem more characteristic of hot spots than of normal mid-ocean ridges. The volume of material they have ejected greatly exceeds the norm for mid-ocean ridges; that is why they have been built up into islands while the rest of the ridge crest has remained submerged. More important, the lavas of these regions are the alkali-rich basalts that are rare at plate margins but that are typical of hot spots.

Our census probably underestimates the number of hot spots. There are domes or rises on some plates that are not capped by volcanoes; in spite of similarities in shape and geophysical properties we have not included them. There are probably also small, active volcanoes on the ocean floor that remain to be discovered. Finally, we have not attempted to include hot spots on con-

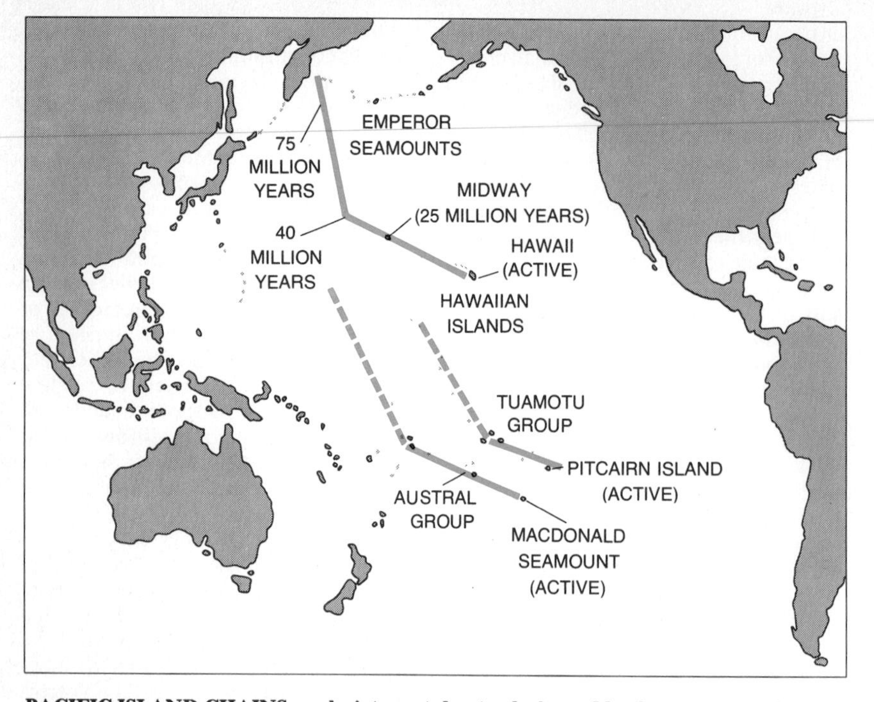

PACIFIC ISLAND CHAINS can be interpreted as tracks formed by the movement of the sea floor over stationary hot spots. The Hawaiian Islands grow older toward the northwest, beginning with Hawaii itself. Two other chains parallel to the Hawaiian Islands display a similar pattern of ages. They are the Austral group, which begins with the MacDonald Seamount, and the Tuamotu group, which begins with Pitcairn Island. All three chains could be generated by the same clockwise rotation of the Pacific plate. The age sequence of the Hawaiian Islands is continued in the Emperor Seamounts, which strike northward from a bend formed 40 million years ago. The change in direction implies that until then the rotation of the plate was centered on a different pole. Seamounts also extend to the north from the Austral and Tuamotu groups (*broken lines*), but there is no convincing evidence their ages form a linear sequence.

verging plate margins. In these areas volcanic activity is both abundant and complex, and it would be difficult to isolate the contribution of hot spots from other sources of volcanism. It should be noted, however, that basalts rich in alkali metals are found in some converging-plate zones.

Of the 122 hot spots we have identified, 53 are in ocean basins and 69 are on continents. Among the oceanic hot spots there is a tendency to congregate on mid-ocean ridges: 15 lie on the crests of ridges and nine others are near the crests. The greatest concentration, however, is in Africa. The African plate has 25 hot spots on land, eight at sea and 10 more on or near the surrounding ocean ridges, for a total of 43.

Even allowing for possible errors in our census, the inhomogeneity is striking. The African plate constitutes 12 percent of the world's surface area, but it has 35 percent of the hot spots. The large-scale topography of the African continent is also unusual. It is characterized by basins and swells, and in recent epochs South and East Africa have been greatly uplifted to produce highlands and the Great Escarpment. The topography and the abundance of hot spots are almost certainly related. Both can be explained by the hypothesis that Africa has come to rest over a population of hot spots.

The most compelling evidence that Africa is stationary is that at some hot spots lavas of several ages are superposed. If the continent were moving, of course, these lavas would be spread out in a chronological sequence. A few hot spots in the vicinity of Cameroon seem at first to be aligned like the island chains of the Pacific. It has been found, however, that these volcanoes are not arranged in chronological sequence. Their alignment is presumably a coincidence; it cannot have been caused by the motion of the plate.

Africa's basin-and-swell topography and the uplifting of large regions could be a direct result of the continent's immobility. Seismic studies have shown that the mantle is not homogeneous, and if there are variations in composition, there may be local concentrations of radioactive elements. The decay of these elements, which contributes a major part of the heat generated in the interior of the earth, would heat and expand some parts of the mantle more than others. The effects of the expansion might uplift regions of a stationary continent, but on a moving continent the uplift would be smeared out and would not be detected.

It is tempting to generalize from these observations, and it does seem there is a relation between the number of hot spots on a continent and the speed with which the continent is moving over the mantle. Antarctica, China and Southeast Asia, like Africa, have relatively large numbers of hot spots on land. Rates of sea-floor spreading imply that if Africa is stationary, these other regions are moving only slowly. In contrast, on rapidly moving plates, such as the North and South American ones, hot-spot volcanism is uncommon.

Opening of the Atlantic

Dated sequences of rock imply that Africa had numerous active volcanoes until the breakup of Gondwanaland 120 million years ago. The volcanic activity then ceased, and it did not resume until 30 million years ago. The two periods of activity and the long intermission between them can be read as signposts indicating the stages in the formation of the Atlantic Ocean.

The earlier episode of volcanic activity suggests that when Africa was a component of Gondwanaland, it was stationary over the mantle. When the supercontinent fractured along the present line of the Mid-Atlantic Ridge, Africa moved east. The motion over the mantle extinguished the volcanism for the next 90 million years. It is convenient to assume that the developing mid-ocean ridge was then stationary and that the two continents spread symmetrically away from it. They rotated in opposite directions around a pole near Cape Farewell on the coast of Greenland.

About 30 million years ago the African plate came to rest; volcanic activity on the continent resumed, and it has continued to the present. Although the African plate had stopped, sea-floor spreading had not. As a result the Mid-Atlantic Ridge was forced to begin drifting west. The relative motion of Africa and South America was unchanged, but the speed of the South American plate with respect to the mantle was doubled. When the Mid-Atlantic Ridge began its migration, the hot spots on the crest were left behind. Today a row of hot spots, which includes Tristan da Cunha and Ascension Island, is found a few

ATLANTIC HOT SPOTS also record the passage of the lithospheric plates. Several of these hot spots are on or near the Mid-Atlantic Ridge; a notable example is Iceland, which has been built up from the massive eruptions of volcanoes on the ridge crest. From some hot spots transverse ridges of volcanic rock extend back to the continental margins, indicating that the mid-ocean ridge developed over these volcanoes and that they were already active when the continents separated.

ARCTIC CIRCLE
JAN MAYEN ISLAND
ICELAND
FAROE ISLANDS
?
?
45° NORTH
AZORES
COLORADO SEAMOUNT
MADEIRA
MID-ATLANTIC RIDGE
CANARY ISLANDS
CAPE VERDE ISLANDS
NORTH BRAZILIAN RIDGE
ST. PAUL'S ROCKS
FERNANDO PO
PRINCIPE
SÃO TOMÉ
ANNOBÓN
EQUATOR
FERNANDO DE NORONHA
ASCENSION ISLAND
ST. HELENA
TRINIDADE
WALVIS RIDGE
RIO GRANDE RIDGE
TRISTAN DA CUNHA
NIGHTINGALE
GOUGH ISLAND
DISCOVERY SEAMOUNT
BOUVET ISLAND
ANTARCTIC CIRCLE

hundred kilometers east of the crest on lithosphere 30 million years old.

The evidence for the mobility of the Mid-Atlantic Ridge lies on the sea floor. From Tristan da Cunha a range of volcanic debris called the Walvis Ridge extends to the northeast. It is believed to be the track of the hot spot during the earlier part of the expansion (when the crest was fixed and Africa was moving), since it extends to lavas on the African coast that date from the disintegration of Gondwanaland. On the other side of the Mid-Atlantic Ridge another line of volcanic debris, the Rio Grande Ridge, extends to the Brazilian coast. There is no hot spot at its seaward terminus, which is separated from the mid-ocean ridge by a gap equivalent to 30 million years.

The disposition of these surface features can be explained by assuming that when the Atlantic was born, Tristan da Cunha was already an active volcano lying directly on the rift that opened to form the ocean. Lava from the hot spot overflowed onto both sides of the ridge and was rafted away by the spreading plates; continued eruption formed a *V*-shaped pair of tracks. When the mid-ocean ridge began to move west, the hot

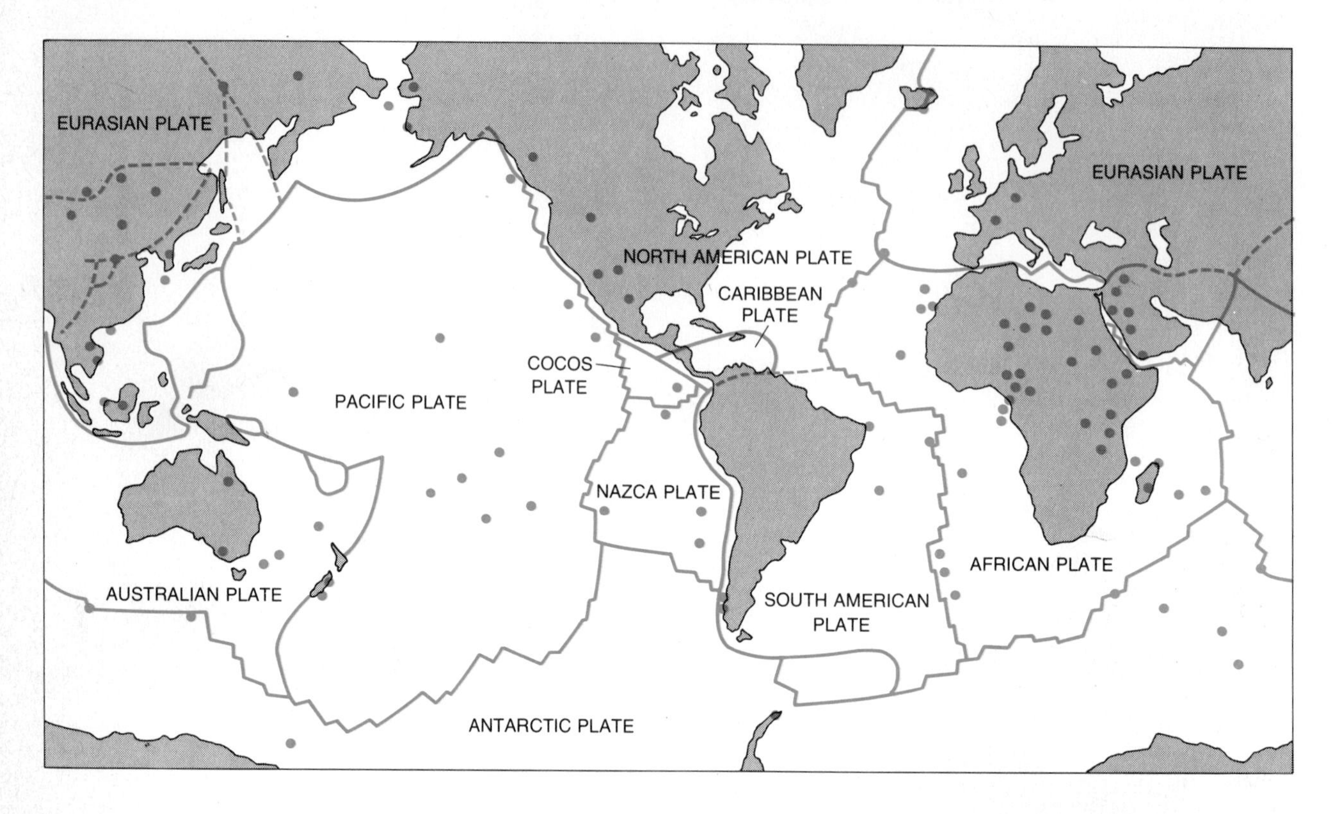

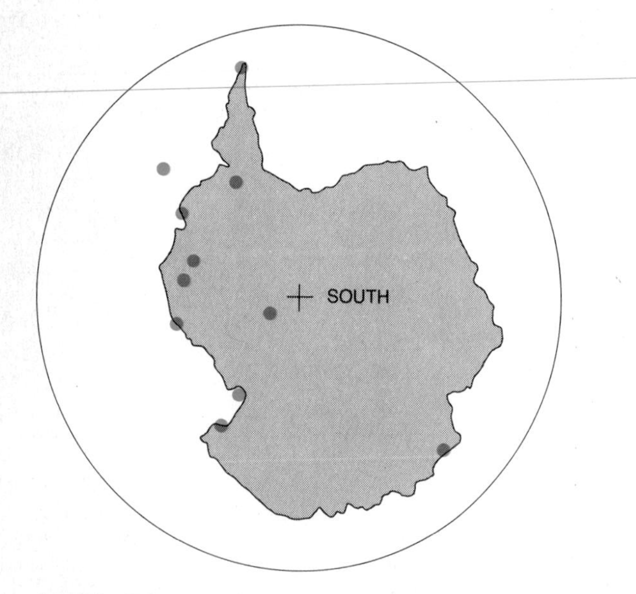

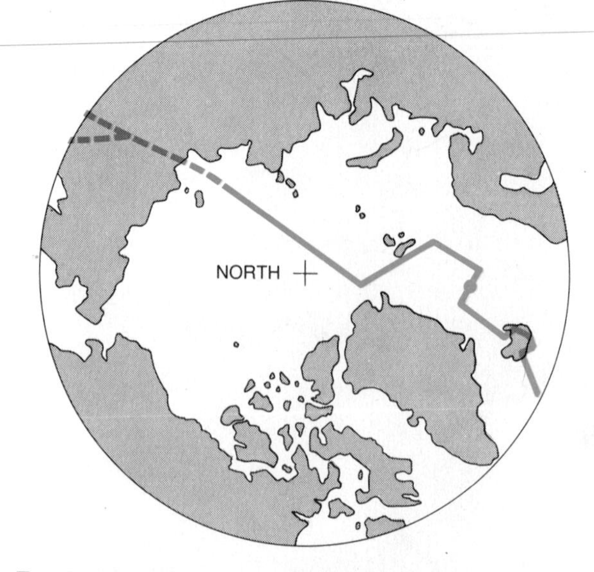

POPULATION OF HOT SPOTS includes at least 122 that have been active in the past 10 million years. They are found on all the major plates and on both oceanic and continental crust, but their distribution is decidedly nonuniform. There is a concentration along mid-ocean ridges, and in particular along the Mid-Atlantic Ridge; what is even more conspicuous, of the 122 hot spots 43 are on the African plate. Together with other evidence, this abundance of hot spots suggests that the African plate is stationary over the mantle. If the African plate is adopted as a frame of reference, other areas that have many hot-spot volcanoes, such as Antarctica and Southeast Asia, are found to be moving only slowly; on fast-moving plates hot-spot volcanism is rare. The map is based on one prepared by W. S. F. Kidd.

spot was left behind on the stationary African plate. It could no longer produce a lateral ridge; instead its successive lava flows simply piled up, one on top of the other. Today at Tristan da Cunha young volcanic rocks are found along with lavas at least 18 million years old. Since lava was no longer deposited on the American plate, the Rio Grande Ridge was also terminated.

Because hot spots are particularly common along ridges and seem to exercise some control over their location, it is reasonable to suppose that the crest of the Mid-Atlantic Ridge will someday jump back to the hot spots it has abandoned. If it does, the 30-million-year gaps in the Walvis Ridge and the Rio Grande Ridge will remain as a record of the interlude.

In the North Atlantic a somewhat different history can be read from the sea floor. The North Atlantic formed by the rotation of the Eurasian and North American plates around a pole in the Arctic Ocean. As we noted above, however, the American plate was already rotating around a pole in Greenland, near Cape Farewell, as a result of its separation from Africa. A single plate cannot rotate around two poles that are both fixed, and in this case the Arctic pole was itself in motion. The result was a shift in the position of the northern Mid-Atlantic Ridge.

When the North Atlantic began to open 80 million years ago, the locus of sea-floor spreading was west of Greenland. Spreading continued there until 50 million years ago and created Baffin Bay. An extinct hot spot left a pair of lateral ridges that trace this movement, extending to Disko Island in Greenland and to Cape Dyer on Baffin Island. Meanwhile 60 million years ago a new mid-ocean ridge developed east of Greenland. The continents have continued to diverge along that line since then.

Motion at Subduction Zones

We have seen that hot spots provide a method for translating the relative motion of lithospheric plates into motion with respect to the mantle. This frame of reference has been employed in clarifying an important aspect of the behavior of the plates that had been imperfectly understood.

When an oceanic plate collides with a continental one, the oceanic plate usually dives toward the mantle and is subducted. That is because the continental plates are thicker and more buoyant. The partial melting of the sinking plate leads to volcanic activity above the subduction zone, but this activity can have two quite different surface expressions. In some cases an island arc forms offshore. The most prominent examples of this process are in South Asia, where

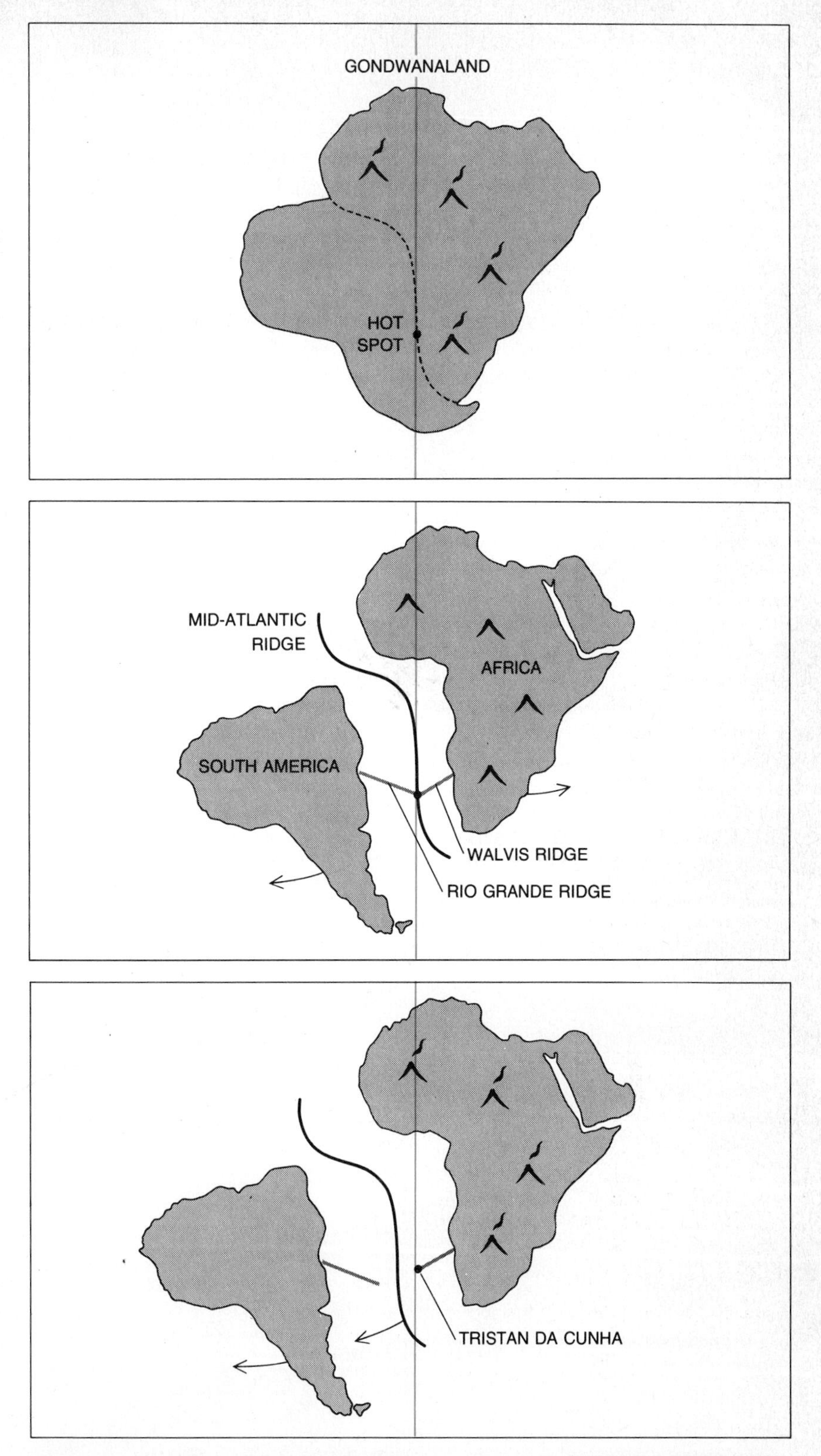

OPENING OF THE SOUTH ATLANTIC began 120 million years ago, when the great southern continent Gondwanaland broke apart. Until then hot-spot volcanoes had been abundant in Africa, suggesting that the continent was stalled over the mantle. When a fissure separated the continents, they receded symmetrically from the developing mid-ocean ridge and the motion of the African plate extinguished the hot spots. About 30 million years ago Africa came to rest and the present era of volcanism on the plate began. Because sea-floor spreading continued at the Mid-Atlantic Ridge, the ridge itself was forced to move west and the speed of South America was doubled. The ridge formed along a line that included several hot spots (*only one is shown*). As long as the ridge was stationary these hot spots generated trails of volcanic rock that extended back to the continental shores. When the mid-ocean ridge began its migration, the hot spots "fell off" the ridge crest and are now isolated on the African plate.

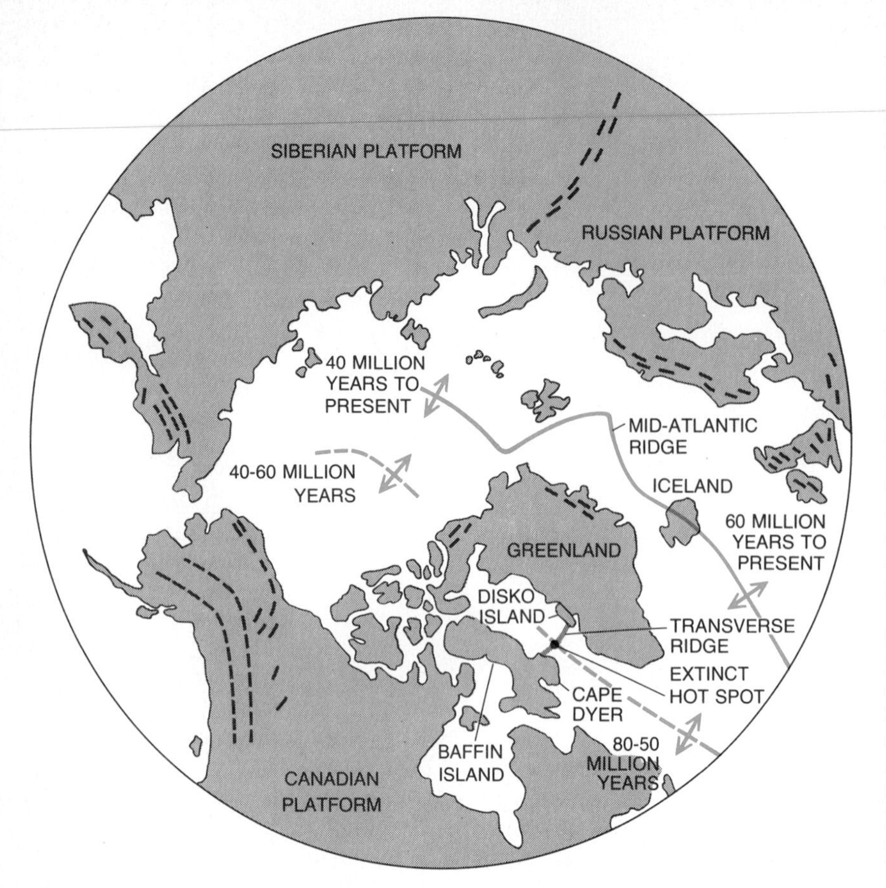

RIDGES IN THE NORTH ATLANTIC and Arctic oceans suggest that the locus of sea-floor spreading in this region shifted between 50 and 60 million years ago. Initially the continents separated along a ridge west of Greenland, opening Baffin Bay. An extinct hot spot has left a record of this movement in trails of volcanic rock that extend from the dormant ridge to Cape Dyer on Baffin Island and to Disko Island in Greenland. About 60 million years ago sea-floor spreading began at the present site of the Mid-Atlantic Ridge, which passes east of Greenland.

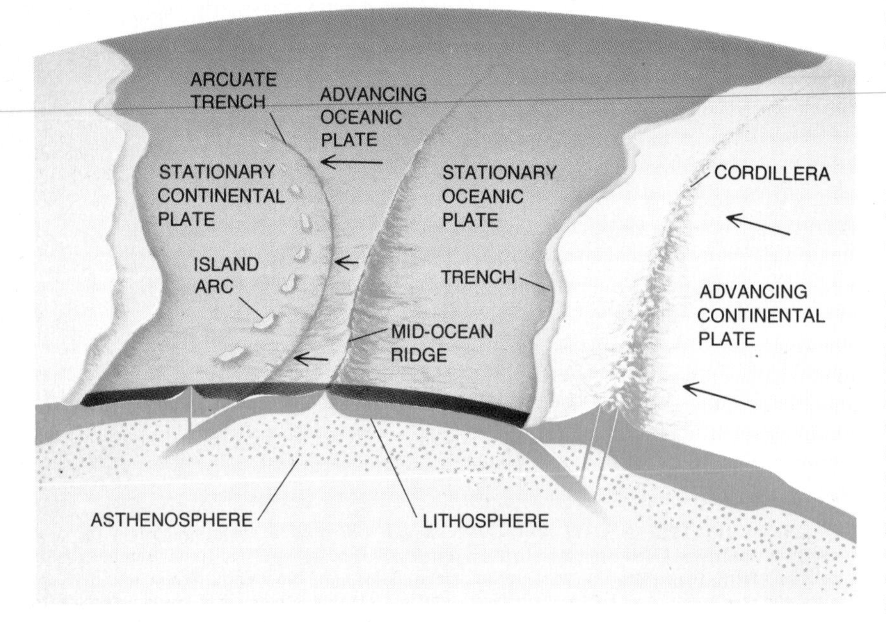

CONVERGENT PLATE MARGINS can assume two different forms. Where an oceanic plate is advancing on a stationary continent, the thin and flexible sea floor buckles offshore in a characteristic arcuate pattern; the volcanoes rising above the subduction zone create an island arc like the arcs of Japan and Indonesia. When a moving continent overrides a stationary oceanic plate, the descending slab of lithosphere is forced to bend at the coastline; as a result the volcanoes rise through the continent, forming a mountain system such as the Andes.

the subduction of the Indian-Australian plate has generated the Indonesian archipelago, and in East Asia, where the sinking Pacific and Philippine plates have produced the islands of Japan and the Philippines. In other cases the volcanic activity appears on the continental landmass. The Andes, for example, were thrown up by the subduction of the Nazca plate, and the Sierra Nevada of California and the Coast range of British Columbia derive from the subduction of the Pacific plate.

It has not been clear why the same process should have two dissimilar manifestations. By referring the plate motions to the hot spots we have attempted to resolve the question. Island arcs form when the continent is stationary over the mantle and the ocean floor moves under it; coastal mountain ranges are raised up when the continent overrides a stationary oceanic plate.

The only plausible explanation for the regular shape of island arcs has been proposed by F. C. Frank of the University of Bristol. He has pointed out that a flexible but inextensible thin spherical shell can bend in on itself only along a circular bend or fracture. This can readily be demonstrated by denting a ping-pong ball. It is suggested that where the ocean floor is moving and free to adopt the preferred shape offshore islands are formed in the characteristic arcuate pattern. When the continent is advancing, on the other hand, the oceanic plate is submerged before it can develop an offshore arc. The known motions of the plates in the Pacific region support this conjecture. The oceanic plates of the Pacific are advancing on Eurasia and underthrusting it, but they are being overtaken by the Americas.

Doming and Rifting

So far we have considered hot spots mainly as indicators of plate motion. They may also act to initiate cycles of tectonic activity.

When a continent comes to rest, the dome that swells up over a hot spot is subject to fracturing. When a rift appears, it very often has a characteristic three-armed pattern. Forty years ago Hans Cloos, a German geologist, recognized the prevalence of such three-armed rifts and showed that they are often related to doming of the continental crust. We would suggest that these rifts are often the seed from which an ocean grows. It follows that the ultimate cause of the rupturing of a continent may be the continent's coming to rest over the mantle. The hot spots appear to guide the fracturing, although they are not necessarily its only cause.

The observed concentration of hot spots on mid-ocean ridges would be accounted for if this mechanism is com-

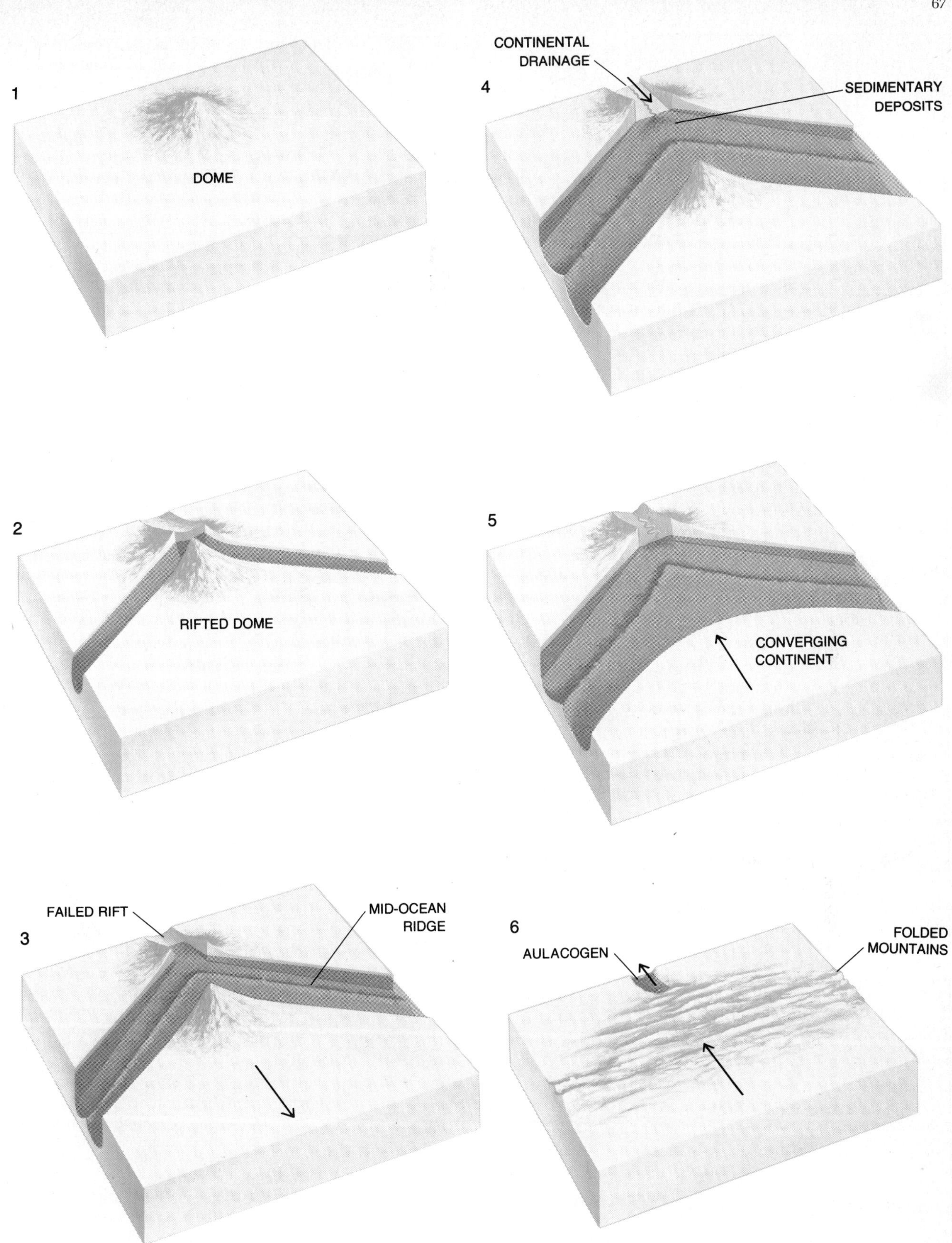

DOMES AND RIFTS associated with hot spots may be involved in the fracturing of continents and the opening of oceans. A dome, often capped with volcanoes (*1*), forms on a continent that is at rest over the mantle. Rifts develop in the dome (*2*), frequently in a three-armed pattern. Two of the arms widen and eventually become the basin of an ocean (*3*), but the third arm fails to develop further. This failed rift may become a major river valley draining the continent and transporting sediments to the new sea (*4*). Later another continent approaches the site of the original rift and closes the ocean (*5*). The collision pushes up a belt of folded mountains, reversing the drainage pattern and carrying sediments back into the failed arm of the rift. Eventually the rift is filled; what remains is a trough of deep sediments roughly perpendicular to the mountain belt (*6*). Nicholas Shatsky of the U.S.S.R. has named such features aulacogens.

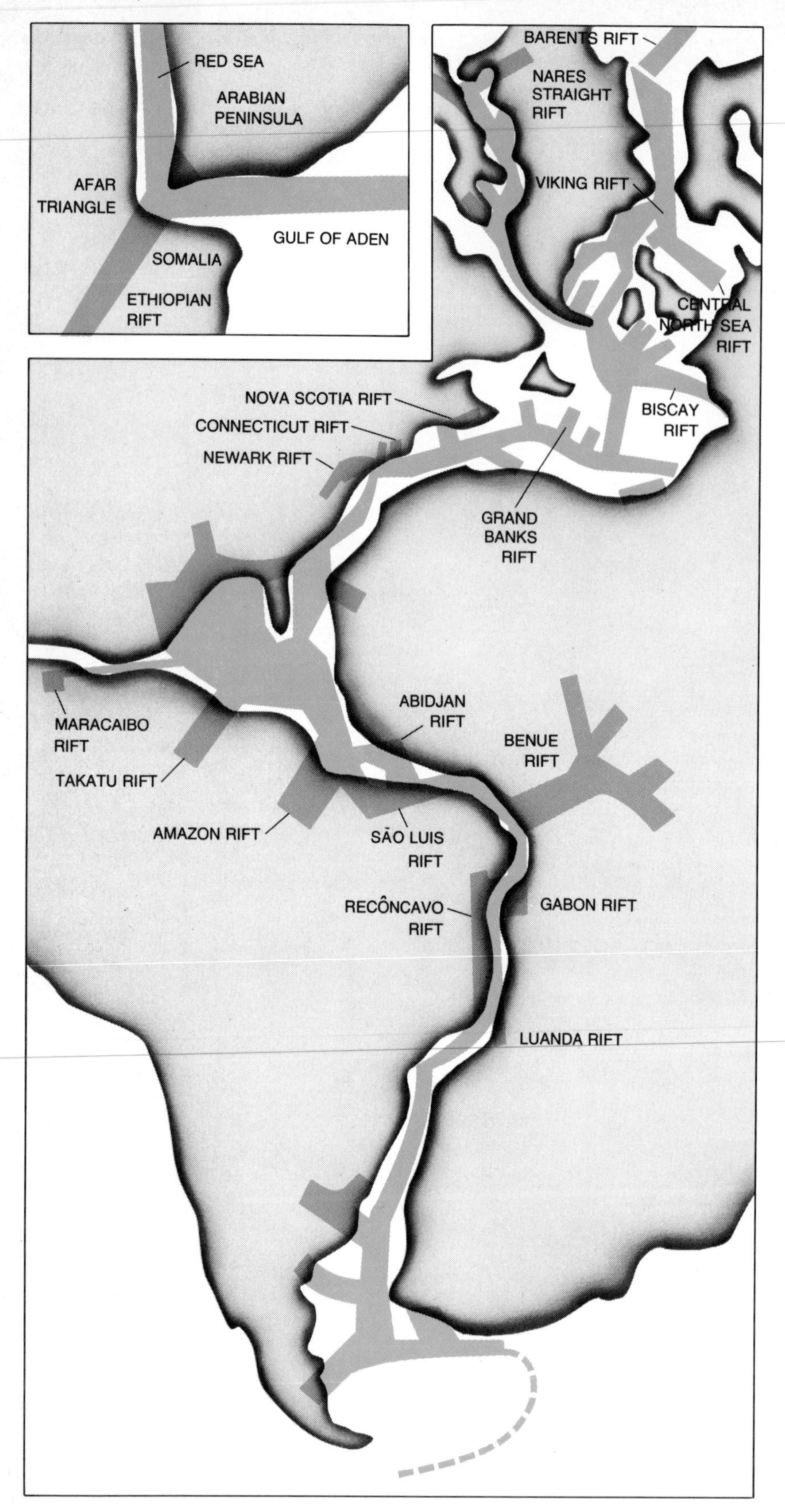

PREVALENCE OF THREE-ARMED RIFTS is revealed by reassembling the continents surrounding the Atlantic. In most cases two of the arms were incorporated into the Atlantic while the third remained a blind rift. A similar process can be observed today where the Arabian Peninsula is splitting away from Africa (*upper left*). The Gulf of Aden and the Red Sea form two branches of a rift; the third extends from the Afar Triangle into Ethiopia and Somalia.

monly involved in the formation of oceans. The breakup of Gondwanaland accords well with this interpretation; it will be remembered that Africa was stationary until the disintegration began.

Typically two arms of the rift open to form an ocean basin, but the third arm fails and remains as a fissure in the continental landmass. By restoring the margins of the Atlantic Ocean to their positions before Pangaea split apart, an abundance of three-armed rifts is revealed. The successful arms merged to create the ocean, whereas the unsuccessful ones remained as rifts extending into the continents. The best example of such a failed rift on the Atlantic coasts is the Benue Rift, which strikes away from the Gulf of Guinea into equatorial Africa.

A much more recent and more conspicuous example can be observed today where the Arabian Peninsula is splitting away from Africa. The Red Sea and the Gulf of Aden both represent arms of a three-armed rift. The third, dry arm strikes into Ethiopia from the Afar Triangle. The symmetry of the pattern is remarkable. The fact that Africa has been stationary over the mantle for 30 million years and that it bears extensive evidence of doming and rifting suggests that we could be witnessing the early stages in the disintegration of the African continent.

Aulacogens

The present cycle of tectonic activity, which dates from the breakup of Pangaea, is not the only one in the earth's history. The recognition that hot spots, domes and rifts form a sequence in the fragmentation of continental landmasses has led to the discovery of a clue that could prove valuable in attempts to reconstruct the earlier wanderings of the lithospheric plates.

In 1941, as German forces threatened the main oil-producing area of the U.S.S.R., Nicholas Shatsky, a Russian stratigrapher, began a search for sedimentary basins that might contain new oil reserves. From the stratigraphic sequences compiled by Shatsky and his colleagues a previously unrecognized pattern emerged. Over much of the Russian and Siberian platforms the sedimentary layer is about a kilometer thick, but they found several narrow troughs up to 800 kilometers long where the sequence is three times the normal thickness. They named these formations aulacogens, from Greek words meaning "born of furrows." Aulacogens are rifts that extend from belts of folded mountains into continental platforms.

Aulacogens can now be recognized as failed arms of three-armed rift systems. When the two successful arms opened to form an ocean, the failed arm remained as a rift valley running inland from the new seacoast. The rift became a feature of the drainage pattern of the continent, accumulating a thick deposit of sediments. Later another continent approached the coast, closing the ocean and blocking the rift. Compressional forces generated by the collision pushed up a chain of folded mountains. The remnant of the rift was a deep bed of sediments striking almost perpendicularly to the mountain chain.

The aulacogens that Shatsky recognized in the U.S.S.R. were of Paleozoic age (between 225 and 600 million years old). Paul Hoffman of the Geological Survey of Canada has since described a formation called the Athapuscow aulacogen that is two billion years old; it underlies the eastern arm of Great Slave Lake in northern Canada. Shatsky himself recognized what is probably the best-developed aulacogen in North America. It is a bed of sediments 15 kilometers deep in southern Oklahoma, parallel to the Texas border. It formed as a rift 600 million years ago when an ocean opened up roughly where the North Atlantic is today. The closing of that ocean was responsible for the building of the Caledonian, Appalachian and Ouachita mountains.

These ancient aulacogens are evidence that the cycle of continental disintegration and reassembly has been going on for at least two billion years. The development of domes and rifts in continents that come to rest over hot spots may have been a part of the process throughout this period.

AULACOGEN in southern Oklahoma is a remnant of an earlier cycle of continental drift. The photograph is a false-color image made in December, 1972, by the LANDSAT satellite. The aulacogen starts in the belt of flatland in the lower half of the photograph and extends another 400 kilometers to the west. To the north are the Ouachita Mountains. When a sea opened up to the south and east 600 million years ago, the aulacogen was the failed arm of a three-armed system of rifts. The closing of that sea raised up the Ouachita range as well as the Appalachians. Erosion of the Ouachitas has added to sediments already in the rift to form a layer of sediments 15 kilometers deep.

II

SEA-FLOOR SPREADING, TRANSFORM FAULTS, AND SUBDUCTION

SEA-FLOOR SPREADING, TRANSFORM FAULTS, AND SUBDUCTION

II

INTRODUCTION

The plates into which the earth's surface layer is divided move relative to one another in three ways: sea-floor spreading where plates separate, transform faulting where they slide past one another, and compression where they meet. In this section, four articles provide details about these processes, one article about each of the first two, and two about the last, which is discussed from two points of view: a geologist's ideas about the surface features over the collision boundary between two plates and a geophysicist's concepts about subduction, where one plate collides with and overrides another, which sinks into the mantle.

"Sea-floor Spreading," by J. R. Heirtzler, begins with the statement of the theory of that name in 1960 by H. H. Hess and with the discovery, about the same time, of regular patterns of magnetic anomalies on the ocean floor. At first no one realized that there was any connection between the two phenomena. Between 1958 and 1961 Victor Vacquier, Arthur D. Raff, and Ronald G. Mason published maps of a regular pattern of magnetic anomalies off the west coast of the United States, extending for hundreds of kilometers as elongated parallel stripes. At intervals, large topographic disturbances offset the stripes by hundreds of kilometers. These disturbances had already been noted and named fracture zones by H. W. Menard, who had concluded from their straightness that they were the traces of great strike-slip faults. A curious feature of these faults is that they stop at coasts and do not extend into continents.

In 1963 F. J. Vine and D. H. Matthews observed and then interpreted a similar pattern in the Indian Ocean. Vine and Matthews used an earlier discovery that the geomagnetic field reverses at intervals of a few tens—or hundreds—of thousands of years. During any one interval, they suggested, the lava forming at the crest of the spreading mid-ocean ridge is magnetized uniformly in one direction, thus forming one stripe in the anomaly pattern. When the earth's field reverses, this stripe is split along the crest of the ridge, which is also the center line of the stripe. As new lava wells up, it cools and is magnetized in the reversed direction to form a new reversed anomaly within the previous one. The two equal halves of the old anomaly move out on both sides. With successive reversals, a symmetrical pattern of stripes is built up parallel to the crest. If the times of these reversals can be ascertained, the rate of spreading can be calculated. Two years later, F. J. Vine and J. Tuzo Wilson showed how the anomalies off the west coast of North America could be interpreted to show the movements of the Pacific plate and of the San Andreas Fault. They used the time scale of reversals that A. Cox at Stanford University and I. D. McDougall in Australia had worked out with their colleagues.

J. R. Heirtzler and his colleagues at Columbia University's Lamont Geological Observatory (the name has since been changed to Lamont-Doherty Geological Observatory) had been working along similar lines, investigating the Reykjanes Ridge, a part of the Mid-Atlantic Ridge south of Iceland. A great body of magnetic observation was available at the observatory, and Heirtzler's paper shows how the interpretation of this data provided a time scale of 171 reversals in the past 76 million years. His demonstration that this same pattern marks all the earth's ocean floors gave great support to the concept of sea-floor spreading.

When this time scale of reversals was introduced, Cox and McDougall had dated the reversals back only 3.5 million years. Heirtzler extrapolated to the older ages on the assumption that sea-floor spreading had proceeded at a constant rate in the area of the South Atlantic Ocean upon which the time scale was based. Although such a large extension might seem arbitrary and unreliable, subsequent drilling, recovery, and dating of basalt samples from the ocean floor by the ship *Glomar Challenger* has shown that the assumptions were warranted and the time scale is broadly accurate.

Don L. Anderson, in "The San Andreas Fault," points out that this fault through California forms part of the boundary between the Pacific plate and the North American plate, The former is moving northwestward past the American continents at a rate of two and a half inches per year and is carrying with it the coastal strip of California lying to the west of the fault. Although the plates move steadily, their adjoining edges have a tendency to lock together. When this happens the plates continue to move, but the parts of each of them close to the fault become distorted and compressed like springs. In this way energy may build up and be stored for many years, until the edges unlock and jump ahead—sometimes moving several feet in opposite directions on either side of the fault. This shakes the ground and produces an earthquake.

This article gives a good account of Californian earthquakes. It discusses their relative intensity and the scales by which the magnitude and effects of earthquakes are assessed. It also discusses the geological history of California in terms of plate tectonics, and it shows that the intersection of the San Andreas Fault with the south end of the Sierra Nevada north of Los Angeles is important to an interpretation of the earthquakes and the geology of California. It concludes that nothing can be done to stop the motion of the plates, and that California will continue to experience earthquakes, just as it has for millions of years.

The prediction of earthquakes is still in its infancy, but Anderson suggests that a dense network of tiltmeters might be of some use. Although earthquakes cannot be stopped, some very recent work has opened the possibility of artificially triggering many small earthquakes, which would prevent energy from being stored up for release in less frequent but larger shocks. It is possible that water pumped into the ground in the proper places would have the effect of lubricating the fault plane, but little has been learned to show how practical this proposal is.

Since the paper was published a few earthquakes have indeed been predicted and a few others have been predicted in error, for they did not materialize. The subject is under intense study, especially by American, Chinese, Japanese, and Soviet seismologists. Although prediction is not yet on a firm basis, the progress already made offers some hope that the time and place of earthquakes may in some cases be predicted. This raises great social problems as well as scientific ones. Less progress has been made in the amelioration of shocks by lubricating faults. Until this very difficult problem has been solved more earthquakes can be expected, and at long intervals great disasters can be expected, but none of them are likely to be more cataclysmic than those that have occurred in the past.

Robert S. Dietz's "Geosynclines, Mountains and Continent-building" relates the new concepts of sea-floor spreading and plate tectonics to the much older realization that there are great prisms of sedimentary rock in the earth's crust. He shows that, after these had accumulated by the growth and coalescence of deltas into continental shelves, subduction zones could have formed beneath them, causing them to be buckled into folded mountains.

Although Dietz suggests that the shelves provide most of the material of which mountains are made (with a moderate addition of magmatic granites), others emphasize the role of volcanoes, which add much andesitic lava derived from subduction zones deep in the mantle. In that connection, it is important to note that off some coasts, such as that of east Asia, volcanoes form island arcs, whereas along other coasts, such as that of Chile, volcanoes lie on the continent. Dietz describes at some length the processes of accumulation of sediments and their conversion into folded and metamorphosed rocks of mountains, using particularly a comparison between the Appalachian Mountains and the present coastal shelf along the east coast of the United States. Consequently he expands the concept of continental drift and plate tectonics from a single phenomenon of the Mesozoic and Cenozoic eras into a continuing, cyclical phenomenon responsible for inactive, ancient mountains as well as modern ones.

In the final article in this section, "The Subduction of the Lithosphere," M. Nafi Toksöz begins by pointing out the vast scale on which subduction is occurring. No part of the present ocean floor is more than 200 million years old, so that all the oceans have opened during that time. Since the earth has not changed much in size, an equivalent area of former ocean floors must have been subducted. The volume of this lithosphere must have been about 20 billion cubic kilometers of rock; in other words, on the average, 100 cubic kilometers of rock is recycled back into the earth's interior each year.

This process involves the penetration of cool slabs of rock from the surface into the deformable asthenosphere. The exact cause of this is still uncertain. Toksöz suggests that in certain areas convection currents in the interior appear to be driving the plates, but in other areas plate motion may be driving the currents. This is compatible with the view already expressed that, because of an increase in density with age, older lithosphere may sink by its own weight, whereas young lithosphere may have to be driven. Most of his article describes what is known of the sinking parts of lithospheric plates.

6 Sea-Floor Spreading

J. R. Heirtzler
December 1968

Comprehensive new theories that rationalize large numbers of observations and explain major aspects of the physical world are rare in any field of investigation. Such a synthesis may be within reach in geophysics. The past few years have seen the emergence of a new theory concerning systematic movements of the sea floor. It deals with vast and formerly unsuspected forces that churn the interior of the earth and account for the arrangement of ocean basins and land masses as we know them today. The theory is based on a variety of observations and hypotheses concerning the topography of the sea floor and the distribution of its sediments, the occurrence of faults and earthquakes, the internal structure of the earth, its magnetic field and periodic reversals of the field. It neatly supports the developing theory of continental drift. Together these theories have already been successful in explaining many surface features of the earth and providing information on internal earth processes. And it is possible that their full importance has yet to be appreciated—that they point toward a major synthesis relating the internal dynamics of the earth, its magnetic field and the dynamics of its orbital motions.

History of the Theory

The stage was set for the discovery of sea-floor spreading by the long debate over continental drift [see "Continental Drift," by J. Tuzo Wilson, beginning on p. 19, and "The Confirmation of Continental Drift," by Patrick M. Hurley, Scientific American Offprint 874]. Evidence from the shape, geological structure and paleontology of various continents and, within the past 20 years, studies of the "paleomagnetism" frozen into volcanic rocks had suggested that the continents have drifted to their present locations from appreciably different positions in the course of millions of years. Even after the possibility of such drifting began to be recognized, it was not at all clear what forces could have caused great land masses to move over the surface of the globe.

By the late 1950's oceanographers had discovered that a continuous range of undersea mountains twists and branches through the world's oceans, that this ridge is usually found in the middle of the ocean and that earthquakes are associated with it. Marine geologists were aware too of the striking youth of the ocean floor: no bottom samples were ever found to be older than the Cretaceous period, which began some 135 million years ago. About 1960 Harry H. Hess of Princeton University proposed that the ocean floor might be in motion. He suggested a kind of convective movement that forced material from deep in the earth to well up along the axis of the mid-ocean ridges, to spread outward across the ocean floor and to disappear into trenches at the edges of continents. (The hypothesis seemed particularly attractive in the case of the Pacific Ocean, which is bordered by trenches, but it was less satisfactory for other oceans, which lack them.)

At about the same time Ronald G. Mason, Arthur D. Raff and Victor Vacquier of the Scripps Institution of Oceanography discovered that the ocean floor off the west coast of North America exhibited a remarkably regular striped pattern of variations in magnetic intensity [see "The Magnetism of the Ocean Floor," by Arthur D. Raff; Scientific American, October, 1961]. The pattern suggested great filamentary magnetic bodies, oriented north-south and offset at intervals along distinct lines running approximately at a right angle to the lineations. No structural features that could explain such a pattern had ever been observed. The origin of these unique magnetic bodies remained a mystery for nearly five years. In 1963, after it had been noted that a distinct magnetic body could often be detected at the axis of a

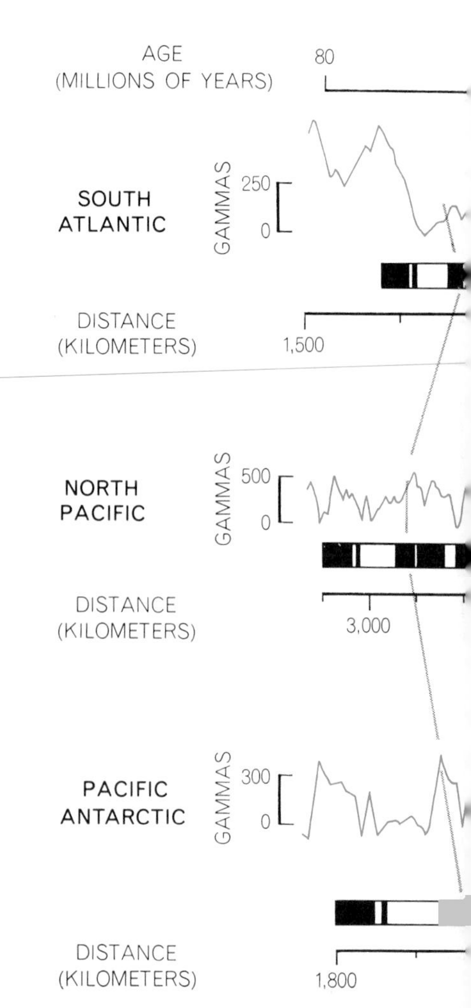

MAGNETIC ANOMALIES (*color*) recorded in all the world's oceans reveal the same succession of magnetic bodies (*black and white bands in strips*) parallel to the mid-ocean ridge. The bodies represent rock that

mid-ocean ridge, F. J. Vine and D. H. Matthews of the University of Cambridge proposed a convincing test of the hypothesis advanced by Hess. It was based on the discovery (which was just then being confirmed in detail) that the earth's magnetic field had reversed direction a number of times in past ages. They reasoned that if molten rock were pushed up along the axis of the mid-ocean ridge, it would become magnetized in the direction of the earth's prevailing magnetic field as it cooled. If the newly cooled material was subsequently pushed out away from the ridge, it would form strips of alternately "normal" and "reversed" magnetism, depending on the polarity of the earth's magnetic field when the rock solidified. A magnetometer at the surface of the ocean should detect these strips as positive or negative anomalies in the earth's smooth field.

Confirmation

As the Vine-Matthews proposal was being published, I was engaged, with colleagues from the Lamont Geological Observatory of Columbia University and the U.S. Naval Oceanographic Office, in a careful magnetic survey of the Reykjanes Ridge, a section of the Mid-Atlantic Ridge south of Iceland that was known to have large magnetic anomalies. We found that the anomalies were linear and symmetrically distributed parallel to the axis of the ridge. This strongly supported the idea of sea-floor spreading from the ridge and the formation of magnetic anomalies, just as Vine and Matthews had suggested. A little later Vine and J. Tuzo Wilson of the University of Toronto pointed out that the recent reversals of the field matched, one for one, part of the extensive pattern of magnetic lineations recorded just off the west coast of North America by Mason, Raff and Vacquier.

By 1965 it was clear to us and others that magnetism could be the key to reconstructing the history of the ocean floor and the movements of the continents. In only three years a great deal has been learned. Indeed, so many different workers have made significant contributions that it is impossible to name them all in a brief review or even always to know who was the first to make a new observation or propose a new model.

Pioneering efforts to measure the earth's magnetic field at sea had begun at Lamont about 20 years ago. Simple and precise instruments were designed to be towed behind ships and in time efficient techniques were devised for recording, storing and interpreting the data to delineate sea-floor structures hidden under layers of sediments. In 1965, when the new importance of magnetic anomalies became apparent, we had a large stock of data from all the oceans of the world and computer techniques with which to process the data. Examining the data in the light of the new hypotheses, we were able to recognize the same sequence of magnetic bodies extending away from the axis of the mid-ocean

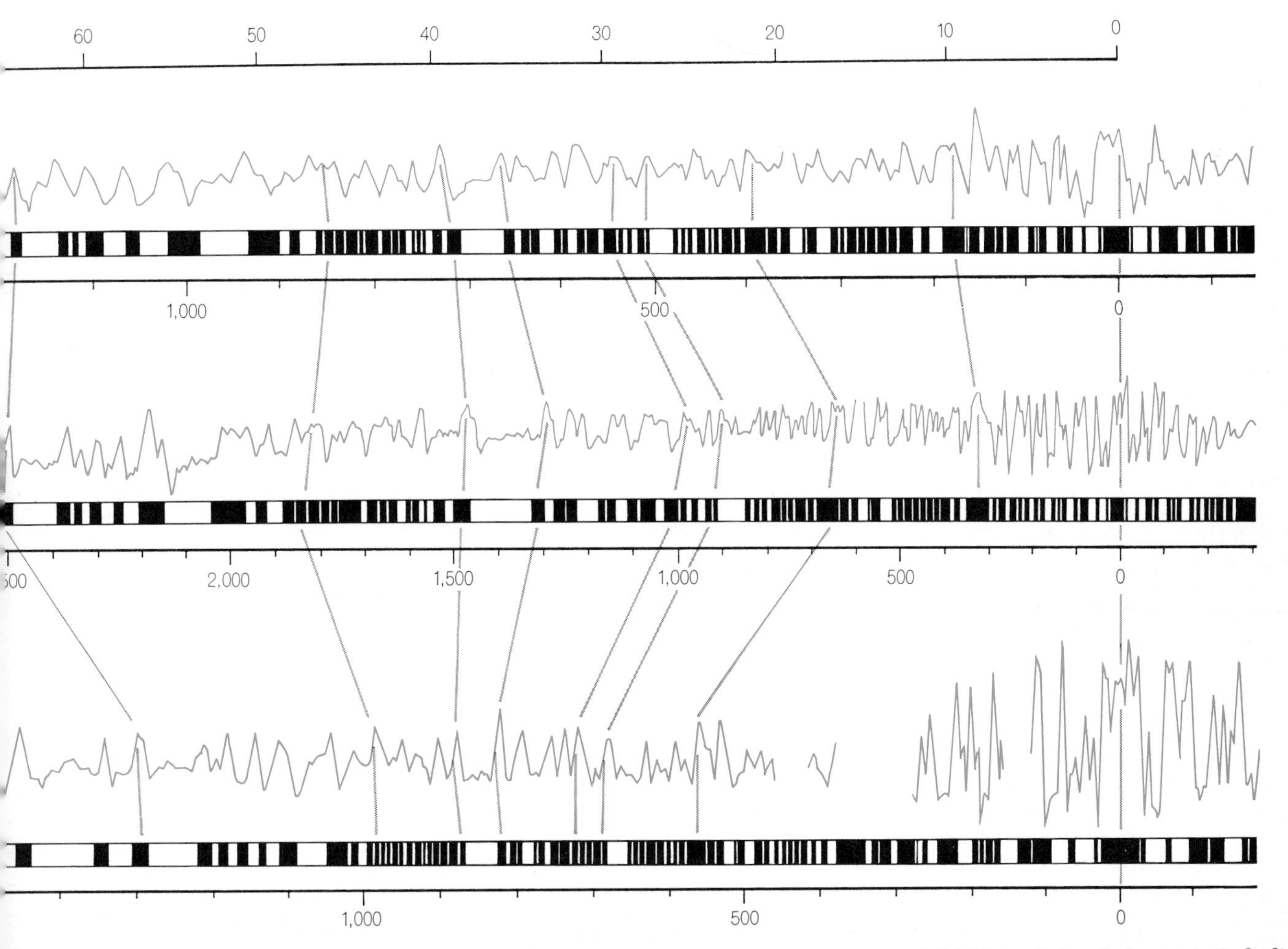

welled up along a "spreading axis" at the ridge during successive periods when the earth's magnetic field was "normal" as it is today or "reversed." The rock was magnetized in the ambient field and later forced out from the axis by subsequent flows. Here three magnetic traces are shown, from three oceans. The anomalies (in gammas, a measure of field intensity) and the magnetic bodies associated with each are spaced differently in each ocean because the spreading rates were different (the rate in the South Atlantic is believed to have been the most constant), but each ocean has the same sequence of 171 reversals extending back 76 million years.

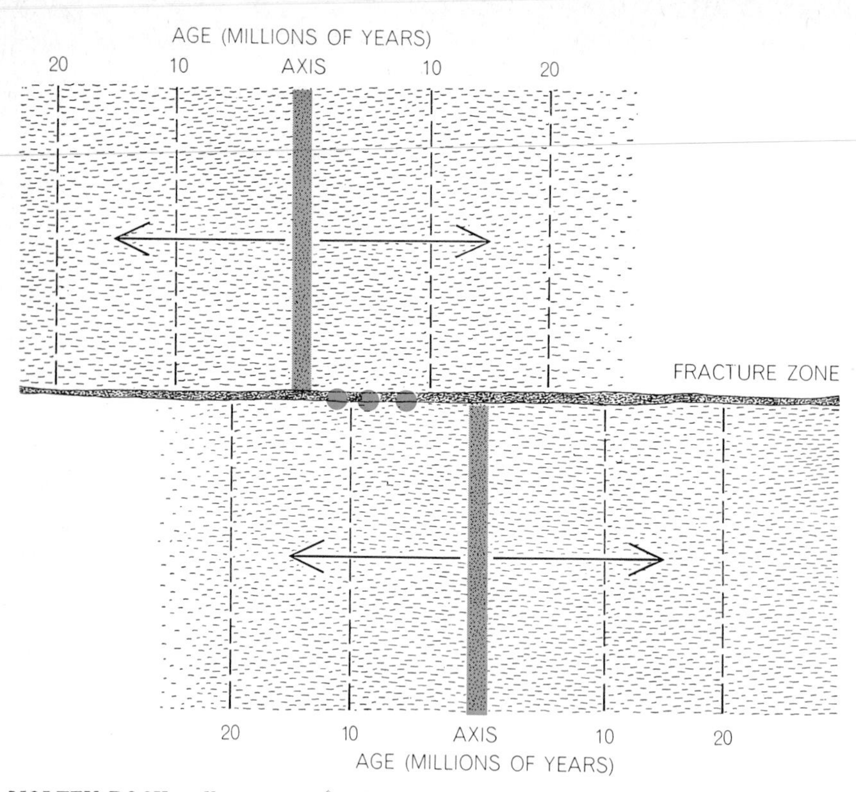

MOLTEN ROCK wells up from the deep earth along a spreading axis, solidifies and is pushed out (*arrows*) by subsequent upwelling. The axis is offset by a fracture zone. Between two offset axes material on each side of the fracture zone moves in opposite directions and the friction between two blocks of the crust causes shallow earthquakes (*colored disks*).

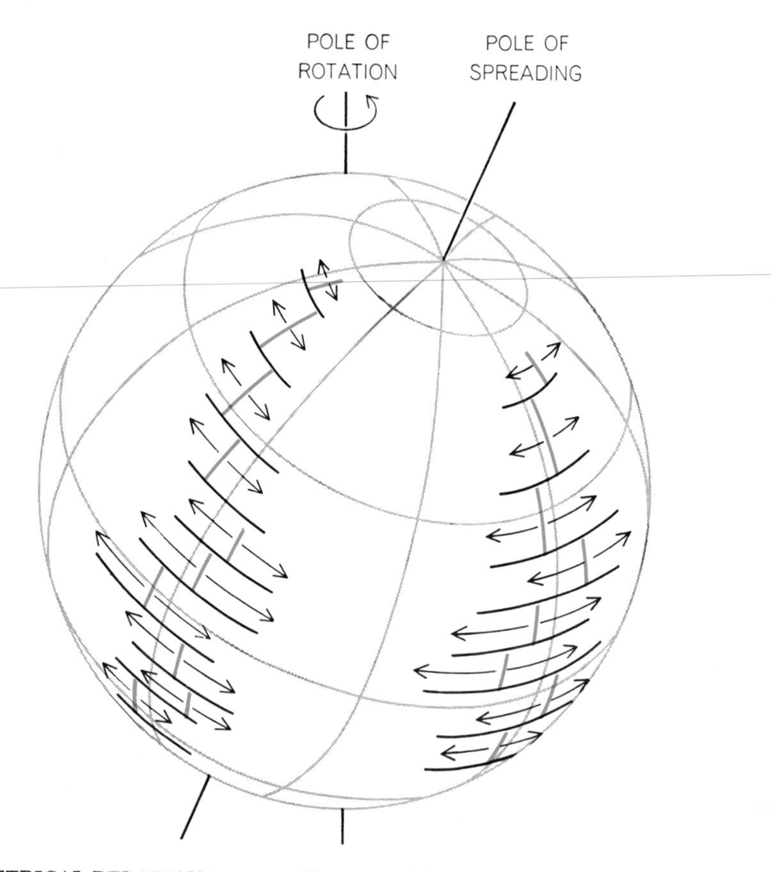

GEOMETRICAL RELATION between ridge axis and fracture zone becomes evident if one conceives of lines of latitude and longitude drawn about a "pole of spreading" rather than the pole of rotation. In each ocean the fracture zones are perpendicular to the spreading axis, and the rate of spreading (*arrows*) varies directly with distance from the equator.

ridge in the South Pacific, South Atlantic and Indian oceans.

Further examination has revealed a worldwide pattern of sea-floor spreading that makes sense of a wide variety of observations. It seems to explain the occurrence of many earthquakes. It establishes a detailed time scale for magnetic reversals and accounts for the direction and rate of continental drift. However, the precise geological events involved in the upwelling along the ridge and the downwelling at the edges of continents are still not understood in detail.

The Worldwide Pattern

At this point we know the main features of the spreading pattern in about half of the world's ocean areas, for the most part those that are adjacent to the better-explored sections of the mid-ocean ridge. The areas where the spreading pattern is unknown either are unexplored or simply resist explanation. For example, many oceans are not wide enough to show the very oldest magnetic bodies, and so not enough magnetic anomalies can be recorded to establish a rate of spreading. Some areas of the ocean floor show no magnetic anomalies. This may be because they are actually not part of the moving floor or because they were created in ancient geologic ages when the earth's field may not have been reversing. There may be places where the spreading is so slow that magnetic bodies crowd one another and confuse the magnetic-anomaly pattern. We have even found places where the axis of spreading is apparently under the continents or does not coincide with the axis of the mid-ocean-ridge system.

The mid-ocean ridge, the axis of the sea-floor spreading, does not meander smoothly from ocean to ocean; instead it is abruptly offset at many points. The offsets, or fracture zones, often extend to great distances on each side of the axis and are usually marked by some irregularity in the topography of the sea floor. Since material wells up at the axis and moves outward on both sides of a fracture zone, it is clear that there will be substantial friction along the zone between offset sections of the axis, where the material is moving in opposite directions [*see top illustration at left*]. Earthquakes, of course, are generated by just such rubbing and scraping between blocks of the earth's crust. Seismologists are now able to pinpoint the location and depth of an earthquake to within a few miles and also to identify the direction of initial motion of the crustal blocks in-

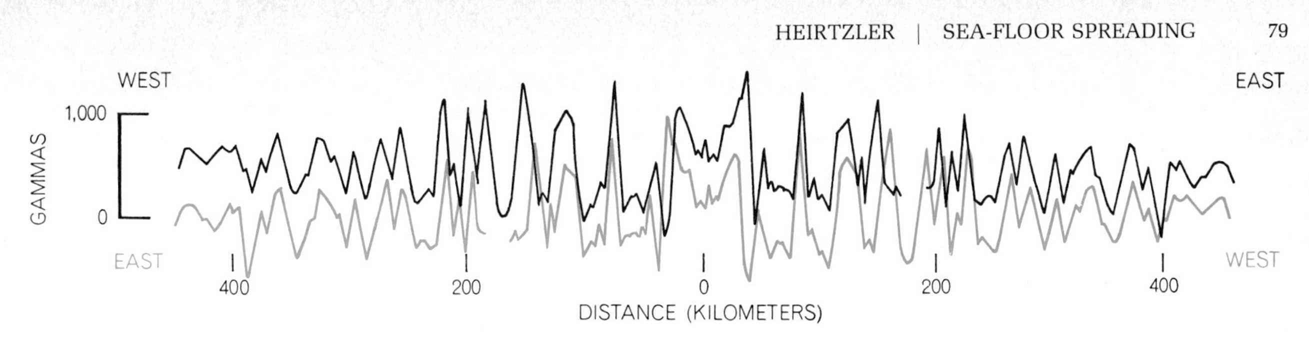

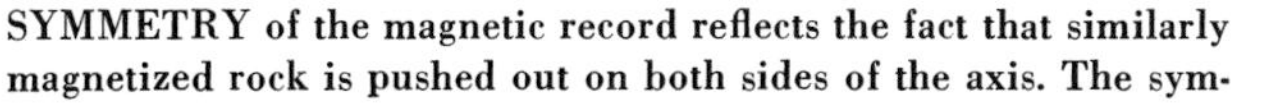

SYMMETRY of the magnetic record reflects the fact that similarly magnetized rock is pushed out on both sides of the axis. The symmetry is demonstrated by reversing a record covering about 1,000 kilometers (*color*) and superposing it on the original (*black*).

volved. Such measurements confirm the theory of sea-floor spreading: Most earthquakes along the ridge system occur in fracture zones between offset sections of the spreading axis—and the direction of first motion is just what one would predict on the basis of the sea-floor motion.

Although it does not appear so on a map with the usual Mercator projection, an axis of spreading and its associated fracture zones are at right angles. In the case of the Pacific and South Atlantic oceans the relation is particularly clear and includes even the rate of spreading. If one conceives of latitude and longitude lines drawn not about the earth's pole of rotation but about a "spreading pole," then the axis of spreading is parallel to the new lines of longitude and the fracture zones are perpendicular to the axis and parallel to the lines of latitude [*see bottom illustration on opposite page*]. Furthermore, the fastest spreading occurs along the equator of this new coordinate system and the rate decreases regularly with distance from the equator—as if giant splits in the floor of the oceans were taking place along the axes. The spreading poles vary for different oceans. For the Pacific and South Atlantic they appear to be very close to the earth's magnetic poles: near Greenland and in Antarctica south of Australia. Spreading in the Indian Ocean seems to be related to a different pair of poles, which can be located, with less certainty, in North Africa and in the Pacific north of New Zealand. Present information indicates that spreading in the North Atlantic occurs about a third set of poles, and fracture zones in older parts of the northeast Pacific floor seem to have been associated with yet another set of poles. There are indications that at least some of and perhaps all the spreading axes are themselves in motion, so that the location of their poles must be changing.

Some workers, notably W. Jason Morgan of Princeton, have attempted to generalize the motions of the earth's crust still further, suggesting that the crust of

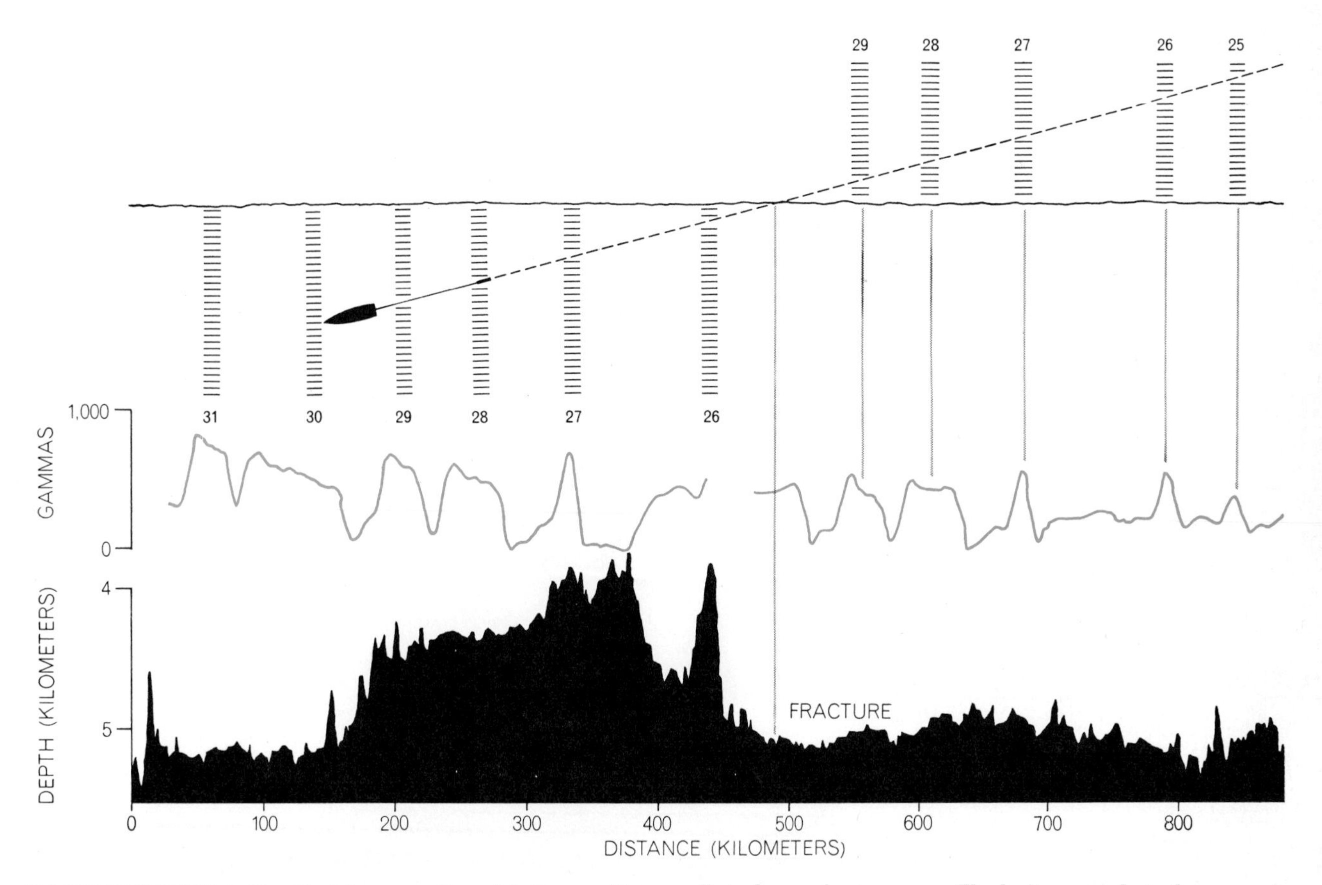

FRACTURE ZONE is identified by magnetic and bottom-profile data. A magnetometer was towed along a course (*broken line at top*) that twice crossed the same magnetic bodies (*Nos. 26–29*), offset along a fracture zone. The bodies caused similar magnetic anomalies, recognizable in the magnetic record (*middle*). Soundings revealed a prominent feature near the fracture zone (*bottom*).

the earth is divided into perhaps six rigid plates. These plates grow by the addition of new crust as new material wells up at the spreading axis; at their outer edges they override or are overridden by other plates. The concept of such strong and rigid plates is appealing to the seismologist since, as will be explained later, deep earthquakes seem to be located at the proposed edges of many of them.

The Geomagnetic Time Scale

The magnetized bodies in the ocean floor have provided an amazingly complete history of magnetic-field reversals that now extends back about 76 million years, into the Cretaceous period. We began with a time scale that had been established by workers at Stanford University and the U.S. Geological Survey,

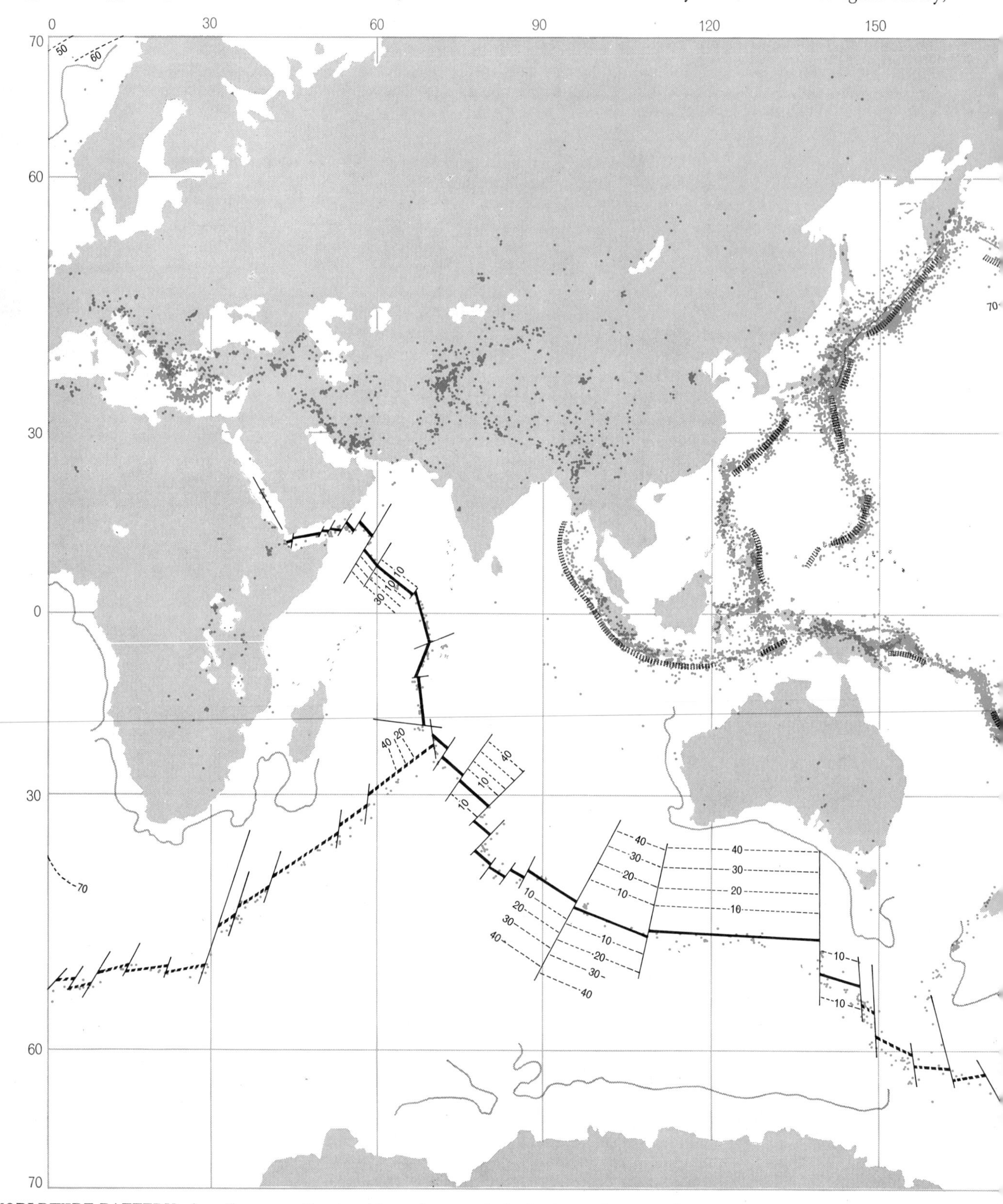

WORLDWIDE PATTERN of sea-floor spreading is evident when magnetic and seismic data are combined. Mid-ocean ridges (*heavy black lines*) are offset by transverse fracture zones (*thin lines*). On the basis of spreading rates determined from magnetic data the author and his colleagues established "isochrons" that give the age of the sea floor in millions of years (*broken thin lines*). The edges

who correlated the magnetism and radioactive-decay dates of rock samples going back some 3.5 million years [see "Reversals of the Earth's Magnetic Field," by Allan Cox, G. Brent Dalrymple and Richard R. Doell; SCIENTIFIC AMERICAN, February, 1967]. By comparing the ages they had assigned to specific reversals of the geomagnetic field with the distance of the corresponding anomalies from the ridge axis, we were able to extend the observations of Vine and Wilson over much of the ocean floor and thus to determine the rate at which the floor had spread in the various oceans.

The rates were found to be different for different sections of the ridge but seemed to have been remarkably steady over the ages in many areas. They vary from about half an inch to a little more

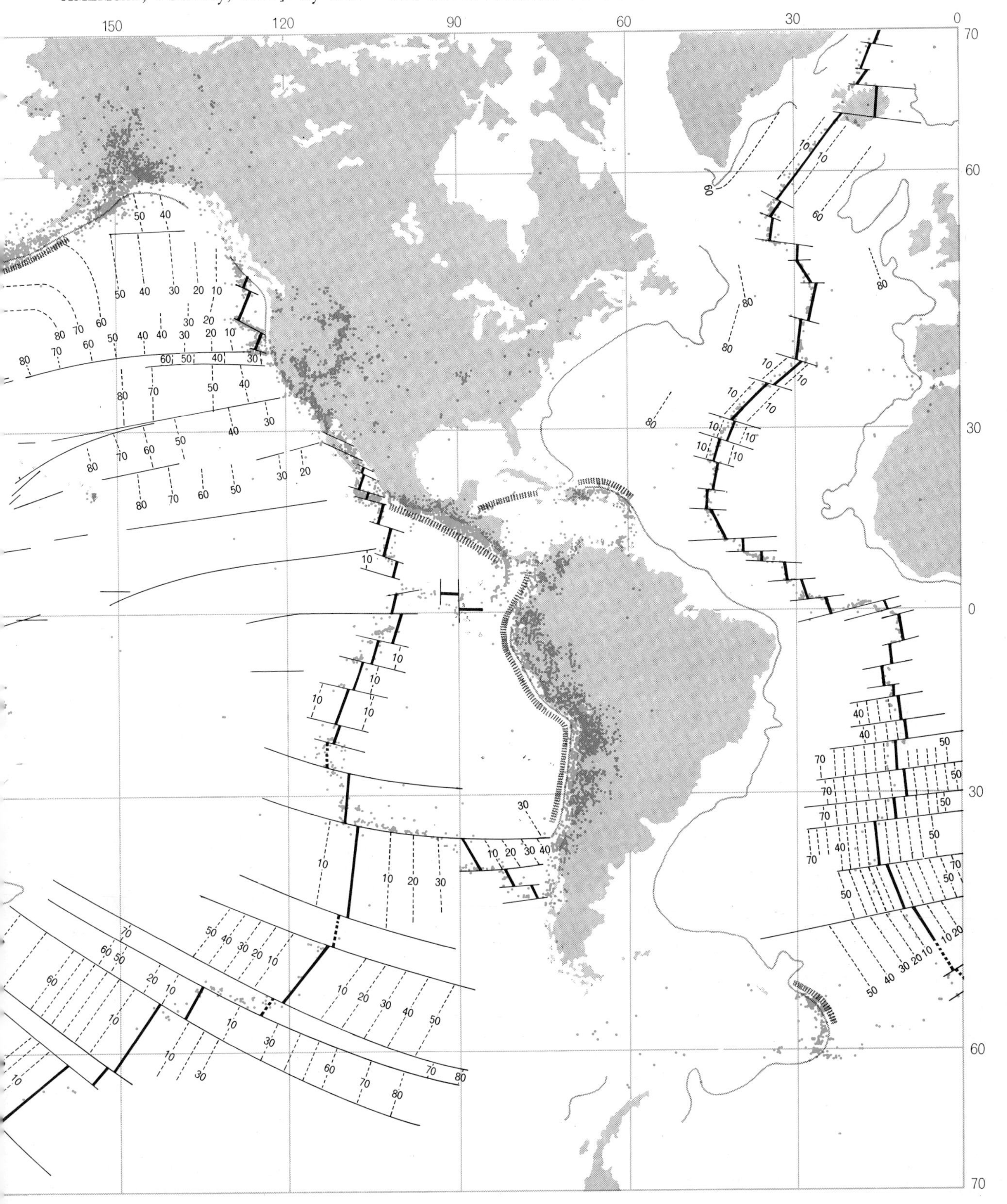

of many continental masses (*gray lines*) are rimmed by deep ocean trenches (*hatching*). When the epicenters of all earthquakes recorded from 1957 to 1967 (plotted by Muawi Barazangi and James Dorman from U.S. Coast and Geodetic Survey data) are superposed (*colored dots*), the vast majority of them fall along mid-ocean ridges or along the trenches, where the moving sea floor turns down.

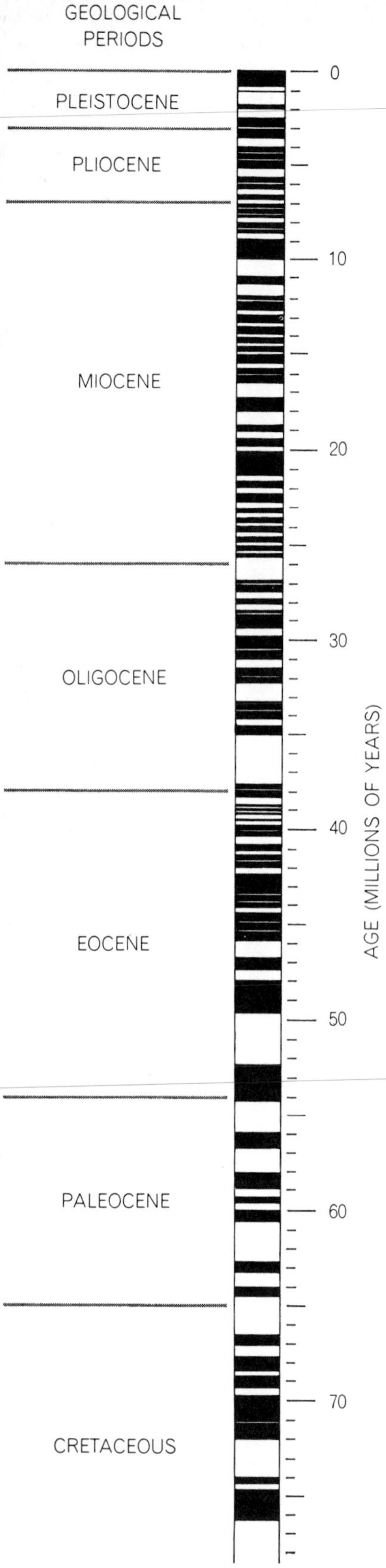

CHRONOLOGY of geomagnetic-field reversals was worked out by extrapolation, beginning with the dates that had been established for recent reversals and assuming a constant rate for spreading within each ocean. The dates were confirmed by geological data.

than two inches per year away from the axis of spreading. Although these rates of movement are small by everyday standards, they are large on a geological time scale. Their magnitude came as a surprise to geophysicists even though comparable rates of slippage had been observed along the San Andreas fault, an exposed fracture zone in southern California.

In several areas we saw that there was no obvious hiatus in spreading. We therefore felt free to assume a constant spreading rate and on that basis to assign dates to geomagnetic field reversals extending far beyond the 3.5-million-year scale. This extrapolation of spreading rates may not seem justified at first, but it is supported by consistency from ocean to ocean and by agreement with other geophysical evidence, such as radioactive and paleontological dating of rock samples. Within the probable errors of the methods no discrepancies have been found. We have now identified 171 reversals of the earth's magnetic field over the 76 million years and believe that the dates assigned to each are quite accurate. (Of course, if spreading stopped abruptly all over the world for a certain time, our time scale before the spreading stopped would be too young by that number of years; this is improbable but cannot be entirely ruled out.) On the whole the evidence for the correctness of the time scale is so strong that it can now be used in turn to study variations in spreading rates over the ages in certain disturbed areas.

The average "normal" interval (when the magnetic field was oriented as it is today) works out to 420,000 years and the average "reversed" interval to 480,000 years. The closeness of the two numbers means that the earth is just about as likely to be found in one state of magnetic polarity as in the other. The present era of normal polarity has lasted 700,000 years. Are we due for a change? Only 15 percent of the normal intervals have been longer than that, although some were apparently as long as three million years. The shortest intervals seem to have been less than 50,000 years, but brief instants of geologic time are difficult to confirm by absolute dating methods. This suggests one of the drawbacks of the magnetic scale: simply because it is so detailed, it is unlikely soon to be proved wrong; by the same token, however, it is difficult to use to date a piece of magnetized earth material. Just as one may need a "finder" telescope to orient a high-powered telescope, so one must start by knowing the approximate age of materials in order to utilize the geomagnetic time scale.

The geomagnetism I have been discussing is frozen into igneous rock of the basaltic type that wells up from deep within the earth. Over much of the ocean floor, of course, this bedrock is not exposed but is covered by varying thicknesses of accumulated sediments. These sediments can also be magnetized, since the tiny particles of which they are composed can become oriented in the earth's field as they settle slowly to the ocean floor. (They are magnetized only one ten-thousandth as strongly as the basalt, however, so that even a thick layer of sediment does not interfere with measurements of anomalies caused by the underlying basalt.) By dropping a hollow cylindrical pipe into the bottom mud one can bring to the surface long "cores" of successive layers of sediment, each constituting an undisturbed record of magnetic-field reversals. Workers at Scripps and Lamont have recently developed sensitive techniques for measuring these weakly magnetized specimens, and the record of geomagnetic reversals has been verified back about 10 million years. The ability to correlate sedimentary layers of the same age in different oceans has proved of immense value to marine geologists, who often lack good paleontological or other indicators of geological horizons. In making detailed comparisons of magnetic reversals and the populations of microscopic animal fossils in such cores, investigators have noted a striking correlation between reversals and major changes in microfaunal species. It has been proposed that such changes are the result of mutations caused by increased exposure to cosmic rays if the earth's protective magnetic field is largely attenuated during a reversal. The evidence for this is not strong, however, and an alternative explanation can be put forward, as I shall suggest later in this article.

Footprints of Continents

The indicated movements of the ocean floor are of the right size and in the right directions to account for continental drift. Topographic and geological evidence had pointed to the probable existence some 200 or 300 million years ago of a single large land mass in the Southern Hemisphere (named Gondwanaland for a key geological province in India) that included Africa, India, South America, Australia, New Zealand and Antarctica, and another mass in the Northern Hemisphere called Laurasia. The positions of the present continents within these land masses were not clear, nor was it possible to trace in detail the sequence of events in their breakup.

The magnetic lineations in the ocean floor serve as "footprints" of the continents, marking their consecutive positions before they reached their present locations. We found that the slow but steady and prolonged rates of motion were sufficient to separate South America from Africa—thus creating the entire South Atlantic Ocean—in about 200 million years and to separate Australia from Antarctica in about 40 million years. As more of the sea floor was dated we could establish more exactly just when the various continents separated and how they moved [*see illustration at right*]. It is now possible to reconstruct the original positions of the continents and the shallow continental shelves, so that land geologists can go into the field and check the continuity of ancient geological structures that were torn apart when the separations occurred.

Although it is possible to tell when and how the continents pulled away from one another, it is not always entirely clear whether or not—and for how long—a continent may have stood still. The impression is that both Africa and Antarctica have remained fixed with respect to the rotational axis for the last 100 million years, or since Africa split off from the remainder of Gondwanaland. If this is true, then the fact that the spreading axis between South America and Africa remains about midway between them indicates that the axis itself must be moving too.

Source and Sink

Until the theory of sea-floor spreading was advanced the only known sources of deep-earth material at the surface were volcanic eruptions. It now appears, however, that most currently active volcanoes are not at the sites of upwelling along the mid-ocean ridge but rather in areas where the moving sea floor turns down under the continents. Recent volcanic eruptions in the Philippines, Mexico and Guatemala are examples, as is the continuing intermittent activity of volcanoes in western North America, Alaska and Japan. The upwelling that initiates sea-floor spreading must therefore represent some unfamiliar geophysical phenomenon, and investigators have concentrated much attention on the axis of the mid-ocean ridge where it occurs. The axis has several unusual properties: a large heat flux, a concentration of shallow earthquakes, unusual seismic-wave velocities, a lack of sediment and eroded rocks, and the presence of a prominent magnetic anomaly. While most of these unusual conditions extend over a distance of from tens to hundreds of miles on both sides of the crest of the ridge, the magnetic anomaly is sharply localized; by studying the symmetry of the magnetic pattern it is possible to locate the spreading axis to within a few miles. The terrain of the ridge is rugged and it is not possible to associate the spreading axis with any single topographic feature,

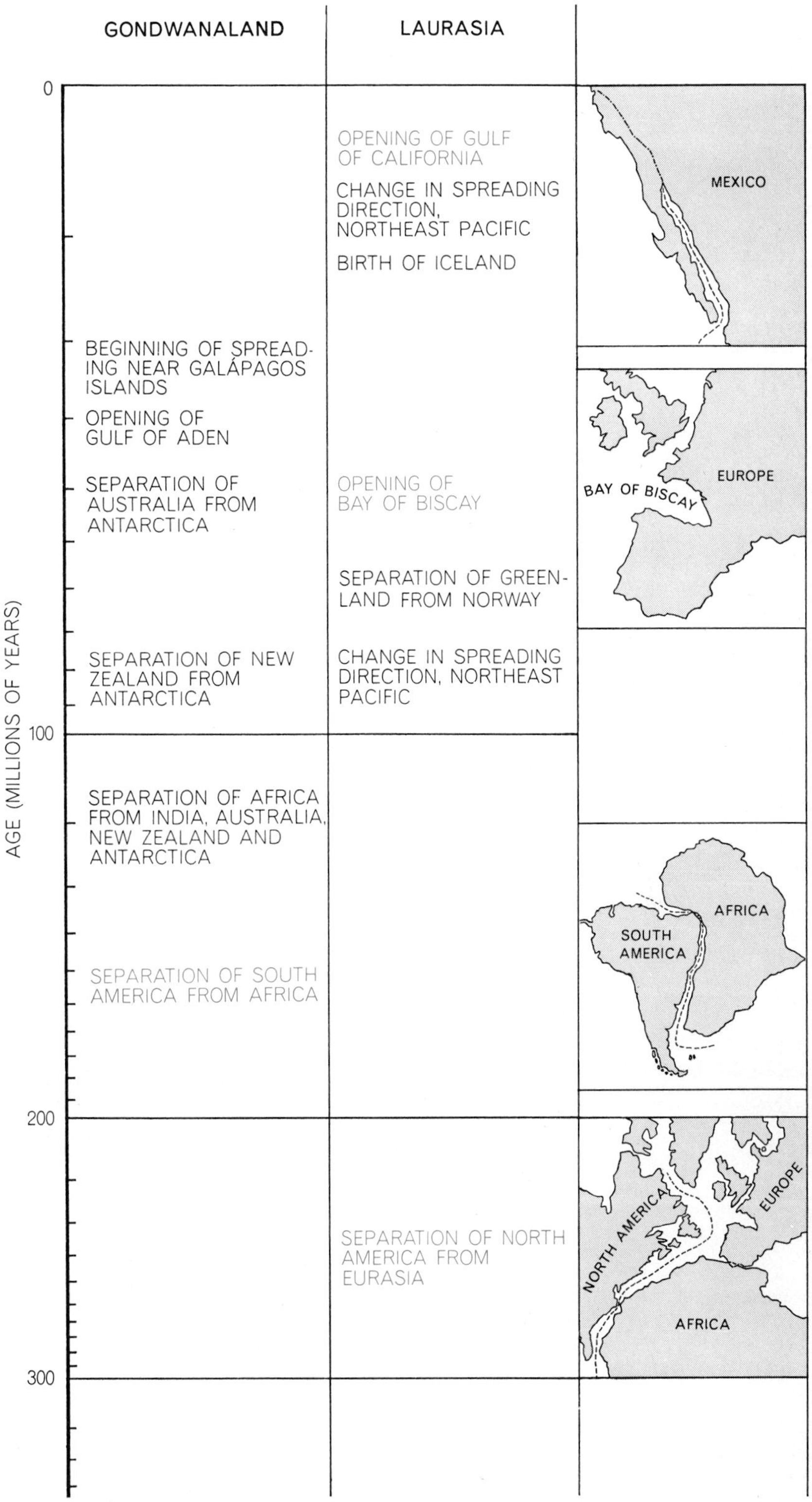

GONDWANALAND
LAURASIA
AGE (MILLIONS OF YEARS)
0
OPENING OF GULF OF CALIFORNIA
CHANGE IN SPREADING DIRECTION, NORTHEAST PACIFIC
BIRTH OF ICELAND
MEXICO
BEGINNING OF SPREADING NEAR GALÁPAGOS ISLANDS
OPENING OF GULF OF ADEN
SEPARATION OF AUSTRALIA FROM ANTARCTICA
OPENING OF BAY OF BISCAY
BAY OF BISCAY
EUROPE
SEPARATION OF GREENLAND FROM NORWAY
SEPARATION OF NEW ZEALAND FROM ANTARCTICA
CHANGE IN SPREADING DIRECTION, NORTHEAST PACIFIC
100
SEPARATION OF AFRICA FROM INDIA, AUSTRALIA, NEW ZEALAND AND ANTARCTICA
SEPARATION OF SOUTH AMERICA FROM AFRICA
SOUTH AMERICA
AFRICA
200
SEPARATION OF NORTH AMERICA FROM EURASIA
NORTH AMERICA
EUROPE
AFRICA
300

TIME SCALE for continental movements and other changes can now be established, since isochrons show the age of the ocean floor and the direction of spreading at any time.

although one can say that when the ridge contains a median "rift valley," as many do, the axis lies within it.

The magnetized plate is not very thick. Analyses of the observed magnetic anomalies suggest that the thickness is from a half-mile to a few miles; studies of the transmission of seismic waves and the distribution of heat flux, on the other hand, indicate a thickness of a few tens of miles or perhaps 100 miles for the moving plate. The precise linearity of the magnetized bodies shows that the upwelling material was not tumbled about after it became cooled and magnetized; in this it is different from the usual folded rocks seen on land or the irregular flows that surround a volcanic eruption. Attempts to locate the axis with a magnetometer submerged near the sea floor show that there are a number of very magnetic bodies where the axis would be expected.

The evidence suggests to most investigators that the upwelling mechanism is an injection of molten deep-earth material by linear intrusions called dikes. Such bodies have a high probability of being injected along the line of the spreading axis. Each injection may be quite localized rather than greatly elongated along the axis. The new material is hot enough to reheat adjacent rock so that both the new material and the slightly older material nearby is magnetized in the direction of the ambient field before being quenched by the cold seawater. This explanation does not indicate anything precise about the thickness of the moving layer; it does account for the lack of tumbling of magnetized blocks, since a dike would tend to push the older material horizontally away from the ridge.

The deep oceanic trenches found around the periphery of the Pacific are thought to be places where surface material returns to the deep earth; so, very likely, are smaller trenches in the North Atlantic and South Atlantic and the Indian Ocean. In many parts of the world where plates of moving sea floor have been identified the outer edge of the plate has not been located, and so we probably do not know where all the sinks are. In most places the spreading sea floor seems to be turning down under the continents, but in some places it seems to be pushing continents ahead of it or even tearing continents apart; we know of no place where the spreading floor is overriding a continent.

If the epicenters of deep earthquakes are mapped, they are almost all found to be on a plane that starts at the floor of a trench and makes an angle of about 45 degrees with the horizontal [*see top illustration on page 85*]. The slippage of earth material parallel to the plane and extending hundreds of miles below the surface is the cause of the earthquakes. These earthquake planes almost certainly define regions of downwelling; studies of first motion in such areas show that the sea floor moves down with respect to the adjacent continent. A thin crustal layer with a characteristic speed of seismic-wave propagation has been shown to underlie the ocean sediments and turn down at 45 degrees to great depths.

The magnetic evidence for what happens at the trenches is ambiguous, however. The sea-floor magnetic pattern is altered suddenly. Measurements at the Aleutian Trench, for example, show a

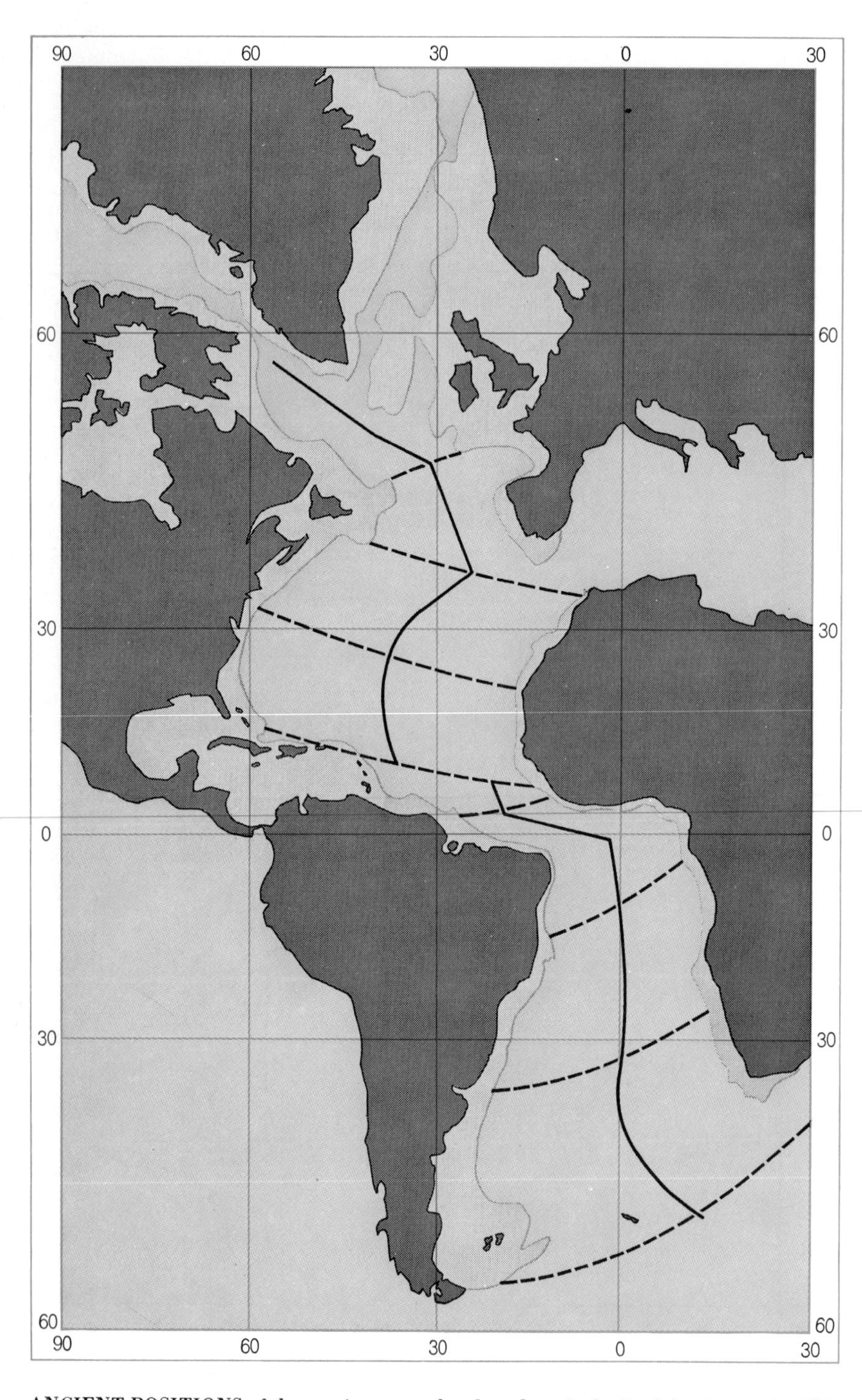

ANCIENT POSITIONS of the continents can be plotted on the basis of the time scale. This map shows the relation of the Americas to Eurasia and Africa 70 million years ago. The continents fitted together generally along the edges of continental masses (*light grey*) rather than of present dry-land areas (*dark gray*). Broken lines trace directions of movement.

sharp discontinuity [*see bottom illustration on this page*]. There is no sign of a magnetic body that should (on the basis of measurements made elsewhere) be located about three kilometers below the trench floor. Its absence might be explained by heating or mechanical deformation, but neither sufficiently high temperatures nor enough seismic activity to cause deformation are indicated so close to the trench floor. Another problem at the Aleutian Trench is that the magnetic bodies seem to be in the wrong sequence. If the trenches are sinks, the older bodies should be in the trench and the younger ones to seaward; the opposite seems to be the case! Moreover, south of the Alaska Peninsula bodies oriented essentially north-south turn and run east-west, parallel to the coast. This complexity is difficult to explain if trenches are sinks. It is one of the most awkward sticking points in the theory.

Speculations

The feeling among many geophysicists that we are on the brink of even more comprehensive theories about the earth stems from the striking geographic or temporal coincidence of certain geophysical phenomena, many having to do with the dynamics of our planet. Existing theories suggest no clear causal relation among these phenomena, and so one hopes for some higher-order synthesis that will establish such a relation.

What are some of these coincidences? The present pole of spreading (for several of the oceans, at least) is near the geomagnetic axis; in Cretaceous times the spreading pole for the North Pacific was near the Cretaceous geomagnetic pole. There have often been significant changes in the microfaunal population of the sea at magnetic reversals. A major meteorite (tektite) fall occurred just at the time of the last reversal [see "Tektites and Geomagnetic Reversals," by Billy P. Glass and Bruce C. Heezen; SCIENTIFIC AMERICAN, July, 1967]. Some authors have speculated recently on a relation between mountain-building activity and magnetic reversals; others see a relation between changes in sea-floor spreading and mountain-building. Mechanisms have been suggested whereby the earth's magnetic field could be generated by convective motion caused in turn by irregularities in the earth's orbit. There has been a revival of a 30-year-old theory that the glacial ages were caused by changes in the tilt of the earth's axis. Finally, there is clear evidence that large earthquakes occur

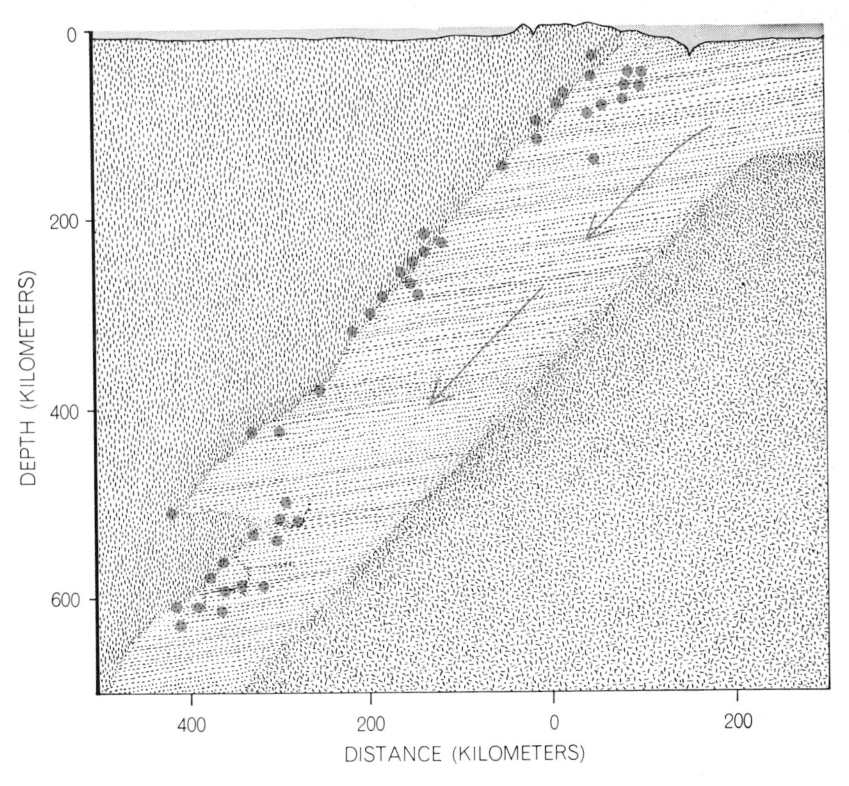

MOVING SEA FLOOR turns down at the trenches rimming the Pacific Ocean. Deep earthquake epicenters line the downward path where floor material moves under land masses. Symbols (*color*) show epicenters recorded at the Tonga Trench in the South Pacific.

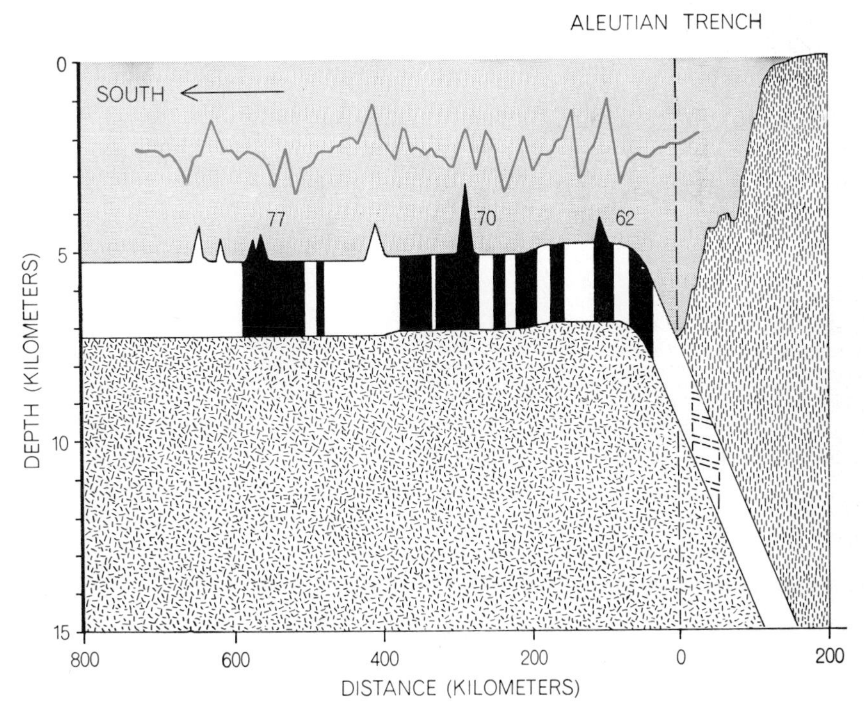

ALEUTIAN TRENCH data fail to fit the model, however. The magnetic record (*color*) shows the expected sequence of normal and reversed bodies (*black and white bands*) approaching the trench, but there is no evidence of the next body, which should appear about three kilometers below the trench (*hatched band*). Moreover, the magnetic bodies seem to be younger (*77 to 62 million years*) as one approaches the trench rather than older.

at about the same time as certain changes in the earth's rotational motion.

This article is not the place for a full evaluation of all these developments. It is interesting to note, however, that a common thread running through these and many other proposals at the frontier of geophysical research is the role played by displacements of the earth's axis of rotation. It seems that rather minor variations can affect to a surprising extent both the climate at the surface of the earth and forces and stresses within the earth.

The intimate relation between the spreading pole and the magnetic pole suggests that the convective motion within the earth and the earth's field may have a common cause. It could be due to the proposed effect of orbital irregularities or to some dynamo action within the earth. Since the direction of convection does not reverse when the field reverses, it is clear that the convective motion does not simply generate the field, nor is it likely that the reversing field could "pump" convection cells. Whatever the driving force is for these two phenomena, it would seem to be related to the motion of the earth. Recently it has been shown that earthquakes of a magnitude of 7.5 or greater on the Richter scale either cause or are caused by changes in the earth's "wobble," a small circular motion described by the north pole of rotation [see "Science and the Citizen," SCIENTIFIC AMERICAN, November, 1968]. Whatever the mechanism of these changes, it is not hard to believe that similar changes in the earth's axial motion in times past could have caused major earthquake and mountain-building activity and could even have caused the magnetic field to flip.

To summarize: Every few months there are changes in the earth's rotational motion that affect sea-floor spreading and cause the earthquakes associated with it. If such a change is large enough, it may even reverse the earth's magnetic field. (Both the presence of a geomagnetic field and the spreading of the sea floor seem to be due to the mere fact that the earth is rotating; it is only the changes in motion that are associated with certain large earthquakes and reversals of the field.) Changes in the microfaunal population of the sea are related to changes in the climate, which are related in turn to variations in the earth's motions.

Such speculation cannot yet be confirmed but neither can it be firmly denied; indeed, it is no more outlandish than theories of sea-floor spreading seemed to be a few years ago.

The San Andreas Fault

7

Don L. Anderson
November 1971

The San Fernando earthquake that occurred at sunrise on February 9, 1971, jolted many southern Californians into acute awareness that California is earthquake country. Although it was only a moderate earthquake (6.6 on the Richter scale), it was felt in Mexico, Arizona, Nevada and as far north as Yosemite National Park, more than 250 miles from San Fernando. It was recorded at seismic stations around the world. In spite of its relatively small size the San Fernando earthquake was extremely significant because it happened near a major metropolis and because its effects were recorded on a wide variety of seismic instruments. Within hours the affected region was aswarm with geologists mapping faults and seismologists installing portable instruments to monitor aftershocks and the deformation of the ground. It was immediately clear from data telemetered to the Seismological Laboratory of the California Institute of Technology in Pasadena from the Caltech Seismic Network that the earthquake was not centered on the much feared San Andreas fault or, for that matter, on any fault geologists had labeled as active. The faults in the area, however, are all part of the San Andreas fault system that covers much of California.

The San Andreas fault system (and its attendant earthquakes) is part of a global grid of faults, chains of volcanoes and mountains, rifts in the ocean floor and deep oceanic trenches that represent the boundaries between the huge shifting plates that make up the earth's lithosphere. The concept of moving plates is now fundamental to the theory of continental drift, which was long disputed but is now generally accepted in modified form on the basis of voluminous geological, geophysical and geochemical evidence. The theory had received strong support from the discovery that the floors of the oceans have a central rise or ridge, often with a rift along the axis, that can be traced around the globe. Within the rift new crustal material is continuously being injected from the plastic mantle below, forming a rise or ridge on each side of the rift. The newly formed crustal material slides away from the ridge axis. Since the magnetic field of the earth periodically reverses polarity, the newly injected material "freezes" in stripes parallel to the ridge axis, whose north-south polarity likewise alternates. By dating these stripes one can estimate the rate of sea-floor spreading.

The San Andreas fault system forms the boundary between the North American plate and the North Pacific plate and separates the southwestern part of California from the rest of North America. In general the Pacific Ocean and that part of California to the west of the San Andreas fault are moving northwest with respect to the rest of the continent, although the continent inland at least as far as Utah feels the effects of the interactions of these plates.

The relative motion between North America and the North Pacific has been estimated in a variety of ways. Seismic techniques yield values between 1½ and 2½ inches per year. The ages of the magnetic stripes on the ocean floor indicate a rate of about 2⅓ inches per year. Geodetic measurements in California give rates between two and three inches per year. The ages of the magnetic anomalies off the coast of California indicate that the oceanic rise came to intersect the continent at least 30 million years ago. Geologists and geophysicists at a number of institutions (notably the University of Cambridge, Princeton University and the Scripps Institution of Oceanography) have proposed that geologic processes on a continent are profoundly affected when a continental plate is intersected by an oceanic rise. At the rates given above the total displacement along the San Andreas fault amounts to at least 720 miles if motion started when the rise hit the continent and if all the relative motion was taken up on the fault. Displacements this large have not been proposed by geologists, but the critical tests would involve correlation of geology in northern California with geology on the west coast of Baja California, an area that has only recently been studied in detail. One can visualize how the west coast of North America may have looked 32 million years ago by closing up the Gulf of California and moving central and northern California back along the San Andreas fault to fit into the pocket formed by the coastline of the northern half of Baja California. This places all of California west of the San Andreas fault south of

DISPLACEMENT ALONG SAN ANDREAS FAULT is clearly visible in the aerial photograph on the next page of a region a few miles north of Frazier Park, Calif., itself 65 miles northwest of Pasadena, where the fault runs almost due east and west. This east-west section of the San Andreas fault is part of the "big bend," where the fault appears to be locked. The photograph is reproduced with north at the right. The hilly region to the left (south) of the fault line is moving upward (westward) with respect to the flat terrain at the right, causing clearly visible offsets in the two largest watercourses as they flow onto the alluvial plain.

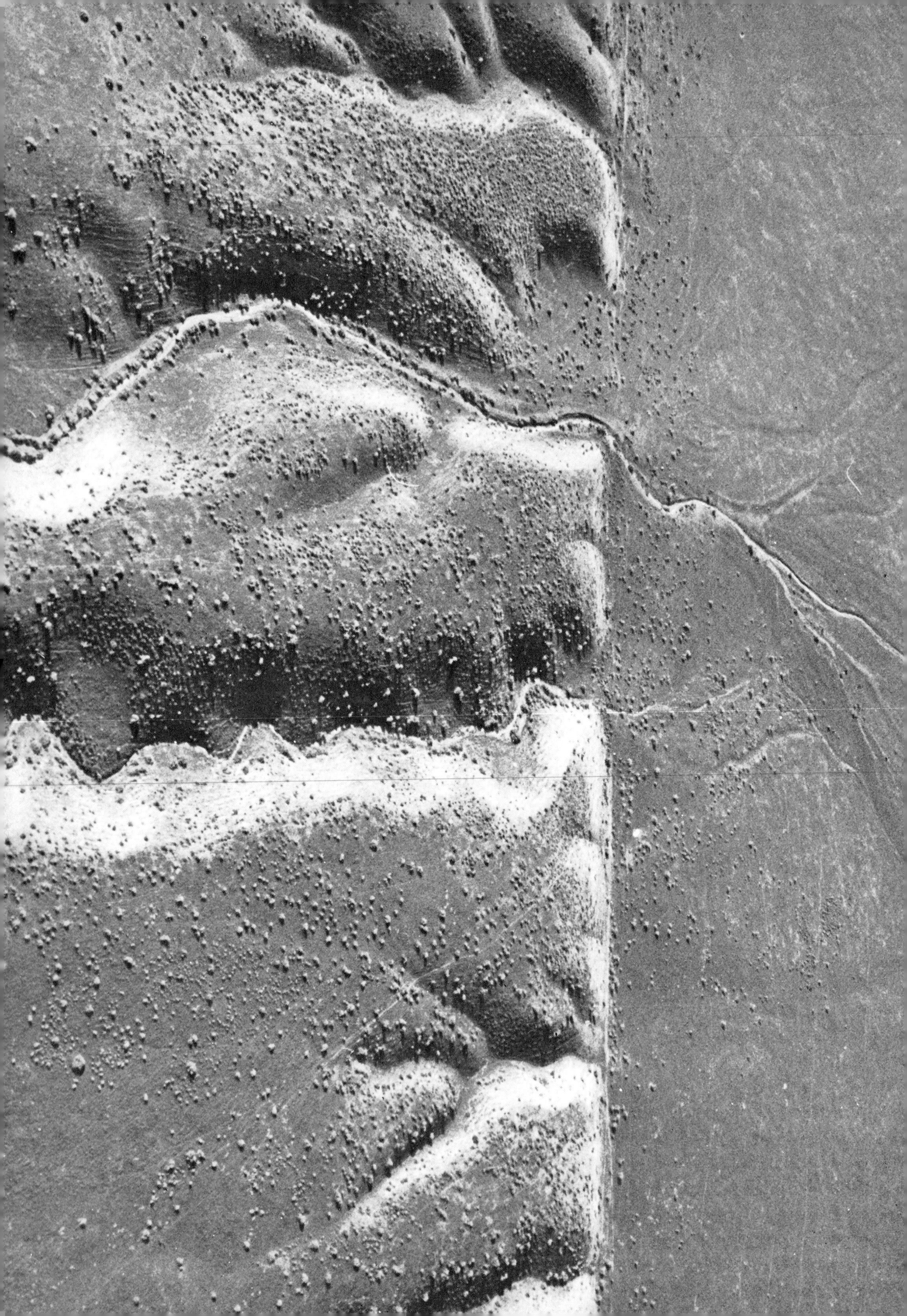

the present Mexican border [*see illustration on page 93*].

California is riddled with faults, most of which trend roughly northwest-southeast, like most of the other tectonic and geologic features of California (such as the Sierra Nevada and the Coast Ranges). The prominent exceptions are the east-west-trending transverse ranges and faults that make up a band some 100 miles wide extending inland from between Los Angeles and Santa Barbara. The San Gabriel Mountains, which form the rugged backdrop to Los Angeles, are part of this complex geologic region, and it was here that the San Fernando earthquake struck. The northeast-trending Garlock fault and the Tehachapi Mountains, which separate the Sierra Nevada and the Mojave Desert, also cut across the general grain of California. The area to the west of most of the northwest-trending faults is moving northwest with respect to the eastern side. This is called right-lateral motion. If one looks across the fault from either side, the other side is moving to the right.

Motion on the Garlock fault is left-lateral, which, combined with the right-lateral motion on the San Andreas fault, means that the Mojave Desert is moving eastward with respect to the rest of California. Parts of the faults that have been observed or inferred to move as a result of earthquakes in historic times are shown in the illustration at the left. Also shown are the dates of the earthquakes and the magnitude of some of the more important ones. In general both the length of rupture and the total displacement are greater for the larger earthquakes. Horizontal displacements as great as 21 feet were observed along the San Andreas fault after the San Francisco earthquake of 1906, which had a magnitude of 8.3 on the Richter scale. (The Richter scale, devised by Charles F. Richter of Cal Tech, is logarithmic. Although each unit denotes a factor of 10 in ground amplitude, or displacement, the actual energy radiated by an earthquake is subject to various modifications.) The San Fernando earthquake produced displacements of six feet, whose direction was almost equally divided between the horizontal and the vertical.

The trend of the San Andreas fault system is roughly northwest-southeast from San Francisco to the south end of the Great Central Valley (the San Joaquin Valley) and again from the north of the Salton Sea depression to the Mexican border. The motion along the faults

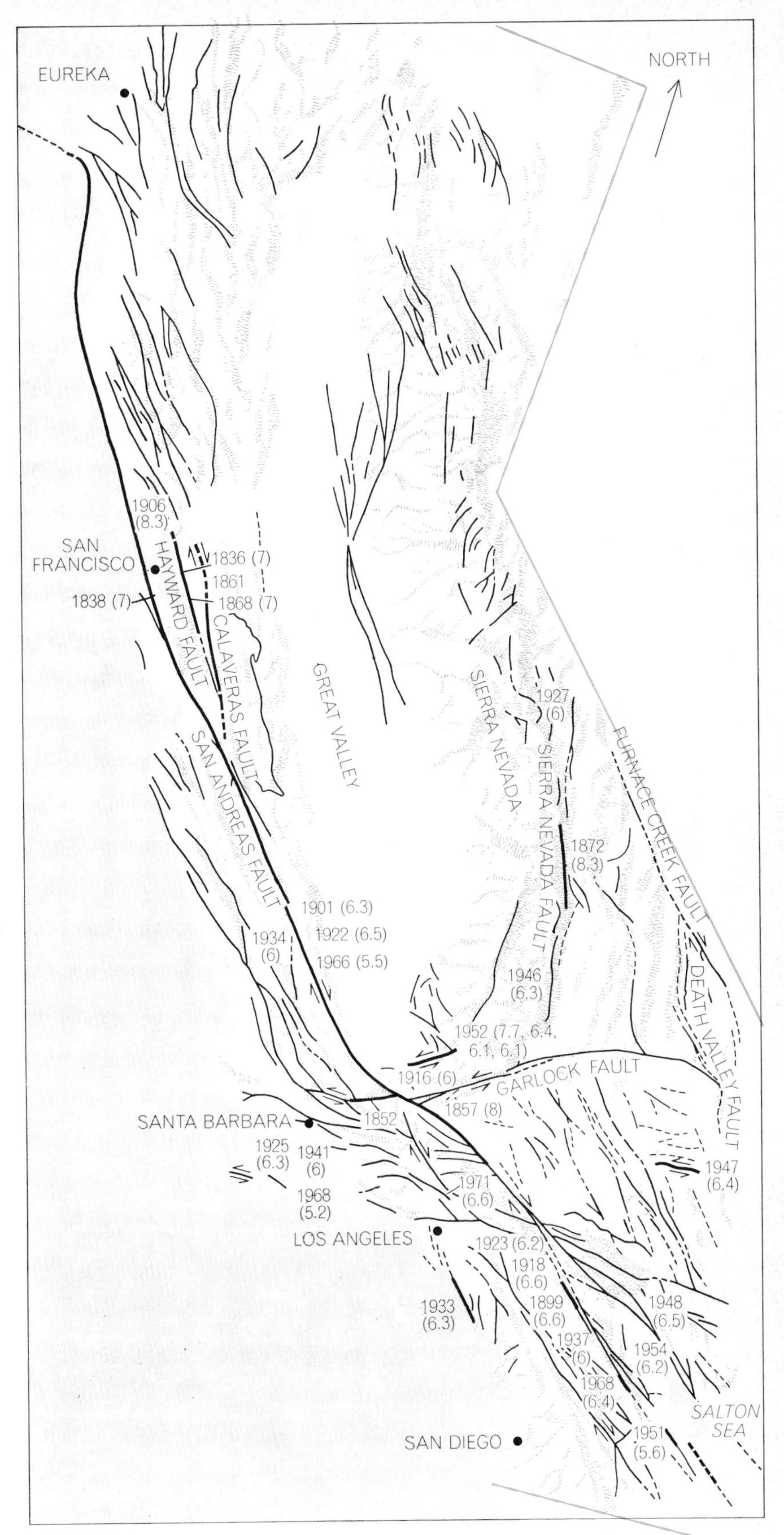

SIMPLIFIED FAULT MAP of California identifies in heavy black lines the faults that have given rise to major earthquakes since 1836. The magnitude of all but two of the earthquakes is given in parentheses next to the year of occurrence. For events that predated the introduction of seismological instruments the magnitudes are estimated from historical accounts. For two major events, the earthquakes of 1852 and 1861, information is too sparse to allow a magnitude estimate. Arrows parallel to the faults show relative motion.

in these areas is parallel to the fault and is mainly strike-slip, or horizontal. Between these two regions, from the south end of the San Bernardino Mountains to the Garlock fault, the faults bend abruptly and run nearly east and west, producing a region of overthrusting and crustal shortening [*see illustration below*]. The attempt of the southern California plate to "get around the corner" as it moves to the northwest is responsible for the complex geology in the transverse ranges, for the abrupt change in the configuration of the coastline north of Los Angeles and ultimately for the recent San Fernando earthquake. The big bend of the San Andreas fault is commonly regarded by seismologists as being locked and possibly as being the location of the next major earthquake. Much of the motion in this region, however, is being taken up by strike-slip motion along faults parallel to the San Andreas fault and by overthrusting on both sides of the fault. The displacements associated with the larger earthquakes in southern California in the vicinity of the big bend have averaged out to about 2½ inches per year since 1800. The Kern County earthquake of 1952 (magnitude 7.7) apparently took care of most of the accumulated strain, at least at the north end of the big bend, that had built up since the Fort Tejon earthquake of 1857 (magnitude 8).

The San Andreas fault system cannot be completely understood independently of the tectonics and geology of most of the western part of North America and the northeastern part of the Pacific Ocean. This vast region is itself only a part of the global tectonic pattern, all parts of which seem to be interrelated. The earthquake, tectonic and mountain-building activities of western North America are intimately related to the relative motions of the Pacific and North American plates. Just as it is misleading to think of the San Andreas fault as an isolated mechanical system, so it is misleading to think of the entire San Andreas fault as a single system. The part of the fault that lies in northern California was activated earlier and has moved farther than the southern California section. The northern portion is less active seismically than the southern section and seems to have been created in a different way. It is also moving in a slightly different direction.

Measuring Displacements

There are several ways to measure displacements on major faults. Fairly recent displacements are reflected in offset stream channels [*see illustration on page 88*]. Many such offsets measured in thousands of feet are apparent across the San Andreas fault in central California, some of which can be directly related to earthquakes of historic times.

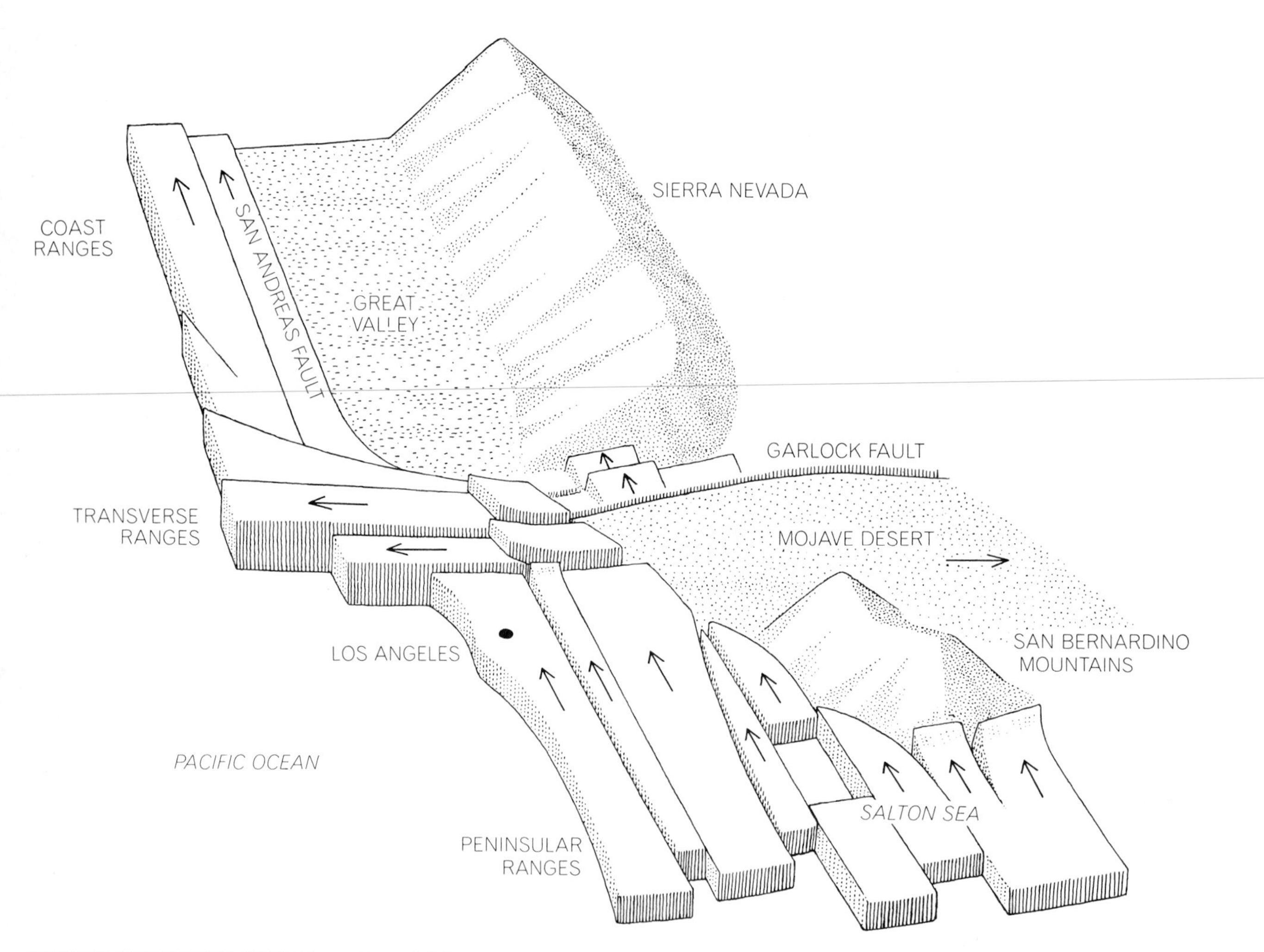

MOTION OF EARTH'S CRUST in southern California is generally northwest except where the lower group of blocks encounters the deep roots of the Sierra Nevada. At this point the blocks are diverted to the left (west), creating the transverse ranges and a big bend in the San Andreas fault system. Above the bend the blocks continue their northwesterly march, carrying the Coast Ranges with them. The Salton Sea trough at the lower right evidently represents a rift that has developed between two blocks.

Erosion destroys this kind of evidence very quickly. By matching up distinctive rock units that have been broken up and moved with respect to each other it is possible to document offsets of tens to hundreds of miles. A sedimentary basin often holds debris that could not possibly have been derived from any of the local mountains; matching up these basins with the appropriate source region on the other side of the fault can provide evidence of still larger displacements. When these various kinds of information are combined, one obtains a rate of about half an inch per year for motion on the San Andreas fault in northern and central California over the past several tens of millions of years.

This is much less than the 2½ inches per year that is inferred for the rate of separation of Baja California and mainland Mexico and the rate that is inferred from seismological studies in southern California. There are several possible explanations for the discrepancy. Northern and southern California may be moving at different rates; this seems unlikely since they are both attached to the same Pacific plate. On the other hand, part of the compression in the transverse ranges may result from a differential motion between the two parts of the state. Another possibility is that all of the relative motion between the North American plate and the Pacific plate is not being taken up by the San Andreas fault or even by the San Andreas fault system but extends well inland. The fracture zones of the Pacific seem to affect the geology of the continent for a distance of at least several hundred miles.

The Great Central Valley and the Sierra Nevada lie between two major fracture zones that abut the California coast: the Mendocino fracture zone and the Murray fracture zone. The transverse ranges, the Mojave Desert and the Garlock fault are all in line with the Murray fracture zone. Recent volcanism lines up with the extensions of the Clarion fracture zone and the Mendocino fracture zone. The basins and range geological province of the western U.S., a region of crustal tension and much volcanism, may represent a broad zone of deformation between the Pacific plate and the North American plate proper. Seismic activity is certainly spread over a large, diffuse region of the western U.S.

Although the subject has been quite controversial, most geologists are now willing to accept large horizontal displacements on the faults in California, particularly the San Andreas. Displacements as large as 450 miles of right-lat-

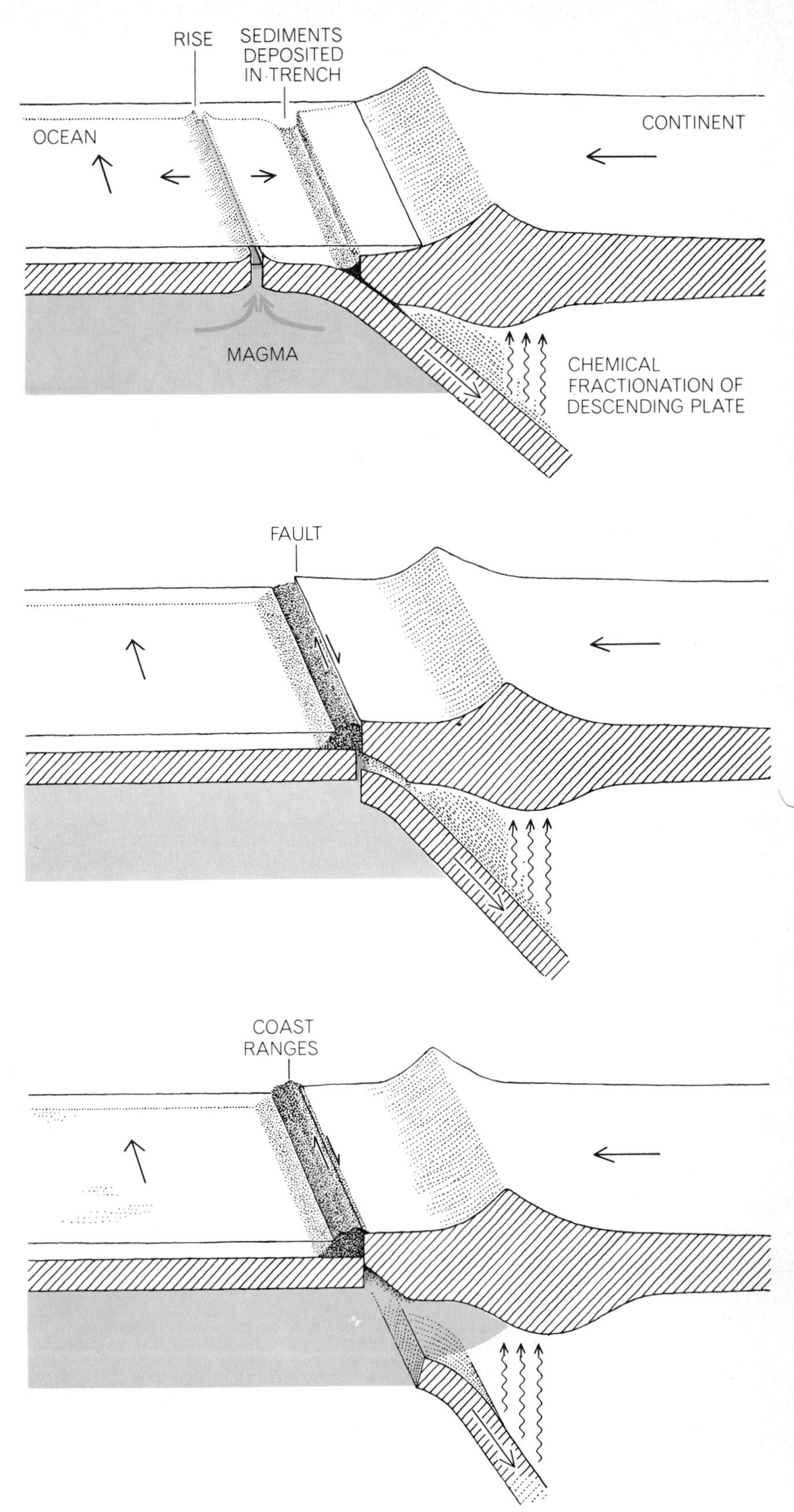

INTERACTION BETWEEN RISE AND TRENCH leads to mutual annihilation. The trench, formed as the oceanic plate dives under the continental plate, slowly fills with sediments carried by rivers and streams (*top*). Meanwhile the melting of the descending slab adds new material to the continent from below. When the axis of the rise reaches the edge of the continent, the flow of magma into the rift is cut off and trench sediments are scraped onto the western (that is, left) part of the oceanic plate (*middle*). The descending plate disappears under the continent and the sediments travel with the oceanic plate (*bottom*). The northern part of the San Andreas fault may have been formed in this way.

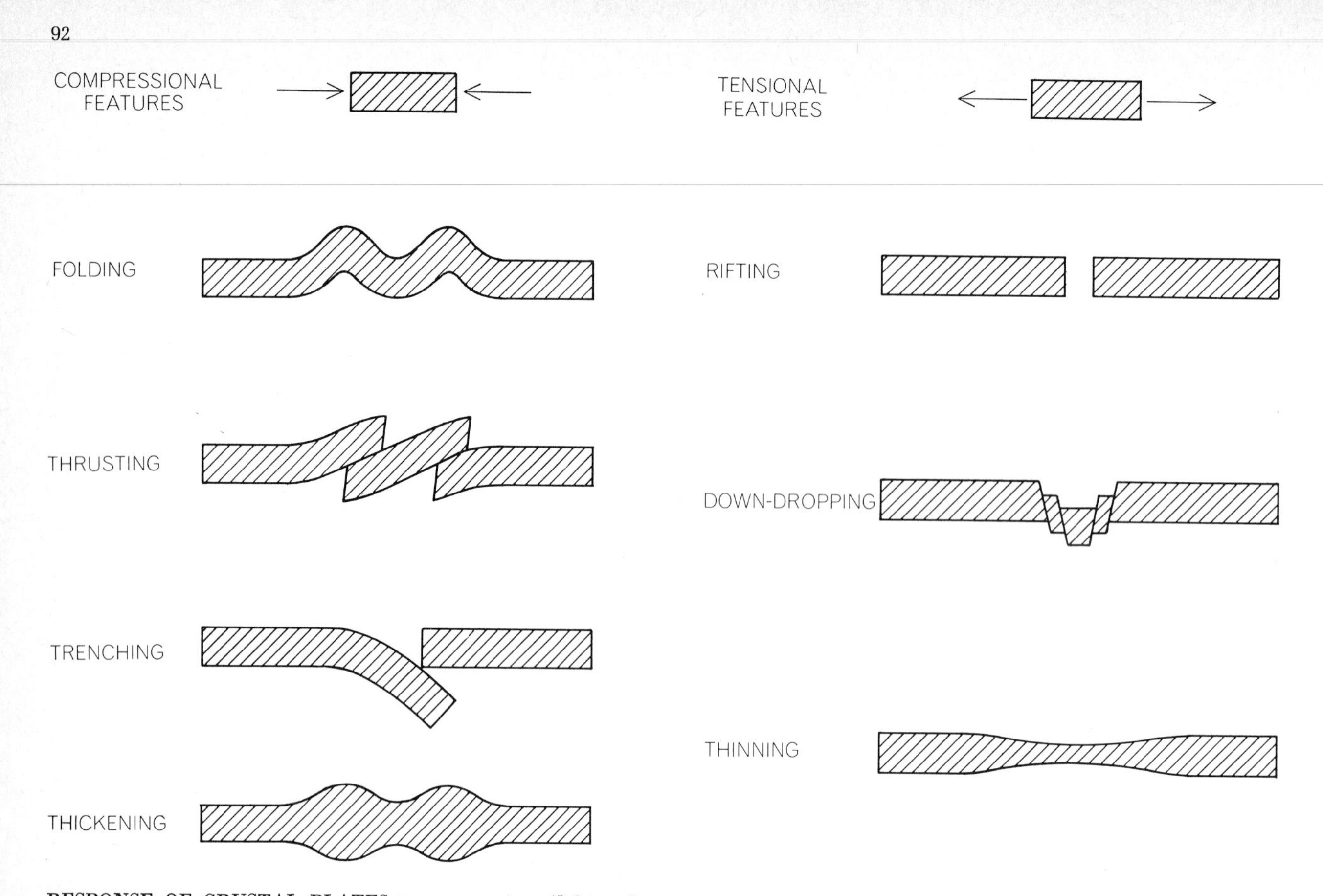

RESPONSE OF CRUSTAL PLATES to compression (*left*) and tension (*right*) accounts for most geologic features. According to the recently developed concept of plate tectonics, the earth's mantle is covered by huge, rigid plates that can be colliding, sliding past one another or rifting apart. The rifting usually occurs in the ocean floor. The San Andreas fault marks the location where two plates are sliding past each other. Plate tectonics helps to explain how the continents have drifted into their present locations.

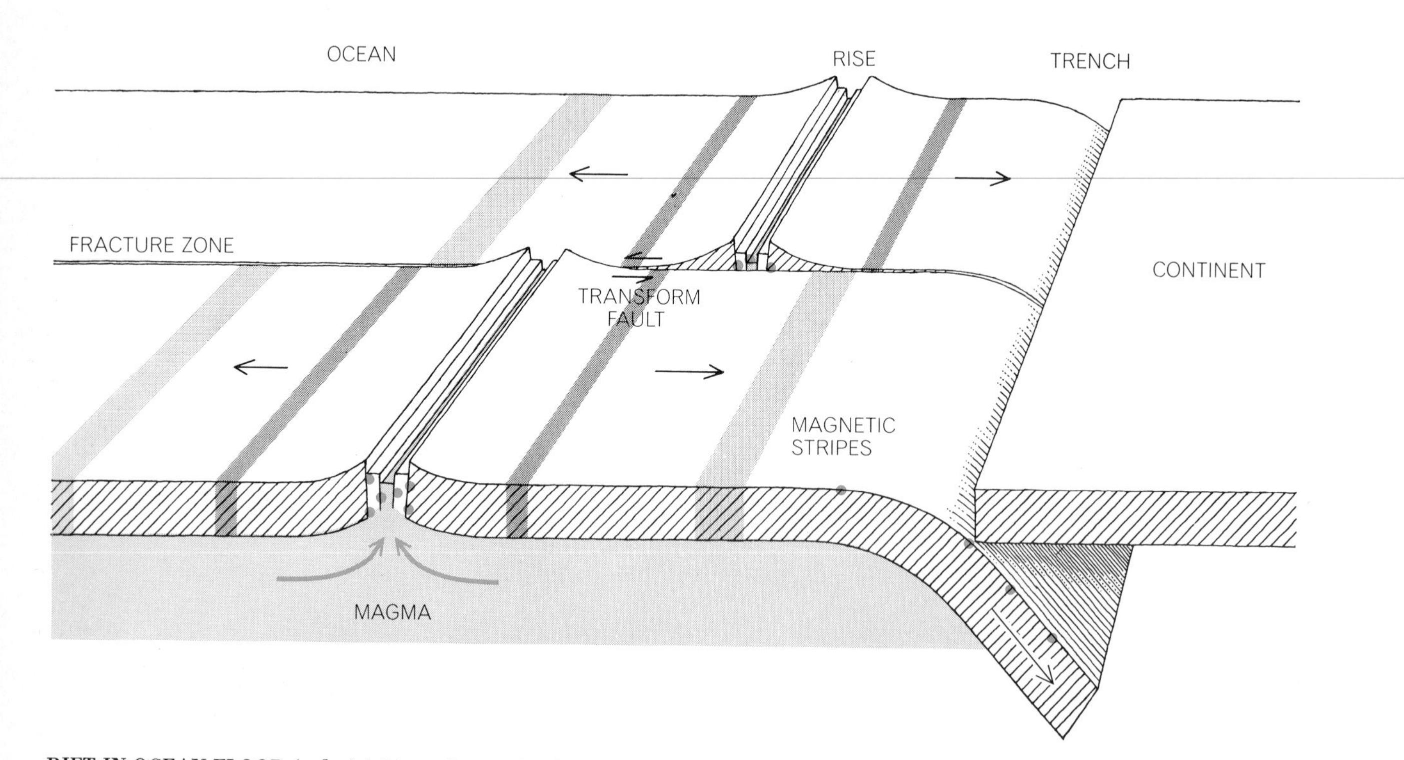

RIFT IN OCEAN FLOOR (*color*) initiates three major features of oceanic plate tectonics. The rift is bordered by a rise or ridge created by magma pushed up from the mantle below. The magma solidifies with a magnetic polarity corresponding to that of the earth. When, at long intervals, the earth's polarity reverses, the polarity of newly formed crust reverses too, resulting in a sequence of magnetic "stripes." A trench results when an oceanic plate meets a continental plate. A fracture zone and transform fault result when two plates move past each other. Earthquakes (*dots*) accompany these tectonic processes. The earthquakes in the vicinity of a rise and along a transform fault are shallow. Deep-focus earthquakes occur where a diving oceanic plate forms a trench.

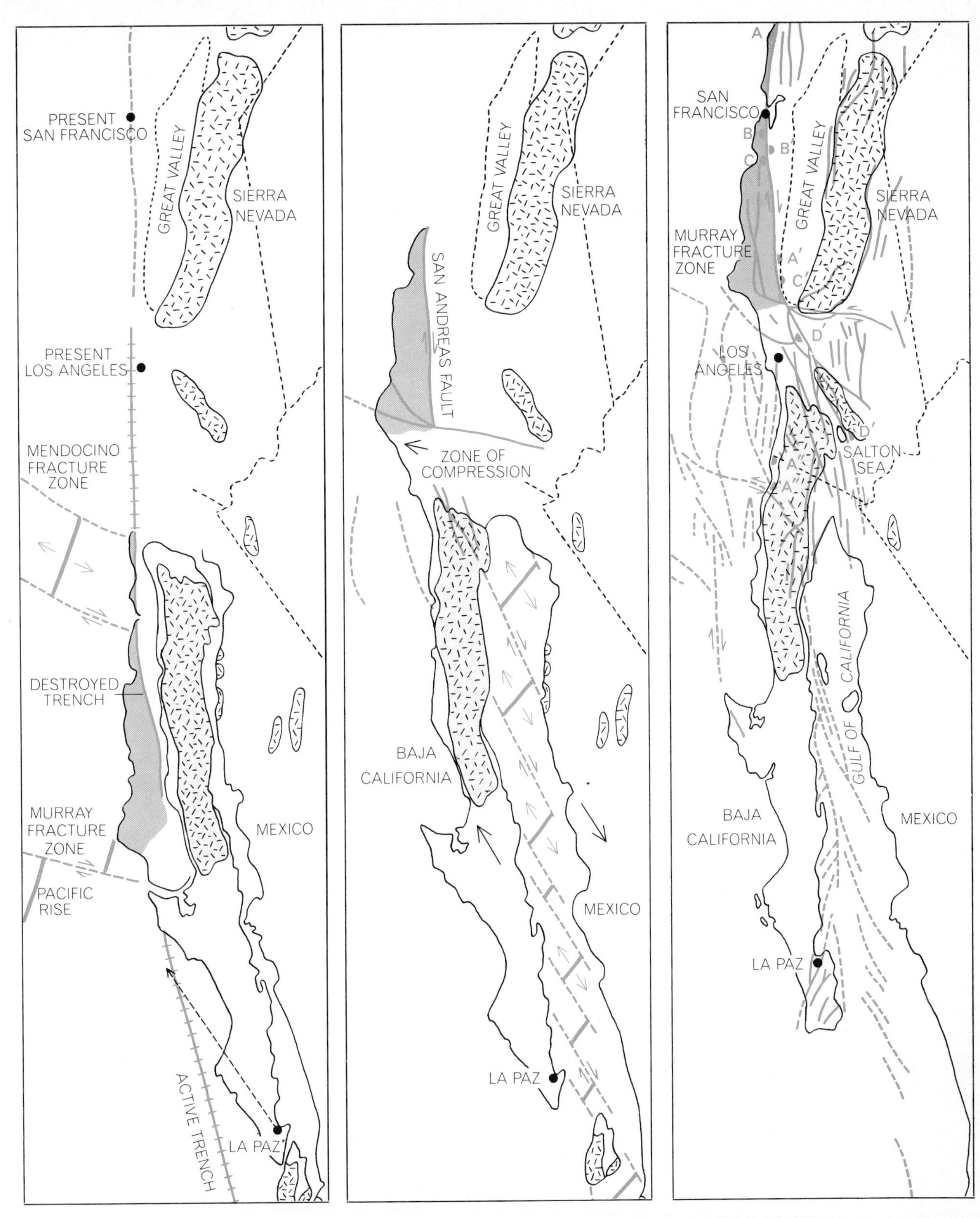

EARLY AND LATE STAGES in the history of the San Andreas fault are depicted. Twenty-five million years ago (*left*) Baja California presumably nestled against mainland Mexico. The first section of oceanic rise between the Murray fracture zone and the Pioneer fracture zone has just collided with the continent. Trench deposits are uplifted and become part of the Coast Ranges of California. The block containing the present San Francisco area (*dark color*) is about to start its long northward journey. A block immediately to the east (*light color*) becomes attached to the Pacific plate and eventually is jammed against the San Bernardino Mountains. Three million years ago (*middle*) the Gulf of California has started to open. As the peninsula moves away from mainland Mexico a series of rifts appear, fill with magma and are offset by numerous fractures. Baja California may have been torn off in one piece or in slivers. Southwest California and Baja California today (*right*) continue to slide northwest against the North American mainland. The illustration shows major fault systems and offshore fracture zones. On the basis of unique rock formations geologists infer that the Los Angeles area has moved northwest about 130 miles (*D′* to *D*) in the past 20 million years or less. Other studies indicate that the Palo Alto region has been carried about 200 miles (*C′* to *C*). Coastal rocks to the north of San Francisco have been displaced at least 300 miles (*A′* to *A*) and perhaps as much as 650 miles (*A″* to *A*) in the past 30 million years.

eral slip have been proposed for the northern segment of the fault. Displacements on the southern San Andreas fault are put at no more than 300 miles. This discrepancy has been puzzling to geologists. My own conclusion is that the part of northern and central California west of the San Andreas fault has moved northwest more than 700 miles and that the southern San Andreas fault has slipped about 300 miles, which makes the apparent discrepancy even worse. The discrepancy disappears if one drops the concept of a single San Andreas fault and admits the possibility that the two segments of the fault were initiated at different times.

The two-fault hypothesis is supported by straightforward extrapolation of the record on the ocean floor. The two San Andreas faults formed at different times, in different ways and may be moving at different rates. The record indicates that the western part of North America caught up with a section of the East Pacific rise somewhere between 25 million and 30 million years ago. Before the collision a deep oceanic trench existed off the coast such as now exists farther to the south off Central America and South America. The trench had existed for many millions of years, receiving sediments from the continent; subsequently the sediments were carried down into the mantle by the descending oceanic plate, which was diving under the continent. Based on what we know of trench areas that are active today one can assume that the plate sank to 700 kilometers and that the process was accompanied by earthquakes with shallow, intermediate and deep foci.

The Origin of Continents

Let us examine a little more closely what happens when an oceanic rise, the source of new oceanic crust, approaches a trench, which acts as a sink, or consumer of crust. Evidently the rise and the trench annihilate each other. The oceanic crust and its load of continental debris, which was formerly in the trench, rise because the crust is no longer connected to the plate that was plunging under the continent. The trench deposits are so thick they eventually rise above sea level and become part of the continent. The deposits are still attached to the oceanic plate, however, and travel with it [*see illustration on page 91*].

In the case of the Pacific plate off California the deposits move northwest with respect to the continent. This is the stuff of coastal California north of Santa Barbara, particularly the Coast Ranges. According to this view, the northern segment of the San Andreas fault was born at the same time as northern California. The rise and the trench initially interacted near San Francisco, which then was near Ensenada in Baja California. Ensenada in turn was near the northern end of the Gulf of California, which was then closed.

The tectonics and geologic history of California, and in fact much of the western U.S., are now beginning to be understood in terms of the new ideas developed in the theories of sea-floor spreading, continental drift and plate tectonics. Many of the basic concepts were laid down by the late Harry H. Hess of Princeton and Robert S. Dietz of the Environmental Science Services Administration. Tanya Atwater of the University of California at San Diego and Warren Hamilton of the U.S. Geological Survey and their colleagues have made particularly important contributions by applying the concepts of plate tectonics to continental geology. We now know that the outer layer of the earth is immensely mobile. This layer, the lithosphere, is relatively cold and

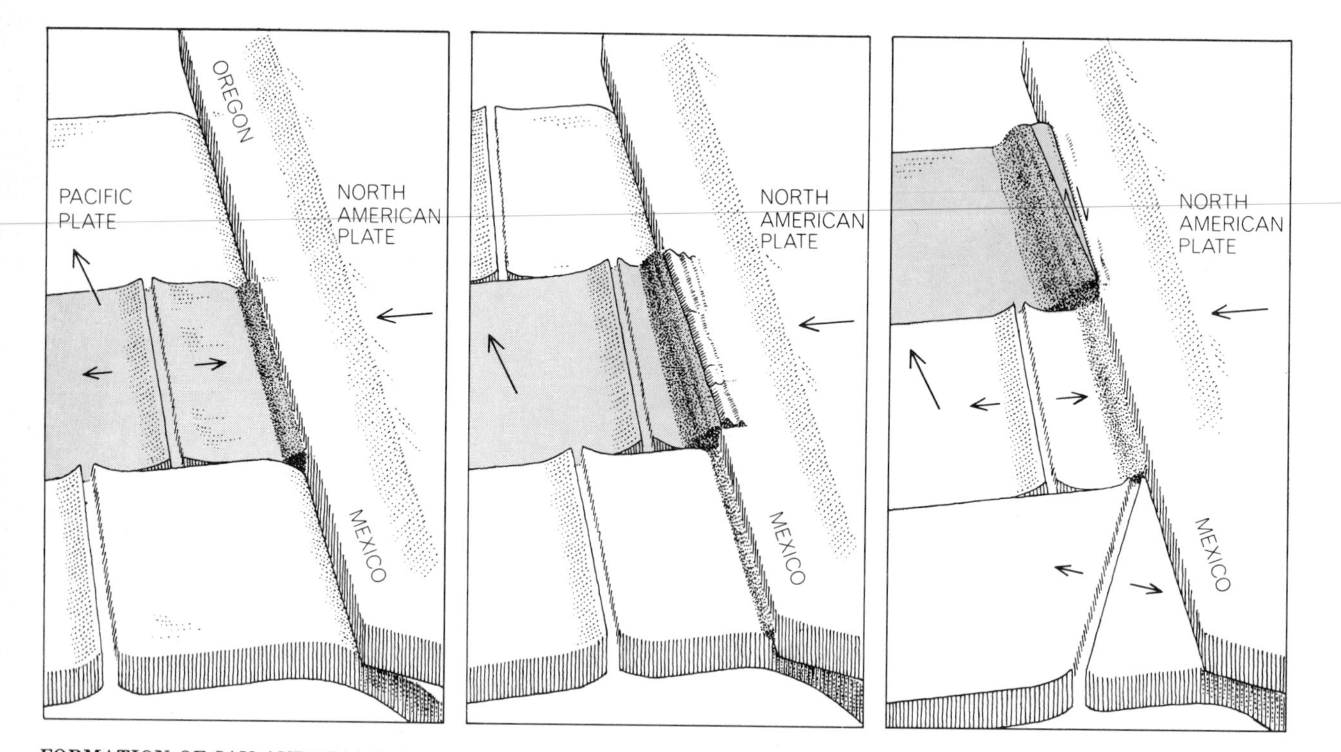

FORMATION OF SAN ANDREAS FAULT SYSTEM is depicted schematically in the six diagrams on these two pages. Some 30 million years ago (*left*) an oceanic rise system lay off the west coast of North America, which was carried by a plate moving toward the rise crests. The continental plate overrides the Pacific plate, producing a long trench. Meanwhile the entire Pacific plate is moving northwest. After a few million years (*right*) the rise nearest the continent is shut off. The trench by now has been filled with material eroded from the continent. These deposits will later become the California Coast Ranges.

NORTHERN SECTION of San Andreas fault is created when the former trench deposits become attached to the northward-moving Pacific plate (*left*). The San Andreas fault lies between the two opposed arrows indicating relative plate motions. Meanwhile to the south a tilted rise crest

rigid and slides around with little resistance on the hot, partially molten asthenosphere.

Where the crust is thick, as it is in continental regions, the temperatures become high enough in the crust itself to cause certain types of crustal rocks to lose their strength and to offer little resistance to sliding. There is thus the possibility that the upper crust can slide over the lower crust and that the moving plate can be much thinner than is commonly assumed in plate-tectonic theory. The molten fraction of the asthenosphere, called magma, rises to the surface at zones of tension such as the mid-oceanic rifts to freeze and form new oceanic crust. The new crust is exposed to the same tensional forces (presumably gravitational) that caused the rift in the first place; therefore it rifts in turn and subsequently slides away from the axis of the rise. In addition to providing the magma for the formation of new crust, the melt in the asthenosphere serves to lubricate the boundary between the lithosphere and the asthenosphere and effectively decouples the two. The rise is one of the types of boundary that exist between lithospheric plates and is the site of small, shallow tensional earthquakes.

When two thin oceanic lithospheric plates collide, one tends to ride over the other, the bottom plate being pushed into the hot asthenosphere. The boundary becomes a trench. When the lower plate starts to melt, it yields a low-density magma that rises to become part of the upper plate; this magma becomes the rock andesite, which builds an island arc on what is to become the landward, or continental, side of the trench. (The rock takes its name from the Andes of South America. Mount Shasta in California is primarily andesite, as are the island arcs behind the trenches that surround the Pacific.) The thickness of the crust is essentially doubled as a result of the underthrusting. The material remaining in the lower plate is now denser than the surrounding material in the asthenosphere, both because it has lost a low-density fraction and because it is colder; thus it sinks farther into the mantle. In some parts of the world the downgoing slab can be tracked by seismic means to 700 kilometers, where it seems to bottom out. By this process new light material is added to the crust and new dense material is added to the lower mantle. A large part of what comes up stays up; a large part of what goes down stays down.

The introduction of chemical fractionation and a mechanism for "unmixing" makes the process different from the one customarily visualized, in which gigantic convection cells carry essentially the same material in a continuous cycle. The new process is able to explain in a convincing way how continents are formed and thickened. As the continent thickens and rises higher because of buoyancy, erosional forces become more effective and dump large volumes of continental sediments into the coastal trenches. A portion of the sediments is ultimately dragged under the continent to melt and form granite. The light granitic magma rises to form huge granitic batholiths such as the Sierra Nevada. A batholith is a large mass of granitic rock formed when magma cools slowly at great depth in the earth's crust. It is carried to the surface by uplifting forces and exposed by erosion.

The concept of rigid plates moving around on the earth's surface and interacting at their boundaries has been remarkably successful in explaining the evolution of oceanic geology and tectonics. The oceanic plates seem to behave rather simply. Tension results in a rise, compression results in a trench and lateral motion results in a transform fault

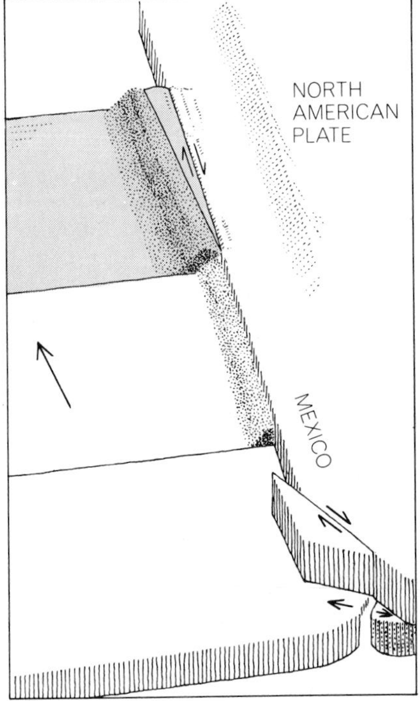

(not yet visible in the first pair of diagrams) is ready to encounter the continent end on at a break in the coastline south of Baja California. The collision (*right*) breaks off a part of the Baja California peninsula, which becomes attached to the Pacific plate and starts its journey to the northwest.

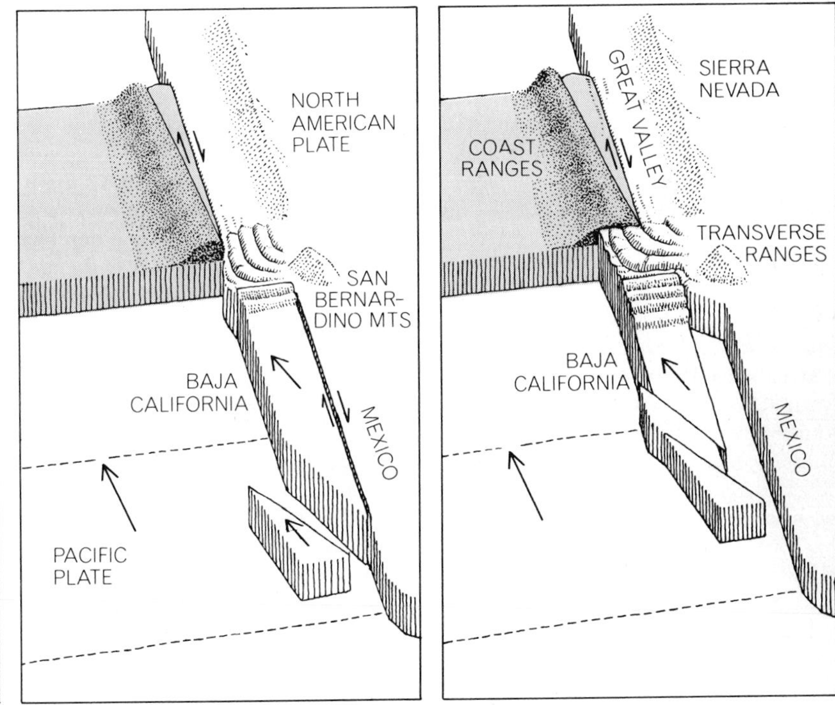

SOUTHERN SECTION of San Andreas fault is now fully activated (*left*) as the Baja California block begins sliding past the North American plate and collides with deeply rooted structures to the north, the Sierra Nevada and San Bernardino Mountains, which deflect the block to the west. More of Baja California breaks loose, opening up the Gulf of California. As Baja California continues to move northwestward (*right*) the Gulf of California steadily widens. The compression at the north end of the Baja California block creates the transverse ranges, which extend inland from the vicinity of present-day Santa Barbara.

and a fracture zone [*see bottom illustration on page 92*]. The interaction of oceanic and continental plates or of two continental plates is apparently much more complicated, and this is one reason the new concepts were developed by study of the ocean bottom rather than continental geology.

The boundary between two oceanic plates can be a deep oceanic trench, an oceanic rise or a strike-slip fault depending on whether the plates are approaching, receding from or moving past each other. The forces involved are respectively compressional, tensional and shearing. When a thick continental plate is involved, compression can also result in high upthrust and folded mountain ranges. The Himalayas resulted from the collision of the subcontinent of India with Asia. I shall show that the transverse ranges in California were formed in a similar way. Tension can result in a wide zone of crustal thinning, normal faulting and volcanism; it can also create a fairly narrow rift of the kind found in the Gulf of California and the Red Sea [*see top illustration on page 92*].

The interaction of western North America with the Pacific plate has led to large horizontal motions along the San Andreas fault, to concentrated rifting as in the Salton Sea trough and the Gulf of California, to diffuse rifting and normal faulting accompanied by volcanism in the basins and range province of California, Nevada, Utah and Arizona, to large vertical uplift by overthrusting as in the transverse ranges north and west of Los Angeles, to the generation of large batholiths such as the Sierra Nevada and to the incorporation of deep-sea trench material on the edge of the continent. Ultimately the geology, tectonics and seismicity of California can be related to the collision of North America with the Pacific plate.

Most of the Pacific Ocean is bounded by trenches and island arcs. Trenches border Japan, Alaska, Central America, South America and New Zealand. Island arcs are represented by the Aleutians, the Kuriles, the Marianas, New Guinea, the Tongas and Fiji. The arcs are themselves bordered by trenches. All these areas are characterized by andesitic volcanism and deep-focus earthquakes. Western North America is lacking a trench and has only shallow earthquakes, but the geology indicates that there was once a trench off the West Coast, and in fact there was once a rise. The present absence of a rise and a trench, the absence of deep-focus earthquakes and the existence of uplifted deep-sea sediments are all related.

Tracing back the history of the interaction of the Pacific plate with the North American plate, one is forced to conclude that the northern part and the southern part of the San Andreas fault originated at different times and in different ways. The northern part was evidently formed about 30 million years ago when a portion of rise between the Mendocino fracture zone and the Murray fracture zone approached an offshore trench bordering the southern part of North America. At that time the west coast of North America resembled the present Pacific coast of South America: there was a deep trench offshore, high mountains paralleled the coastline and large underthrusting earthquakes were associated with the downgoing lithosphere.

Origin of the Fault

As the rise approached the continent both the geometry and the dynamics of interaction changed [*see illustrations on pages 94 and 95*]. Depending on the spreading rate of the new crust generated at the rise and the rate at which the rise itself approaches the continent, the relative motion between the rise and the continental plate will decrease, stop or reverse when the rise hits the trench. The forces keeping the trench in existence will therefore decrease, stop or reverse, leading to uplift of the sedimentary material that has been deposited in the trench. In classical geologic terms these are known as eugeosynclinal deposits. Although they have been exposed to only moderate temperatures, they have been subjected to great pressures, both hydrostatic (owing to their depth of burial) and directional (owing to the horizontal compressive forces between the impinging plates). Eugeosynclinal sediments are therefore strongly deformed and become even more so as they are contorted and sheared during uplift. Much of the western edge of California and Baja California is underlain by this material, called the Franciscan formation. The formation is physically attached to the Pacific plate and is therefore moving northwest with respect to the rest of North America. The present boundary is the northern part of the San Andreas fault. Today this section of the San Andreas system extends from Cape Mendocino, north of San Francisco, to somewhere south and east of Santa Barbara, near the beginning of the great bend of the San Andreas fault, where the San Andreas and the Garlock faults intersect.

Meanwhile, 30 million years ago, another section of the rise south of the Murray fracture zone was still offshore, together with an active trench. Baja California was still attached to the mainland of Mexico and the Gulf of California had not yet opened up. The southern part of the San Andreas fault had not yet been formed.

The abrupt change in the direction of the coastline south of the tip of Baja California suggests that here the rise approached the continent more end on than broadside. A sliver of existing continent was welded onto the Pacific plate and rifted away from Mexico, thus forming Baja California and the Gulf of California. Thereafter Baja California participated in the northwesterly motion of the Pacific plate, with the result that the Gulf of California widened progressively with time.

About five million years were required for northern California, which had broken off from Baja California, to be carried about 200 miles to the northwest. At the end of that time the Gulf of California and the Salton Sea trough had not yet opened. The faults that delineate the major geologic blocks in southern California had not yet been activated. The block bearing the San Gabriel fault, now north of San Fernando, occupied the future Salton Sea trough. The transverse ranges will eventually be formed from the Santa Barbara, San Gabriel and San Bernardino blocks by strong compression from the south when Baja California breaks loose from mainland Mexico. This also opens up the Gulf of California and the Salton Sea trough.

As northern California is being carried away from Baja California by the Pacific plate another segment of oceanic rise south of the Murray fracture zone approaches the southern half of Baja California, where the situation described above is repeated except that the rise crest encounters a sharp bend in the coastline and the trench hits just south of the tip of the peninsula. Now instead of approaching the continent more or less broadside the rise approaches the continent end on. Mainland Mexico is still decoupled from that part of the Pacific plate to the west of the rise by the rise and the trench. Baja California, however, is now coupled to the northwestward-moving Pacific plate and Baja California is torn away from the mainland. This happened between four and six million years ago. Magma from the upper mantle wells up into the rift, forming a new rise that works its way north into the widening gulf. Alternatively, the entire peninsula of Baja California could have broken off from the mainland at the same time. As the peninsula, including parts of southern California, moves north it collides with parts of the continent that are still attached to the main North American plate. This results in compression, overthrusting and shearing and the eventual formation of the transverse ranges.

The southern part of the San Andreas fault system was therefore formed by the rifting off of a piece of continent. Today it represents the boundary between two parts of the continental plate that are moving with respect to each other. This part of the San Andreas fault was formed well east, or inland, of the southward projection of the northern San Andreas.

The northerly march of southern California and Baja California seems to have been blocked when the moving plate encountered the thick continental crust to the north, particularly the massive granitic San Bernardino mountain range, which includes the 11,485-foot San Gorgonio Mountain. Since large and high mountain ranges have deep roots, the crust in this region is probably much thicker than normal, perhaps as thick as 50 kilometers. Earthquakes in this region are all shallower than 20 kilometers, which may be the thickness of the sliding plates. The blocks veer westward and are strongly overthrust as they attempt to get around the obstacle; this movement generates the big bend in the San Andreas fault system. The deflected blocks eventually join up with the northern California block.

Earthquake Country

From a social and economic point of view earthquakes are one of the most important manifestations of plate interaction. From a scientific point of view they supply a third dimension to the study of faults and the nature of the interactions between crustal blocks, including the stresses involved and the nature of the motions.

Seismologists at the University of California at Berkeley and at the Cal Tech Seismological Laboratory have been keeping track of earthquake activity in California for more than 40 years. Both groups have installed arrays of seismometers that telemeter seismic data to their laboratories for processing and dissemination to the appropriate public agencies. During the 36-year period 1934 through 1969 there were more than 7,300 earthquakes with a Richter magnitude of 4 or greater in southern California and adjacent regions [*see illustration on page 99*]. Many thousands more earthquakes of smaller magnitude are

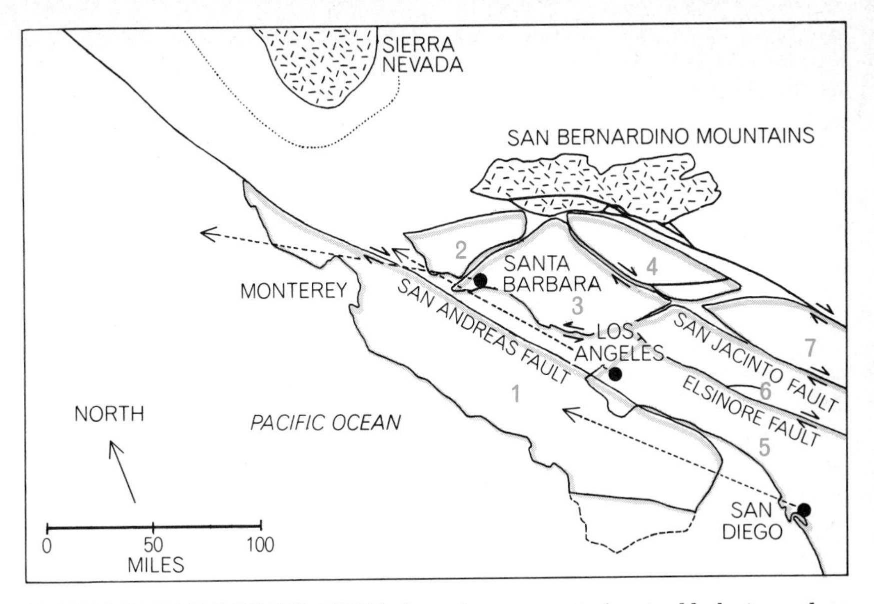

SEQUENCE OF SIMPLIFIED VIEWS shows the movement of major blocks in southern California over the past 12 million years. In the first view (*above*) the Gulf of California has not yet appreciably opened but the block carrying the Coast Ranges (*1*) has started to move rapidly northwest with activation of northern portion of San Andreas fault. Dots show origin and arrows show displacement of San Diego, Los Angeles and Santa Barbara.

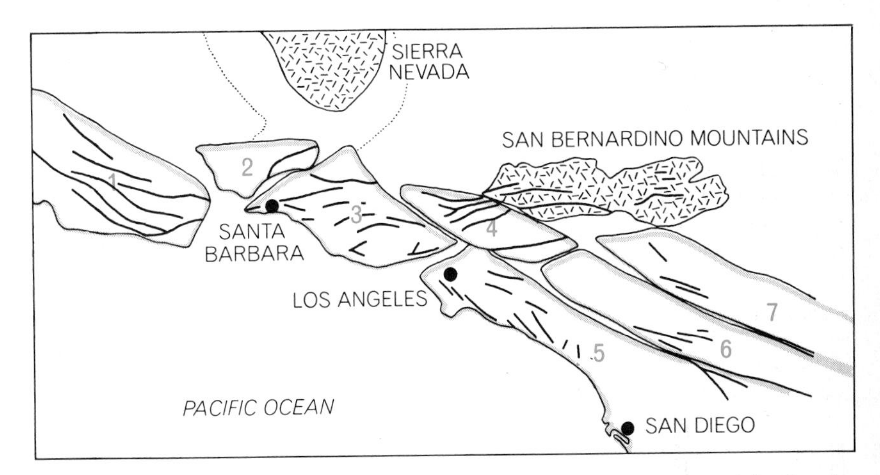

TWO MILLION YEARS AFTER ACTIVATION of the southern portion of the San Andreas fault four blocks (*2, 3, 4, 7*) have been forced against the deep roots of the Sierra Nevada and San Bernardino Mountains. Compressive forces create the transverse ranges. Meanwhile the block carrying the Coast Ranges (*1*) has been carried far to the northwest.

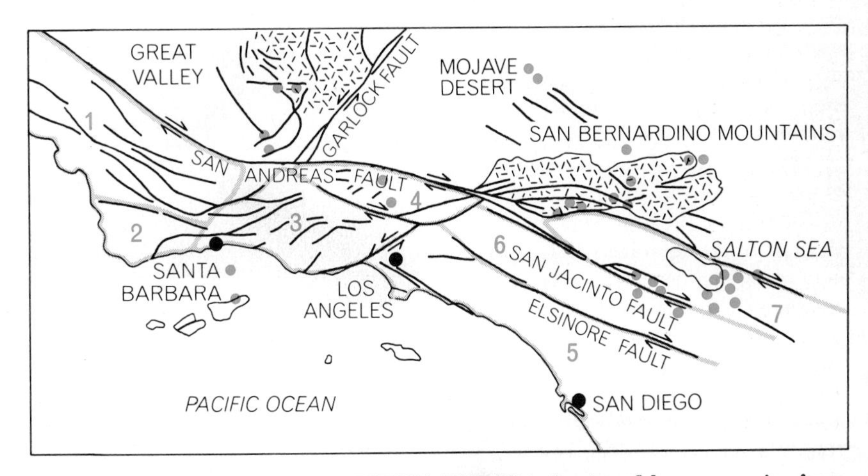

GEOLOGY OF SOUTHERN CALIFORNIA TODAY is dominated by compressive forces operating in the big bend of the San Andreas fault, which connects the southern and northern parts of the system. Colored dots show the location of earthquakes in the recent past.

routinely located and reported in the seismological bulletins. Although damage depends on local geological conditions and the nature of the earthquake, a rough rule of thumb is that a nearby earthquake of magnitude 3.5 or greater can cause structural damage. The average annual number of earthquakes of magnitude 3 or greater in southern California recorded since 1934 is 210; the number in any one year has varied from a low of 97 to a high of 391. The strongest earthquake in this period was the Kern County event of magnitude 7.7 in 1952. The aftershocks of that event increased the total number of events for several years thereafter.

In general the larger the earthquake, the greater the displacement across a fault and the greater the length of fault that breaks. The great earthquakes of 1906 and 1857 respectively caused large displacements across the northern and central parts of the San Andreas fault and relieved the accumulated strain in these areas. The accumulation of strain in southern California is relieved mainly by slip on a series of parallel faults and by overthrusting on faults at an angle to the main San Andreas system; that is what happened in the Kern County and San Fernando earthquakes. The unique east-west-trending transverse ranges were formed in this way. In the process deep-seated ancient rocks were uplifted and exposed by erosion.

Another seismically active area associated with major faults is south of the Mojave Desert near San Bernardino, where the faults show a sudden change in direction. The central part of the Mojave Desert is also moderately active. This is consistent with the idea that the sliding lithosphere is diverted by the San Bernardino Mountains. Faults and evidence of relatively recent volcanic events abound in the area. The northern part of Baja California is also quite active. An interesting feature of seismicity maps of southern California is the alignment of earthquakes in zones that trend roughly northeast-southwest, approximately at right angles to the major trend of the San Andreas system.

The map on the next page shows that the San Andreas fault itself has played only a small role in the seismicity of southern California over the past 30-odd years. One must not forget, however, that the great earthquake of 1857 probably broke the San Andreas fault for about 100 miles northwest and southeast from the epicenter. That epicenter is thought to have been near Fort Tejon, which is close to the projected intersection of the Garlock and San Andreas

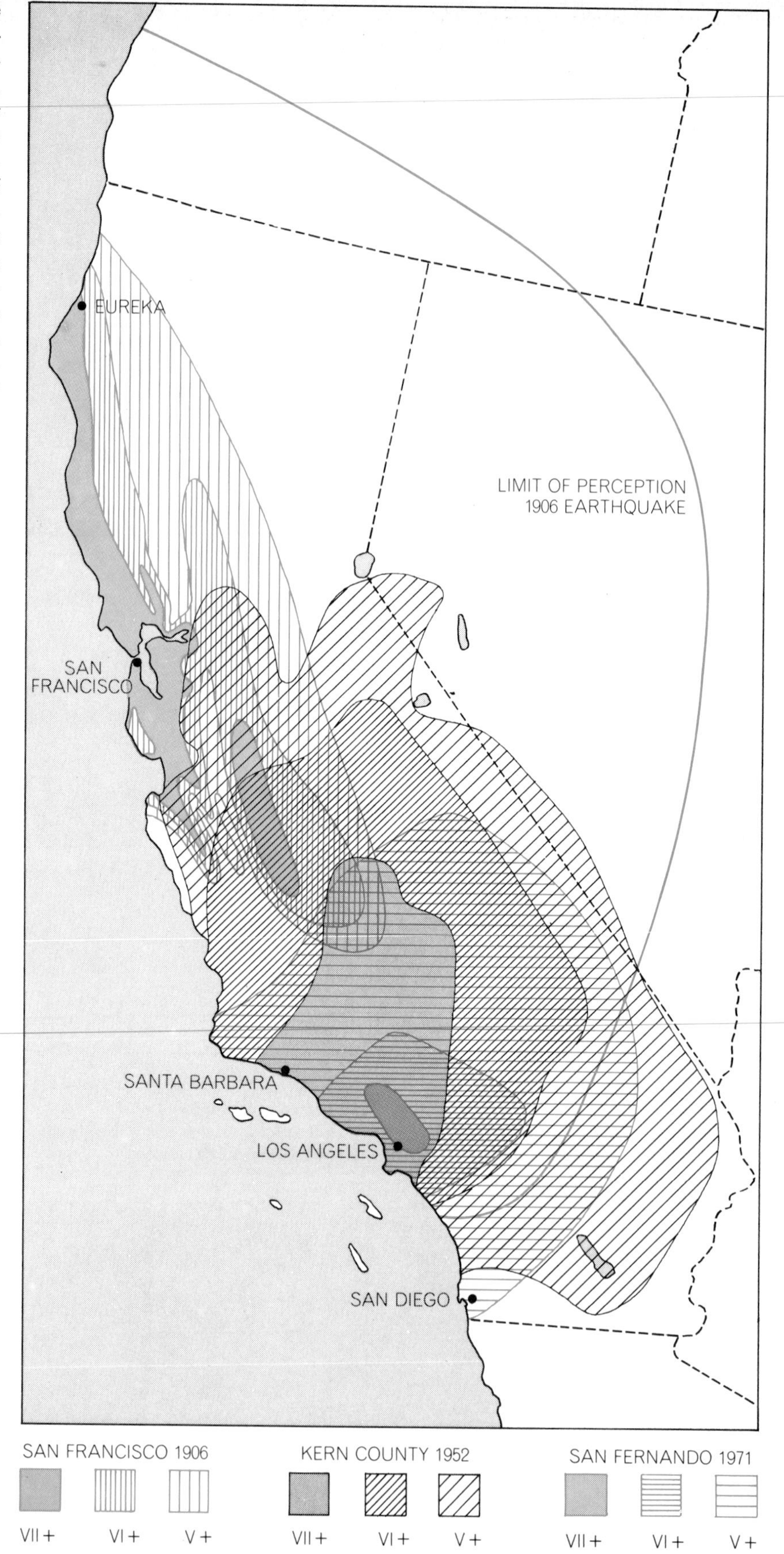

ISOSEISMAL CONTOUR MAP shows the pattern and intensity of ground-shaking produced by the 1906 San Francisco earthquake of magnitude 8.3, the 1952 Kern County earthquake of magnitude 7.7 and the 1971 San Fernando earthquake of magnitude 6.6. The Roman numerals indicate levels of perceived intensities as defined by the modified Mercalli scale. A short description of each level in the scale appears in the text on page 100.

faults; the actual location of the epicenter is uncertain by hundreds of miles because there were no seismic instruments in those days. Since that time this part of southern California has been remarkably quiet and seems to be locked, generating neither earthquakes nor creep. Activity along the San Andreas fault picks up near Coalinga, which is about midway between Bakersfield and San Francisco. Alignments of earthquakes are apparent along the San Jacinto and Imperial faults in the Salton Sea trough near the Mexican border. Although these faults lie west of the main San Andreas fault, they are part of the San Andreas system. The White Wolf fault, which is northwest of and parallel to the Garlock fault, has also been quite active, particularly after the Kern County earthquake, which occurred on this fault. The White Wolf fault lines up with the Santa Barbara Channel area, which has similarly been quite active.

One way to quantify the seismicity of southern California is to count the number of earthquakes per year per 1,000 square kilometers and compare this figure for the world as a whole. For example, southern California averages one earthquake of magnitude 3 or greater per year per 1,000 square kilometers. Thus within the entire region there are about 200 such earthquakes per year. The rate for earthquakes of magnitude 6.6, the size of the San Fernando earthquake, is about one every five or six years. The actual rates, however, vary considerably from year to year and depend somewhat on the time interval of the sample. The number of earthquakes decreases rapidly with size, and the average recurrence interval is not well established for the larger earthquakes. Southern California is about 10 times more active seismically than the world as a whole, which is simply to say that California is earthquake country.

Although certain areas in southern California are relatively free of earthquakes, none is immune from their effects. One of the largest quiet areas is the western part of the Mojave Desert wedge. This is surprising because the region is bounded on the northwest and southwest by areas that are obviously under large compression, as is shown by the upthrust mountains in the transverse and Tehachapi ranges and the large overthrust earthquakes that occurred in Kern County and San Fernando. It appears that the region is being protected from the northwesterly march of the southern California–Baja California block by the San Bernardino batholith and may represent a stagnation area in the lee of the mountains. Only a small number of earthquakes are centered near San Diego, although the larger earthquakes in northern Baja California and in the mountains between San Diego and the Salton Sea are felt in San Diego. The Great Central Valley north of Bakersfield and the eastern part of the Sierra Nevada are fairly inactive, as is a large area north of Santa Barbara in the Coast Ranges.

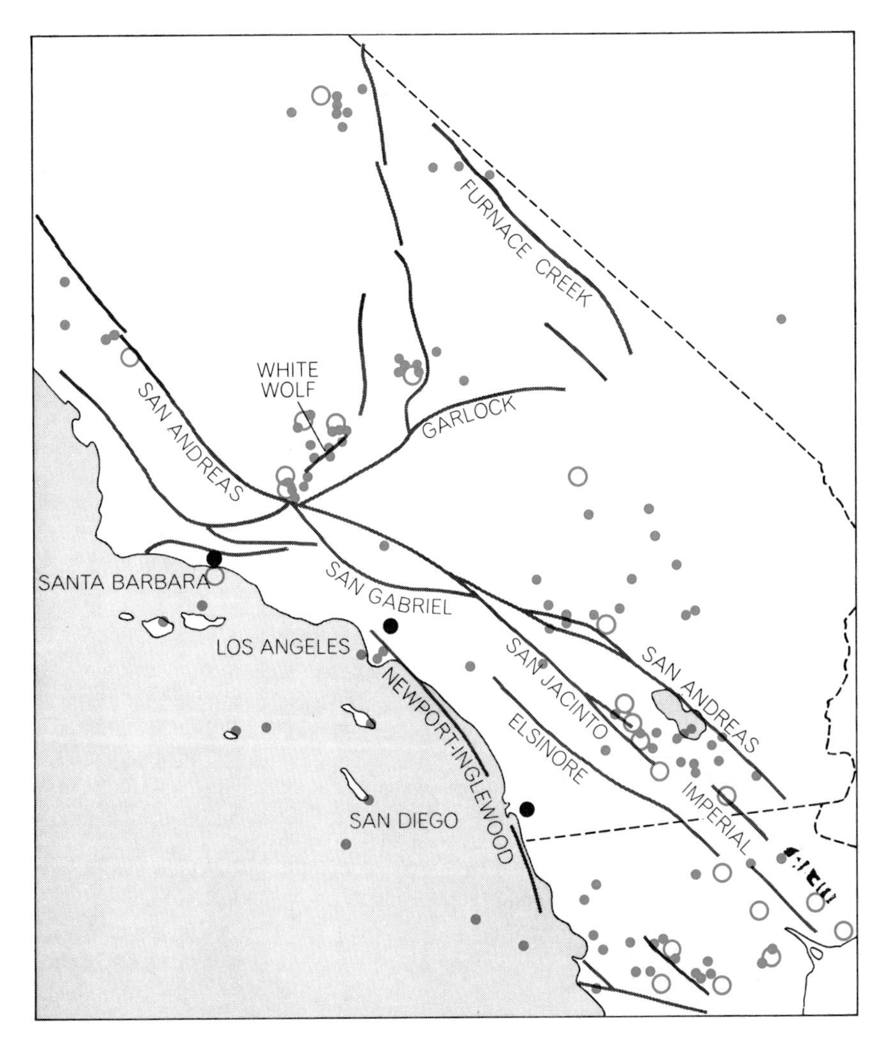

THIRTY-SIX-YEAR EARTHQUAKE RECORD shows the epicenters of all events of magnitude 5 or greater recorded in southern California and in the northern part of Baja California from 1934 through 1969. The epicenter is the point on the earth's surface above the initial break. Dots show earthquakes between 5 and 5.9 in magnitude. Open circles indicate earthquakes of magnitude 6 or greater. The hypocenter, the point of the initial break in the earth's crust, is often many miles below the surface in thrust-type earthquakes, a type frequently observed in this region. In the 36-year period southern California and adjacent regions experienced more than 7,300 earthquakes with a magnitude of 4 or more. Earthquakes are about 10 times more frequent in this area than they are in the world as a whole.

Magnitude and Intensity

It is somewhat deceptive to plot earthquakes as small points on a map. The points represent the epicenter: the point on the surface above the initial break. Once the break is started it can continue, if the earthquake is a major one, for hundreds of miles. Earthquakes of the thrust type, which result from a failure in compression, typically first break many miles below the surface; the surface break and maximum damage can be 10 miles or more from the epicenter. The distance over which strong shocks were felt during three large California earthquakes in this century (1906, 1952 and 1971) can be represented by plotting isoseismals: lines of equal intensity [*see illustration on facing page*]. The shape of the pattern varies with the type of earthquake and with the nature of the local geology; structures on deep sedimentary basins or on uncompacted fill get a more intense shaking than structures on bedrock. The isoseismals of the

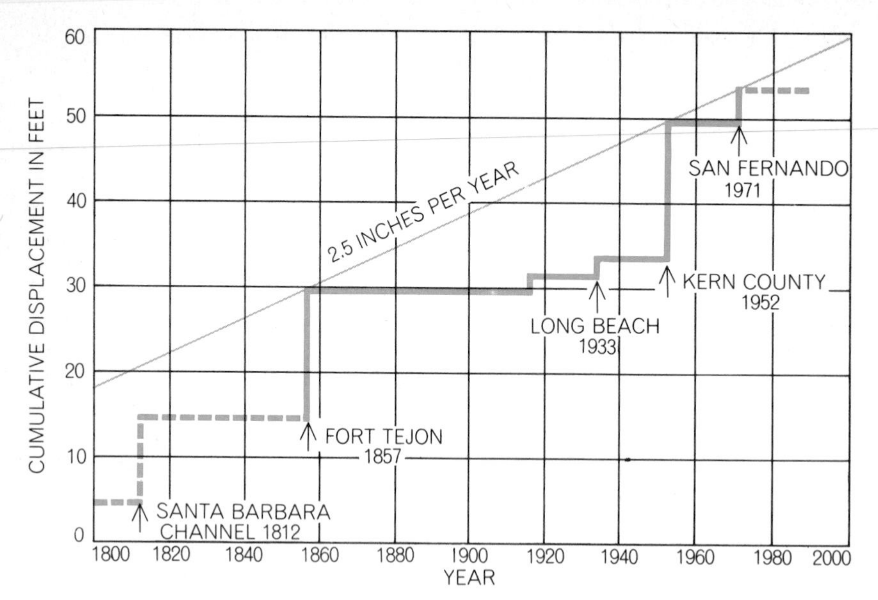

CUMULATIVE DISPLACEMENTS directly related to earthquakes indicate that southern California west of the San Andreas fault system is sliding northwestward at an average rate of 2½ inches per year. Major earthquakes relieve stresses that have built up over decades.

San Francisco earthquake are long and narrow, both because of the orientation of the fault and the length of the faulting and because of the northwest-southeast trend of the valleys. The orientation of the valleys in turn is controlled by the orientation of the San Andreas fault.

The public and the news media are confused about the various measures of the size of an earthquake. There are many parameters associated with an earthquake; they are usually regarded as fault parameters. They include the length, depth and orientation of the fault, the direction of motion, the rupture velocity, the radiated energy, the causal stresses and their orientation, the stress drop (which is related to the strength or the friction along the fault), the energy spectrum, the amount of offset or displacement and the time history of the motion. Most of these parameters can be estimated from seismic records, even from signals recorded several thousand kilometers from the earthquake. To obtain high precision, however, one needs records from many well-distributed seismic stations together with field observations at the site of the earthquake.

The magnitude on the Richter scale is a number assigned to an earthquake from instrumental readings of the amplitude of the seismic waves on a standard seismometer, the Wood-Anderson torsion seismometer. The amplitude must be suitably corrected for spreading and attenuation in the earth, and for instrumental response if a non-standard instrument is used. The magnitude is closely related to the energy of the earthquake, the single most important quantity by which earthquakes can be ranked one against another. If all the corrections are adequately made, a seismologist anywhere in the world will assign the same magnitude. In practice, because of the complicated radiation pattern of earthquakes and because of the distortion of the waves traveling through the earth, the initial magnitude assigned by various observatories may differ slightly. The magnitude scale is logarithmic and is open-ended at both ends. It is not a scale with a maximum value of 10, as is often reported in the press, and negative magnitudes are routinely measured by seismologists working on microearthquakes.

The intensity scale was developed for engineering purposes and is a qualitative measure of the intensity of ground vibration and structural damage. These qualitative assessments are assigned Roman numerals from I to XII. Unlike the magnitude of an earthquake, the intensity varies with distance and depends on the nature of the local ground. In general alluvial valleys, soft sediments and areas of uncompacted fill will magnify ground-shaking and will register higher intensities than adjacent areas on solid rock.

The intensity scale in common usage today is the Modified Mercalli Intensity Scale. The following characterizations of intensity, abridged from longer descriptions, indicate the kind of observations on which the Mercalli scale is based:

I. Not felt except by a very few under special circumstances. Birds and animals are uneasy; trees sway; doors and chandeliers may swing slowly.

II. Felt only by a few persons at rest, particularly on the upper floors of buildings.

III. Felt indoors, but many people do not recognize as an earthquake. Vibrations like the passing of light trucks. Duration of the shaking can be estimated.

IV. Windows, dishes and doors rattle. Walls make creaking sounds. Sensation like the passing of heavy trucks. Felt indoors by many, outdoors by few.

V. Felt by nearly everyone; many awakened. Small unstable objects are displaced or upset; plaster may crack.

VI. Felt by all; many are frightened and run outdoors. Some heavy furniture is moved; books are knocked off shelves and pictures off walls. Small church and school bells ring. Occasional damage to chimneys, otherwise structural damage is slight.

VII. Most people run outdoors. Difficult to stand up. Noticed by drivers of automobiles. Damage is negligible in buildings of good design and construction, slight to moderate in well-built ordinary structures, considerable in poorly built or badly designed structures. Waves on ponds and pools.

Intensity VII corresponds to the general experience within five or 10 miles of the surface faults associated with the San Fernando earthquake of last February. The following intensity levels were experienced in a small area of the northern San Fernando Valley and would be widely experienced in more severe earthquakes.

VIII. Steering of automobiles affected. Frame houses move on foundations if not bolted down; loose panel walls are thrown out. Some masonry walls fall. Chimneys twist or fall. Damage is slight in specially designed structures, great in poorly constructed buildings. Heavy furniture is overturned.

IX. General panic. Damage is considerable in specially designed structures; partial collapse of substantial buildings. Serious damage to reservoirs and underground pipes. Conspicuous cracks in the ground.

X. Most masonry and frame structures are destroyed with their foundations. Some well-built wooden structures are destroyed. Rails are bent slightly. Large landslides.

XI. Few, if any, masonry structures remain standing. Bridges are destroyed. Broad fissures in the ground. Rails are bent severely.

XII. Damage is nearly total. Objects are thrown into the air.

It is clear that the Mercalli intensity scale is people-oriented; anyone can estimate the intensity from his own experience during an earthquake. The National Oceanic and Atmospheric Administration compiles information on intensities by mailing out questionnaires to a sample of the population living in an area that has experienced a sizable earthquake.

In order to obtain more exact information about the ground motions involved in earthquakes engineers have developed strong-motion accelerometers that automatically trigger and start to record when shaken severely. Most of these instruments are installed in the seismic areas of the U.S., with a particularly heavy concentration in and around Los Angeles. The instruments are expensive and must be located very close to an earthquake to provide useful data. More than 250 of the instruments were triggered during the San Fernando earthquake, and a wealth of engineering data will be provided by these records.

A strong-motion instrument records ground acceleration as a function of time. Accelerations are commonly reported as fractions of a g, the acceleration due to gravity at the earth's surface. One g is roughly 10 meters per second per second. In designing a building to withstand moderate earthquakes, engineers are concerned chiefly with the maximum accelerations, the period or frequency of shaking and the duration of shaking. Buildings in earthquake-hazard regions with stringent building codes are usually designed to withstand at least .1 g of acceleration; this corresponds to an intensity of about VII on the Mercalli scale.

Although there is no direct correlation between intensity and magnitude, the zone of destruction increases as the magnitude increases for shallow-focus earthquakes. In general the larger the magnitude of an earthquake, the longer the fault length, the larger the displacement across the fault and the longer the duration of shaking. The longer fault length alone accounts for much of the increased area of destruction. For example, the San Francisco earthquake of 1906 had an intensity of VII or greater out to a distance of 500 miles from the epicenter, and this may not have been the largest California earthquake in historic times. The San Francisco earthquake had a magnitude of 8.3. The 1952 Kern County earthquake (magnitude 7.7) had an intensity of VII or greater out to 50 miles. The recent San Fernando earthquake (magnitude 6.6) damaged older structures out to 25 miles. An earthquake of magnitude 5.5, the Parkfield earthquake of 1966, produced comparable damage to a distance of 10 miles.

The February Earthquake

The San Fernando earthquake occurred in the San Gabriel Mountains just north of the San Fernando Valley, a densely populated northern suburb of Los Angeles. The San Gabriel Mountains are part of the structural province of the transverse ranges: the band of east-west-trending mountains, valleys and faults that is characterized by strong and geologically recent tectonic deformation. Geologists recognize that the region is one of recent crustal shortening caused by north-south compression. The mountains, produced by buckling and thrusting, are one result of this crustal shortening. They have been thrusting over the valleys to the south for at least five million years along fault planes that dip to the north or northeast.

Although many faults are known to have been active in this area in the past several thousand years, the San Fernando earthquake produced the first historic example of surface faulting. The San Gabriel Mountains rise abruptly some 5,000 feet above the San Fernando Valley and the Los Angeles basin to the south. During the earthquake of February 9 a wedge-shaped prism of the crystalline basement rock comprising the San Gabriel Mountains was thrust over the San Fernando Valley to the southwest, thereby raising the elevation of a section of the San Gabriel Mountains and sliding it slightly to the west. The displacement is consistent with the motions that have been occurring for millions of years, as one can infer from geologic offsets and uplifts. It also agrees with the general picture presented here, namely that the transverse ranges were formed by the collision of the southern and Baja California block with the central and northern California block, and with the concept that the southern California block is being diverted to the west by the massive San Bernardino batholith. One can infer that the thickening of the crust involved in the overthrusting and uplift of the San Gabriel Mountains made this region an additional obstacle to the northwesterly march of

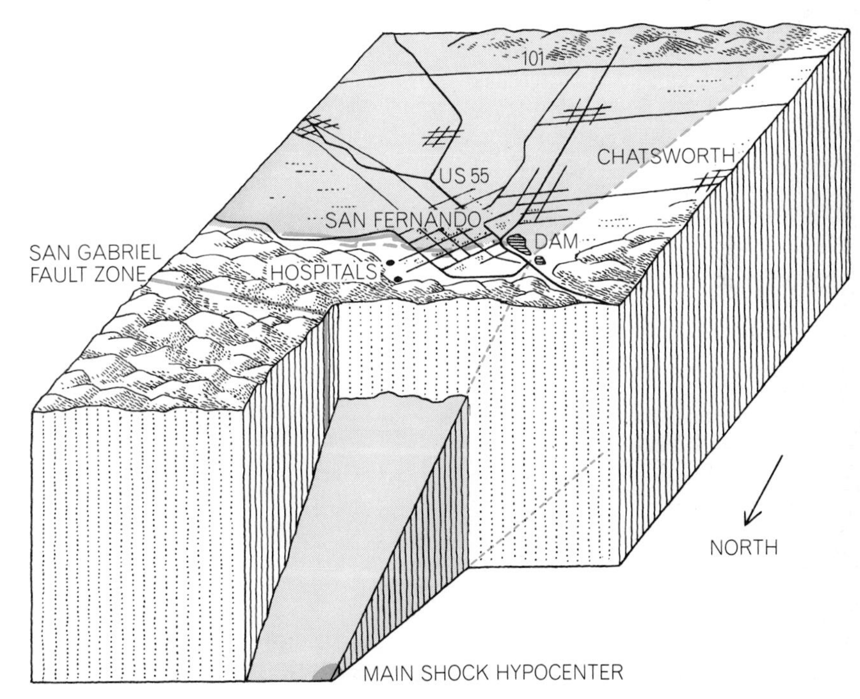

HYPOCENTER OF SAN FERNANDO EARTHQUAKE (*dark color*) **of last February was 13 kilometers deep and 12 kilometers north of the area where the principal ground-shaking occurred. The earthquake collapsed sections of two hospitals in the San Fernando Valley, taking 64 lives, and so seriously weakened the earthen wall of the Van Norman Dam at the northern end of the San Fernando Valley that 80,000 people living below the dam had to leave their homes until the water level in the reservoir could be lowered. Total damage caused by the earthquake is estimated at $500 million to $1 billion. This three-dimensional view is based on a drawing prepared by two of the author's colleagues, Bernard Minster and Thomas Jordan, who worked with information supplied by geologists and geophysicists of the California Institute of Technology. The view is looking toward Los Angeles.**

the southern California block. If it did, this would lend additional support to the notion that the plates in California are only 15 to 20 kilometers thick. An intriguing possibility is that the upper part of the crust is sliding with relatively little friction on a layer of rock rich in the mineral serpentine.

The hypocenter, or point of initial rupture, of the San Fernando earthquake was at a depth of 13 kilometers under the San Gabriel Mountains. The fault motion was propagated to the surface along a fault inclined northward at an angle of 45 degrees and broke the surface near the cities of San Fernando and Sylmar, at the boundary between the crystalline rocks of the mountains and the sediments of the valley [*see Illustration, on page 101*]. Two heavily damaged hospitals were between the epicenter and the surface break and were therefore on the upthrust, or elevated, block. The hundreds of aftershocks following the earthquake covered an area of approximately 300 square kilometers; the total volume of rock lifted up was about 2,500 cubic kilometers.

Even though the elevation difference between the peaks of the San Gabriel Mountains, such as Mount Wilson and Mount Baldy, and the floors of the adjacent valleys is impressive, it does not represent the total uplift. Erosion removes material from the mountains and deposits it in the valleys. The total amount of differential vertical motion probably exceeds two and a half miles, and horizontal displacements in the transverse ranges probably exceed 25 miles. Many thousands of earthquakes of the San Fernando type must have occurred in the area over the past several million years.

Seismic surveillance of the region with instruments dates back only four decades. In this period the northern San Fernando Valley was less active seismically than many other parts of the greater Los Angeles area, although it was comparable to the average for all southern California. On the basis of the seismic data there was no reason to believe the San Fernando area was any more or less likely than any other region of recent mountain-building in southern California to experience a large earthquake. On the other hand, the thrusting and bending associated with the geologic processes in the region, and the tilting that was associated with the earthquake and its aftershocks, suggest that a dense network of tiltmeters could provide a warning of the next large earthquake here.

Geosynclines, Mountains, and Continent-Building

8

Robert S. Dietz
March 1972

A geosyncline is a long prism of sedimentary rock laid down on a subsiding region of the earth's crust. It has long been recognized that geosynclines are fundamental geologic units. Furthermore, it has been a dictum of geology that they eventually evolve into mountains consisting of folded sedimentary strata. The laying down of such sediments and their subsequent folding constitute a basic geologic cycle that requires a few hundred million years. Until recently the original nature of geosynclines has been inferred only by studying folded mountains. It was commonly believed that there are no nascent (unfolded) geosynclines in the world today, but this would defy another geologic dictum: that the present is the key to the past.

In recent years the study of marine geology has been revolutionized by the concept of plate tectonics, which holds that the earth's crust is divided into a mosaic of about eight rigid but shifting plates in which the continents are embedded and drift along as passive passengers. With this concept the evolution of ocean basins has been rather clearly resolved. The question arises: Must plate tectonics stay at sea, or is it also the prime mover of the geosyncline mountain-building cycle? In other words, can it account for the collapse of geosynclines and the growth of continents? I am among those who believe it can. Some notable advocates of this new concept of continental evolution are John Dewey and John M. Bird of the State University of New York at Albany, Andrew Mitchell and Harold Reading of the University of Oxford and William R. Dickinson of Stanford University.

When one examines the structure of ancient folded mountains, one finds that the classic geosyncline is divided into a couplet: two adjacent and parallel structures consisting of a eugeosyncline (true geosyncline) and a miogeosyncline (lesser geosyncline), often shortened to eugeocline and miogeocline. Now that the ocean floor is becoming better known, one need not look far to find an example of the geosynclinal couplet in process of formation. A probable example of a "living" eugeocline is the continental rise that lies seaward of the continental slope off the eastern U.S. Landward of the rise and capping the continental shelf is a wedge of sediments that becomes progressively thicker as it extends toward the shelf edge. This wedge seems to be a living miogeocline.

In dimensions and in the overall character of its rocks and stratigraphy the modern continental-rise prism closely matches typical ancient eugeoclines. It parallels the Atlantic seaboard for 2,000 kilometers, forming an apron 250 kilometers wide from the continental slope to the abyssal plain [*see illustrations on pages 108 and 109*]. Seismic studies reveal that the rise is the top of a huge plano-convex lens of sediments whose maximum depth is about 10 kilometers. The sediments are turbidites, deposited by the muddy suspensions known as turbidity currents. Such suspensions periodically cascade down submarine canyons and pour across the continental rise, depositing sedimentary fans that eventually coalesce into an apron. Turbidites consist of thin graded beds of poorly sorted particles of silt and sand in which coarse material is at the base and finer material is at the top. The gradation in particle size reflects the differential rate of settling from a single injection of muddy sand. Interlayered with the graded beds are fine clays (pelagites) that slowly settle from the overlying water as a "gentle rain" between major influxes of turbidity currents.

Collapsed eugeoclines in ancient folded mountains are similarly composed of thick and repetitive sequences of turbidites; these strata are usually termed flysch or graywacke. Mixed with the graywackes are thin limestones, ironstones and cherts formed from the skeletons of radiolarians, indicating that the sediments were deposited in deep water. True fossils are sparse, but many eugeoclinal sequences of the lower Paleozoic era contain graptolites: extinct plantlike animals that settled down from the surface.

Close examination of the graded beds also reveals what are called sole mark-

COLLISON OF CONTINENTS is depicted in the view, shown on the next page, of the Zagros Mountains in Iran along the Persian Gulf taken from the spacecraft *Gemini 12* in November, 1966. The mountains are uplifted folds of sedimentary strata, originally deposited as a geosyncline, whose cores have been exposed by erosion. The foldbelt has apparently been thrown up by the collision of the Arabian block, rotating counterclockwise, with the Eurasian block, rotating clockwise. Since the Arabian block is part of the African block, the folding represents the collision between Africa and Eurasia. The Zagros Mountains and the shallow Persian Gulf are both part of the Arabian block that extends to the Red Sea. The suture between the Arabian block and the Eurasian block is marked by a major thrust fault that passes through the upper right corner of the photograph just beyond the mountain chains.

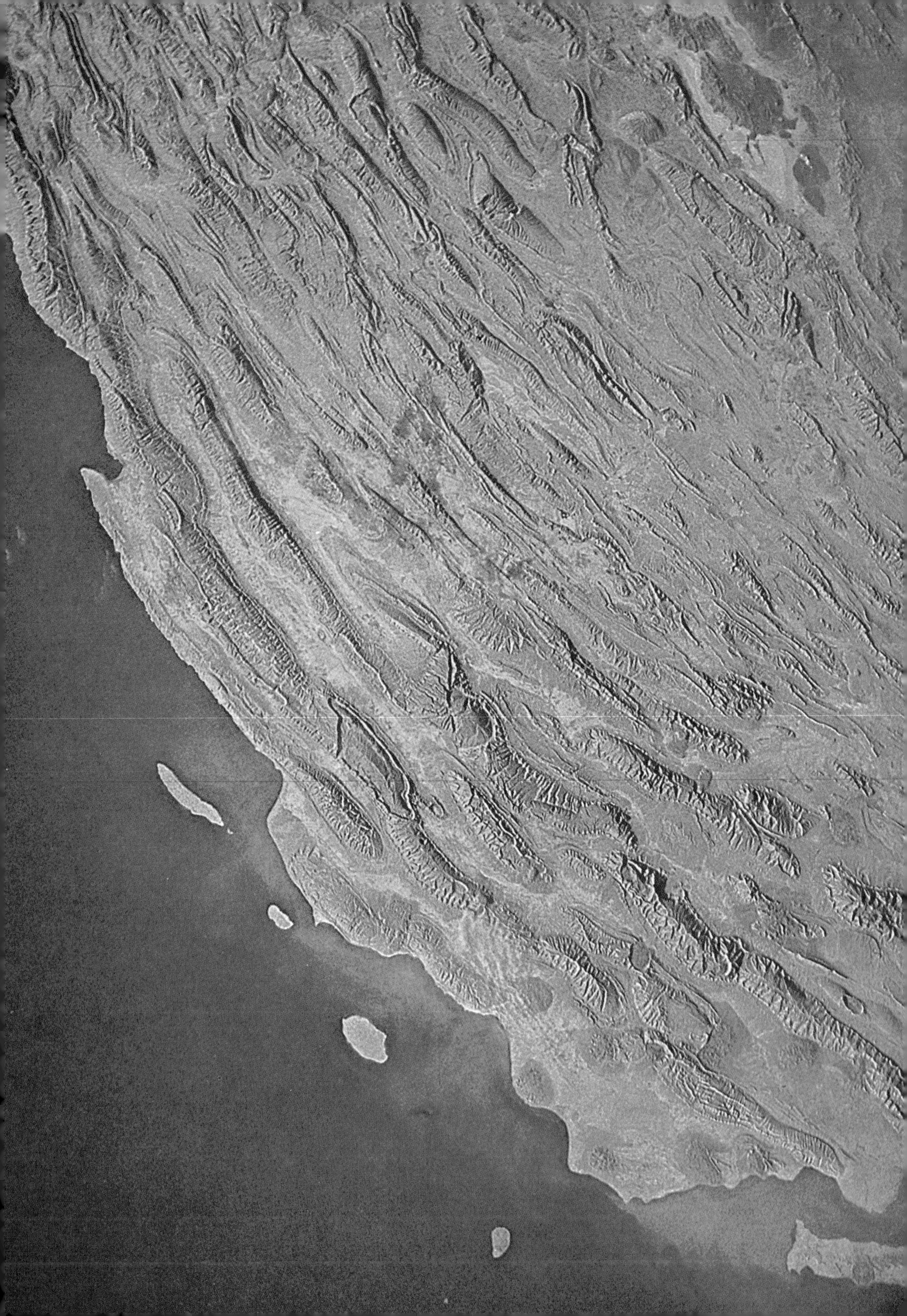

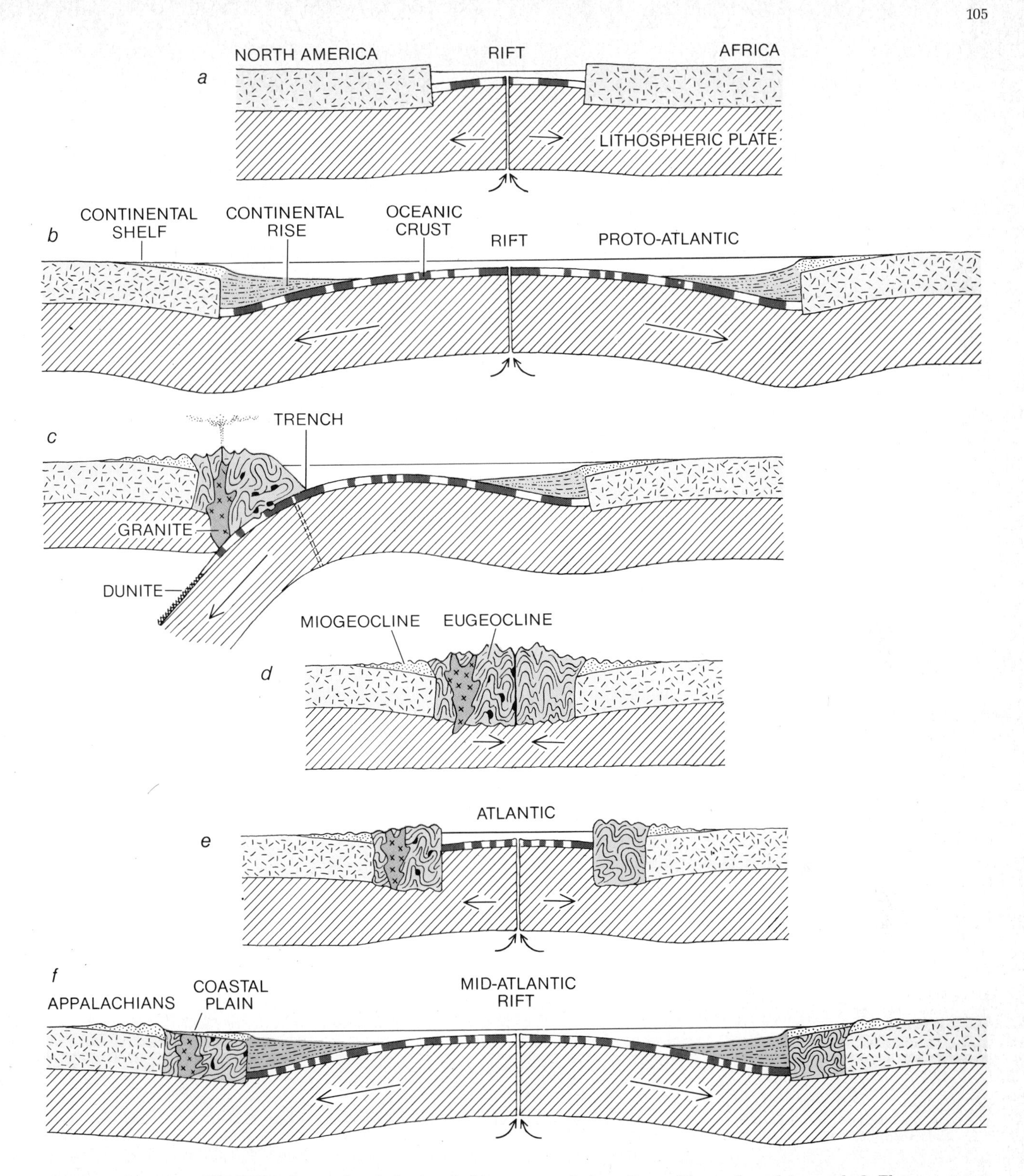

MECHANISM OF CRUMPLING that produced the Appalachian foldbelt is depicted on the hypothesis that the Atlantic Ocean has opened, closed and reopened. In the late Precambrian (*a*), North America and Africa are split apart by a spreading rift, which inserts a new ocean basin. By the process of sea-floor spreading (*b*) the ancestral Atlantic Ocean opens. New oceanic crust is created as the plates on each side move apart. As the crust cools, its direction of magnetization takes the sign of the earth's magnetic field; the field periodically reverses, and the reversals are represented by the striped pattern. On the margin of each continent sediments produce the geosynclinal couplet: miogeocline on the continental shelf, eugeocline on the ocean floor itself. The ancestral Atlantic now begins to close (*c*). The lithosphere breaks, forming a new plate boundary, and a trench is produced as the lithosphere descends into the earth's mantle and is resorbed. The consequent underthrusting collapses the eugeocline, creating the ancient Appalachians. The eugeocline is intruded with ascending magmas that create plutons of granite and volcanic mountains of andesite. The proto-Atlantic is now fully closed (*d*). The opposing continental masses, each carrying a geosyncline couplet, are sutured together, leaving only a transform fault (*vertical black line*). The shear contains squeezed-up pods of ultramafic mantle rock. Sediments eroded from the mountain foldbelt create deltas and fluvial deposits collectively called molasse. North America and Africa were apparently joined in this way between 350 million and 225 million years ago. About 180 million years ago (*e*) the present Atlantic reopened near the old suture line. Today (*f*) the central North Atlantic is opening at the rate of three centimeters per year, creating new geosynclines.

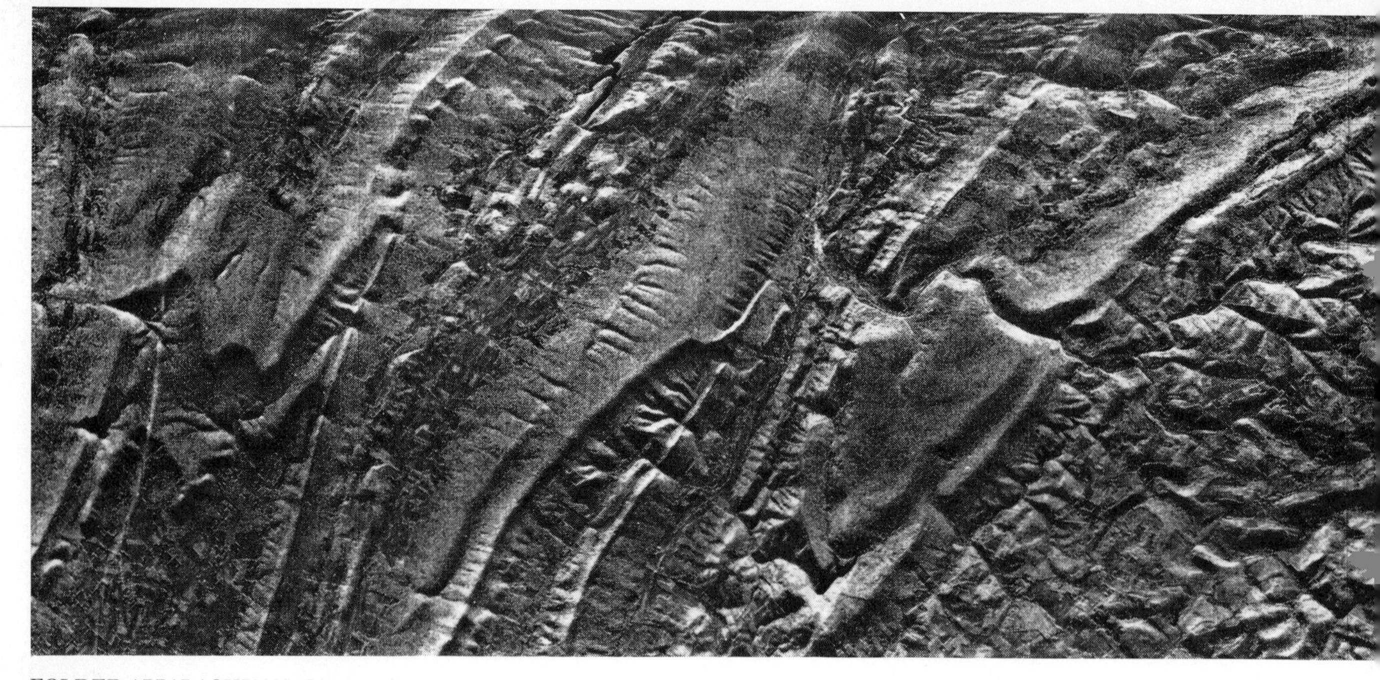

FOLDED APPALACHIAN MOUNTAINS in western Pennsylvania are depicted in this image produced by side-looking radar. The picture covers a region 25 miles long parallel to the Maryland border, centered approximately at 78 degrees 45 minutes west longitude. The picture is printed with north at the bottom so that the landscape appears to be lighted from the top. (The illuminant, of course, is the radar beam transmitted from an airplane.) If the picture is inverted, features that are actually elevated appear depressed and vice versa. This image and the sequence of three views at the bottom of these two pages were made by the National Aeronautics

ings or "flysch figures," for example ripple marks of a kind that could have been produced by turbidity currents. There can be little doubt that most of these sequences are the uplifted and eroded remnants of former continental-rise prisms. The crystalline Appalachians, which are that part of the Appalachians lying seaward of the Blue Ridge Mountains and equivalent ranges to the north and south, bear the clear imprint of being a collapsed continental-rise prism laid down in the early Paleozoic some 450 to 600 million years ago. The original prism has been much altered by intrusions and metamorphism.

The sedimentary wedge that underlies the coastal plain and continental shelf along the Atlantic seaboard appears to be an actively growing miogeocline. The wedge thickens as it progresses seaward, attaining a total thickness of between three and five kilometers along the shelf edge. Laid down on a basement of Paleozoic rocks, the wedge is composed of well-sorted shallow-water sediments deposited during the past 150 million years under conditions much like those of today. The stratified beds exhibit characteristics indicating they were deposited across the continental shelf in alluvial plains, in lagoons, along shorelines and offshore. Taking into account expected changes in the pattern of sedimentation over geologic time, the present Atlantic marine

APPALACHIAN FOLDBELT north of Harrisburg, Pa., is an extension of the foldbelt shown at the top of the page. In these side-looking radar views north is at the right. The three pictures cover a distance of 75 miles from just south of Mechanicsburg to the vicinity of a town called Jersey Shore on the West Branch of Susquehanna River. The Susquehanna River itself appears in the first frame at the left. The folded Appalachians were probably created in a late compressional stage of the collision between Africa and North America more than 450 million years ago, which caused "rugfolds" in the strata of sedimentary rock that formed part of a

and Space Administration in collaboration with the Remote Sensing Laboratory of the University of Kansas. The *K*-band radar system that produced the images was built by the Westinghouse Electric Corporation.

deposits closely resemble the ancient miogeoclinal foldbelts of the Paleozoic era and earlier. For example, the modern sedimentary wedge is much like the one found in the folded Appalachians of Pennsylvania. Both wedges are characterized by "thickening out," signifying that they grow steadily thicker toward the east before they abruptly terminate.

If the foregoing analysis is correct, one must conclude that geosynclines are actively forming along many continental margins today: eugeoclines at the base of the continental slope and miogeoclines capping the continental shelves. It remains to be shown, however, that the crustal shifting associated with plate tectonics can convert these sedimentary prisms into the mountainous foldbelts that make up the fabric of the continents, mostly as ancient eroded mountain roots rather than as modern mountain belts. In order to examine this possibility we must first summarize some of the basic concepts of plate tectonics.

The approximately eight rigid but shifting plates into which the earth is currently divided are thought to be about 100 kilometers thick. Most of the plates support at least one massive continental plateau, often referred to as a craton. We can visualize the ideal plate as being rectangular, although only the plate supporting the Indian craton approaches this simple shape. Along one edge of a crustal plate there is a subduction zone, usually marked by a trench, where the plate dives steeply into the earth's mantle, attaining a depth as great as 700 kilometers before being fully absorbed into the mantle. On the opposite side of the plate from the subduction zone is a mid-ocean rift, or pull-apart zone. As the rift opens, the gap is quickly healed from below by the inflow of liquid basalt and quasi-solid mantle rock. The other two opposed sides of the plate, connecting the rifts to the trenches, are shears called transform faults.

Hence three types of plate boundary are possible: divergent junctures (the mid-ocean rifts where new ocean crust is created), shear junctures (the transform faults where the plates slip laterally past one another, so that crust is conserved) and convergent junctures (trenches where two plates collide, with one being subducted and consumed). Only the last of the three, the convergent juncture, can help to explain how the sedimentary prism of a submarine geosyncline might be collapsed into a folded range of mountains. As the plate carrying a prism collides with a plate carrying a continental craton one would expect the prism to be compressed into folds. Thrusting and crustal thickening would follow, assisted by isostatic forces that act to keep adjacent crustal masses in balance. Such forces would cause the collapsed prism to be uplifted. The entire process would be accompanied by the generation and intrusion of magma, together with extensive metamorphism of the crustal rocks.

A grand theme of plate tectonics is that ocean basins are not fixed in size or shape; they are either opening or closing. Today the Atlantic Ocean is opening and the Pacific Ocean is closing. The drifting of the continents is another theme; every continent must have a leading edge and a trailing edge. For the past 200 million years the Pacific coast of North America has been the leading edge and the Atlantic coast the trailing edge. The trailing margin is tectonically stable, and since the continental divide is near the mountainous Pacific rim, most of the sediments are ultimately dumped into the Atlantic Ocean, including the Gulf of Mexico. Therefore it is primarily along a trailing edge that the great geosynclinal prisms are deposited.

Consider, however, what would hap-

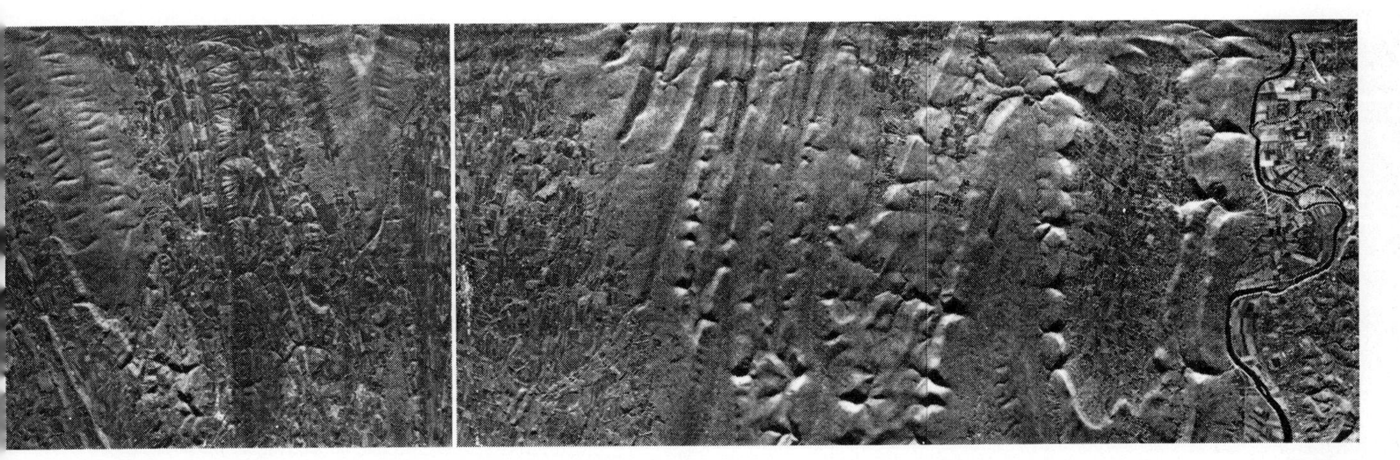

Paleozoic geosyncline (*see illustration on page 110*). After folding the region was eroded to a level plain and then uplifted. The modern mountains were subsequently etched out according to the hardness of the various strata. Thus the ridges are composed mainly of dense sandstone and can be either synclines (troughs) or anticlines (arches). The V-shaped chevrons in the first frame are synclines that plunge to the northeast. The Susquehanna River established its course when the entire region was reduced to a level surface, so that its course has been superimposed on the folded structure, thereby cutting directly across the folds and creating water gaps.

pen if, with the changing patterns of plate motions, a subduction zone (a trench) were created along a former trailing edge, forming a new plate boundary. The Atlantic would be transformed into a closing ocean with its geosynclinal prisms riding toward the trench. The continental margin and the trench would eventually collide, collapsing the eugeocline into a contorted mountainous foldbelt and also folding the miogeocline to a much lesser extent. Before that happened the continental margin would encounter and incorporate an island arc, similar to the island arcs found along the perimeter of the western Pacific. These arcs are created by tectonic and magmatic activity triggered by the plunging crustal plate. It is also quite possible that the Atlantic Ocean would close entirely, causing North Africa to collide with eastern North America. The collision of India with the underbelly of Asia, throwing up the Himalaya rampart, would be a present-day analogy. One can imagine many possible scenarios, depending on the geometry of plate boundaries and other variables.

The creation of a eugeoclinal foldbelt is of course considerably more than simply the accordion-like collapse of a continental-rise prism. The foldbelt is sheared into thrust faults and the landward edge of the eugeocline is commonly thrust onto the adjacent miogeocline. The descending crustal plate is not entirely consumed within the earth's hot mantle, with the result that low-density magmas buoyantly rise and invade the eugeocline. This leads to intrusions of granodiorite (a granite-quartz rock) and the growth of volcanic mountains consisting of andesite (the rock characteristic of the Andes). This lava is highly explosive because it is charged with water sweated out of the descending plate. Magma is not generated from the plunging lithosphere until it has reached a considerable depth. As a result the eugeocline can be subdivided into two parallel geologic belts. Toward the sea one finds sedimentary rock transformed at high pressure and low temperature; farther inland the sedimentary rock has been altered predominantly at low pressure and high temperature by the numerous intrusions of magma. From the new marginal mountain range, delta and river deposits sweep back across the continent, covering the miogeocline with a suite of continental shales and conglomerates collectively called molasse.

The concept that the geosynclinal cycle is controlled by plate tectonics provides some new answers to old questions about geosynclines. For example, is mountain-building periodic and worldwide or is it random in space and time? The answer must be both yes and no. On the one hand, the crustal plates are highly intermeshed; the drift of any one plate has global repercussions, giving rise to synchronous mountain-building. Any brief interval of rapid plate motion would also be one of widespread mountain-building. On the other hand, the rate of plate convergence is highly dependent on the latitudinal distance from the relative pole of rotation of that plate and on the particular geometry of the plate boundary.

A law of plate tectonics states that sea-floor spreading (injection of new ocean crust) proceeds at right angles to a rift; the crustal plate, however, may be subducted into a trench at any angle. The rate of subduction and the attendant distorting of the crust therefore vary greatly from place to place, as can be observed on the perimeter of the Pacific today. Thus it would seem that although mountain-building over the span of geologic time may reach crescendos, it must also be continuous and random.

The plate-tectonic version of the geosynclinal cycle predicts that miogeoclines are ensialic, or laid down on continental crust (sial), whereas eugeoclines are ensimatic, or deposited on oceanic crust (sima). This differs from the earlier view that all geosynclines are ensialic,

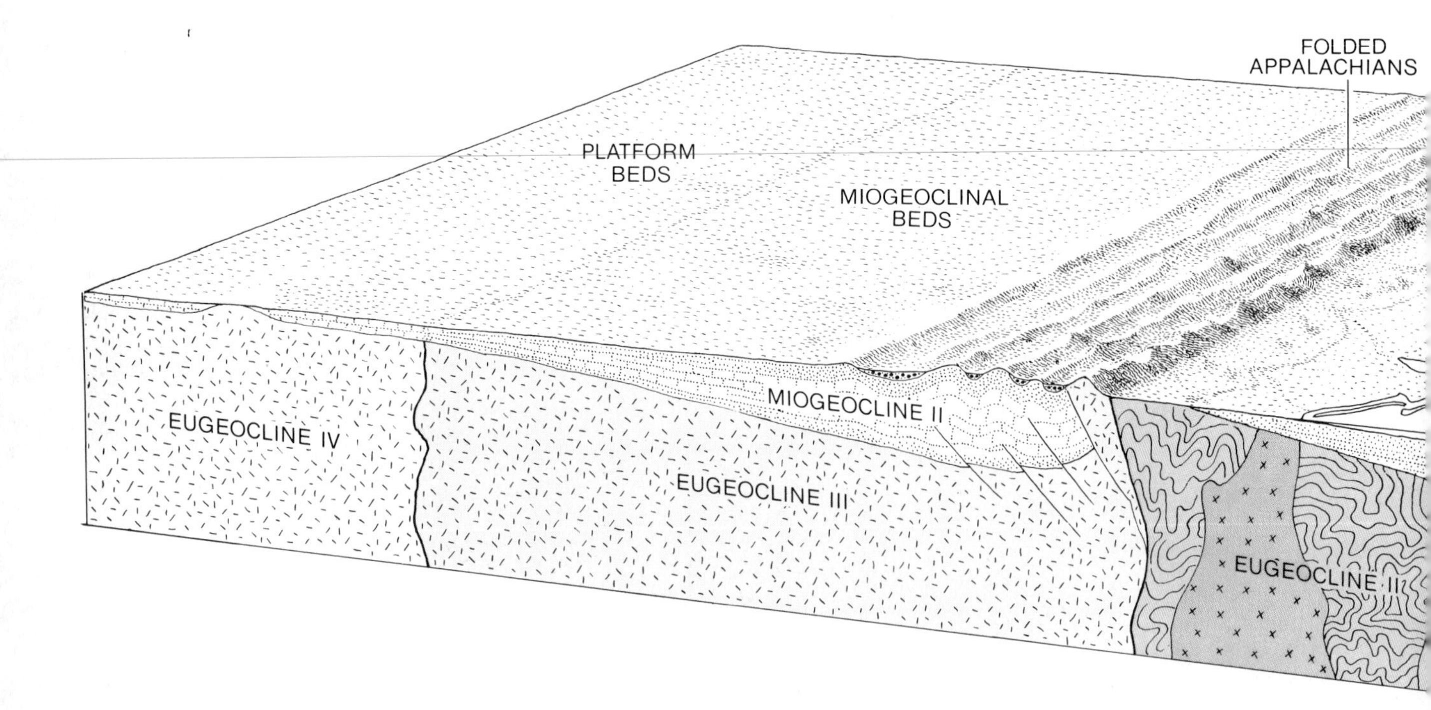

SUCCESSION OF EUGEOCLINES underlies nearly all North America below the relatively undisturbed cover beds. These contorted and intruded prisms constitute the fundamental fabric of continents, known as the basement complex. A new geosynclinal couplet is being deposited today. It consists of a miogeocline (lesser geosyncline) of shallow water beds that caps the coastal plain and continental shelf paralleled by a eugeocline (true geosyncline) that is formed at the base of the continental slope by detritus

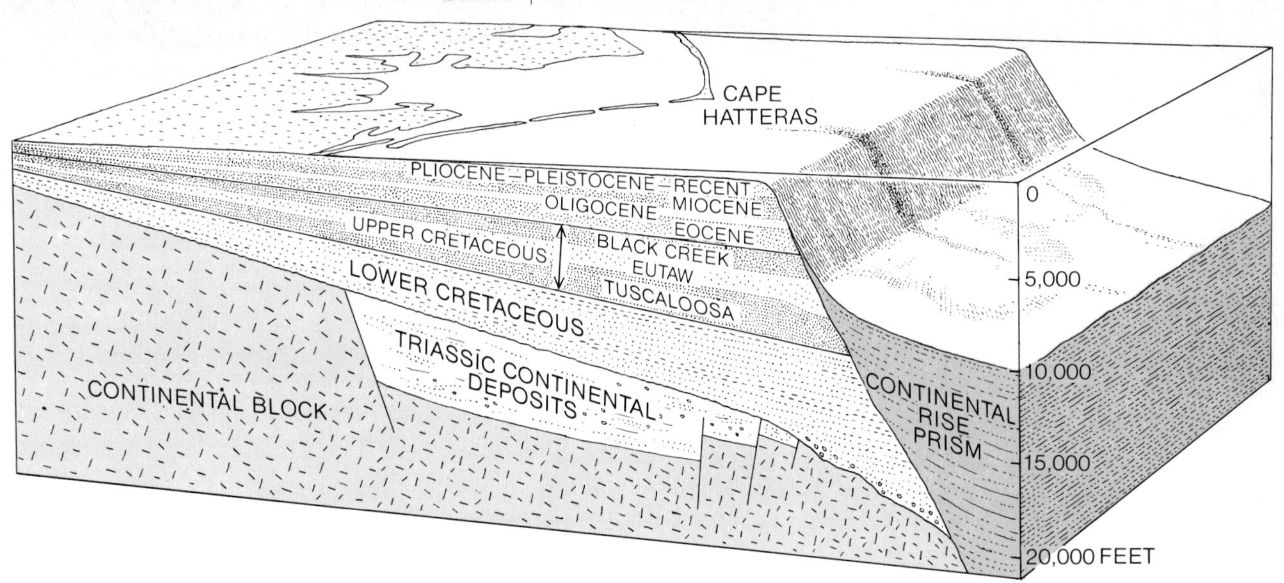

LIVING GEOSYNCLINAL COUPLET off the Atlantic coast of the U.S. consists of a miogeocline, strata laid down on the shallow continental shelf during the past 150 million years, and a eugeocline prism *(dark color)*, consisting of thin beds of sand and mud deposited by turbidity currents flowing down the continental slope. The material in the Triassic basin represents continental deposits laid down before the foundering of the continental margin under tension 190 million years ago, prior to opening of Atlantic Ocean.

which is certainly incorrect. Early investigators observed that a granitic basement is invariably present under miogeoclines and evidently reasoned that a similar sialic basement, although it was unseen, must also be present under eugeoclines. A collapsed eugeocline is as thick as the continental plate, about 35 kilometers, so that its basement is beyond the depth of even the deepest boreholes. We can infer that the ultimate basement is simatic, however, by observing detached fragments that are caught up in the contorted mélange of the eugeoclinal pile. These fragments include samples of oceanic crust (for example radiolarian cherts and sodium-altered lavas) and upper-mantle rocks (for example serpentinites and peridotites).

The ensimatic location of eugeoclines can also account for their tectonic style. They are tightly folded, faulted, tumbled and dynamically metamorphosed into an almost unmappable mélange. This contorted state is understandable, since the ocean floor is shearing under the eugeocline and thrusting the sedimentary pile against the continental slope. Extensive tectonic thickening and interleaving must occur before the pile will rise to mountainous altitudes. On the other hand, the miogeocline beds are protected by the stable continental slab, so that they are simply thrown into a series of loose, open, ruglike folds.

It now seems amusing to recall that 19th-century geologists, using a wrinkled apple as an analogy, interpreted folded mountains to mean that the earth was shrinking. Today it seems clear that eugeoclines are deposited at the edge of a continent on oceanic crust seaward of the continental slope, so that folded mountains really show that the continents are growing larger through marginal accretion. Mountain-building is therefore evidence of an even more fundamental geologic process: the growth of continents. The continents grow not as a layer cake but as a craton that is divided vertically into zones with an old nucleus and young margins.

An important aspect of geosynclines requiring explanation is that they are laid down on foundations that are continuously subsiding. This aspect is par-

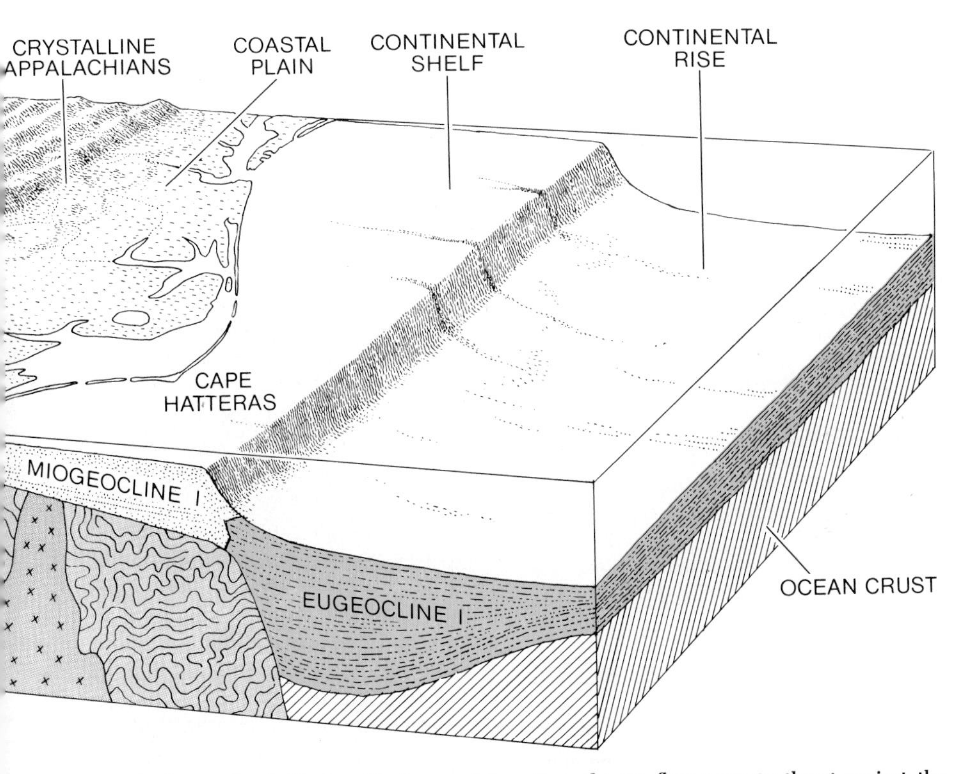

washed over the shelf edge. If at some future time the sea floor were to thrust against the continent, the modern eugeocline (*1*) would collapse into a new foldbelt like the earlier ones. The hypothetical mechanism that creates foldbelts is shown on page 105. This diagram and others are based on drawings by the author's colleague John C. Holden.

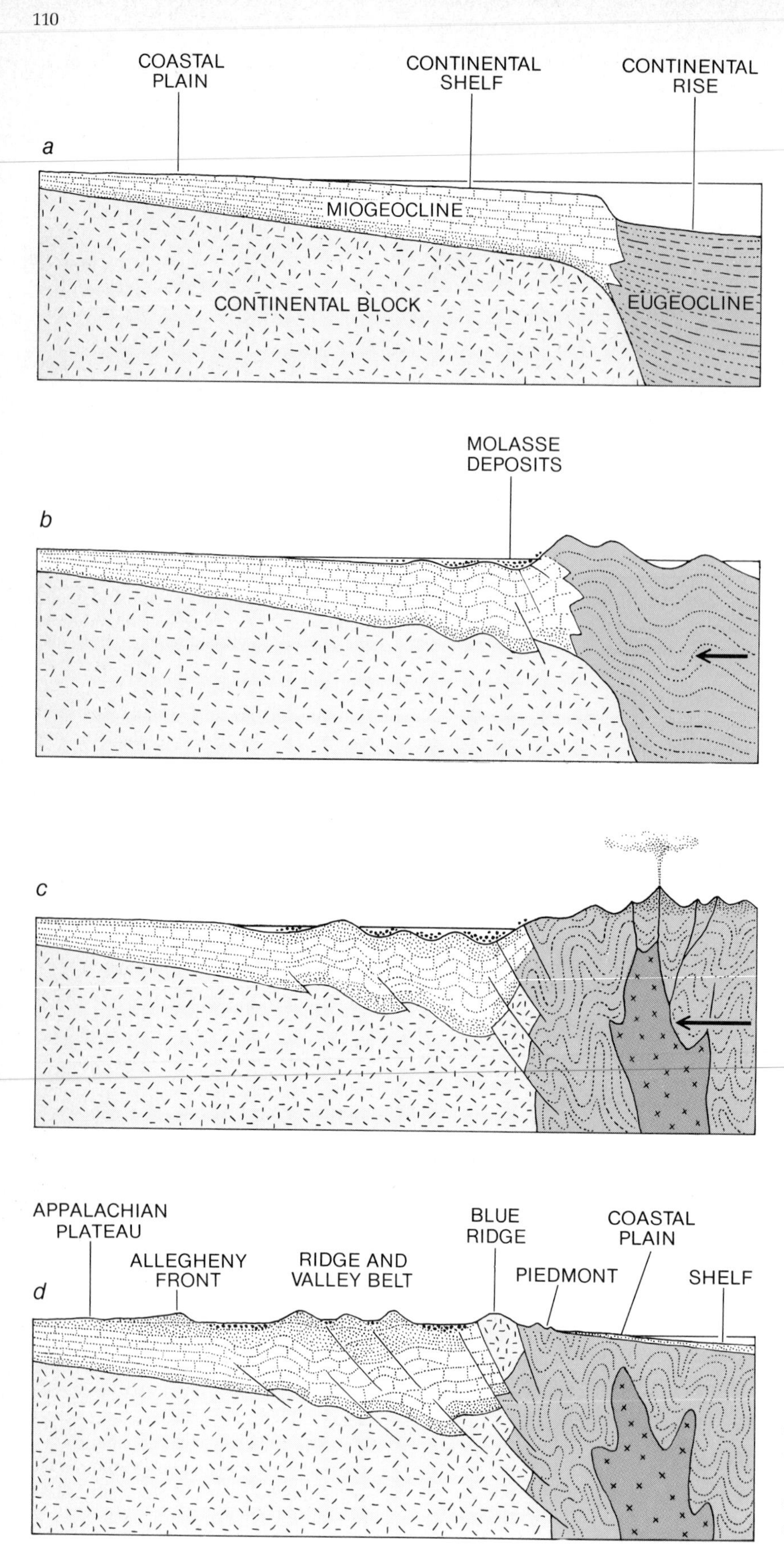

CRUMPLING OF EARLIER GEOSYNCLINAL COUPLET, apparently laid down in late Precambrian and early Paleozoic time more than 450 million years ago, produced the Appalachian foldbelt. The four-part sequence shows how the miogeocline, or western part of the geosynclinal couplet, was folded into the series of ridges between the Blue Ridge line and the Allegheny front. The eugeocline, altered by heat, pressure and volcanism, formed a lofty range of mountains, now almost completely eroded, east of the Blue Ridge line.

ticularly evident in miogeoclines, which can attain a total thickness at their seaward edge of five kilometers even though they are entirely composed of beds deposited in shallow water. This phenomenon is nicely accounted for by plate tectonics: the margins of rift oceans inherently have, as one geologist has expressed it, a "certain sinking feeling."

Let us take as an example the Atlantic Ocean between the U.S. and the bulge of Africa. This new ocean basin was created about 180 million years ago by the insertion of a spreading rift that split North America away from Africa [*see illustration on page 105*]. Attendant swelling of the mantle arched the continents upward along the rift line by about two kilometers. Erosion then beveled the raised edges, thinning the margins of the two continental plates.

A modern example of crustal arching associated with incipient rifting of the crust can be observed in the high dorsal of Africa from Ethiopia southward. The Red Sea provides a more advanced stage of a newly opening ocean basin. Along the flanks of this linear trough crustal arching has stripped away young rocks, exposing "windows" of Precambrian basement.

With the insertion of new oceanic crust by sea-floor spreading, the ocean grew ever wider. In the process the continental edges subsided, as is demonstrated by the sloping flanks of the mid-Atlantic ridge today. When the ocean was smaller, the continental edges had to ride down a similar slope. Eventually the inflated mantle under the ridge reverts to normal mantle, but this takes 100 million years or more. Therefore as a geosyncline is laid down on the trailing edges of a drifting continent it slowly subsides for reasons external to the sedimentary deposit itself.

Additional subsidence, however, is caused by the steadily growing mass of the sedimentary apron, which must be isostatically compensated because the earth's crust is not sufficiently strong to sustain the load. For every three meters of sediment deposited the crust sinks about two meters. This crustal failure, however, is spread over a large geographic area, so that the growth and subsidence of a huge continental-rise prism causes a sympathetic downward flexing of the adjacent continental margin. As the continental shelf slowly tilts, wedges of shallow-water sediments are deposited.

Over the millenniums, as the shoreline transgresses and regresses repeatedly across the shelf, a large composite megawedge of shallow-water sediments caps the shelf. The abundant supply of sediment usually ensures that the top of the prism is maintained close to sea level. Excess detritus bypasses the shelf, is temporarily dumped on the continental slope and is then carried onto the continental rise by turbidity currents. The shelf edge and the continental-rise prism comprise a couplet within which there is constant interplay.

Like the sedimentary wedge under the Atlantic coastal plain today, the early Paleozoic Appalachian miogeocline thickens in the seaward direction. The abrupt termination of this miogeocline was long a mystery to early geologists who mapped it. They suggested that a missing seaward limb had been thrust upward and completely eroded away or that it had foundered into an ancient oceanic basin. This hypothetical land mass of Appalachia was the geological equivalent of the legendary Atlantis. The wedgelike structure of the existing continental-plain prism provides a satisfactory solution to the puzzle: the hypothesized seaward limb never existed. We now see that the thickening out of sedimentary deposits at the shelf edge is a normal mode of sedimentation. One way in which this may happen is that reefs of coral and algae build up along the margin of the continental shelf, creating a carbonate dam behind which other shelf sediments accumulate.

The mechanism of building continents by the peripheral accretion of collapsed continental rises seems also to ensure that the sedimentary deposits become dry land. (We take for granted that continents are above sea level, but it should be remembered that the mid-ocean ridge system, which approaches the continents in importance as a topographic feature, almost never rises above the sea surface.) The sedimentary apron gradually thickens until it approaches the height of the continental slope (about five kilometers), but upward growth ceases once the slope is completed and sea level is attained.

As we have seen, however, isostasy is at work, causing the oceanic crust to subside under the sedimentary load. The result is that a fully developed sedimentary prism attains an overall thickness of about 15 kilometers. When the prism is subsequently collapsed into a eugeosynclinal foldbelt, it becomes thicker still. The attendant metamorphism and granitic intrusion (which increases the total mass of rock) give rise to a monolithic structure that is more than 35 kilometers thick, thicker than a continental plate. Thus new foldbelts not only rise above sea level but also throw up rugged mountain ranges.

The hypothesis that geosynclines are deposited along a continental margin and then crushed against the continent as a result of plate tectonics seems to explain satisfactorily how geosynclines are transformed into folded mountains. The close relation between eugeoclines and foldbelts is not one of cause and effect but a simple consequence of location: geosynclines are laid down along continental margins and such margins are the locus of interaction between continents and subduction zones.

In spite of the vast span of geologic time and the rigors of erosion, the continents remain in a good state of health. We can predict that they always will be: detritus lost to the oceans is eventually carried back to the continents and collapsed into accretionary belts that also incorporate new igneous rock. Although the earthquakes that punctuate mountain-building are sometimes disastrous to man's culture, they are acts of continental construction. The great flood—the complete inundation of the erosionally leveled continents—will always threaten but will never come to pass.

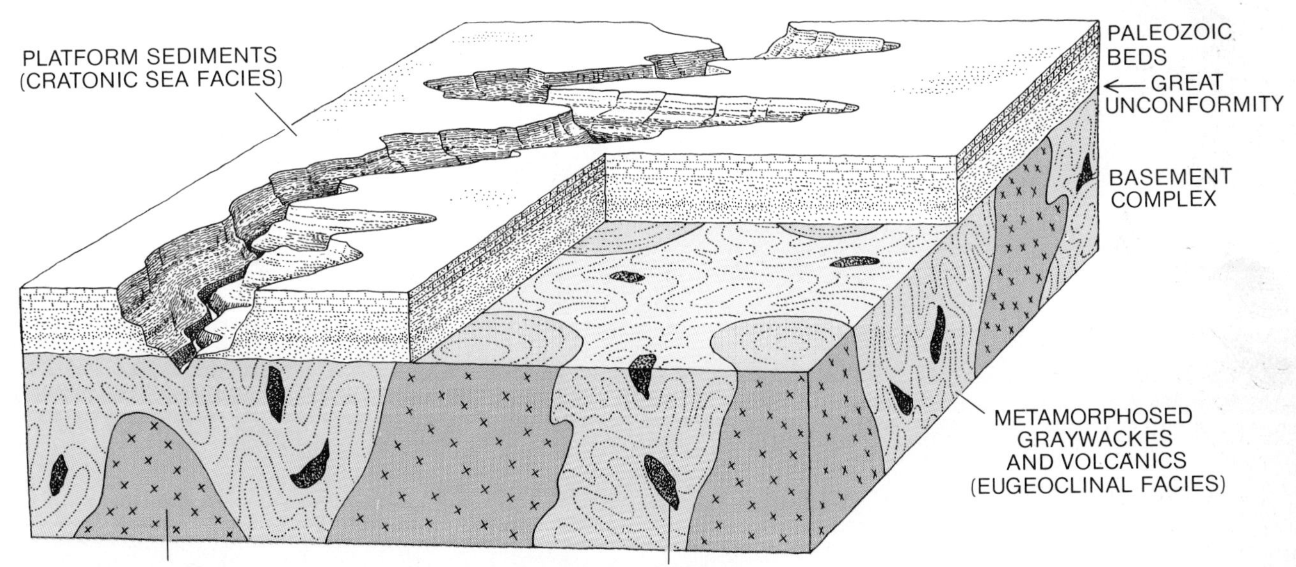

DEEP FABRIC OF CONTINENTS, the basement complex, is the fundamental rock unit of the continental plateaus, or cratons. This complex is usually obscured from view by miogeoclinal beds or by the coating of shallow-sea deposits that have invaded the continents from time to time. (Hudson Bay is a modern example of such an invading shallow sea.) Long a puzzle to geologists, the basement is composed of eugeoclinal foldbelts that have undergone intensive folding, metamorphism and intrusion. Geologists once thought that these "roots of mountains" indicated that the earth had contracted while cooling. The folds were likened to the skin of a dried apple. The present interpretation is that the eugeoclinal facies was laid down on the ocean floor and subsequently was crumpled against the continental margin, building up an onion-like vertically zoned craton. On this view the continents have grown larger rather than smaller with time. Moreover, the basement complex need not be Archean (composed of the oldest rocks) as formerly supposed, because the high degree of metamorphism does not necessarily indicate great antiquity and repeated mountain-building events. Instead it reflects the intensity of the collapsing process; once accreted, the foldbelt is not usually mobilized again. It has long been known that the granites of the basement complex are always younger than the metamorphosed sediments they intrude. On the other hand, the included pods of ultramafic rocks are always older, since they are detached fragments of the oceanic foundation on which the eugeocline was deposited. If geologists ever find "the original crust of the earth," it will be one of these pods within the oldest foldbelts.

The Subduction of the Lithosphere

M. Nafi Toksöz
November 1975

The lithosphere, or outer shell, of the earth is made up of about a dozen rigid plates that move with respect to one another. New lithosphere is created at mid-ocean ridges by the upwelling and cooling of magma from the earth's interior. Since new lithosphere is continuously being created and the earth is not expanding to any appreciable extent, the question arises: What happens to the "old" lithosphere?

The answer came in the late 1960's as the last major link in the theory of sea-floor spreading and plate tectonics that has revolutionized our understanding of tectonic processes, or structural deformations, in the earth and has provided a unifying theme for many diverse observations of the earth sciences. The old lithosphere is subducted, or pushed down, into the earth's mantle. As the formerly rigid plate descends it slowly heats up, and over a period of millions of years it is absorbed into the general circulation of the earth's mantle.

The subduction of the lithosphere is perhaps the most significant phenomenon in global tectonics. Subduction not only explains what happens to old lithosphere but also accounts for many of the geologic processes that shape the earth's surface. Most of the world's volcanoes and earthquakes, including nearly all the earthquakes with deep and intermediate foci, are associated with descending lithospheric plates. The prominent island arcs—chains of islands such as the Aleutians, the Kuriles, the Marianas and the islands of Japan—are surface expressions of the subduction process. The deepest trenches of the world's oceans, including the Java and Tonga trenches and all others associated with island arcs, mark the seaward boundary of subduction zones. Major mountain belts, such as the Andes and the Himalayas, have resulted from the convergence and subduction of lithospheric plates.

In order to appreciate the gigantic scale on which subduction takes place, consider that both the Atlantic and the Pacific oceans were created over the past 200 million years as a consequence of sea-floor spreading. Thus the lithosphere that underlies the world's major oceans is less than 200 million years old. As the oceans opened, an equivalent area of lithosphere was simultaneously subducted. A simple calculation shows that the process involved the consumption of at least 20 billion cubic kilometers of crustal and lithospheric material. At the present rate of subduction an area equal to the entire surface of the earth would be consumed by the mantle in about 160 million years.

To understand the subduction process it is necessary to look at the thermal regime of the earth. The temperatures within the earth at first increase rapidly with depth, reaching about 1,200 degrees Celsius at a depth of 100 kilometers. Then they increase more gradually, approaching 2,000 degrees C. at about 500 kilometers. The minerals in peridotite, the major constituent of the upper mantle, start to melt at about 1,200 C., or typically at a depth of 100 kilometers. Under the oceans the upper mantle is fairly soft and may contain some molten material at depths as shallow as 80 kilometers. The soft region of the mantle, over which the rigid lithospheric plate normally moves, is the asthenosphere. It appears that in certain areas convection currents in the asthenosphere may drive the plates, and that in other regions the plate motions may drive the convection currents.

The mid-ocean ridges mark the region where upwelling material forms new lithosphere. The ridges are elevated more than three kilometers above the average level of the ocean floor because the newly extruded rock is hot and hence more buoyant than the colder rock in the older lithosphere. As the lithosphere spreads away from the ridge it gradually cools and thickens. The spreading rate is generally between one centimeter and 10 centimeters per year. The higher velocities are associated with the Pacific plate and the lower velocities with the plates bordering the Mid-Atlantic Ridge. At a velocity of eight centimeters per year the lithosphere will reach a thickness of about 80 kilometers at a distance of 1,000 kilometers from the ridge. Under most of the Pacific abyssal plains a thickness of this value has been confirmed by measurements of the velocities of seismic waves.

Where two plates move toward each

HIMALAYAS OF NEPAL, shown in this false-color picture made from the Earth Resources Technology Satellite (ERTS), are a zone in which continental lithosphere is being subducted. In most subduction zones oceanic lithosphere plunges under continental lithosphere. Here the lithosphere of the Indian subcontinent (*bottom*) is being subducted under the snow-covered Himalayas (*top*), raising the mountain range in the process. The area covered by the picture is 125 kilometers (78 miles) across. Mount Everest is one of the peaks on the ridge at the very edge of the picture in the upper right-hand corner. The main boundary fault between the two lithospheric plates runs from left to right in the valley that is marked by two clusters of cloud that are visible at lower center and lower right.

other and converge, the oceanic plate usually bends and is pushed under the thicker and more stable continental plate. The line of initial subduction is marked by an oceanic trench. At first the dip, or angle of descent, is low and then it gradually becomes steeper. Profiles across trenches of the reflections of seismic waves clearly show the downward curve of the top of descending oceanic plates.

Several factors contribute to the heating of the lithosphere as it descends into the mantle. First, heat simply flows into the cooler lithosphere from the surrounding warmer mantle. Since the conductivity of the rock increases with temperature, the conductive heating becomes more efficient with increasing depth. Second, as the lithospheric slab descends it is subjected to increasing pres-

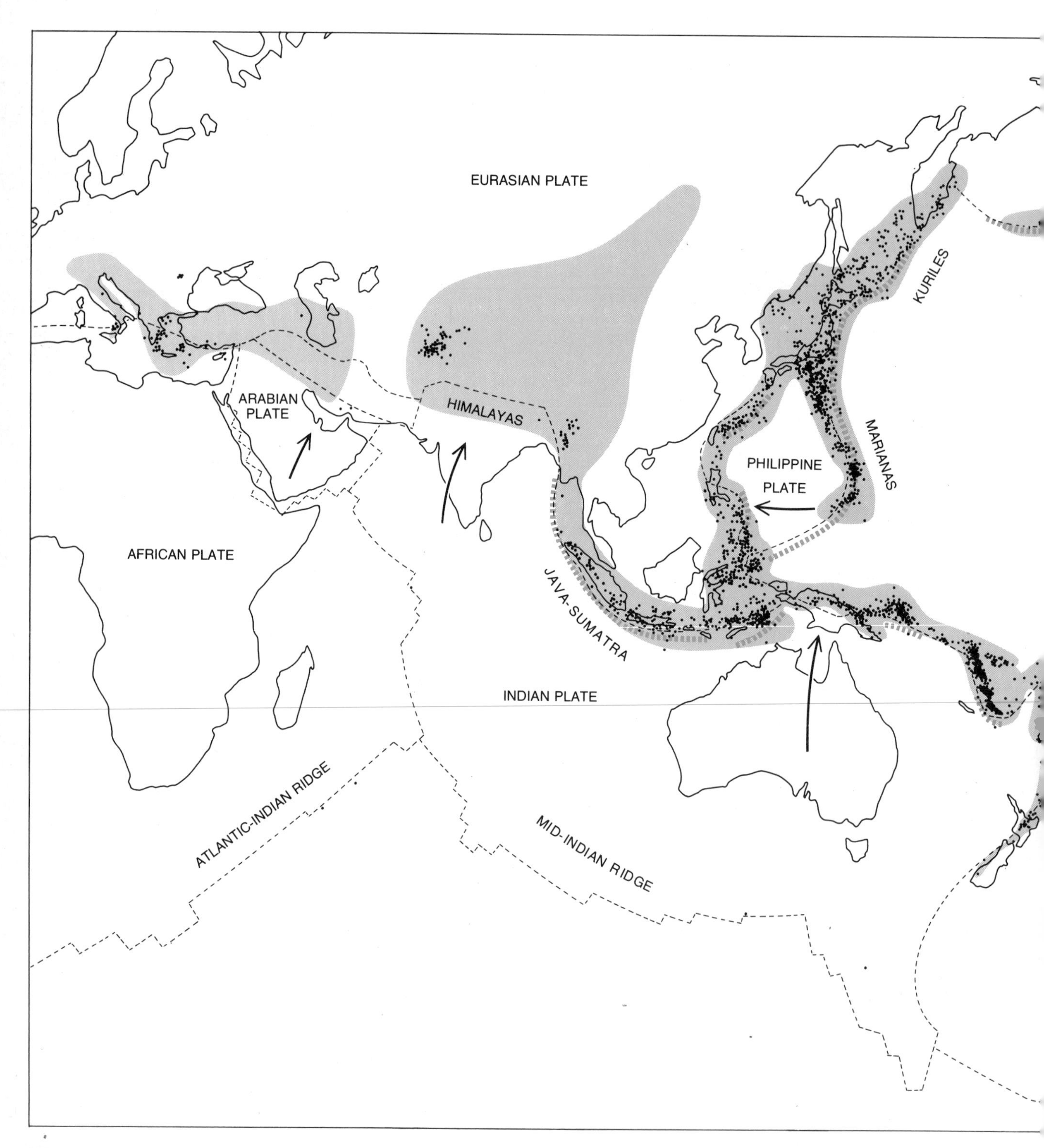

TECTONIC MAP OF THE EARTH depicts the principal lithospheric plates and their general direction of motion (*arrows*). New material is continually added to the plates at mid-ocean ridges by the upwelling and cooling of magma from the earth's mantle. It moves outward and is eventually returned to the mantle by subduction. There it is slowly consumed. The subduction process creates deep oceanic trenches (*broken lines in color*) and island arcs such as those bordering the western and northern Pacific. On the islands of the arcs are many active volcanoes. Young mountain belts in Europe and Asia identify zones where continental litho-

sure, which introduces heat of compression. Third, the slab is heated by the radioactive decay of uranium, thorium and potassium, which are present throughout the earth's crust and add heat at a constant rate to the descending material. Fourth, heat is provided by the energy released when the minerals in the lithosphere change to denser phases, or more compact crystal structures, as they are subjected to higher pressures during descent. Finally, heat is generated by friction, shear stresses and the dissipation of viscous motions at the boundaries between the moving lithospheric plate and the surrounding mantle. Among all these sources the first and fourth contribute the most toward the heating of the descending lithosphere.

The temperatures inside a descending

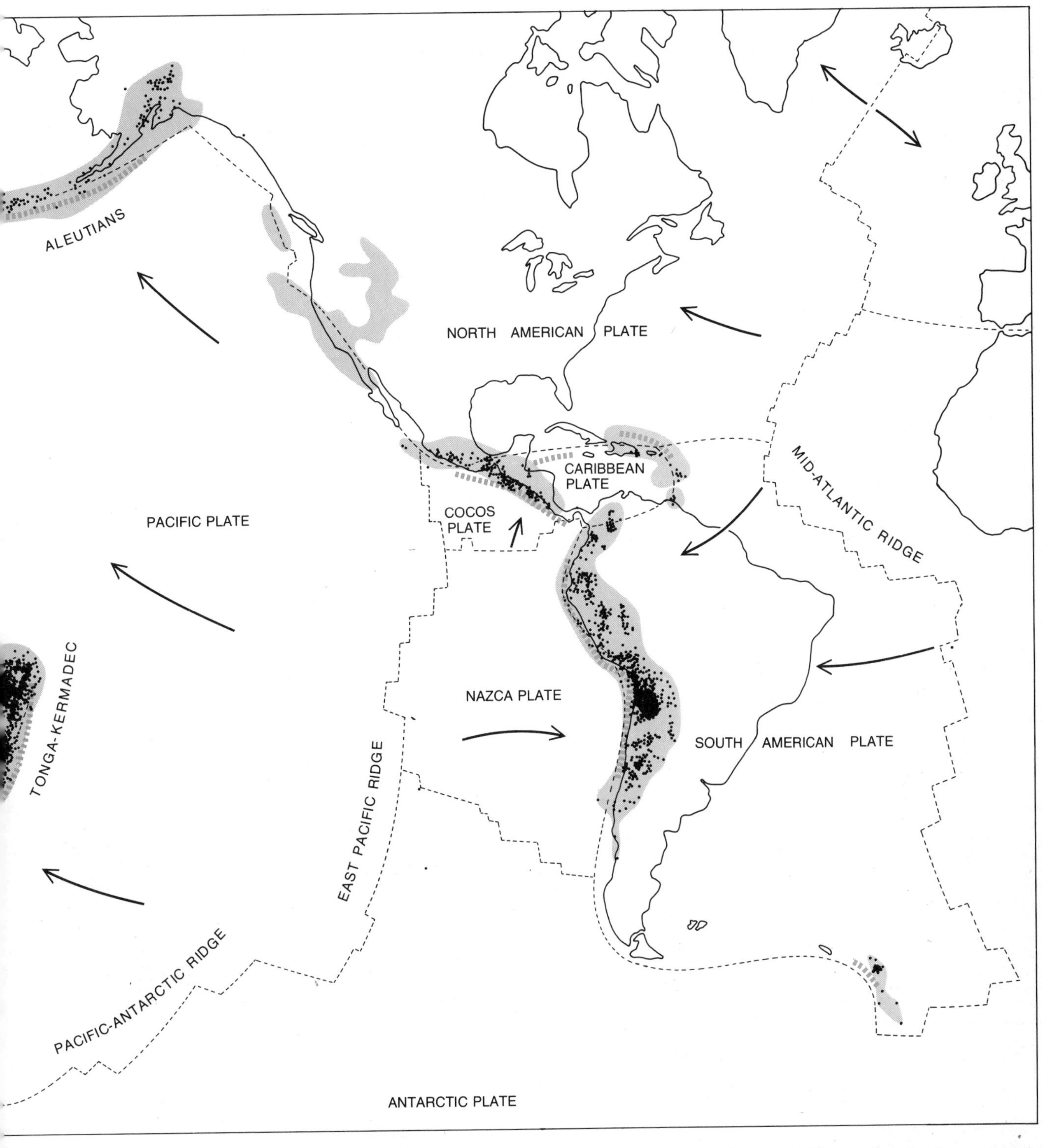

spheric plates converge; around the Pacific young mountain ranges result from the subduction of oceanic plates. The areas in color identify the general location of the great majority of earthquakes that occurred at all depths between 1961 and 1967; they are based on maps made by H. J. Dorman and M. Barazangi of the Lamont-Doherty Geological Observatory. Most earthquakes have a magnitude below 6.5 and occur at a shallow depth (between five and 15 kilometers). The locations of deep earthquakes, those occurring below 100 kilometers, are given by the black dots. All the deep earthquakes take place in cold descending slabs of the oceanic type.

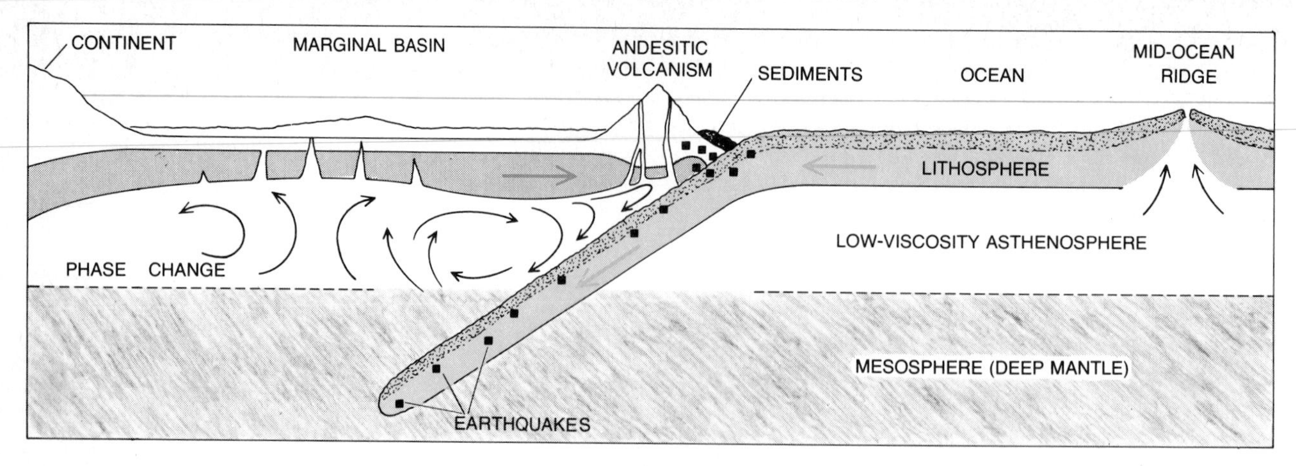

FORMATION AND SUBDUCTION OF LITHOSPHERE are shown in this cross section of the crust and mantle. New lithosphere is created at a mid-ocean ridge. A trench forms where the lithospheric slab descends into the mantle. Earthquakes (*small squares*) occur predominantly in the upper portion of the descending slab. Arrows in soft asthenosphere indicate direction of possible convective motions. Secondary convection currents in asthenosphere may form small spreading centers under marginal basins.

lithospheric plate have been calculated theoretically over the past five years by geophysicists in Britain, Japan and the U.S. Although different approaches were taken in the calculations, the results are in good agreement. For example, our group at the Massachusetts Institute of Technology has computed the progressive heating of plates penetrating into the mantle at various velocities over periods ranging from several hundred thousand years to more than 10 million years. A typical calculation based on our model of the phenomenon shows what happens to a plate descending at the rate of eight centimeters per year (the velocity characteristic of the Pacific subduction zones) at three points in time: 3.6, 7.1 and 12.4 million years after the beginning of subduction [*see illustration on opposite page*].

NAME	PLATES INVOLVED	TYPE	LENGTH OF ZONE (KILOMETERS)	SUBDUCTION RATE (CENTIMETERS PER YEAR)	MAXIMUM EARTHQUAKE DEPTH (KILOMETERS)	TYPE OF SUBDUCTING LITHOSPHERE
KURILES-KAMCHATKA-HONSHU	PACIFIC UNDER EURASIAN	A	2,800	7.5	610	OCEANIC
TONGA-KERMADEC-NEW ZEALAND	PACIFIC UNDER INDIAN	A	3,000	8.2	660	OCEANIC
MIDDLE AMERICAN	COCOS UNDER NORTH AMERICAN	B	1,900	9.5	270	OCEANIC
MEXICAN	PACIFIC UNDER NORTH AMERICAN	B	2,200	6.2	300	OCEANIC
ALEUTIANS	PACIFIC UNDER NORTH AMERICAN	B	3,800	3.5	260	OCEANIC
SUNDRA-JAVA-SUMATRA-BURMA	INDIAN UNDER EURASIAN	B	5,700	6.7	730	OCEANIC
SOUTH SANDWICH	SOUTH AMERICAN SUBDUCTS UNDER SCOTIA	C	650	1.9	200	OCEANIC
CARIBBEAN	SOUTH AMERICAN UNDER CARIBBEAN	C	1,350	0.5	200	OCEANIC
AEGEAN	AFRICAN UNDER EURASIAN	C	1,550	2.7	300	OCEANIC
SOLOMON–NEW HEBRIDES	INDIAN UNDER PACIFIC	D	2,750	8.7	640	OCEANIC
IZU-BONIN-MARIANAS	PACIFIC UNDER PHILIPPINE	D	4,450	1.2	680	OCEANIC
IRAN	ARABIAN UNDER EURASIAN	E	2,250	4.7	250	CONTINENTAL
HIMALAYAN	INDIAN UNDER EURASIAN	E	2,400	5.5	300	CONTINENTAL
RYUKYU-PHILIPPINES	PHILIPPINE UNDER EURASIAN	E	4,750	6.7	280	OCEANIC
PERU-CHILE	NAZCA UNDER SOUTH AMERICAN	E	6,700	9.3	700	OCEANIC

MAJOR SUBDUCTION ZONES and some of their principal characteristics are listed. One of the smallest plates, the Nazca plate, is associated with the longest single subduction zone, embracing almost the entire west coast of South America. It also has the second-highest subduction rate: 9.3 centimeters per year perpendicular to the arc of the earth's surface. In general the more rapidly a plate descends, the greater is the maximum depth of earthquakes associated with it. (A major exception is the subduction zone under the Philippines.) The five principal types of subduction zone (*A—E*) are depicted schematically in the illustration on page 8.

In this model the interior of the descending plate remains distinctly cooler than the surrounding mantle until the plate reaches a depth of about 600 kilometers. As the plate penetrates deeper its interior begins to heat up more rapidly because of the more efficient transfer of heat by radiation. When the plate goes beyond a depth of about 700 kilometers, it can no longer be thermally distinguished as a structural unit. It has become a part of the mantle. Significantly, 700 kilometers is a depth below which no earthquake has ever been recorded. Apparently deep earthquakes cannot occur except in descending plates; therefore the occurrence of such earthquakes implies the presence of sunken plate material.

The descending lithosphere does not always, however, penetrate to 700 kilometers before it is assimilated. A slow-moving plate will attain thermal equilibrium before reaching that depth. For example, at a velocity of one centimeter per year the subducting plate will be assimilated at a depth of about 400 kilometers. If subduction ceases altogether, the subducted segment of the lithosphere will lose its identity and become part of the surrounding mantle in roughly 60 million years. At half that age a stationary plate will already have become too warm to generate earthquakes. These calculations make it clear why we can identify only those subducted plates that are associated with the latest episode of sea-floor spreading. Although there are surface geological expressions of older subduction zones, the plates subducted under these regions cannot be identified in the earth's mantle. The old slabs are lost not only because of the assimilation process but also because of the motion of the surface with respect to the mantle.

So far I have been describing ideal subduction zones without major complications. Such zones are found, for example, under the Japanese island of Honshu, under the Kuriles (extending to the north of Japan) and under the Tonga-Kermadec area (to the north of New Zealand). In many other areas the lithosphere descends in a more complicated manner.

In new subduction areas the descending slab may have penetrated a good deal less than 700 kilometers, as is the case under the Aleutians, the west coast of Central America and Sumatra. In other areas where the subduction rate is low the slab may be assimilated well before it reaches that depth; the subduction

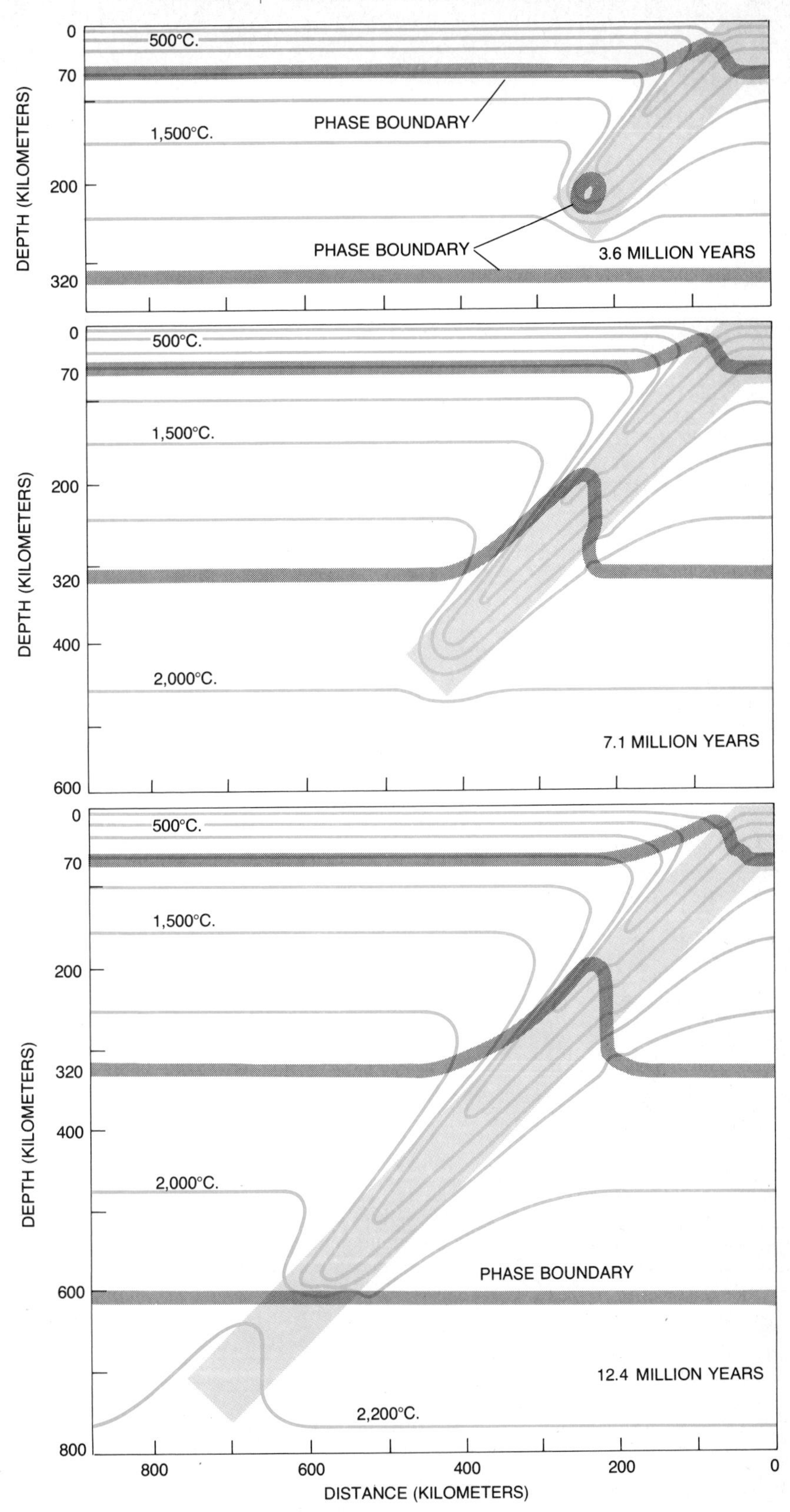

EVOLUTION OF DESCENDING SLAB is described by computer models developed in the author's laboratory at the Massachusetts Institute of Technology. These diagrams depict the fate of a slab subducting at an angle of 45 degrees and at a rate of eight centimeters per year. Phase changes, induced by increasing pressure, normally occur at depths of 70, 320 and 600 kilometers. In the descending slab the first two phase changes occur at shallower depths because of the slab's lower temperature. The phase conversions to denser mineral forms help to heat the slab and to speed its assimilation. When the slab reaches the temperature of the surrounding mantle at a depth of 700 kilometers, it loses its original identity.

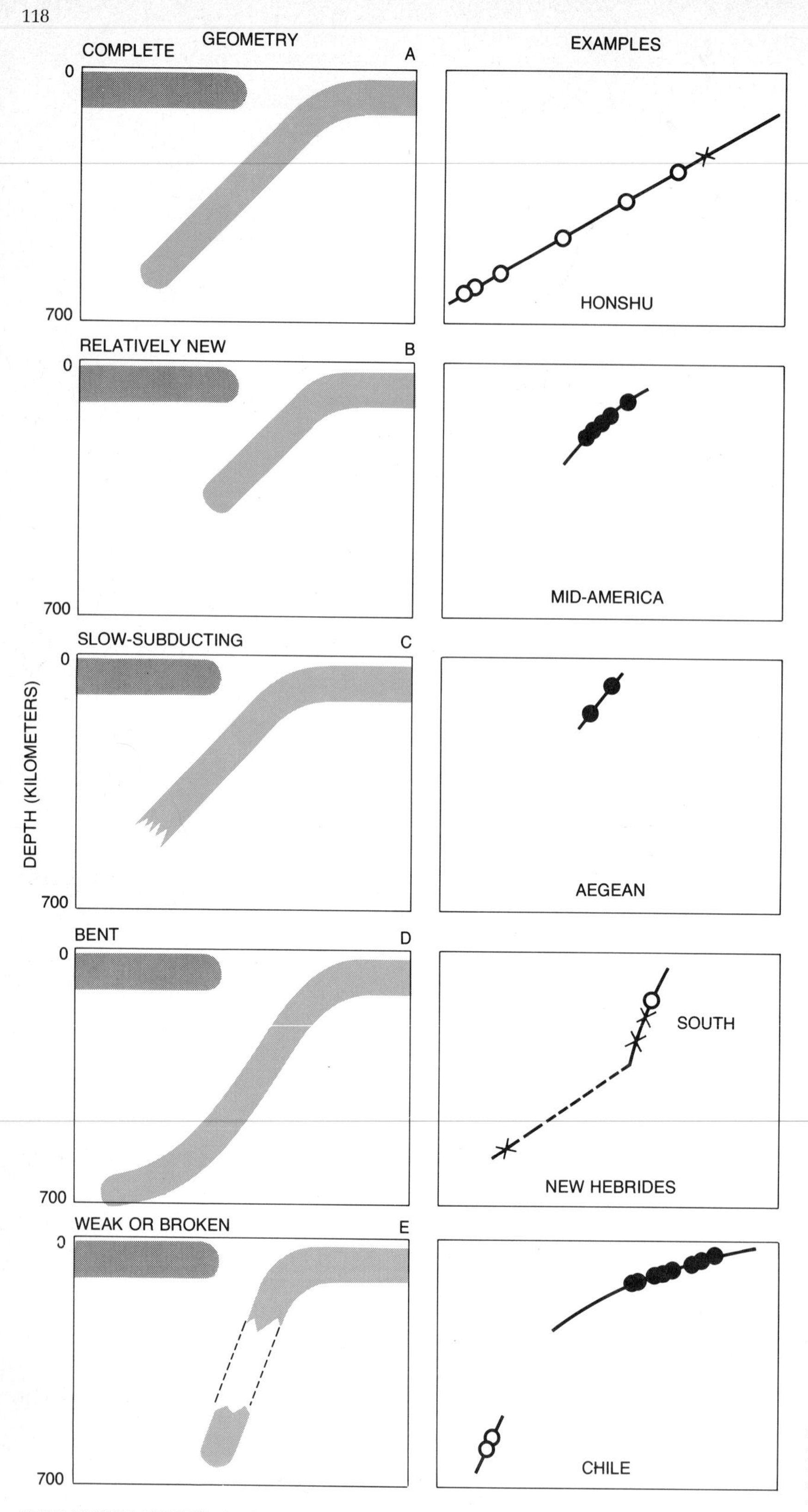

FIVE MAJOR TYPES of subducting oceanic slabs can be identified. Examples of each type are shown to the right of the schematic diagrams. In the examples the solid lines represent the location of all earthquakes projected onto a cross section. The symbols on the lines identify particularly large earthquakes from which the direction of stress was determined. Open circles indicate compression along the length of the slab; filled circles indicate tension along the length of the slab, and crosses show stresses that do not lie in the plane of the cross section. Many subduction zones exhibit a "seismic gap" between 300 and 500 kilometers where no earthquakes occur. It is not known whether this is because the slab is broken (*Type E*) or because stresses are absent at that depth. Examples given are based on a survey conducted by Bryan L. Isacks of Cornell University and Peter Molnar of M.I.T.

of the Mediterranean plate under the Aegean Sea is an example. In still other areas the subduction starts at a shallow angle, gets steeper at intermediate depths and bends again nearly to the horizontal at about 500 kilometers. Such a sigmoid configuration is observed dramatically under the New Hebrides in the South Pacific. The double bend may be attributable to low resistance in the upper asthenosphere and much greater resistance at a depth of 600 kilometers, resulting from an increase either in the density or in the strength of the mantle, possibly both. Another anomalous situation is found under Peru and Chile, where there is a marked absence of earthquakes at intermediate depths, indicating a stress-free zone or possibly a broken slab.

Most frequently the oceanic lithosphere is subducted under an island arc, as is generally the case in the western Pacific. Here, however, there are many other combinations and complications. For example, a small oceanic plate, such as the Philippine plate, may get trapped between two trenches. Or an oceanic plate may be subducted under a continent, as in the case of the Nazca plate, which plunges under the Andes. The Andes can be regarded as being equivalent to an overgrown island arc. Elsewhere transform faults such as the San Andreas fault may interrupt subduction boundaries. In other cases multiple subduction zones may develop within relatively small areas. Finally, subducting plates may bring two continents together, with major tectonic consequences. Continental collisions place major restrictions on plate motions because the buoyancy of the continental crust, which is less dense than the mantle, resists subduction. Collisions of this type create major mountain belts, such as the Alps and the Himalayas.

Continental subduction is qualitatively different from oceanic subduction because it is a transient process rather than a steady-state one. When continental crust moves into a subduction zone, its buoyancy prevents it from being carried down farther than perhaps 40 kilometers below its normal depth. As plate convergence continues the crust becomes detached from the plate and is itself underthrust by more continental crust. That creates a double layer of low-density crust, which rises buoyantly to support the high topography of a major mountain range. It is possible that the long oceanic slab below the surface ultimately becomes detached and sinks; in any case it is no longer a source of earthquakes. After this stage further deformation and compression may take place behind the line of collision, producing a high plateau with surface volcanoes, like the plateau of Tibet. Eventually the plate convergence itself will stop as resisting forces build up. It now seems that continental collisions are probably a major factor in the periodic reorientations of the relative motions of the plates.

It is clear that an understanding of the geological, geochemical and geophysical consequences of lithospheric subduction helps to explain many major features of the earth's surface. At the same time the observable features enable us to test the validity of theoretical subduction models. A wide variety of features can be investigated. For the sake of brevity I shall mention only the geological characteristics of the trench sediments and the subducting crust, the andesitic magmas associated with island-arc volcanoes, and heat-flow and gravity anomalies. The measurable quantities related to these features are primarily sensitive to the properties of subduction down to a depth of about 100 kilometers. The most definitive observations on the deeper parts of subducting plates are seismic observations. The velocity and attenuation of seismic waves, and most significantly the indication the waves give of the locations of deep- and intermediate-focus earthquakes, outline the extent of the relatively cool and rigid zone of the descending lithosphere.

With the passage of time the deep oceanic trenches created by descending plates accumulate large deposits of sediment, primarily from the adjacent continent. As the sediments get caught between the subducting oceanic crust and either the island arc or the continental crust they are subjected to strong deformation, shearing, heating and metamorphism. Profiles of seismic reflections have identified these deformed units. Some of the sediments may even be dragged to great depths, where they may eventually melt and contribute to volcanism. In this case they would return rapidly to the surface, and the total mass of low-density crustal rocks would be preserved.

A prominent feature of subduction zones is volcanism that gives rise to andesite, a fine-grained gray rock. Where the magma for these volcanoes originates is not definitely known. Most geochemical and petrological evidence favors a depth of about 100 kilometers for the magma source. The magma may come from the partial melting of the subducted oceanic crust, as A. E. Ringwood of the Australian National University suggested in 1969. The shearing that takes place at the top of the descending plate may provide the heat required for partial melting. Convective motions in the wedge of asthenosphere above the descending plate may also contribute to magma sources by raising asthenospheric material to a depth where it could melt slightly under lower pressure.

The flow of heat through the earth's surface tells us something about the thermal characteristics of shallow layers. (It is influenced only indirectly by deeper phenomena.) Trenches have low heat flow (less than one microcalorie per square centimeter per second); island arcs generally have a high and variable heat flow because of their volcanism. High heat flow is also associated with the marginal basins behind the island arcs, for example the Sea of Japan, the Sea of Okhotsk, the Lau Basin west of Tonga and the Parece Vela Basin behind the Marianas arc.

These basins are underlain by relatively hot material brought up either by convection currents behind the island arc or by upwelling from deeper regions. The convection is induced in the wedge

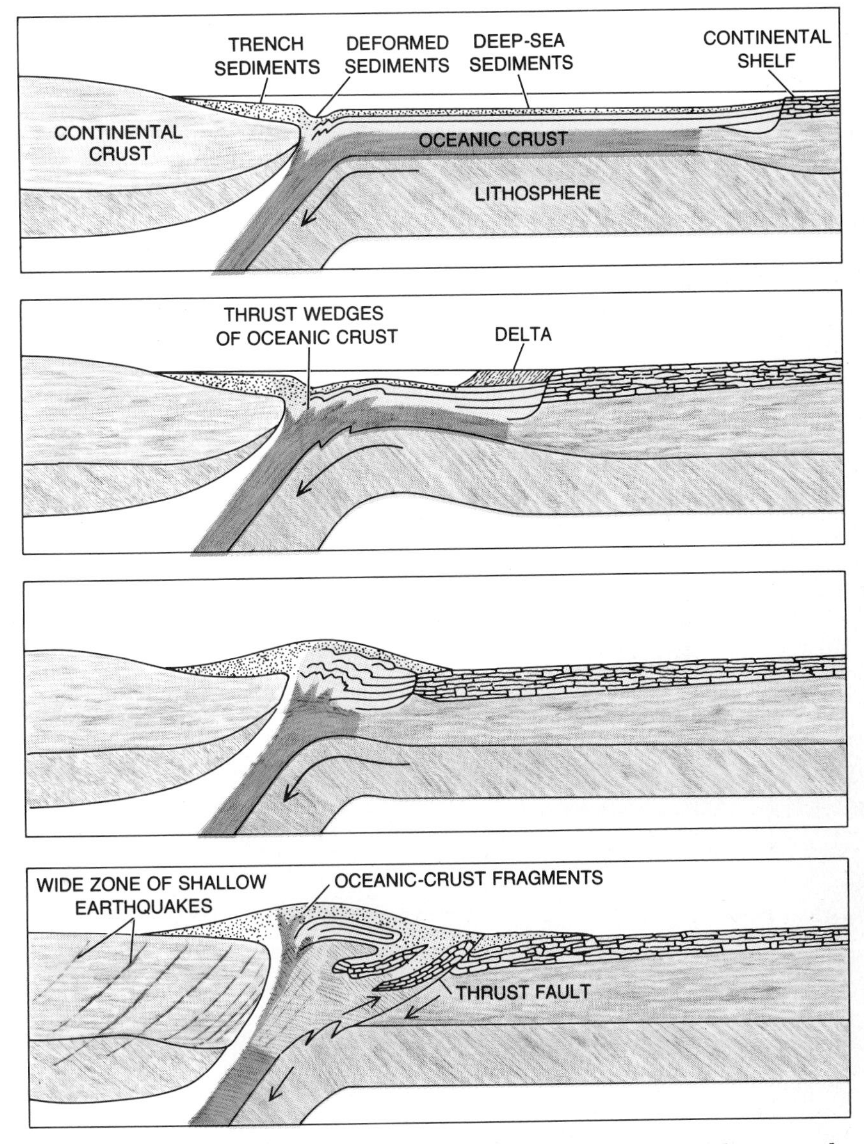

COLLISION OF CONTINENTS occurs when an oceanic slab that is subducting at the edge of one continent (*left*) is itself part of a lithospheric plate bearing a second continent (*right*). Such a collision took place when the Indian lithospheric plate, traveling generally northward for 200 million years, subducted under Eurasian plate. This kind of subduction eventually ends, but not before crust of subducting plate has been detached and deformed and has pushed up a mountain range (in the case of the Indian plate the Himalayas).

of asthenosphere above the descending lithospheric plate by the downward motion of the plate. Since it takes time for such currents to be set in motion, high heat flow would not be expected in basins behind the youngest subduction zones. Indeed, the observed heat-flow values in the Bering Sea behind the Aleutians are normal.

Gravity anomalies associated with subduction zones are large and broad. A descending lithospheric plate is cooler and denser than the surrounding mantle; therefore it gives rise to a positive gravity anomaly. The hot region under a marginal basin would show a density lower than normal and hence would create a negative gravity anomaly. Changes in the character of the crust from the ocean to an island arc or a continent add more anomalies. A combination of all these anomalies is needed to account for the gravity observations that have been made across subduction zones [*see illustration on opposite page*]. The gravity evidence provides strong support for the subduction models, but it is not conclusive because of the uncertainties as to the depth of the masses that give rise to the anomalies.

The most compelling evidence for the subduction of the lithosphere comes from seismology. Most of the world's earthquakes and nearly all the deep- and intermediate-focus earthquakes are associated with subduction zones. The hypocenters of the earthquakes and their source mechanisms can be explained by the stresses in the subducting plate. The models that explain the seismic-wave observations outline the location of the subducted cool lithospheric plates. In some areas (Japan, the Aleutians, the Tonga Trench, South America) the data are abundant and convincing.

The general picture that emerges is as follows. At shallow depths, where the edges of the two rigid lithospheric plates are pressing against each other, there is intense earthquake activity. Many of the world's greatest earthquakes (for example the Chile earthquake of 1960, the Alaska earthquake of 1964 and the Kamchatka earthquake of 1952), as well as many smaller ones, occur along the shear plane between the subducting oceanic lithosphere and the continental or island-arc lithosphere. Some normal-faulting (tensional) earthquakes on the ocean side of a trench are caused by arching of the lithosphere. Other earthquakes result from the tearing of the lithosphere and other adjustments in this zone of intense deformation.

The deep- and intermediate-focus earthquakes generally occur along the Benioff zone, a plane that dips toward a continent. At first this plane was thought to be the shear zone between the upper surface of the descending lithospheric plate and the adjoining mantle. Detailed studies conducted by Bryan L. Isacks of Cornell University and Peter Molnar of M.I.T. and others over the past 10 years have shown, however, that the forces needed to account for the observed earthquakes could not be provided by the shearing process. These studies, combined with more precise determinations of the location of earthquake foci under several island arcs, indicate that the deep- and intermediate-focus earthquakes occur in the coolest region of the interior of the descending plate. The stresses generated by the gravitational forces acting on the dense interior of the slab and the resistance of the surrounding mantle to the slab's penetration are also highest in the coolest region. Moreover, the cool and rigid interior of the slab acts as a channel to transmit stresses. The computed directions of the stresses are consistent with the directions that have been deduced from earthquakes.

These concepts can be tested in areas where detailed studies of earthquakes have been made. Two such regions are the Aleutians and Japan. At Amchitka Island in the central Aleutians the nuclear explosions named Longshot, Milrow and Cannikin provided energy sources with precisely known locations and times. From the travel times of the seismic waves going through the subducting lithosphere the location of the coolest region was determined precisely. The dense network of seismic stations installed in the area also provided precise locations of earthquakes. The shallow earthquakes are concentrated along the thrust plane and the deeper ones along the coolest region [*see illustration at left*].

The islands of Japan constitute probably the most intensively studied seismic belt in the world. The velocities of seis-

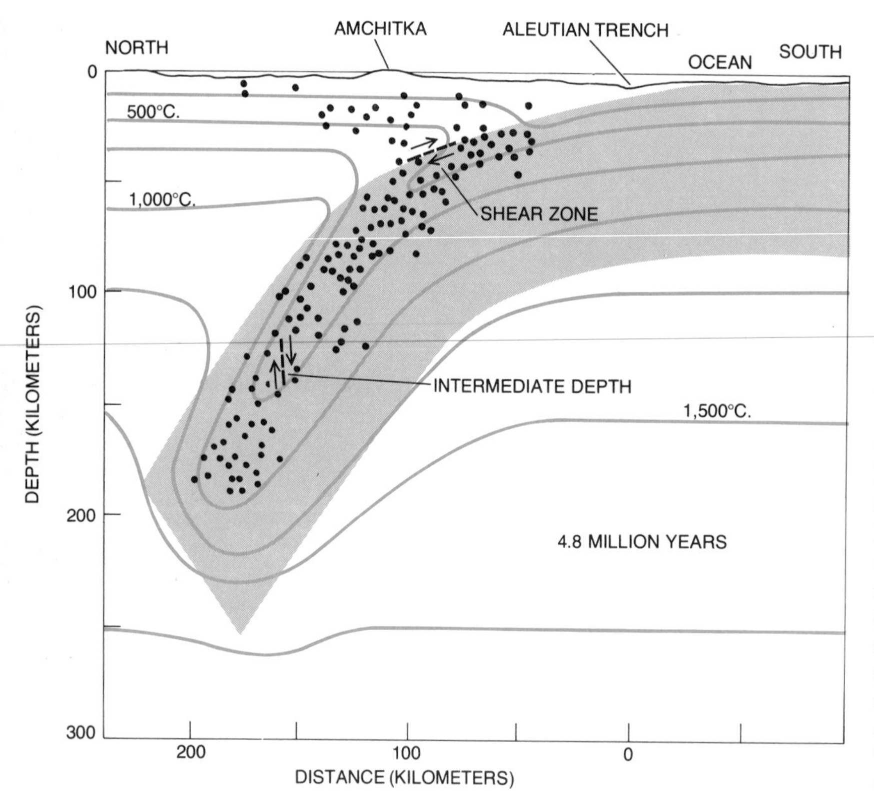

ALEUTIAN EARTHQUAKES mark the general location of the subducting Pacific plate in that region. The precise location of the cold descending slab in relation to the earthquakes was determined with the help of seismic waves from nuclear tests on Amchitka Island, which showed that the waves travel more rapidly through the cold slab than through the surrounding mantle. A computed simulation of seismic records revealed that intermediate-depth earthquakes (*dots*) occur in the cold center of the slab, as shown here, and not, as had been thought, at the shear zone on its upper face. At shallower depths the earthquakes occur in shear zone and in overriding plate. Arrows show the slip planes and the sense of motion.

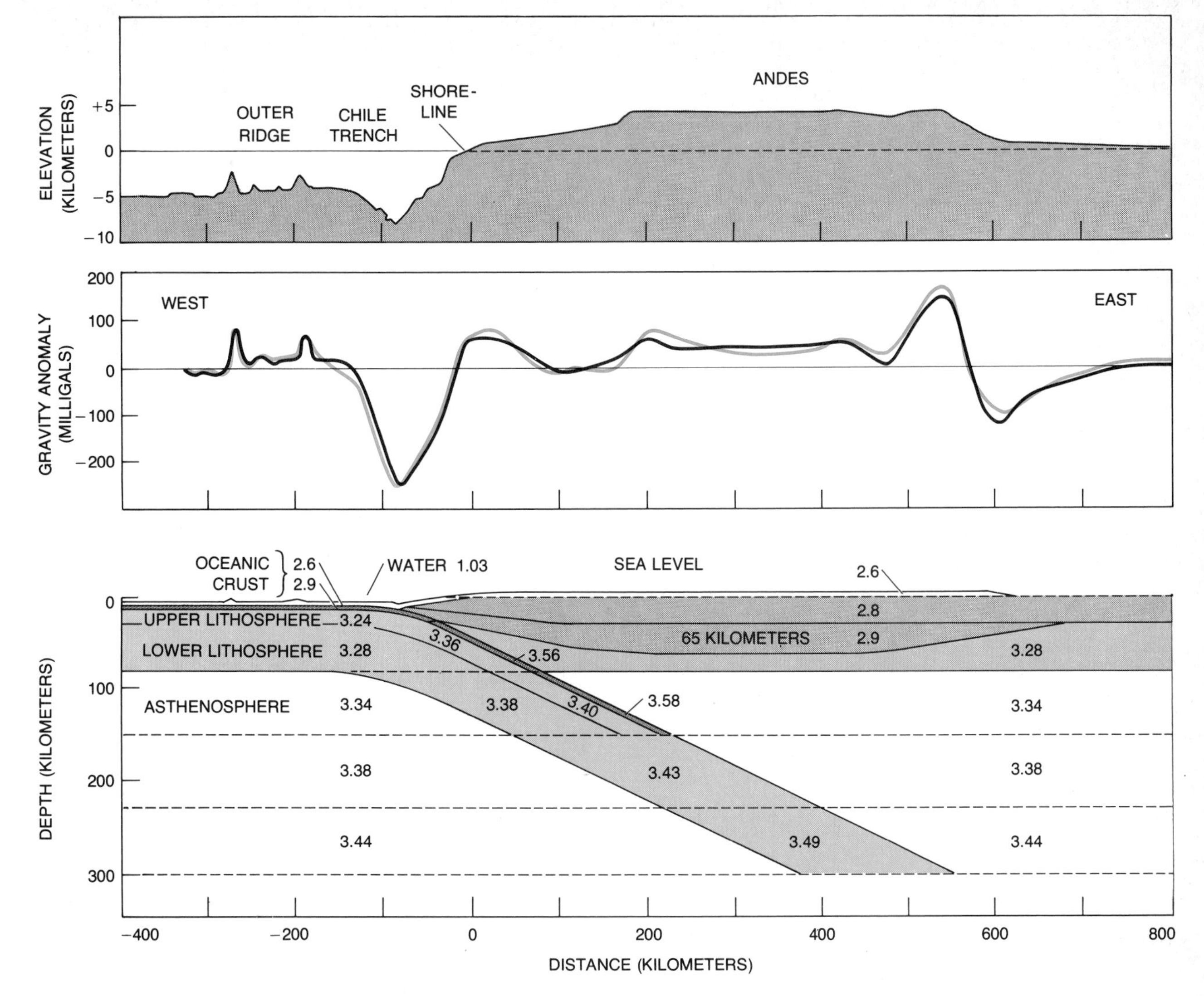

EFFECT OF A SUBDUCTING PLATE ON GRAVITY is clearly represented in the gravity anomaly that has been measured over the west coast of Chile and the Andes. The diagram at the top is a topographical cross section of the region. The observed gravity anomaly, given in milligals, is shown by the black curve in the middle diagram. The colored curve is the anomaly calculated on the basis of the lithospheric model shown at the bottom. (One gal, named for Galileo, is one 980th the normal gravity at the earth's surface; thus an anomaly of —260 milligals over the trench corresponds to a gravity deficit of about .026 percent.) The model includes the trench, which gives rise to the gravity low, and cold dense slab, which has opposite effect. Densities given in model are in grams per cubic centimeter. Model was worked out by J. A. Grow and Carl O. Bowin of the Woods Hole Oceanographic Institution.

mic waves, the characteristics of the waves' attenuation, the precise locations of earthquake hypocenters and the focal mechanisms all fit the subduction model in this region. The descending plate shows high velocities and low attenuation, which is a measure of the nonelastic damping of high-frequency seismic waves. There are numerous shallow earthquakes along and near the boundary where the plates meet near the surface. Deep- and intermediate-focus earthquakes are in the coolest region of the slab where the stresses are highest [*see illustration on next page*]. In other subduction zones the locations of earthquakes are not as precisely known. Nevertheless, wherever adequate data exist, for example for the areas of the Tonga Trench and of Peru and Chile, the deep- and intermediate-focus earthquakes are found to occur in the interior of the subducting plate along the coolest region.

The absence of earthquakes below a depth of 700 kilometers can now be explained. The descending lithosphere heats up below that depth and can no longer behave as a rigid elastic medium susceptible to faulting or brittle fracture. Moreover, below that depth the stresses are small, and they are relieved by slow plastic deformation rather than by the sudden failure associated with an earthquake.

The gravitational energy associated with large masses of subducting cool, dense material is large even in terms of the total energy associated with plate motions. The gravitational forces are largely balanced by the resistance of the mantle to the penetration of the descending lithosphere. The net force acting on the plates in the subduction zone is still enough to play a major role in global plate motions. Other forces that contribute are the horizontal flow of convection currents under the plates and the outward push of the material coming to the surface at the mid-ocean ridges.

Not all the problems of plate motions and subduction have been solved. It is puzzling, for example, that the Pacific plate can move laterally for 6,000 kilometers before it subducts. It is not known why some subduction zones are where they are. It is not clear why plate motions change at certain times. These are minor problems, however, compared with the understanding of continental drift, earthquakes, volcanism and mountain building that has been gained. The theory of plate tectonics is a concept that unifies the main features of the earth's surface and their history better than any other concept in the geological sciences.

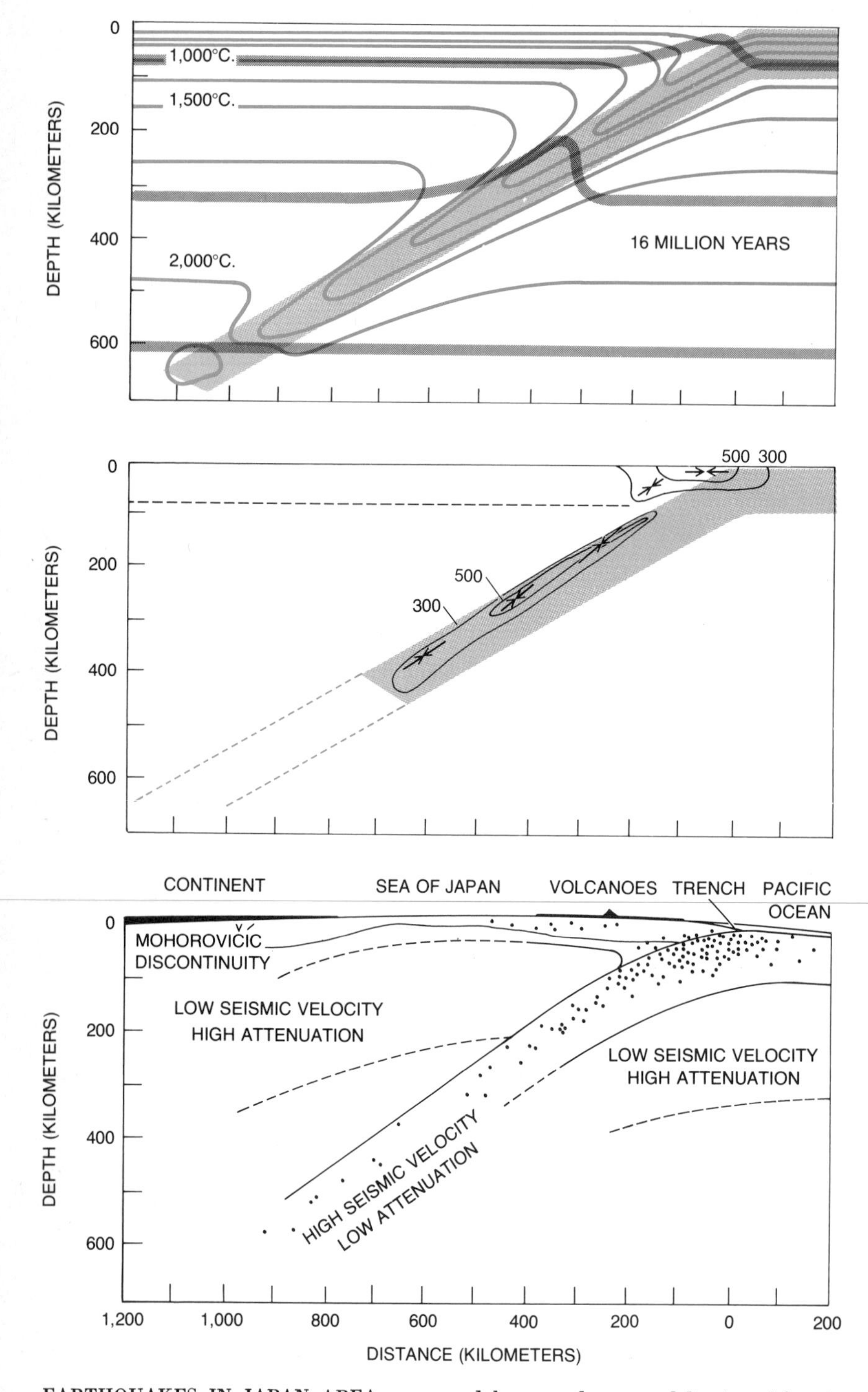

EARTHQUAKES IN JAPAN AREA are caused by several westward-dipping slabs of Pacific lithosphere. The author has calculated a temperature model for a typical Japan slab (*top*). This in turn has been used to calculate the stresses generated within the upper portion of the slab (*middle*). The stresses result from the interaction between the slab's tendency to sink because of its high density and opposing forces: friction near surface and viscous drag in asthenosphere. Nonhydrostatic stresses are computed in bars (one bar is 14.7 pounds per square inch). Arrows show direction of compression. Calculated stresses account well both for distribution of earthquakes (*bottom*) and their mode of initiation.

III

EXAMPLES OF PLATE MOTION

III

EXAMPLES OF PLATE MOTION

INTRODUCTION

The four papers in this section describe in some detail the way in which continents have broken apart and the rifts between them have spread to form ocean basins.

Appropriately enough the first paper, "The Break-up of Pangaea," by Robert S. Dietz and John C. Holden, presents an overview of plate movements during the Mesozoic and Cenozoic eras. After reviewing the principles of plate tectonics, Dietz and Holden consider what the global pattern of continents would have been 200 million years ago. They agree with Alfred L. Wegener that they formed a single supercontinent, for which Wegener coined the name Pangaea, and they provide a map of it. Apparently Pangaea was then on the point of breaking up; by 180 million years ago it had split apart. At that time the eastern shore of Pangaea was deeply indented near the equator by a great seaway called the Tethys Sea, which has since closed to form the Himalayan-Alpine mountain chains and the Mediterranean Sea. This connected with the Caribbean Sea and the Pacific Ocean to form the first opening in the Atlantic (between what is now the Atlantic coast of the United States, on the northwest, and what is now northwest Africa and South America, on the southeast), separating Pangaea into two supercontinents—Laurasia in the Northern Hemisphere and Gondwanaland in the Southern Hemisphere. Gondwanaland also started to break up at about that time. Dietz and Holden illustrate these and later stages (135 and 65 million years ago) by maps. They also provide a map to show what they speculate the world will look like 50 million years from now, if the present directions and rates of motion continue.

Without a doubt this reconstruction of Pangaea contains errors; indeed, the authors have already changed their minds and would omit the Kerguelen–Broken Ridge fragment to the west of Australia because they no longer believe it to be a microcontinent.

Their reconstruction draws attention to a number of points about which there has been debate and of which many are still unsettled. In contrast to the fitting of South America to Africa, about which there is no controversy, the east coast of the United States can be placed against the curved shore of West Africa in a number of ways. Dietz and Holden followed the pattern that E. C. Bullard, A. G. Smith and J. E. Everett had already obtained, with the aid of a computer, as the best fit between the 500-fathom submarine contours of the two continents. It must be remembered, however, that the shape of the continental shelves may have changed as they grew and were modified by ocean currents. Others would rotate North America farther south so that South America fills the Gulf of Mexico. This removes the Bahamas overlap and involves a different arrangement of the ancestral fragments of Mexico and

Central America, which may not have been joined together at that time, and which certainly had not then assumed their present positions. Still others would delay the break-up of the Indian Ocean and place the original position of Madagascar farther south, in the bay into which the Limpopo River debouches.

Many also consider that the wide and shallow shelves in and around the Bering Strait indicate that it has not been a permanent seaway, but that the eastern tip of Siberia is geologically part of North America. They consider that the passage into Sinus Borealis, shown on the first map of Dietz and Holden, lay along the locus of the Verkhoyansk Mountains before that range was thrust up by the union of the tip of Siberia with the rest of Asia.

Dietz and Holden support the concept that hot spots exist as long-lived, stationary upwellings in the deep mantle. They use the one presently beneath the island of Tristan da Cunha as a fixed reference point from which to estimate the past longitude and latitude of the continents. This concept has been developed further in the last article of Section I.

"The Evolution of the Indian Ocean," by D. P. McKenzie and J. G. Sclater, reconstructs the latter part of the history of development of the Indian Ocean, using maps of magnetic anomalies on the ocean floor. The complex pattern contrasts with the simple opening of the South Atlantic Ocean. The reason is probably connected with the need for mid-ocean ridges to adjust their shapes and positions. Consider the location of the ridge when the southern continents were assembled together in Pangaea as Dietz and Holden showed in their article. It must have lain against and have had the shape of the coast of Africa. Today the oceans have grown and the ridge has expanded to quite a different shape. In addition, if we follow Dietz and Holden's argument and agree that the South Atlantic Ridge remained fixed over Tristan da Cunha (at least until later Tertiary time), then Africa must have moved away from the ridge, and in consequence the parts of the ridge through the Southern Indian Ocean would have had to move far over the deeper mantle and greatly change their shape. It is believed that this has led to the abandonment of some old ridges and the formation of some new ones, producing the complex pattern of ridges described by McKenzie and Sclater.

"The Evolution of the Pacific," by B. C. Heezen and I. D. MacGregor, considers the effects of plate motions upon the sedimentary deposits on the floor of the Pacific Ocean. Like the South Atlantic and unlike the Indian Ocean and the Arctic part of the North Atlantic, the Pacific Ocean presents a simple pattern. For more than 100 million years the East Pacific Rise has been spreading from the same location. This has been possible because, unlike the Indian Ocean, the Pacific basin is decoupled from plate movements elsewhere by the subduction zones that form the well-known Ring of Fire surrounding all its margins except the southern.

Heezen and MacGregor point out that upwelling of ocean currents along the equator supports abundant life from which shells rain down to form a deposit of sediments thicker beneath the equator than elsewhere. Drilling in the floor of the ocean has traced these thick deposits, which have been found to indicate that the Pacific plate has moved northward. The indications of motion of the Pacific plate given by the equatorial deposits are compatible with the evidence for its motion relative to the volcanoes of Hawaii, which are held by these authors to have arisen from a hot spot like Tristan da Cunha.

"The Floor of the Mid-Atlantic Rift," by J. R. Heirtzler and W. B. Bryan, gives an account of a joint French-American investigation of the axial rift of the Mid-Atlantic Ridge south of the Azores Islands, and shows in detail how lava wells up along the rift at the crest of the ridge.

10 The Breakup of Pangaea

Robert S. Dietz and John C. Holden
October 1970

The history of science is replete with outrageous hypotheses. They are mostly forgotten, as best they should be, but from time to time one of them turns out to be true. So it was with the concept that the earth is a sphere spinning in space, supported by nothing at all. Now it also seems to be with the theory of continental drift, which in its extreme form holds that all the continents were once joined in a single great land mass. Named Pangaea, this universal continent was somehow disrupted, and its fragments—the continents of today—eventually drifted to their present locations.

Over the past three years geologists and geophysicists have been forced to abandon the old dogma that the crust of the earth is essentially fixed and to accept the new heresy that it is quite mobile. The notion that continents can drift thousands of kilometers in a few hundred million years is now generally accepted. Geology therefore finds itself in much the same position that astronomy was in at the time of Copernicus and Galileo. Textbooks are being rewritten to embrace the new mobilistic viewpoint.

Although the theory of continental drift has triumphed, many of its details remain uncertain. Advocates of drift are challenged to say exactly how the present continents fitted together to form Pangaea, or alternatively to reconstruct the two later supercontinents Laurasia and Gondwana, which some theorists prefer to a single all-embracing land mass. The original concept of Pangaea ("all lands") was proposed in the 1920's by Alfred Wegener. Most attempts to improve on his reconstruction have been rather generalized sketches showing how the continents might have been joined. A few workers have made jigsaw fits with considerable care but without taking advantage of the latest concepts in geotectonics. Recently British theorists have presented detailed reconstructions showing how land masses were juxtaposed before the opening of either the Atlantic or the Indian Ocean, but their solutions show only the relative motions of the masses involved.

In this article we present a reconstruction of Pangaea in which the continents are assembled with cartographic precision. For the first time Pangaea is positioned on the globe in absolute coordinates. This reconstruction is accompanied by four maps that show the breakup and subsequent dispersion of the continents by the end of the four major geologic periods covering the past 180 million years: the Triassic, the Jurassic, the Cretaceous and the Cenozoic.

The guiding rationale for our reconstruction is the drift mechanism associated with plate tectonics and sea-floor spreading [*see illustration on page 128*]. According to this concept the earth has a strong lithosphere, or outer shell of rock, about 100 kilometers thick. Presumably in response to forces generated in the asthenosphere, the weak upper mantle of rock underlying the lithosphere, the shell was broken up into a number of separate plates. There are now some 10 major plates, plus numerous additional subplates. The continents resting on these plates were rafted across the surface of the globe.

The mechanism of plate movement is not yet clear. The plates may be pushed, carried by convection cells in the mantle, driven by gravitational forces or pulled. We prefer a model based on pulling; we suspect that plates are colder and heavier at one boundary than elsewhere and thus dive down into the earth's mantle along "subduction" zones. These zones usually show themselves as deep trenches, which are disposed principally around the periphery of the Pacific. As a result a tear, or rift, widens along the opposite boundary of the plate; this rift is filled by a solid flow of viscous mantle rock and by dikes of molten tholeiitic basalt (a differentiated partial melt of the mantle). Because the mantle rock and its basaltic derivative are both heavier than the granitoid rock of the continents they assume a level about four kilometers below sea level. Consequently such a pulled-apart region always becomes new ocean floor. As two adjacent plates continue to pull apart, basaltic dikes continue to pour into the suboceanic rift, which remains midway between the two plates. This highly symmetrical process, which creates new ocean basins or continuously repaves old ocean floors, is termed sea-floor spreading. The rate of spreading, measured from the mid-ocean rift to either plate, is from one centimeter per year (10 kilometers per million years) to several times that figure. This is remarkably rapid by geological standards, being many times faster than mountains are elevated by tectonism or leveled by erosion. For example, the North American plate is moving westward the length of one's body in a lifetime.

The discovery of a mid-ocean ridge system some 40,000 kilometers long, winding through all the ocean basins, was an important prelude to the sea-floor-spreading hypothesis. It was soon recognized that the ridge has a fossa, or axial depression, into which dikes of basalt are continuously being injected. This linear depression in the ridge marks the location of the rift. The term "mid-ocean," although appropriate for the part of the ridge system in the Atlantic

SUBCONTINENT OF INDIA, originally attached to what is now Antarctica, made the longest migration of all the drifting land masses: approximately 9,000 kilometers in 200 million years. This picture, taken at an altitude of 650 kilometers from *Gemini XI* in September, 1966, shows all of the subcontinent. The Himalayan mountains, 3,700 kilometers away, are just visible on the horizon.

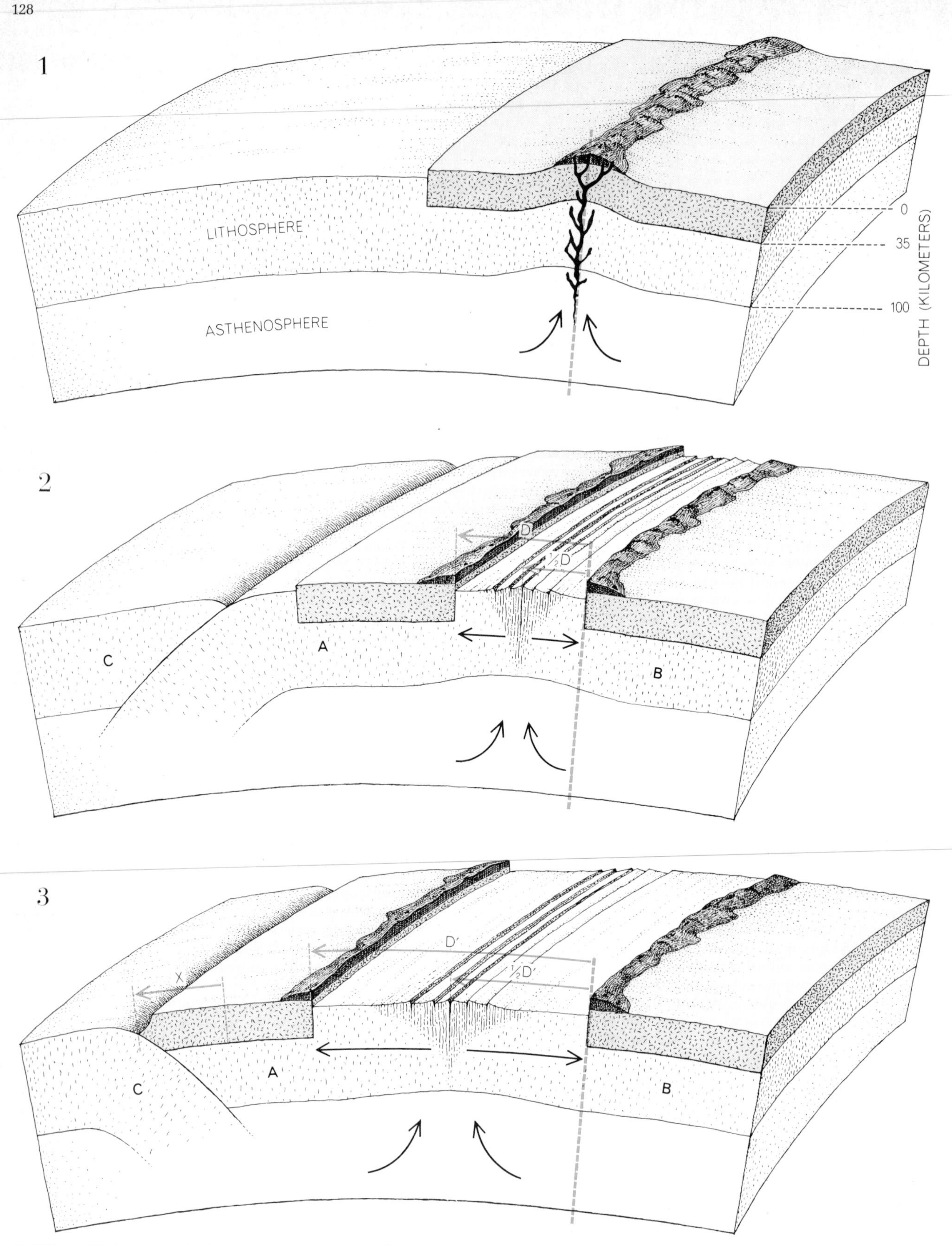

THEORY OF PLATE TECTONICS provides a mechanism for continental drift. The process begins (*1*) when a spreading rift develops under a continent (*color*) that is resting on a single crustal plate. Molten basalt from the asthenosphere spills out. The second simultaneous requirement for continental drift is the formation of a zone of subduction, or trench, into which oceanic crust of the new moving plate (*A*) is pulled and "consumed" (*2*). As the new continent carried by plate *A* is rafted to the left, a new ocean basin is created between the two land masses. In the third stage (*3*) the continent on plate *A* encounters and overrides the trench for some distance (*X*) and eventually reverses, or flips, its direction from west-dipping to east-dipping. Because the continent on plate *B* is here arbitrarily fixed, the mid-ocean rift migrates to the left, remaining in the center of the new ocean basin, whose width is *D′*.

and the Indian Ocean, is a misnomer for the ridge in the Pacific. The Atlantic and the Indian Ocean are rift oceans, formed where continents were once split apart; therefore it is natural for the axis of spreading, marked by the ridge system, to remain in the center of these two oceans. The Pacific, on the other hand, is not a rift ocean; it is clearly the ancestral ocean, and it is becoming smaller as new ocean basins grow. Although the Pacific also has a ridge, it runs north-south well to the east of the ocean's center.

In reality the crustal motions are considerably more complex than the ones we have just outlined. The trenches and rifts apparently migrate, and the opposing plates are also subject to displacements produced by internal shears. The "megashears," the large zones of slippage along plate boundaries, also seem able to accommodate minor amounts of crustal extension or compression. Few of the plates are "ideal" in the sense of being rectilinear, of having a rift matched by an opposing trench and of having these two antithetical zones connected by a megashear. The Antarctic plate, for example, has no trench at all. Perhaps this anomaly is partly explained by the fact that a sphere cannot be covered with rectangles.

We can visualize the continents as being passively rafted over the surface of the globe as embedded plateaus of sialic (granite-like) rock resting on the even larger and thicker crustal plates. The continents have generally maintained their size and shape since the breakup of Pangaea. There have been some accretions with the formation of mountain belts, but these have been mostly confined to the sides of continents facing the Pacific. The sides of continents facing rift oceans (the Atlantic and the Indian Ocean) show little change; hence they can be fitted together almost as neatly as pieces of a jigsaw puzzle.

In contrast, the crustal plates can change in size or shape either by the addition of new ocean floor along the rifts or by the resorption of oceanic crust in trenches. Thus it has been possible for the North American and South American plates moving toward the Pacific to grow larger at first and then smaller as they passed over the great circle of the earth and now converge toward the central Pacific. An even more tortured history is reflected in the complex evolution of the Caribbean Sea region, caught as a "gore" between the North American and South Ameri-

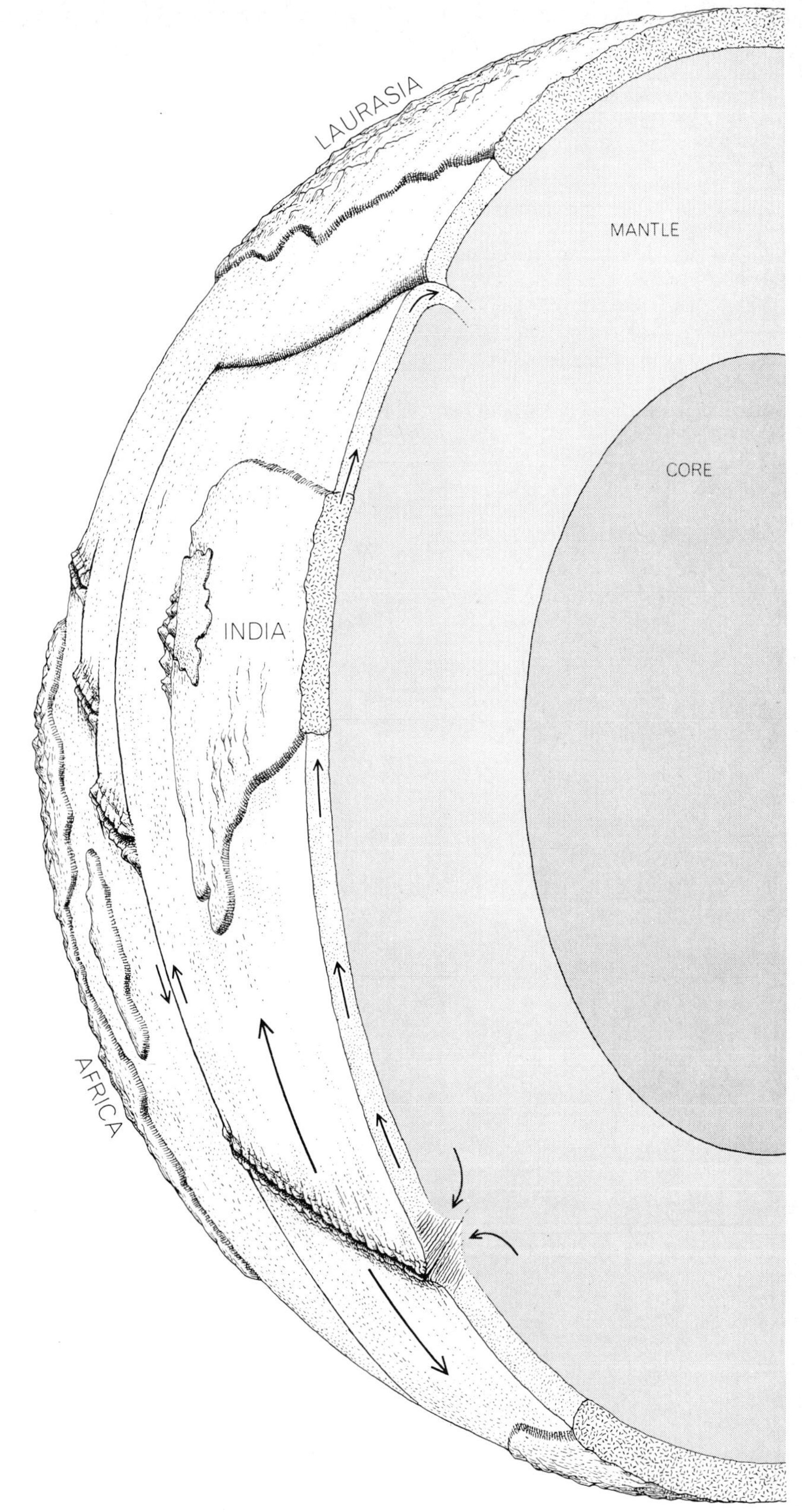

NORTHWARD DRIFT OF INDIA exemplifies how far a land mass can be carried when tectonic conditions are favorable. The plate carrying the Indian land mass is nearly a perfect rectangle, which was sliced away from Antarctica within the primitive universal continent of Pangaea. The plate that rafted India then migrated northward toward and subducted into the Tethyan trench, which ran east-west near the Equator. The plate evidently glided freely along parallel "megashears" on its eastern and western boundaries without interacting with the other crustal plates of the world. India finally collided with and underthrust the southeast margin of Asia, creating the Himalayas, which are thus two plates thick.

can plates, and the Scotia Sea region, similarly trapped between the South American and Antarctica plates. As we shall see, in at least one case two plates evidently collided, producing a mid-continent mountain range: the Himalayas.

In making our reconstruction of Pangaea we selected for fitting not the present coastlines of continents but the contour lines where the continental slope reaches a depth of 1,000 fathoms, or about 2,000 meters [*see illustration below*]. This isobath was selected because it is approximately halfway down the continental slope and thus marks roughly half the height of the vertical walls created when the continents first rifted. On the assumption that these walls subsequently slumped to a condition of stable repose, the 1,000-fathom isobath closely delineates the location of the original break.

For joining the two sides of the Atlantic we have followed, with some modification, the reconstruction proposed by Sir Edward Bullard, J. E. Everett and A. G. Smith of the University of Cambridge. For closing the Indian Ocean we have used the best-fit computer solutions of Walter P. Sproll, a colleague of ours in the Marine Geology and Geophysics Laboratory of the Environmental Science Services Administration. His studies provide precise fits between Australia and Antarctica and between Antarctica and Africa. The three continents together constitute most of Gondwana. Presumably India was also part of the Gondwana complex, but where it was attached remains unclear. Fortunately the pattern of fracture zones in the ocean floor provides crude but useful dead-reckoning tracks showing how the continents drifted. Using such tracks, we have placed the west coast of India against Antarctica rather than against western Australia, the fit that is often proposed.

Another difficult fit is presented by the bulge of Africa and the bight of North America. The areas of mismatch, particularly that caused by the Florida-Bahamas platform, are sufficiently large for one to reasonably argue that Africa and North America were never joined. On this assumption instead of Pangaea one obtains two unconnected supercontinents as the antecedent land masses: Laurasia in the Northern Hemisphere and Gondwana in the Southern. This version of the continental-drift theory has important adherents.

We nevertheless prefer the Pangaea reconstruction; in our view the areas of mismatch can reasonably be regarded as modifications that arose after Africa and North America began drifting apart. We regard the Florida-Bahamas platform as a sedimentary infilling of a small ocean basin that appeared when Africa

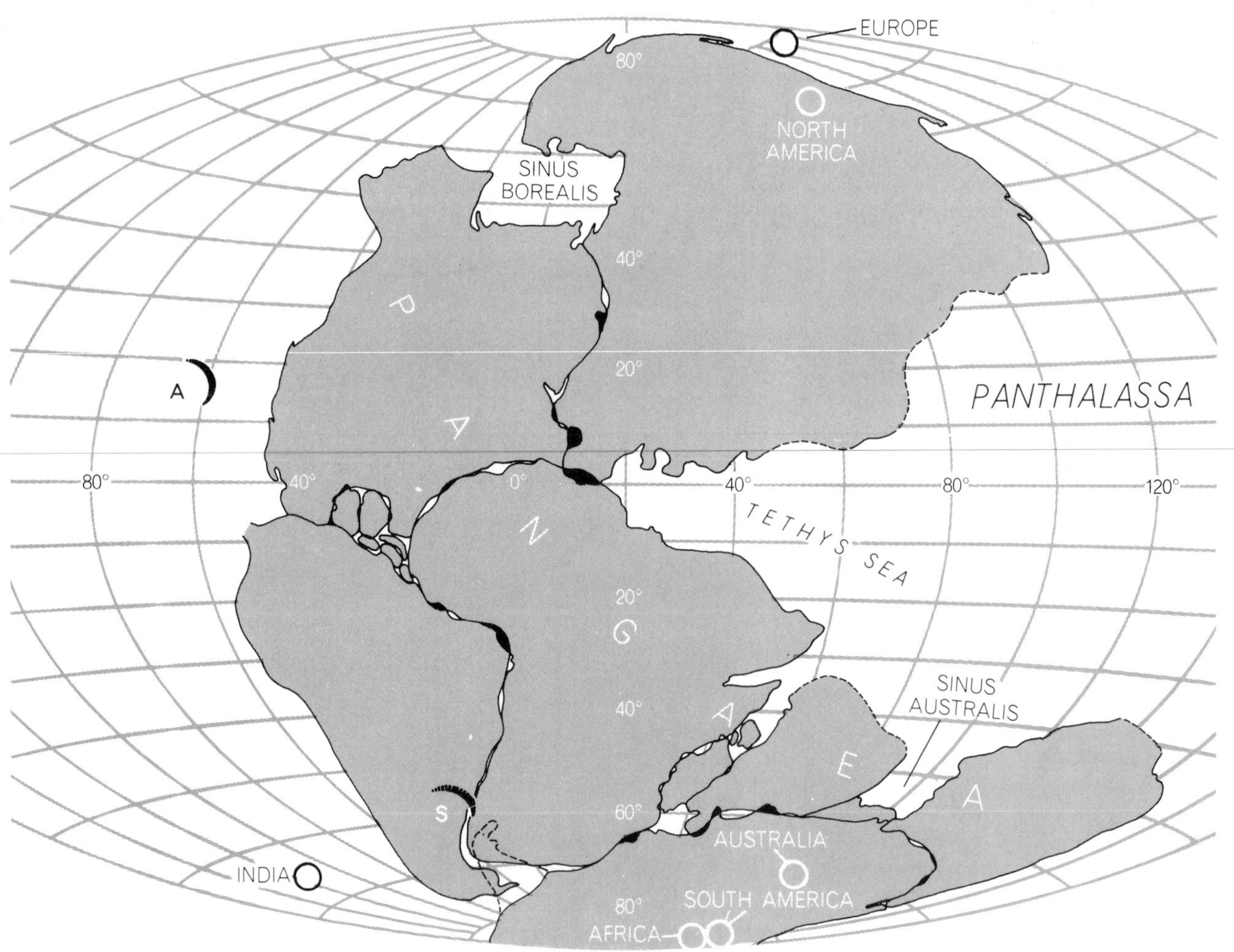

UNIVERSAL LAND MASS PANGAEA may have looked like this 200 million years ago. Panthalassa was the ancestral Pacific Ocean. The Tethys Sea (the ancestral Mediterranean) formed a large bay separating Africa and Eurasia. The relative positions of the continents, except for India, are based on best fits made by computer, using the 1,000-fathom isobath to define continental boundaries. When the continents are arranged as shown, the relative locations of the magnetic poles in Permian times are displaced to the positions marked by circles. Ideally these positions should cluster near the geographic poles. The hatched crescents (*A and S*) serve as modern geographic reference points; they represent the Antilles arc in the West Indies and Scotia arc in the extreme South Atlantic.

and North America first began to pull apart. Without this assumption the platform unaccountably overlays a large portion of the bulge of Africa [*see illustration on page 136*].

According to our reconstruction, Pangaea was a land mass of irregular outline surrounded by the universal ocean of Panthalassa: the ancestral Pacific. The fit between North America and Africa provides the principal connection between the future block of northern continents and the future group of southern ones. On the east the Tethys Sea, a large triangular bight, separated Eurasia from Africa; the present Mediterranean Sea is a remnant of the Tethys. Other major indentations in the outline of Pangaea (adapting terminology from the moon) can be named Sinus Borealis, the ancestral Arctic Ocean, and Sinus Australis, a southern bay off the Tethys separating India from Australia. Our fully closed reconstruction of the Central American region is problematical. An alternate possibility is that the Gulf of Mexico is the remnant of an oceanic arm extending into the Americas from Panthalassa—a Sinus Occidentalis.

When measured down to the 1,000-fathom isobath, the total area of Pangaea was 200,000 square kilometers, or 40 percent of the earth's surface—equal to the area of the present continents measured to the same isobath. When the future continents were still part of Pangaea, they were generally to the south and east of their present location, so that the amount of land in the two hemispheres was almost equally balanced. (Today two-thirds of all the land lies north of the Equator.) The Y-shaped junction connecting North America, South America and Africa was located in the South Atlantic not far from the present position of Ascension Island. If New York had been in existence at the time, it would have been on the Equator and at longitude 10 degrees east (rather than 74 degrees west). Spain would also have been on the Equator, but it would have been near its present longitude. Japan would have been in the Arctic, well north of its position today. India and Australia would have bordered the Antarctic, far to the south of where they are now.

The great event that broke up Pangaea and set its fragments adrift evidently began no more than 200 million years ago, or in the last few percent of geologic time. There may have been—indeed, there probably was—"predrift drift" that assembled Pangaea from two or more smaller land masses. The evidence is still scanty, however, and does not bear directly on this discussion.

We take the immediate prelude to the breakup of Pangaea to be the first large outpourings of basaltic rock along the continental margins being es-

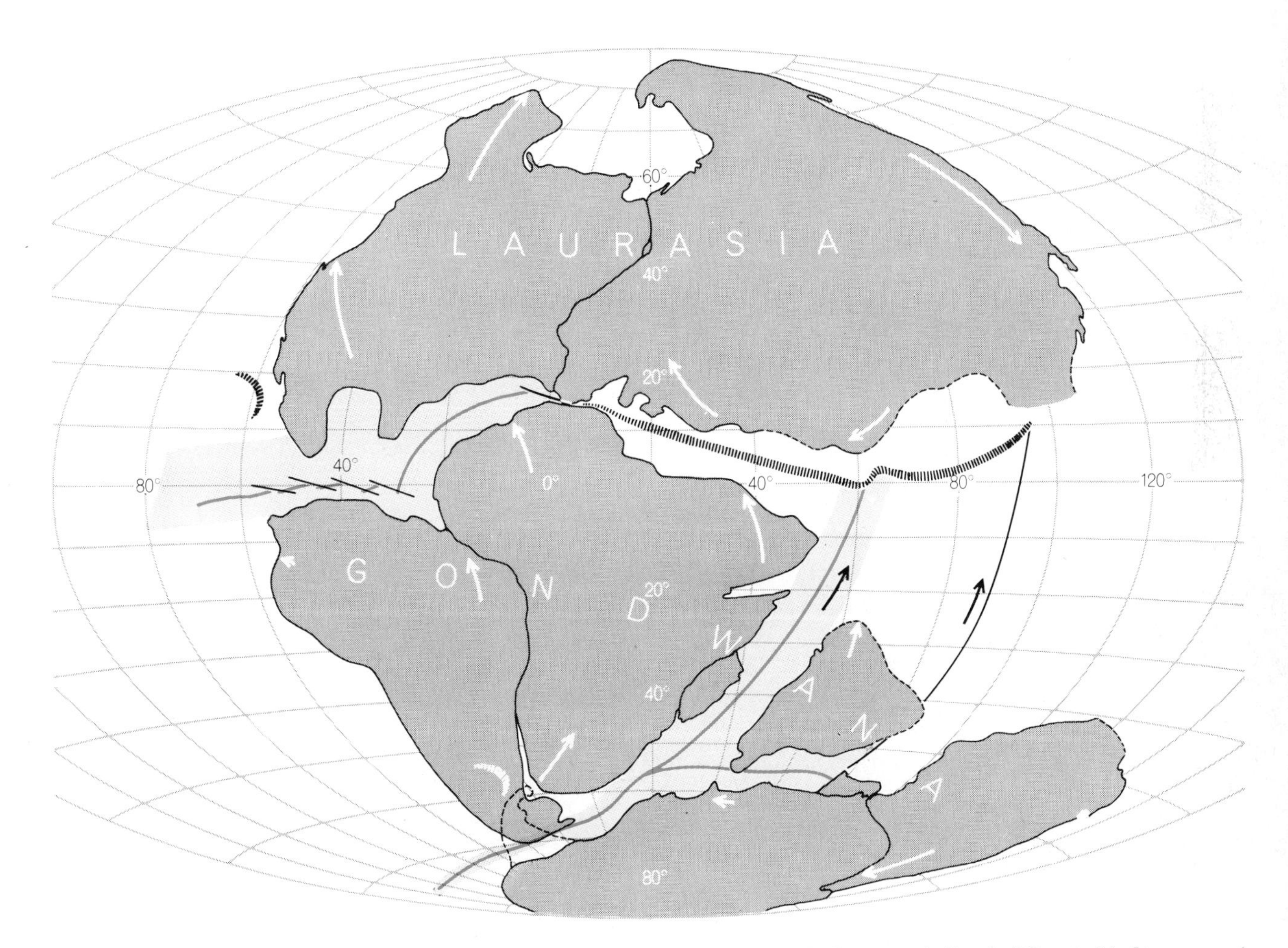

AFTER 20 MILLION YEARS OF DRIFT, at the end of the Triassic period 180 million years ago, the northern group of continents, known as Laurasia, has split away from the southern group, known as Gondwana. The latter has started to break up: India has been set free by a Y-shaped rift (*heavy line in color*), which has also begun to isolate the Africa–South America land mass from Antarctica-Australia. The Tethyan trench (*hatched lines in black*), a zone of crustal uptake, runs from Gibralter to the general area of Borneo. Black lines and black arrows denote megashears, zones of slippage along plate boundaries. The white arrows indicate the vector motions of the continents since drift began. Oceanic areas tinted in color represent new ocean floor created by sea-floor spreading.

tablished by rifting. The Triassic Newark series of basaltic flows along the east coast of the U.S. is a good example. Measurements of radioactivity indicate that the most ancient of these rocks are about 200 million years old, yielding a date that coincides with the middle of the Triassic period. As we interpret the evidence, two extensive rifts were initiated in Pangaea about 200 million years ago, which resulted in the opening of the Atlantic and the Indian Ocean by the end of the Triassic period 180 million years ago [*see illustration on preceding page*]. The northern rift split Pangaea from east to west along a line slightly to the north of the Equator and created Laurasia, composed of North America and Eurasia. The Laurasian land mass evidently rotated clockwise as a single plate around a pole of rotation that is now in Spain, creating a western "Mediterranean" that ultimately became part of the Gulf of Mexico and the Caribbean Sea. The southern rift split South America and Africa as a single land mass away from the remainder of Gondwana, consisting of Antarctica, Australia and India. Soon afterward (if not simultaneously) India was severed from Antarctica by a smaller rift to begin its rapid drift northward.

During the Jurassic period, from 180 to 135 million years ago, the direction of drift established by the Triassic rifts continued, further opening up the Atlantic and the Indian Ocean [*see illustration below*]. As North America drifted to the northwest, the Atlantic became more than 1,000 kilometers wide and probably remained fully connected to the Pacific. The east coast of the present U.S. ran almost east and west at a latitude of about 25 degrees north, so that coral reefs were able to grow all along the edge of the Atlantic continental shelf to the present Grand Banks, off Nova Scotia.

During the 45-million-year Jurassic period the Atlantic rift extended northward, blocking out the Labrador coastline and possibly initiating the opening up of the Labrador Sea between North America and Greenland. The interaction between the African and Eurasian plates forced the region of Spain to rotate counterclockwise 35 degrees, opening up the Bay of Biscay. The Tethys Sea, forerunner of the Mediterranean, continued to close at its eastern end. The Tethys was not only a zone of crustal subduction, or trench, but also a zone of shear along which Eurasia slid westward with respect to Africa. The compression associated with the Tethys trench raised bordering mountains composed of deep-water sediments.

At the close of the Jurassic an incipient rift began splitting South America away from Africa, entering from the south and working only as far north as where Nigeria is today. The tectonic situation first resembled the one now found in the rift zone along the backbone of high Africa (the region from Ethiopia to Tanzania) and then gradually opened farther to form a body of water resembling the Red Sea of today.

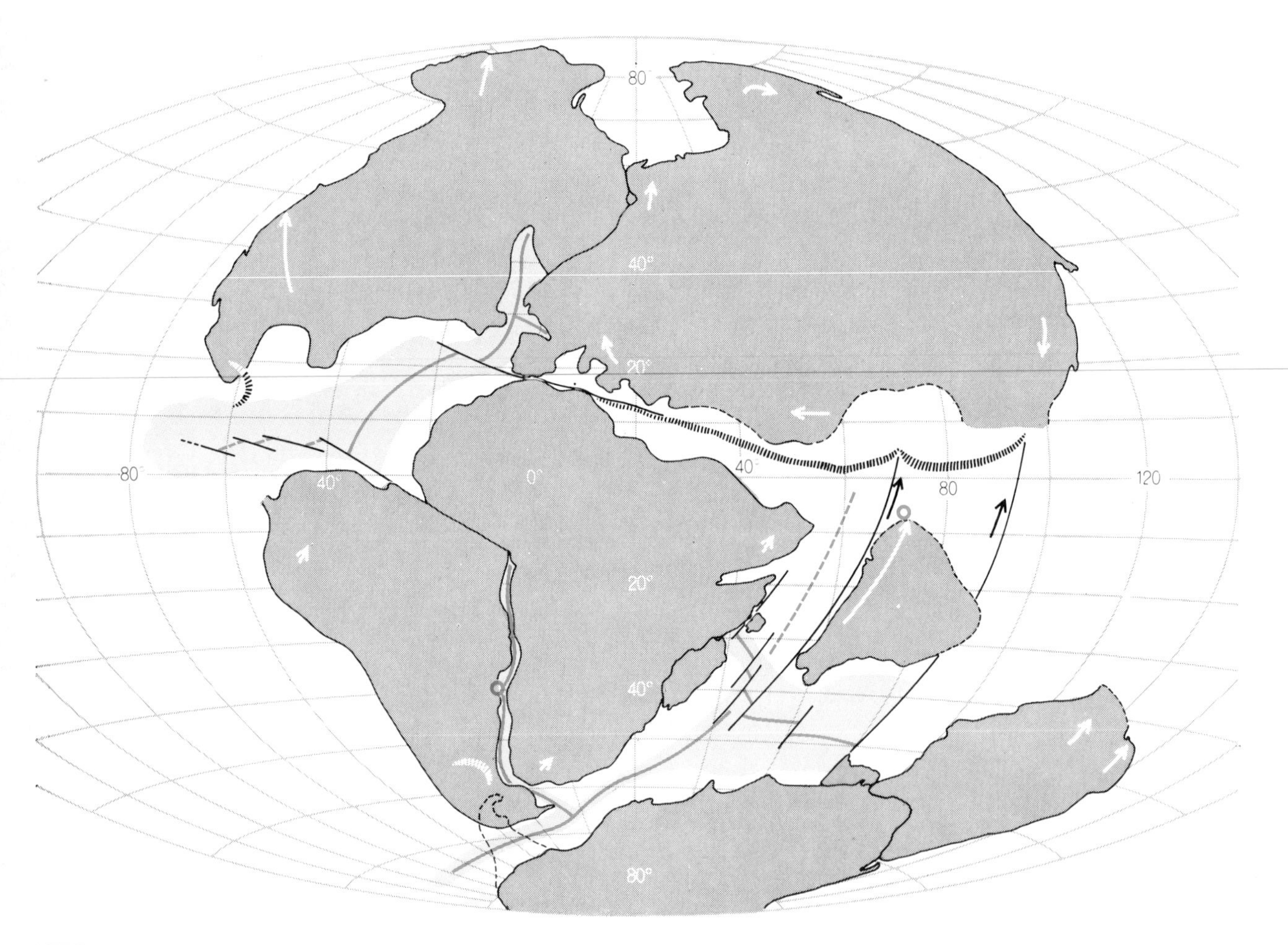

AFTER 65 MILLION YEARS OF DRIFT, at the end of the Jurassic period 135 million years ago, the North Atlantic and the Indian Ocean have opened considerably. The birth of the South Atlantic has been initiated by a rift. The rotation of the Eurasian land mass has begun to close the eastern end of the Tethys Sea. The Indian plate is about to pass over a thermal center (*colored dot*) that will soon pour out basalt to form the Deccan plateau. Later the hot spot will create the Chagos-Laccadive ridge in the Indian Ocean. Similarly, in the South Atlantic the Walvis thermal center (*colored dot*) will create the Walvis and Rio Grande "thread ridges."

At first freshwater sediments created thick deposits in pockets opened by faults; these sediments were overlain by deposits of salt.

By the end of the Cretaceous period, some 70 million years later (and 65 million years ago), the rupture of South America and Africa was complete, and the South Atlantic had widened rapidly to at least 3,000 kilometers [*see illustration below*]. Meanwhile the rift in the North Atlantic had switched from the west side of Greenland to the east side, blocking out its eastern margin (without, however, penetrating to the Arctic Ocean). Africa had drifted northward about 10 degrees and continued its counterclockwise rotation as the Eurasian plate rotated slowly clockwise. These two opposed motions nearly closed the eastern end of the Tethys Sea. The slow westward rotation of Antarctica continued. All the continents were now blocked out except for the remaining connection between Greenland and northern Europe and between Australia and Antarctica.

Although it is not shown on our maps, an extensive north-south trench system must have existed in the ancient Pacific to consume by subduction the rapid westward drift of the two plates carrying North and South America. North America presumably encountered this trench in the late Jurassic and early Cretaceous, with the result that the Franciscan fold belt, the predecessor of the California Coast Ranges, was accreted to the western margin of the continent. It appears that the trench was eventually overridden and "stifled" by North America's continued westward drift. Such trenches have the capacity to resorb ocean crust but not the lighter granitic crust of continents.

At about the same time, or soon afterward, South America first encountered the Andean trench and began to displace the trench westward, without ever overriding it. The early Andean fold belt resulted from this encounter. It seems likely that the trench originally dipped toward the west but was flipped over to its present eastward dip.

In the Cenozoic period (from 65 million years ago to the present) the continents drifted to the positions we observe today. The mid-Atlantic rift propagated into the Arctic basin, finally detaching Greenland from Europe [*see top illustration on next two pages*]. There were three other major developments during the Cretaceous: (1) the two Americas were rejoined by the Isthmus of Panama, created by volcanism and the arching upward of the earth's mantle, (2) the Indian land mass completed its remarkable journey northward by colliding with the underbelly of Asia and (3) Australia was rifted away from Antarctica and drifted northward to its present position.

In the collision of India with Asia the northern margin of the Indian plate was subducted below the Asiatic plate, creating the Himalayas. On India's passage to the north early in the Cenozoic its western margin crossed a fixed source of

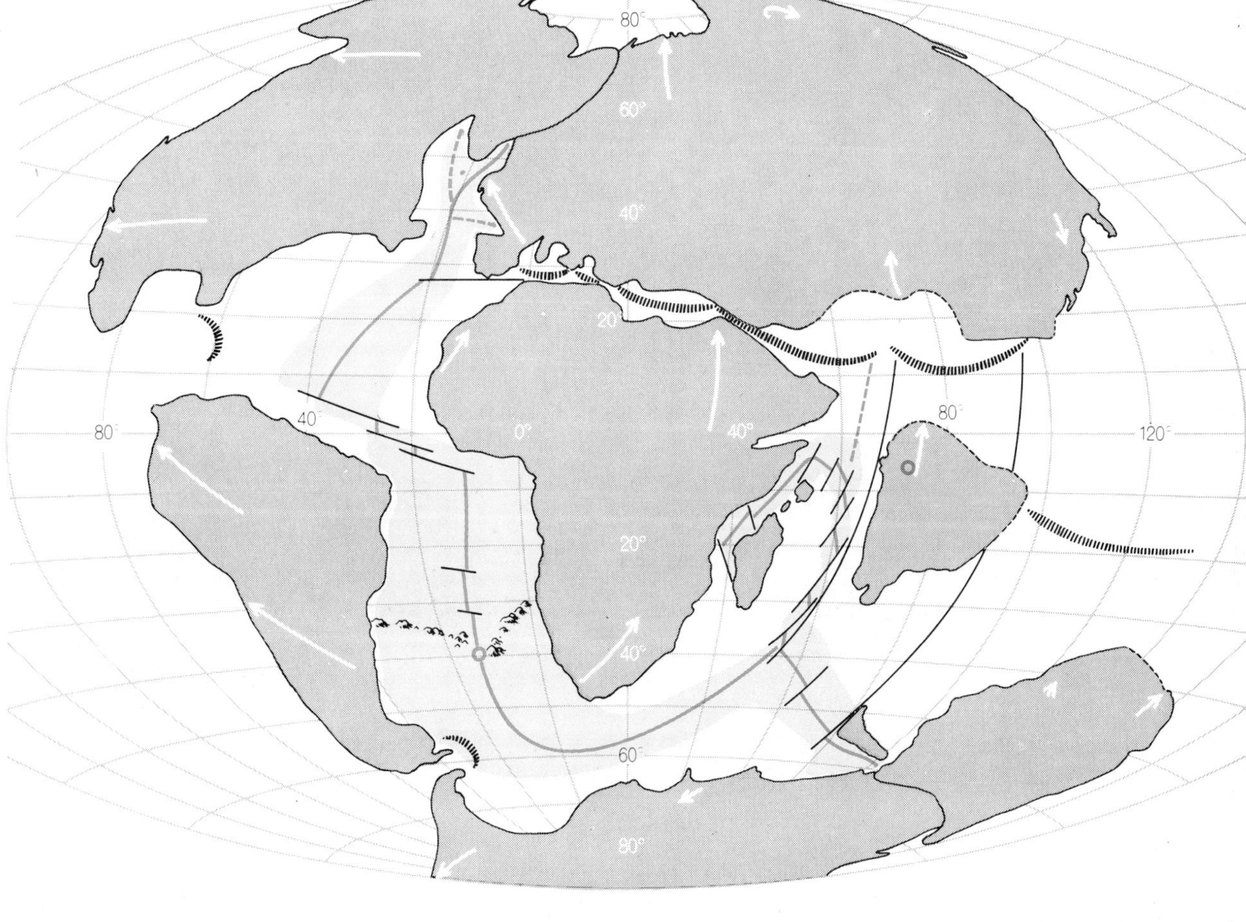

AFTER 135 MILLION YEARS OF DRIFT, 65 million years ago at the end of the Cretaceous period, the South Atlantic has widened into a major ocean. A new rift has carved Madagascar away from Africa. The rift in the North Atlantic has switched from the west side to the east side of Greenland. The Mediterranean Sea is clearly recognizable. Australia still remains attached to Antarctica. An extensive north-south trench (*not shown*) must also have existed in the Pacific to absorb the westward drift of the North American and South American plates. Note that the central meridian in all these reconstructions is 20 degrees east of the Greenwich meredian.

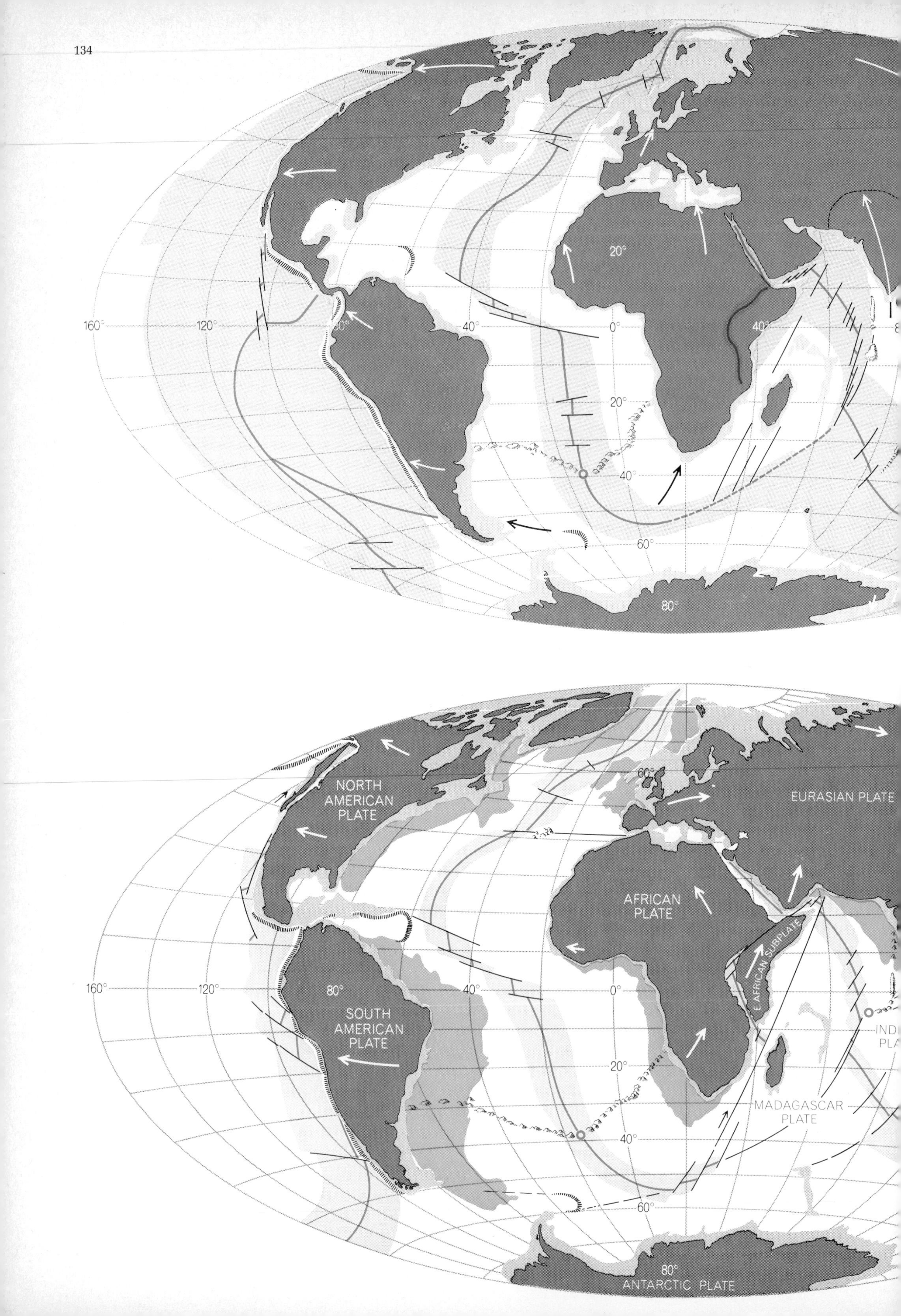
160°
120°
80°
40°
0°
20°
40°
60°
80°
NORTH
AMERICAN
PLATE
EURASIAN PLATE
AFRICAN
PLATE
E. AFRICAN SUBPLATE
SOUTH
AMERICAN
PLATE
MADAGASCAR
PLATE
ANTARCTIC PLATE

WORLD AS IT LOOKS TODAY was produced in the past 65 million years in the Cenozoic period. Nearly half of the ocean floor was created in this geologically brief period, as shown by the areas stippled in color. India completed its flight northward by colliding with Asia and a rift has separated Australia from Antarctica. The North Atlantic rift finally entered the Arctic Ocean, fissioning Laurasia. The widening gap between South America and Africa is closely traced by the thread ridges produced by the Walvis thermal center. The Antilles and Scotia arcs now occupy their proper positions with respect to neighboring land masses.

basaltic magma rising from the earth's upper mantle near the Equator. Molten rock erupted through the crust and poured onto the Indian subcontinent, laying down the basalts of the Deccan plateau. Even after India had left the hot spot behind, magma continued to stream out on the ocean floor, producing the Chagos-Laccadive ridge, which became covered with coral as it subsided into the Indian Ocean. Finally, a branch of the Indian Ocean rift split Arabia away from Africa, creating the Gulf of Aden and the Red Sea, and a spur of this rift meandered west and south into Africa.

Less pronounced changes during the Cenozoic period included the partial closing of the Caribbean region and the continued widening of the South Atlantic as new ocean crust was emplaced by sea-floor spreading. As the Atlantic continued to open in the far north the northwestward movement of the Eurasian land mass was halted and reversed, simultaneously reversing its sense of slippage with respect to Africa. The new direction of shear has been strongly impressed on the tectonic character of the Mediterranean and the Near East. The major north-south rift in the Indian Ocean largely ceased spreading and became instead a megashear that accommodated the counterclockwise and northward rotation of the African plate.

The reader will have observed that our maps of continental drift show more than relative positions and motions; the land masses, beginning with Pangaea itself, are assigned absolute geographic coordinates. Since this has not been attempted before we shall briefly describe how we arrived at our results. In the mobile world of plate tectonics one must assume that all parts of the crust are capable of moving and almost surely have moved.

After an extensive search for some absolute reference point, we finally concluded that the Walvis thermal center, or hot spot, might provide what we sought. In reaching this conclusion we accepted a hypothesis put forward by J. Tuzo Wilson of the University of Toronto. He had suggested that the Walvis ridge and the Rio Grande ridge in the South Atlantic are nemataths, or "thread ridges" of basalt, that had been poured onto the spreading ocean floor from a fixed lava orifice rising from a deep, stagnant region of the mantle. As new floor was carried past the orifice, lava would periodically pour out and form a small volcanic cone. By observing the location of succeeding cones as they merged into a ridge one can establish the absolute direction taken by the crust in that region. A study of the Walvis and Rio Grande ridges enabled us to establish not only the drift of the South American plate with respect to the African plate but also any motion the two plates may have had in some other direction [*see illustration on page 137*].

Unfortunately the Walvis hot spot did not exist earlier than about 140 million years ago, so that its usefulness as a fixed point does not go back earlier than the end of the Jurassic period. To trace crustal motions during the first 60 million years after the breakup of Pangaea one has to rely on dead reckoning. We have made the assumption that Antarctica has moved very little from its original location when it was part of Pan-

WORLD 50 MILLION YEARS FROM NOW may look something like this. The authors have extrapolated present-day plate movements to indicate how the continents will have drifted by the end of what they propose to call the Psychozoic era (the age of awareness). The Antarctic remains essentially fixed but may rotate slightly clockwise. The Atlantic (particularly the South Atlantic) and the Indian Ocean continue to grow at the expense of the Pacific. Australia drifts northward and begins rubbing against the Eurasian plate. The eastern portion of Africa is split off, while its northward drift closes the Bay of Biscay and virtually collapses the Mediterranean. New land area is created in the Caribbean by compressional uplift. Baja California and a sliver of California west of the San Andreas fault are severed from North America and begin drifting to the northwest. In about 10 million years Los Angeles will be abreast of San Francisco, still fixed to the mainland. In about 60 million years Los Angeles will start sliding into the Aleutian trench.

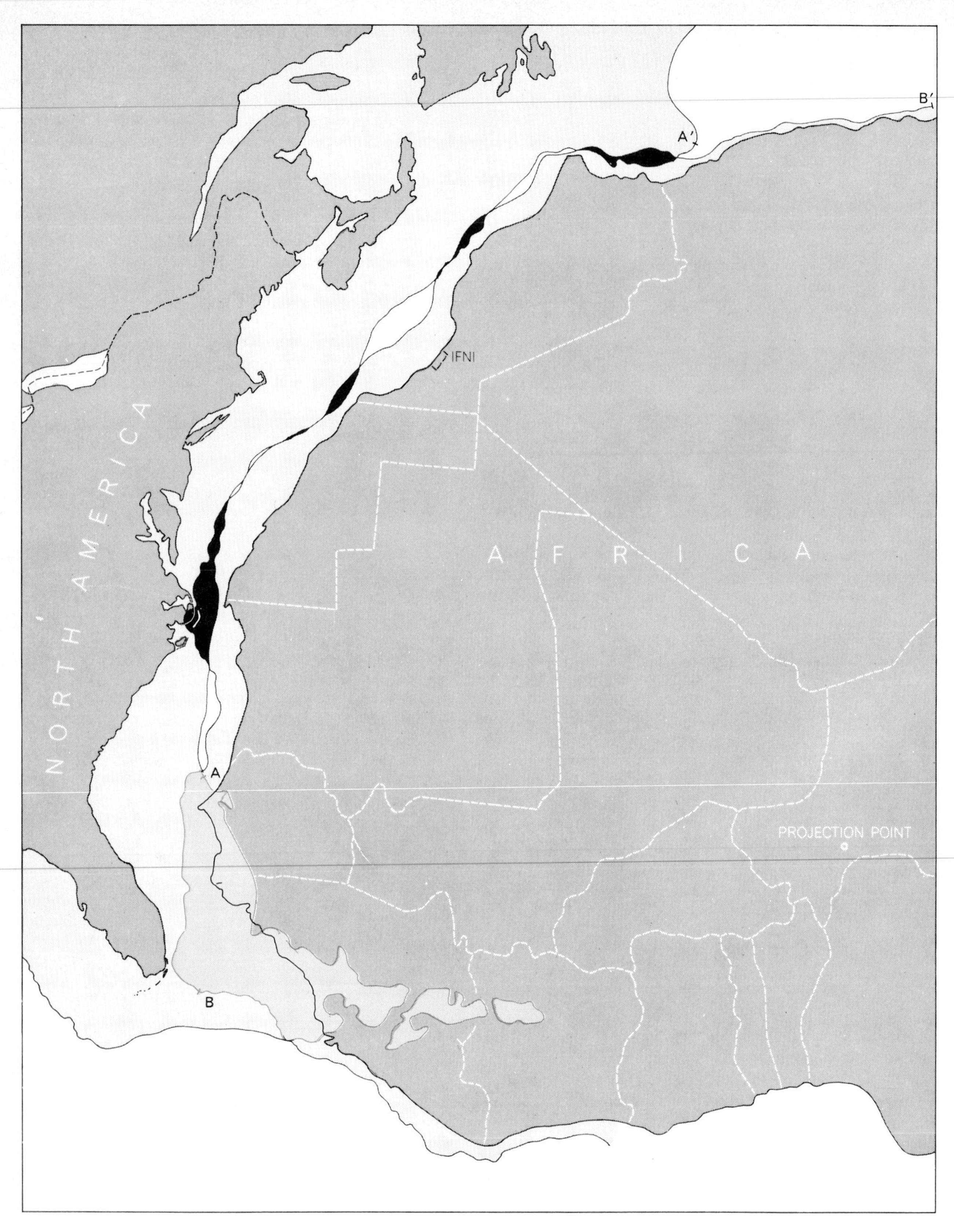

FIT OF AFRICA AGAINST NORTH AMERICA was made by the authors' colleague Walter P. Sproll with the aid of a computer. As in the reconstruction of Pangaea, it is assumed that each continent actually extends out into the ocean and halfway down the continental slope, where the ocean reaches a depth of 1,000 fathoms. The North American "coast" between *A* and *A′* was matched for best fit to the African "coast" between *B* and *B′*. White areas are gaps in the fit; black areas are overlaps. The overlap produced by the Bahamas platform, an enormous area half the size of Texas, is specially depicted in dark color. The authors propose that the platform represents an accumulation of sediments followed by coral growth after the two continents became separated. The largest gap in the proposed fit between the two continents is found off the Spanish enclave of Ifni. The Ifni gap may have been created when a small section of Africa split off and was translated 190 kilometers to the southwest, forming the eastern group of the Canary Islands.

gaea. This seems reasonable because the Antarctica plate is entirely surrounded by a system of rifts and megashears; there is no associated trench toward which the plate would tend to move away from its polar position.

Independent support for this assumption is obtained by plotting the position of the North and South poles before the dispersal of Pangaea. These positions are obtained by studying the direction of magnetization in rocks of the Permian period, as obtained by E. Irving of the Dominion Observatory in Canada and by other workers. We plotted the Permian pole positions with respect to each continent as it exists today and then rotated these pole positions as needed to assemble the continents into our version of Pangaea. By this method the pole positions should ideally cluster at one of the geographic poles. Actually there is some scatter, as can be seen in our reconstruction of Pangaea [*see illustration on page 130*], but all the positions do within either the Arctic Circle or the Antarctic Circle.

We can now summarize how the continents have moved in time and space. The two Americas have drifted a long way, generally westward. North America has drifted more than 8,000 kilometers west northwest; the tip of Florida once lay in the South Atlantic near the present position of Ascension Island. Moving toward the Tethyan trench system, India and Australia were carried far to the north. Africa rotated counterclockwise perhaps 20 degrees as the Eurasian land mass, similarly moving toward the Tethyan trench, rotated clockwise a roughly equal amount. India's remarkable flight is probably attributable to its being rafted on an "ideal" plate. Approximately rectangular, the Indian plate was sliced away from Pangaea by a rift along what is now India's east coast and then was free to move northward toward a major trench. This northward movement was facilitated by two parallel megashears.

Decades ago Wegener proposed that the drift of the continents was vectored by forces he termed *Westwanderung* (westward drift) and *Polarfluchtkraft* (flight from the poles). Although real, these forces are minuscule and not likely to be the underlying cause of drift. Our solution, however, does support Wegener's hypothesis of a westward flight, which, like the slip of the atmosphere, directly opposes the earth's rotation. We have also inferred a latitudinal drift, but from the South Pole only, or, paraphrasing Wegener's terminology, a *Sudpolarfluchtkraft*.

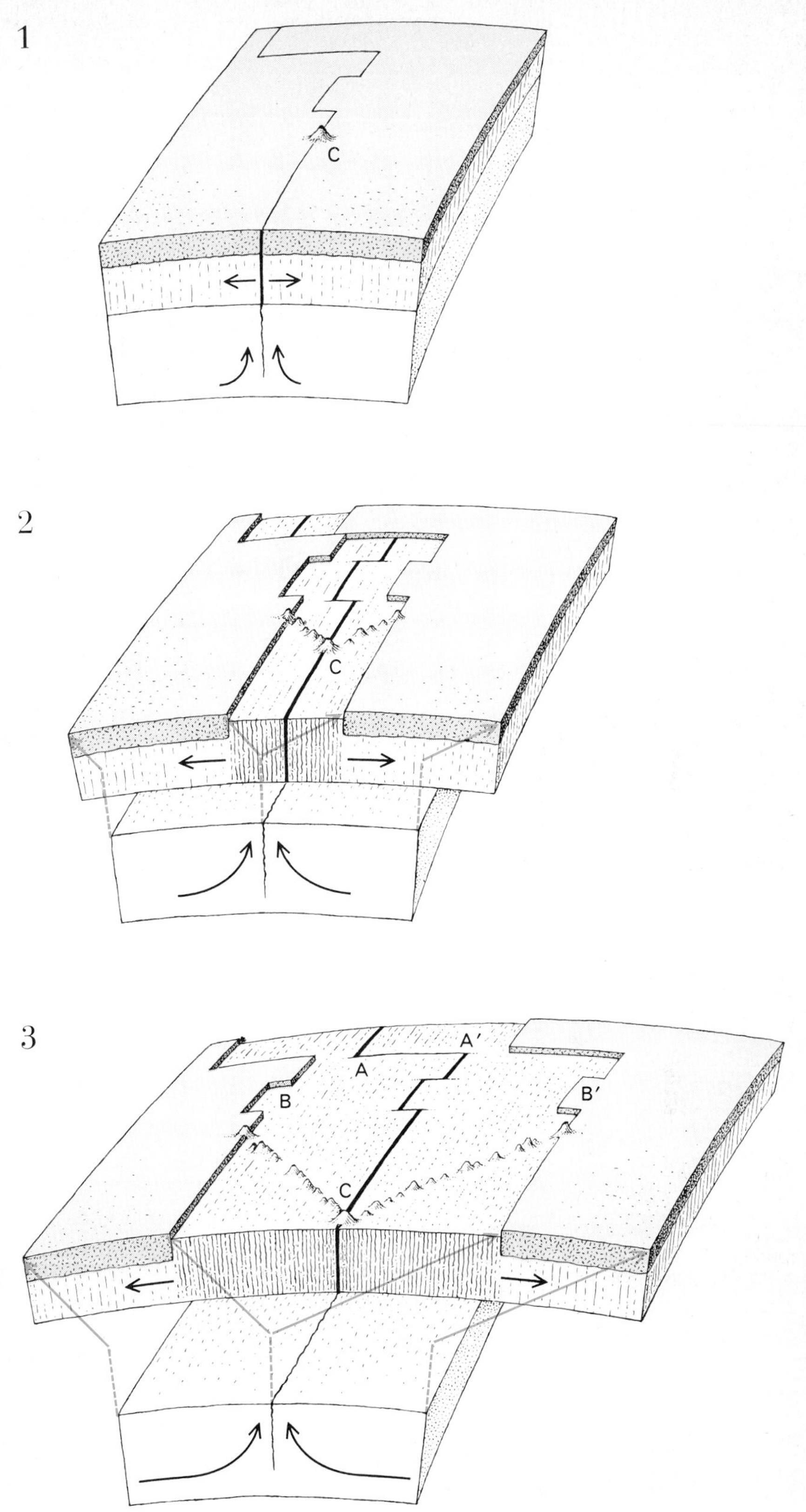

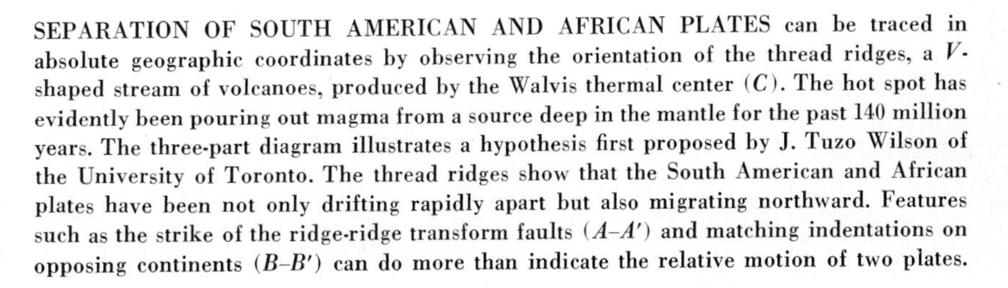
SEPARATION OF SOUTH AMERICAN AND AFRICAN PLATES can be traced in absolute geographic coordinates by observing the orientation of the thread ridges, a *V*-shaped stream of volcanoes, produced by the Walvis thermal center (*C*). The hot spot has evidently been pouring out magma from a source deep in the mantle for the past 140 million years. The three-part diagram illustrates a hypothesis first proposed by J. Tuzo Wilson of the University of Toronto. The thread ridges show that the South American and African plates have been not only drifting rapidly apart but also migrating northward. Features such as the strike of the ridge-ridge transform faults (*A*–*A'*) and matching indentations on opposing continents (*B*–*B'*) can do more than indicate the relative motion of two plates.

The Evolution of the Indian Ocean

11

D. P. McKenzie and J. G. Sclater
May 1973

The creation of the Himalayas, the highest and most dramatic mountain range in the world, long presented a puzzle to geologists. Like the Alps and the heavily eroded Appalachians, it was clear that the Himalayas consisted of sedimentary rocks, laid down over many millions of years in shallow seas, then uplifted and heavily deformed by mighty tectonic forces. But in what sea were the Himalayan rocks deposited and how did they get sandwiched between the subcontinent of India and the great Asian landmass to the north? The geology textbooks of a decade ago could provide no satisfactory answer.

With the rapid development and acceptance of the concept of sea-floor spreading, continental drift and plate tectonics the origin of the Himalayas is no longer a mystery. The vast Himalayan range was created when a plate of the earth's crust carrying the landmass of India collided with the plate carrying Asia some 45 million years ago, having traveled 5,000 kilometers nearly due north across the expanse now occupied by the Indian Ocean. Thus the Indian Ocean and the Himalayas have a common origin.

The principal evidence supporting this theory has come from oceanography rather than from classical geology. A geologist's life is too short for him to study a large part of a major mountain system in any detail. Furthermore, the rocks that form the Himalayas have been so strongly deformed and eroded that it is not possible to discover the extent of the relative motion of India and Asia from a study of the rocks themselves.

That left the ocean as the only other place to look for evidence of a collision. It seemed unlikely that a large continent could cross an entire ocean basin and leave no trace of its passage. Yet as recently as a decade ago oceanographers who had studied the Indian Ocean saw no evidence indicating the passage of a continent and therefore were mostly opposed to the idea of continental drift. The solution to this difficulty and the beginning of our present understanding was provided about 1960 by the late Harry H. Hess of Princeton University. Hess suggested that the continents do not plow their way through the ocean floor but move with it much like coal on a conveyor belt. He proposed that the sea floor is being continuously generated along linear fissures running for thousands of kilometers through the major ocean basins; molten rock wells up into the fissures from the mantle below and quickly hardens [*see top illustration on next page*]. Elsewhere on the earth the ocean floor plunges into a trench and then sinks to great depths in the mantle.

The difference between the concepts of sea-floor spreading and plate tectonics is mostly one of emphasis. Sea-floor spreading is primarily concerned with the processes by which the ocean floor is created and destroyed, and much less with the continents and the passive parts of the ocean floor. The inactive areas, on the other hand, are the principal subject matter of plate tectonics, which starts from the idea that all areas of the earth's surface free from earthquakes are not at present being deformed. The concept is useful because most earthquakes occur within narrow belts that delineate the regions where large plates of the earth's crust are being subjected to irresistible tectonic forces. It turns out that most of the earth's surface can be divided up into only a few large, rigid plates.

Since the plates are rigid, one can work out the relative motions everywhere of any two plates simply by knowing how they move at any two points along a common boundary. One must emphasize "relative motion" because there is as yet no method for defining absolute motion. The simplest type of motion on a plane is when two plates slide past each other with no relative rotation. A more general type of motion also involves rotation, and it must then be everywhere parallel to circles centered on the center of rotation. Wherever plates are moving apart, the separating edges must be growing by the upwelling of material from below. Such accretions almost always add to each retreating plate at the same rate, and commonly they form boundaries known as ridges. Because fresh material usually adds to the two separating plates at the same rate, ridges normally spread symmetrically. The reason for this symmetric behavior is not fully understood. Furthermore, there is no geometric requirement that the ridge should form at right angles to the direction of motion of the plates, but that is usually the case.

When two plates simply slide past

HIMALAYAN RANGE looking to the west was photographed from *Apollo* 7 in October, 1968. The green area to the left of the range is Nepal; the area to the right is mostly Tibet. Mount Everest is among the peaks in the middle of the picture two and a half inches from the bottom. At the lower right edge is Kanchenjunga, the third highest peak in the world.

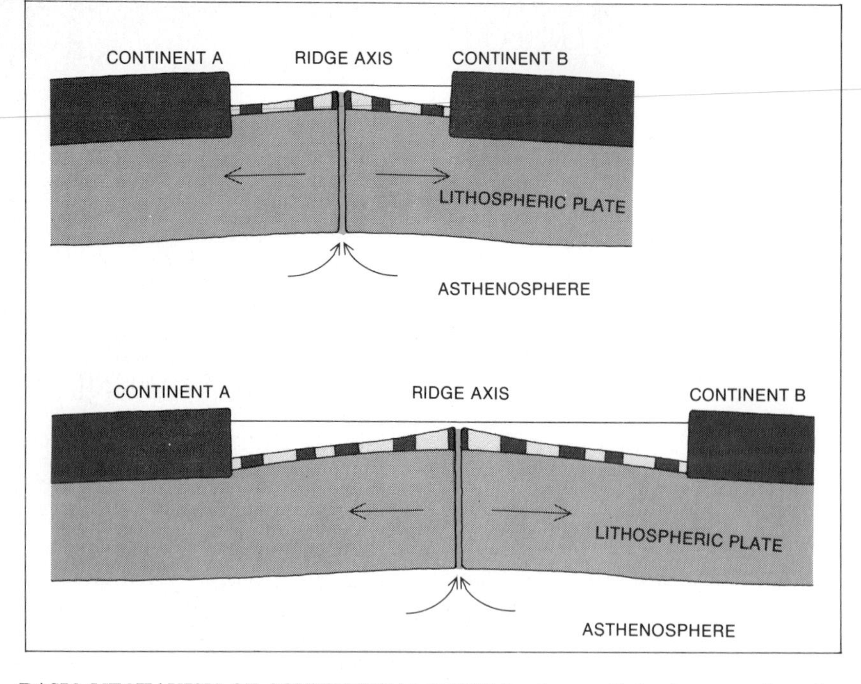

BASIC MECHANISM OF CONTINENTAL DRIFT involves a rift in the ocean floor that allows molten rock to well up from the asthenosphere and form a spreading ridge. The continents are rafted apart at rates up to 20 centimeters per year. As the molten rock solidifies at the ridge axis it becomes magnetized in stripes of alternating polarity (*light and dark bands*) as a consequence of intermittent reversals in the earth's magnetic field.

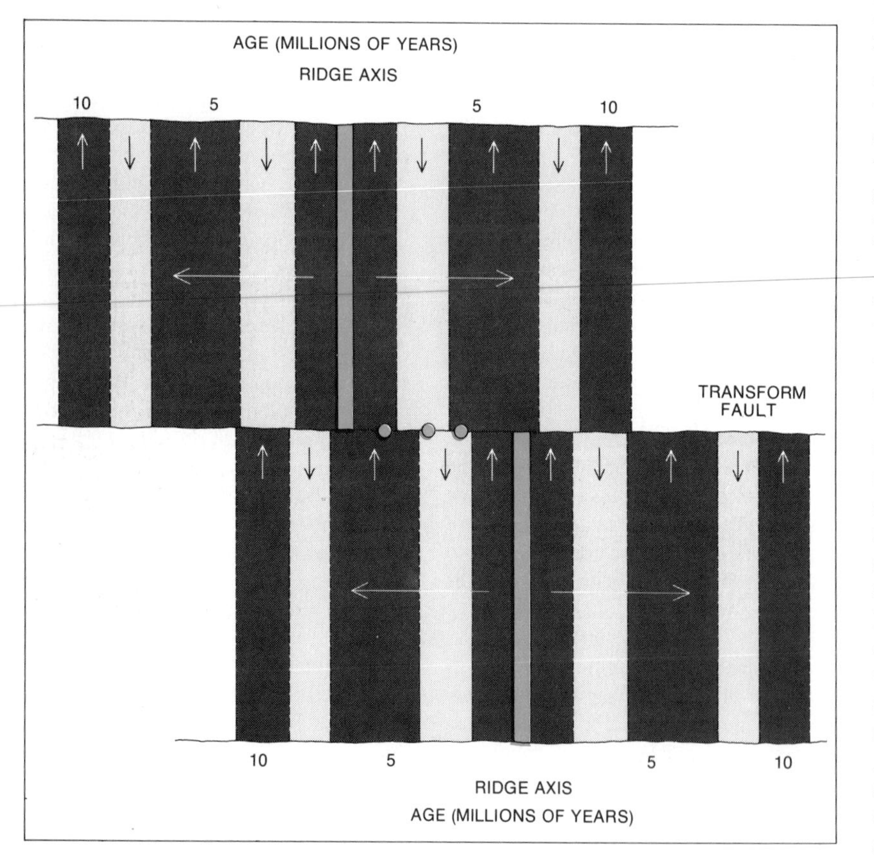

REVERSALS OF EARTH'S MAGNETIC FIELD produce a zebra-striped pattern of magnetization in the crust of the ocean floor. The reversals (*small arrows*) have been dated back about 80 million years by combining radioactivity measurements of rocks with identification of organisms in deep-sea cores. Transform faults often offset the pattern of magnetic anomalies. Friction between the sliding blocks causes shallow earthquakes (*colored dots*).

each other, nothing is added to or subtracted from either plate. Plate boundaries of this type, known as transform faults, are always parallel to the direction of relative motion between the two plates. Such faults must therefore be segments of circles around the center of rotation. When two plates are moving toward each other, one eventually underrides the other and is consumed by plunging into a trench. Simultaneously the edge of the overriding plate is typically uplifted, creating a belt of crumpled and folded rocks parallel to the trench. There is, however, no simple relation between the downward direction of a trench and the relative direction of motion between the plates.

The motion of plates on the earth is slightly more complicated because the earth is spherical. Their relative motion, however, can still be described by a rotation about an axis. Since the plates must remain on the earth's surface, this axis must pass through the earth's center. The motion must again be parallel to circles about the rotation axis. The axis for the opening of the South Atlantic is shown in the illustrations on the opposite page, together with circles drawn with the axis as the center. The earthquake epicenters mark the plate boundary, which is in places a ridge and in places a transform fault parallel to the direction of relative motion marked by the small circles. The original relative position of Africa and South America can be obtained by rotating the two continents toward each other until they meet. The earthquakes have not been moved, but each continent has been rotated through an equal angle toward the ridge axis.

The reason the earthquakes then mark the original break is that the ridge in the South Atlantic has added to the African and South American plates at the same rate. South America and Africa can thus be fitted together rather accurately by a simple rigid rotation. Neither has been deformed during the formation of the South Atlantic, and since both continents consist largely of very ancient rocks, neither has been appreciably deformed in the past 300 million years. Together they form a spherical cap whose diameter is about equal to the radius of the earth. If the earth had expanded appreciably in the past 600 million years, both continents would have been broken into many fragments, since the spherical cap they originally formed would no longer fit on an expanded earth. Plate tectonics therefore shows

AFRICA AND SOUTH AMERICA were locked together as part of the primitive continent of Gondwanaland until about 180 million years ago. The black dots mark the positions of earthquakes that have occurred on the Atlantic mid-ocean ridge in the past 10 years. Their location coincides with the location of the primitive cleavage line between South America and Africa when the two continents are rotated toward each other by an equal amount along the "latitude" lines centered on the axis of rotation (*arrow*). India, Antarctica and Australia were also part of Gondwanaland, but there is still no agreement on how several pieces fitted together.

PRESENT LOCATION of Africa and South America shows the two continents to be equidistant from the still-active spreading ridge (*color*) in the mid-Atlantic. Evidently molten rock flowing into the ridge has increased the size of each plate by an equal amount for the past 180 million years. The two continents are still being pushed apart at the rate of about three centimeters a year. The transform faults developed because the initial break was not everywhere at right angles to the circles of rotation. As required by the theory of plate tectonics, the transform faults lie parallel to the circles.

that geologically important expansion has not occurred, although it does not rule out changes in radius of a few tens of kilometers.

The South Atlantic also shows how transform faults can originate. The original break between South America and Africa was a jagged line that followed old lines of weakness in the Precambrian shield and did not form a plate boundary at right angles to the direction of motion between the plates. After separation had started, the plate boundary changed from a jagged ridge to a series of straight ridge sections joined by transform faults. This change maintained the general shape of the ridge but allowed the ridges to be at right angles to the direction of relative motion.

To a continental geologist it is quite surprising that the simple ideas of plate

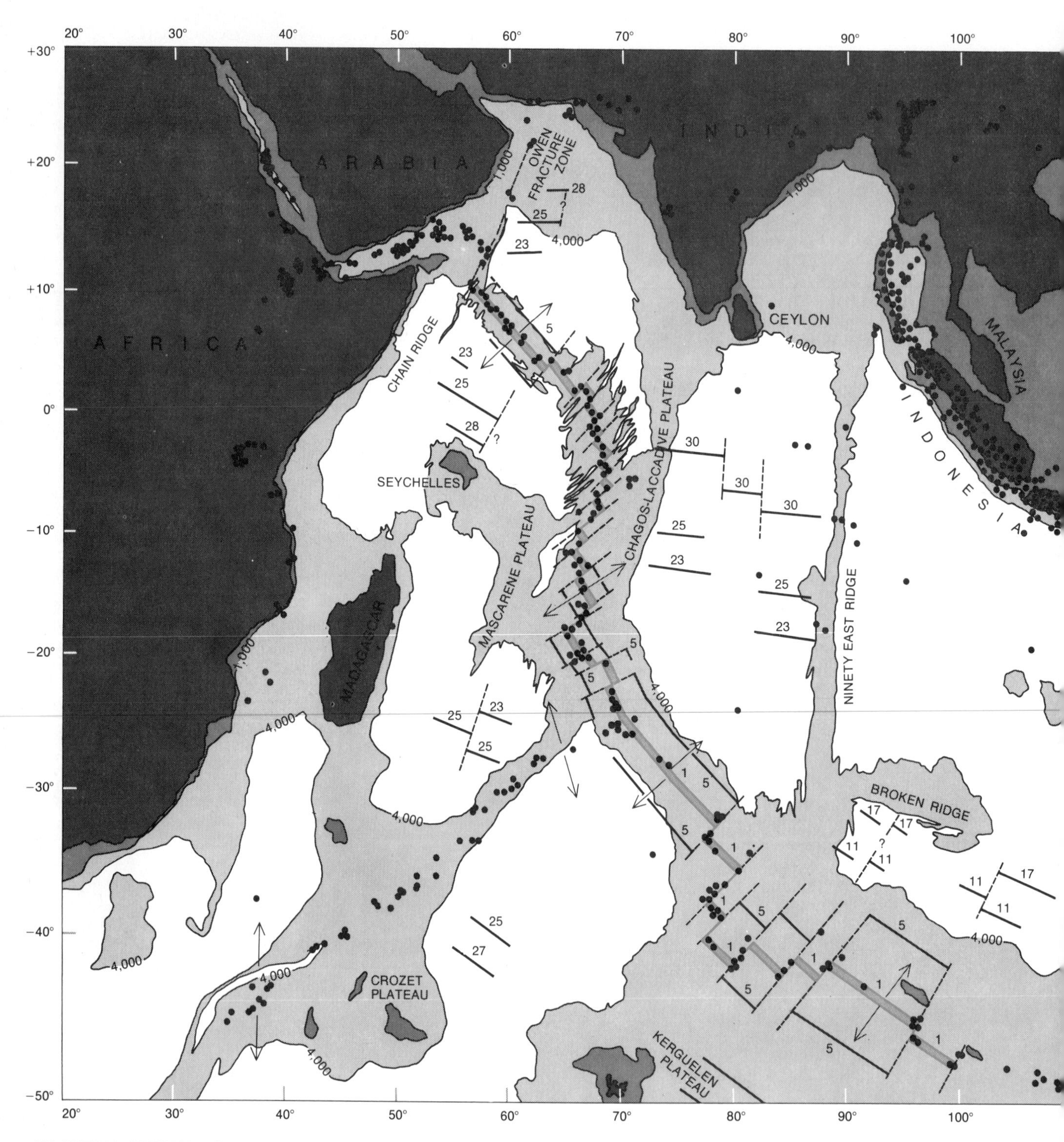

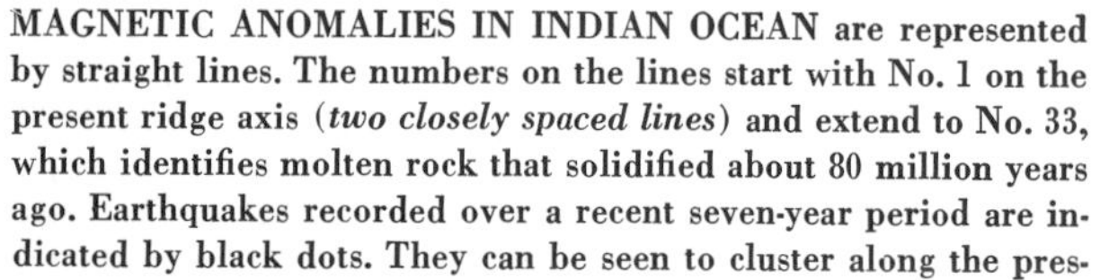
MAGNETIC ANOMALIES IN INDIAN OCEAN are represented by straight lines. The numbers on the lines start with No. 1 on the present ridge axis (*two closely spaced lines*) and extend to No. 33, which identifies molten rock that solidified about 80 million years ago. Earthquakes recorded over a recent seven-year period are indicated by black dots. They can be seen to cluster along the present ridge axis and on lines where the axis is offset by transform faults. The greatest number of earthquakes, however, take place around the island arc between Southeast Asia and Australia where the Indian Ocean plate and the Pacific Ocean plate plunge into deep trenches and are consumed. The major features on the floor of the Indian Ocean are the three large ridges that run north and

tectonics work so well. The rocks he studies were generally severely deformed by plastic flow when they were near a plate boundary. Thus there was opposition to the new ideas from investigators who had worked only on land. That opposition has now largely ceased.

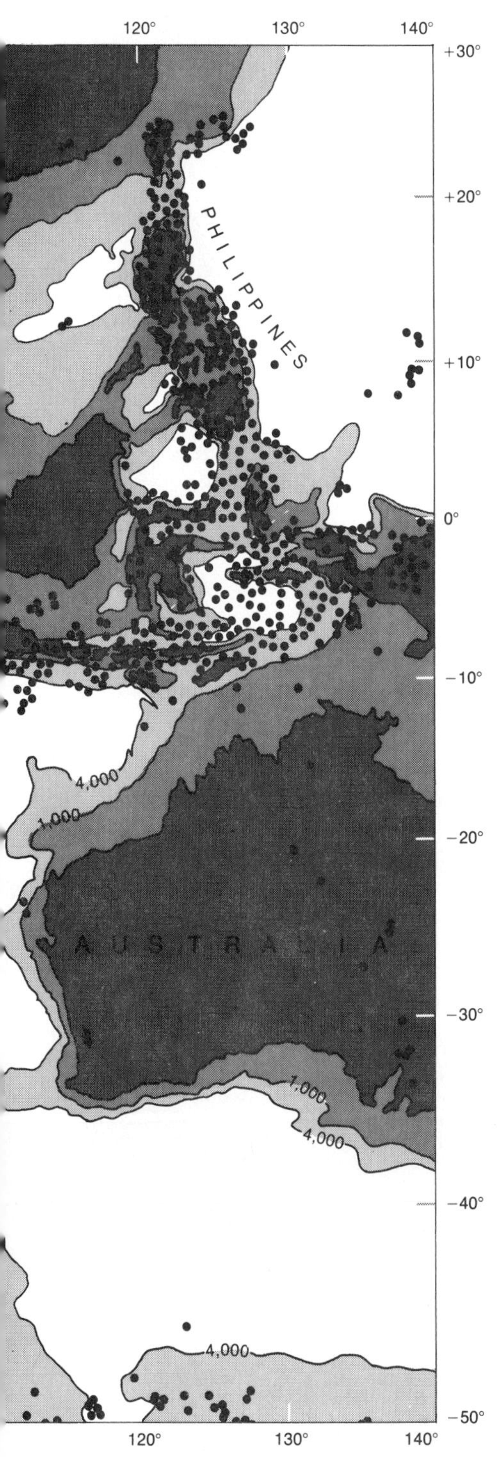

south: Ninety East Ridge, the Chagos-Laccadive Plateau and the Mascarene Plateau. Two parallel ridges, Broken Ridge and the Kerguelen Plateau, run approximately east-west near Australia. Such features aided in reconstructing evolution of Indian Ocean.

The early tests of plate tectonics all involved the present motions of plates measured at plate boundaries as defined by the location of earthquakes. This approach is obviously powerless to determine how plates may have moved in the geologic past. Providentially the rocks of the ocean floor carry a permanent record of their own history. It was noted about 1960 that the oceanic crust is not uniformly magnetized. Instead magnetometer surveys revealed a remarkably regular striped pattern of magnetic variation: stripes of normal polarization (with north pointing to the magnetic north pole) alternate with stripes of reverse polarization.

In 1963 F. J. Vine and D. H. Matthews of the University of Cambridge proposed that when lava wells up in the oceanic ridges, it is magnetized as it cools by the magnetic field of the earth. The alternations of polarity indicate that the magnetic field of the earth reverses at infrequent and irregular intervals. As a result the entire ocean floor is covered with long linear blocks of alternating magnetic polarity [*see bottom illustration on page 140*]. The shape of the blocks is exactly the shape of the ridge that formed them. We are doubly fortunate: the period between reversals of the magnetic field is generally sufficient to produce stripes at least 10 kilometers wide and, equally important, the reversals come at irregular intervals, so that the stripes vary in width. Hence the magnetic anomalies can be matched where they have been shifted by transform faults.

If the present ridge axis is designated No. 1, the magnetic stripes that flank it on each side can be numbered sequentially up to No. 33, formed about 80 million years ago. Older anomalies are known, but their age and relation to the most recent 33 are not yet established. Magnetic surveys can be made with a magnetometer towed behind either a ship or an airplane. The mapping and identification of magnetic anomalies now provides a fast and accurate method of studying the evolution of an ocean.

In regions where there are no magnetic data or where the anomalies are not clear the depth of the sea can often provide a rough estimate of the age of the sea floor. The youngest portion of a plate forms the peak of a ridge system that rises about 3,000 meters above the flatter and older portions of the sea floor. The reason is that the recently intruded rock is hotter and therefore less dense than the older material that has cooled and contracted as it moved away from the plate boundary. Sea-floor profiles calculated on the basis of rock-cooling models agree well with profiles established by accurate depth surveys.

The separation rate recorded by the anomalies indicates the relative motion between two plates. It is also possible to measure the motion between one plate and the earth's magnetic pole. If the magnetic pole is assumed to coincide approximately with the rotational pole, the measurement of plate motion can be used to find the latitude and orientation of the plate on which the observations were made. If the rocks are magnetized near the Equator, the magnetization is horizontal; the closer the rocks are to the poles when they are magnetized, the more vertical is the pitch, or dip, of the magnetization. With many paleomagnetic measurements made over the past two decades one can plot the apparent migration of the pole going back tens of millions of years. One cannot decide, of course, whether the pole itself has moved or whether the pole has remained fixed while the plates have moved. For convenience we take the poles as being fixed and move the plates to a latitude corresponding to the pitch of their magnetization. Naturally any reconstruction of the positions of the plates must be consistent with respect both to latitude and to north-south direction of polarization.

When we started our work on the Indian Ocean, the theory of plate tectonics had been proposed and tested but its success was not yet generally recognized. Like many scientific projects, ours started by chance. When we began in 1968, the Indian Ocean was the only major ocean in which no one was trying very hard to use marine geological and geophysical data to construct a tectonic history. It happened, however, that Robert L. Fisher of the Scripps Institution of Oceanography was doing detailed work on the Central Indian Ridge and had organized an expedition that would spend six months collecting marine-geological and geophysical information in the Indian Ocean. One of us (Sclater) was chief scientist on that expedition for two months and proposed that the other (McKenzie) join the ship for a month, more to show a theoretical geophysicist how cruises are run than with the intention of putting him to work.

The magnetic anomalies observed on that cruise left little doubt in our minds that the plate carrying India had crossed the Indian Ocean. To see if we could strengthen the hypothesis we collected all the available magnetic profiles from ship and airplane traverses of the ocean.

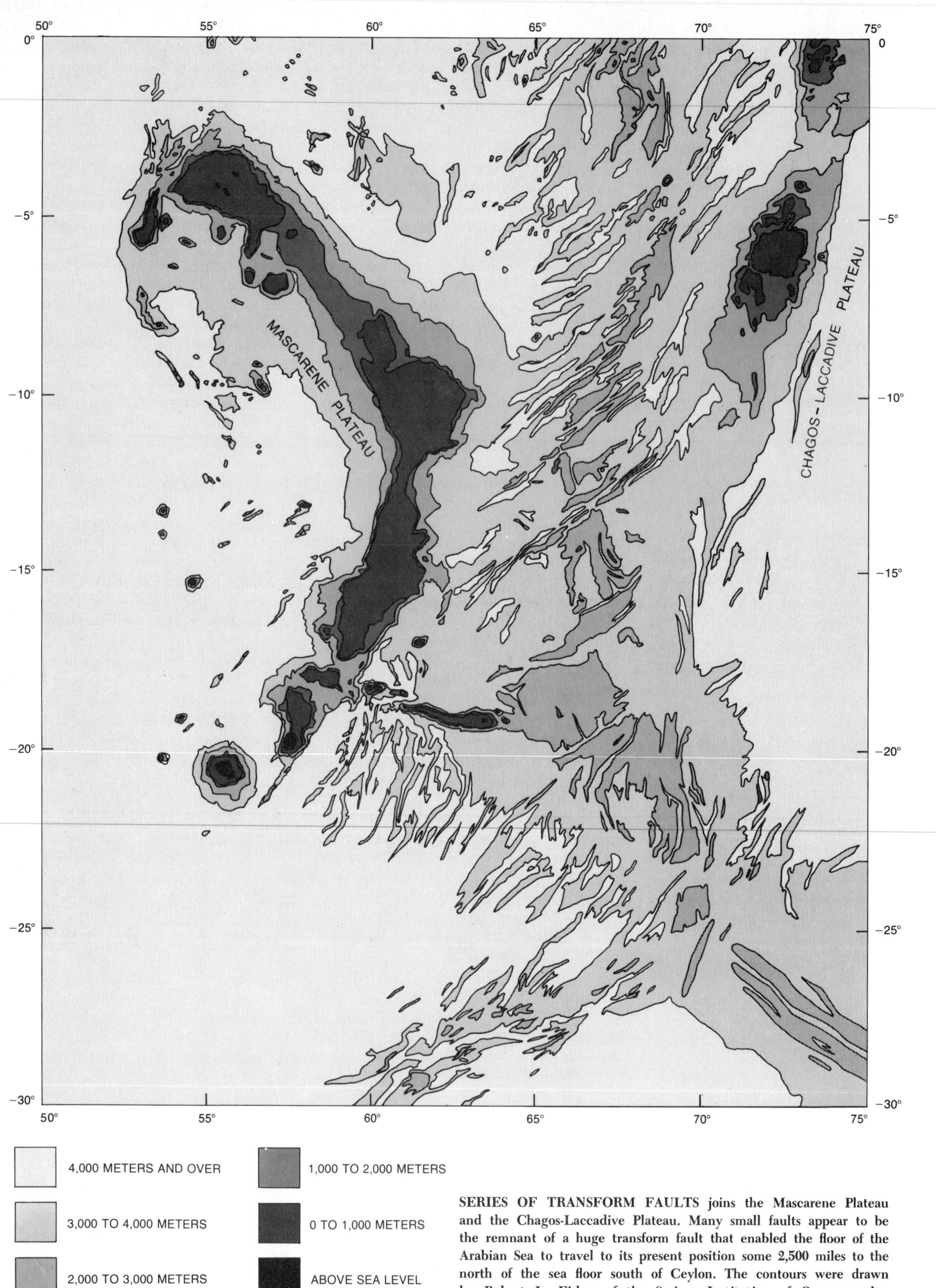

SERIES OF TRANSFORM FAULTS joins the Mascarene Plateau and the Chagos-Laccadive Plateau. Many small faults appear to be the remnant of a huge transform fault that enabled the floor of the Arabian Sea to travel to its present position some 2,500 miles to the north of the sea floor south of Ceylon. The contours were drawn by Robert L. Fisher of the Scripps Institution of Oceanography.

Fortunately the theory of plate tectonics changed very little in the three years that the two of us and three co-workers spent interpreting the observations. When we were done, the evolution of the Indian Ocean emerged clearly.

The shapes of the magnetic anomalies show that the evolution of the Indian Ocean was much more complicated than the evolution of the South Atlantic. The topography of the floor of the Indian Ocean is correspondingly more complex. The floor shows four large north-south ridges, two of which are still actively spreading [see *illustration on pages 142 and 143*]. It is hard to see any relation between the two huge inactive ridges (Ninety East Ridge and the Chagos-Laccadive Plateau) and the ridges that have remained active. It is also difficult to understand how the various ancient continental fragments, such as the island of Madagascar, the Seychelles islands and perhaps also Broken Ridge west of Australia, became isolated from the surrounding landmasses. Islands and ridges of this type are not found in the Atlantic and the Pacific. Their existence in the Indian Ocean is baffling.

A feature of particular interest is the plate boundary that runs north and south between the Chagos-Laccadive Plateau and the Mascarene Plateau in the northwestern part of the Indian Ocean. In this area Fisher's work has shown that the short spreading ridge segments are offset by many transform faults, none of which crosses the two inactive ridges on each side [*see illustration on opposite page*]. These faults therefore cannot have been produced by the shape of the original break, as the faults in the South Atlantic were. Their origin becomes clear only when the anomalies between anomaly No. 23 and anomaly No. 29 are used to map the shape of the ridge axis that produced them.

The anomalies in the Arabian Sea south of Iran run east and west but are about 2,500 kilometers north of the similarly numbered anomalies south of Ceylon. Hence there must at one time have been a huge transform fault joining the ridge in the Arabian Sea to the one south of Ceylon [*see illustration on this page*]. Some 55 million years ago relative motion south of Ceylon slowed from 16 centimeters per year to about six centimeters and probably stopped altogether in the Arabian Sea. Then about 35 million years ago movement resumed in the Arabian Sea, but its direction was no longer parallel to the ancient transform fault. New sea floor was again generated, and the boundary changed from a ridge oblique to the spreading direction to a series of ridge segments that are perpendicular to the direction of motion and are joined by transform faults [*see illustrations on next page*].

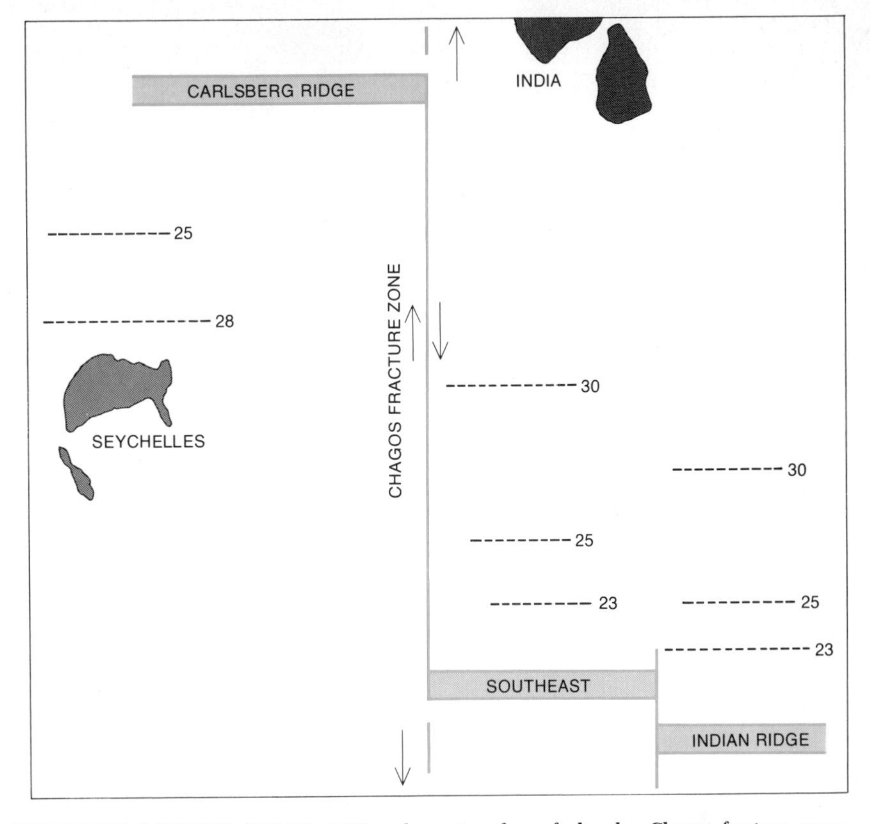

FIFTY-FIVE MILLION YEARS AGO a huge transform fault, the Chagos fracture zone, divided the floor of the Indian Ocean into two large segments. Between 80 million and 55 million years ago the segment carrying India traveled about 16 centimeters a year.

As in the Atlantic, the shape of the present plate boundary preserves the shape of the ancient boundary and the north-south section is the image of the old transform fault preserved by the new structures. This region and the central Atlantic thus show how the complex geometry of plate boundaries often results from the simple behavior of ridges and transform faults. To be sure, not all ridges behave so simply. In some cases a plate breaks and forms a ridge in a new place. Such "jumping" ridges are fortunately rare. Since they produce very muddled patterns of anomalies, they are difficult to identify.

A second area of much interest is the region south of Ceylon. Here we found anomalies from anomaly No. 32 to anomaly No. 21 that were formed between 75 million and 55 million years ago. These anomalies are now just south of the Equator, but the matching anomalies on the other side of the ridge are now in the far southwest Indian Ocean, northeast of the Crozet Islands. The two segments of anomaly No. 32 are now separated by about 5,000 kilometers. India must therefore have traveled this distance at the rate of 7.5 centimeters per year after breaking away from Antarctica 75 million years ago.

The separation of the anomalies between anomaly No. 30 and anomaly No. 22 shows that the separation rate during this period was more than twice as rapid: 16 centimeters per year. At that rate a plate would be carried completely around the world in only 250 million years, which is a very short span of time by geological standards. Even faster rates, however, have been measured on the present spreading ridge in the southeast Pacific, where the plates are moving apart at nearly 20 centimeters per year.

We can now use the magnetic lineations to reconstruct the relative positions of the different plates underlying the Indian Ocean. We can also employ the paleomagnetic observations from the neighboring continent of Australia to place all the plates in their correct position with respect to the magnetic poles, which we believe coincide approximately with the earth's rotational poles. The motion of each plate can be described by a rotation, so that the complete reconstruction process consists of a series

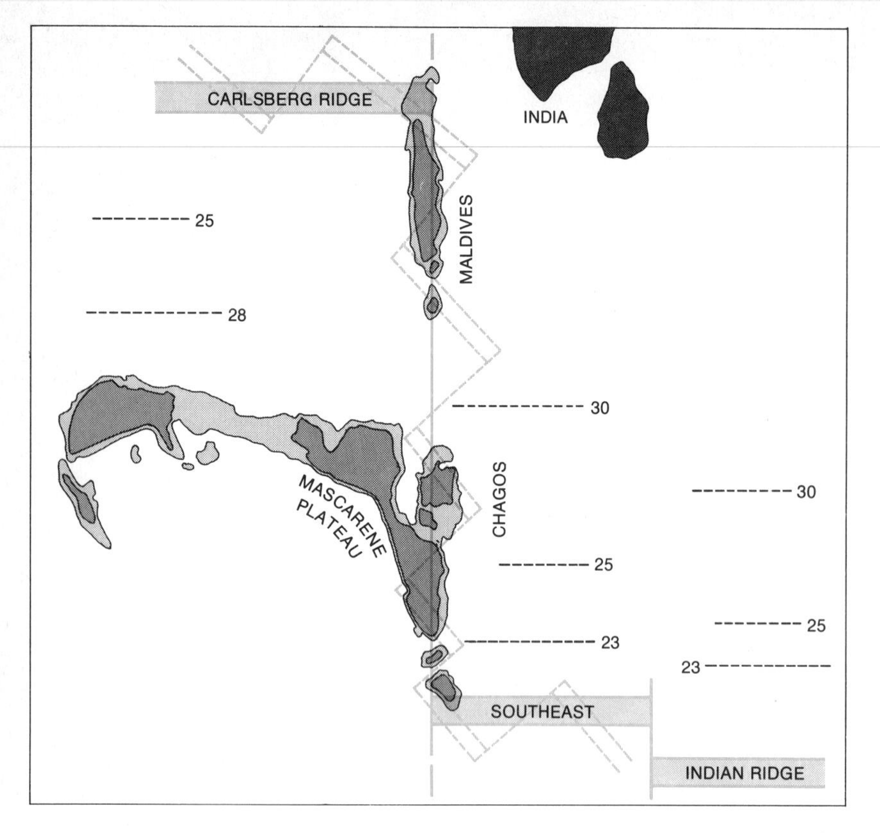

THIRTY-FIVE MILLION YEARS AGO, after a 20-million-year period during which India's northward drift slowed significantly, the single north-south fault of the Chagos fracture zone evidently broke up into a number of ridge segments connected by short faults.

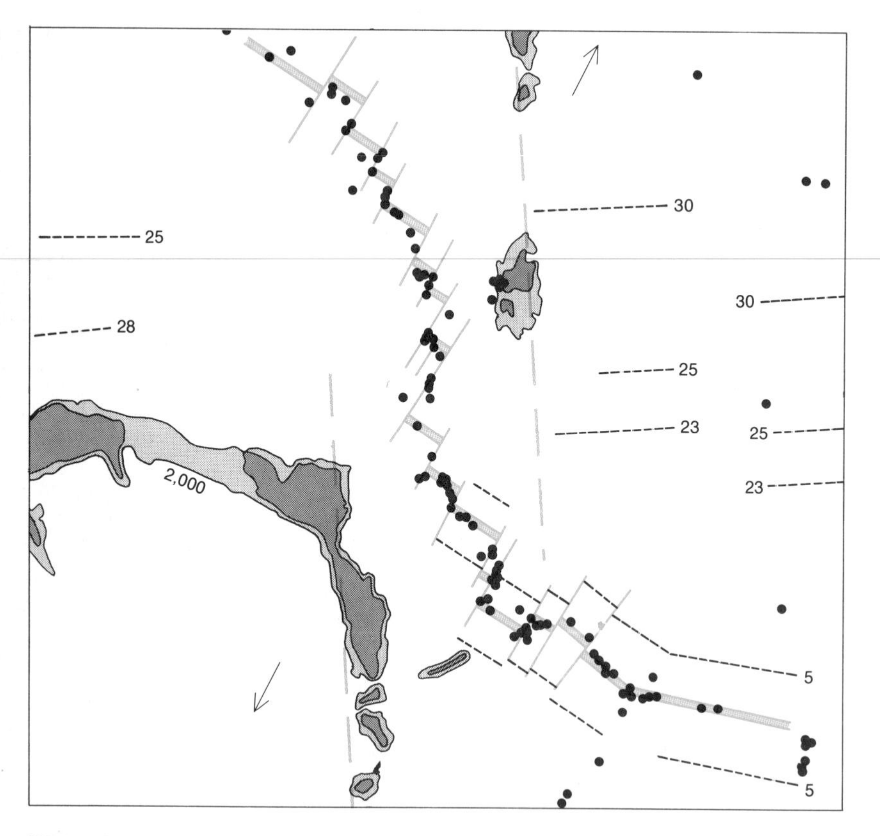

WHEN NORTHWARD MOTION RESUMED about 35 million years ago, new sea floor was extruded along ridges that cut obliquely across the line of the ancient Chagos fault. Thus the direction of sea-floor spreading was rotated about 45 degrees from the original direction. This map repeats a central portion of the ridge-and-fault structure illustrated on page 152.

of rotations. Unlike the single rotation that suffices to close the South Atlantic, however, the closing of the Indian Ocean requires the rotation of several plates that have moved at varying rates and in different directions since India started its rapid movement northward.

The earliest reconstruction we can obtain is governed by the oldest anomaly we can identify: anomaly No. 32, formed about 75 million years ago. At that time the ridge south of Ceylon was spreading in such a way as to propel India generally a little east of north, and Ninety East Ridge was an active transform plate, facilitating the northward slippage of the Indian plate [*see illustration on opposite page*]. Since Asia was evidently more or less stationary, or at least not moving north nearly as fast as India was approaching, the northern edge of the Indian plate must have been consumed in a trench lying somewhere to the south of Asia, but little is known about the position or history of that trench. Australia was then still joined to Antarctica, forming a single continent; Africa and South America had already separated.

During the next 20 million years India moved rapidly northward between two great transform faults: Ninety East on the east and a fault now called the Owen fracture zone on the west. Some 55 million years ago the relative movement between plates slowed down or stopped. Only the ridge between India and Antarctica continued to produce new plate, and the rate of separation dropped from 16 centimeters per year to six centimeters or less. Forty-five million years ago there was a rupture between Australia and Antarctica, setting Australia free for its northward migration. Soon afterward all the plate boundaries in the Indian Ocean became active again, but they were now moving in new directions [*see illustration on page 148*].

The result was the complicated pattern of ridge segments and transform faults that is still active in the central part of the western Indian Ocean. Motion on the Ninety East plate boundary ceased about 45 million years ago. The most recent events in the history of the Indian Ocean have been the formation of the Red Sea and the Gulf of Aden when Arabia separated from Africa, probably about 20 million years ago.

Beginning about 45 million years ago the northern edge of the Indian plate must have begun crumpling up the many layers of shallow-water sediments, known as a geosyncline, laid down over millions of years on the continental shelf that bordered the southern edge of Asia.

The result was the upthrusting of the Himalayas. The application of plate-tectonic concepts to both the evolution of the sea floor and the building of mountains is encouraging. It should eventually be possible to tie the two phenomena together by relating the composition and historical movement of rocks in the mountain ranges to records left in the sea floor.

Many attempts have been made to guess precisely how South America, Africa, India, Antarctica and Australia were once joined to form the primitive continent known as Gondwanaland. There is as yet no general agreement as to how this should be done. The fit between South America and Africa, as is well known, is excellent. The fit between Australia and Antarctica is good. The arrangement of all five major units, however, is controversial, and the original position of Madagascar is unknown.

The principal difficulty is that no magnetic lineations have yet been discovered on the older parts of the floor of the Indian Ocean between the continents. We therefore cannot continue to reassemble the continents by the same methods we have used to trace the movement of

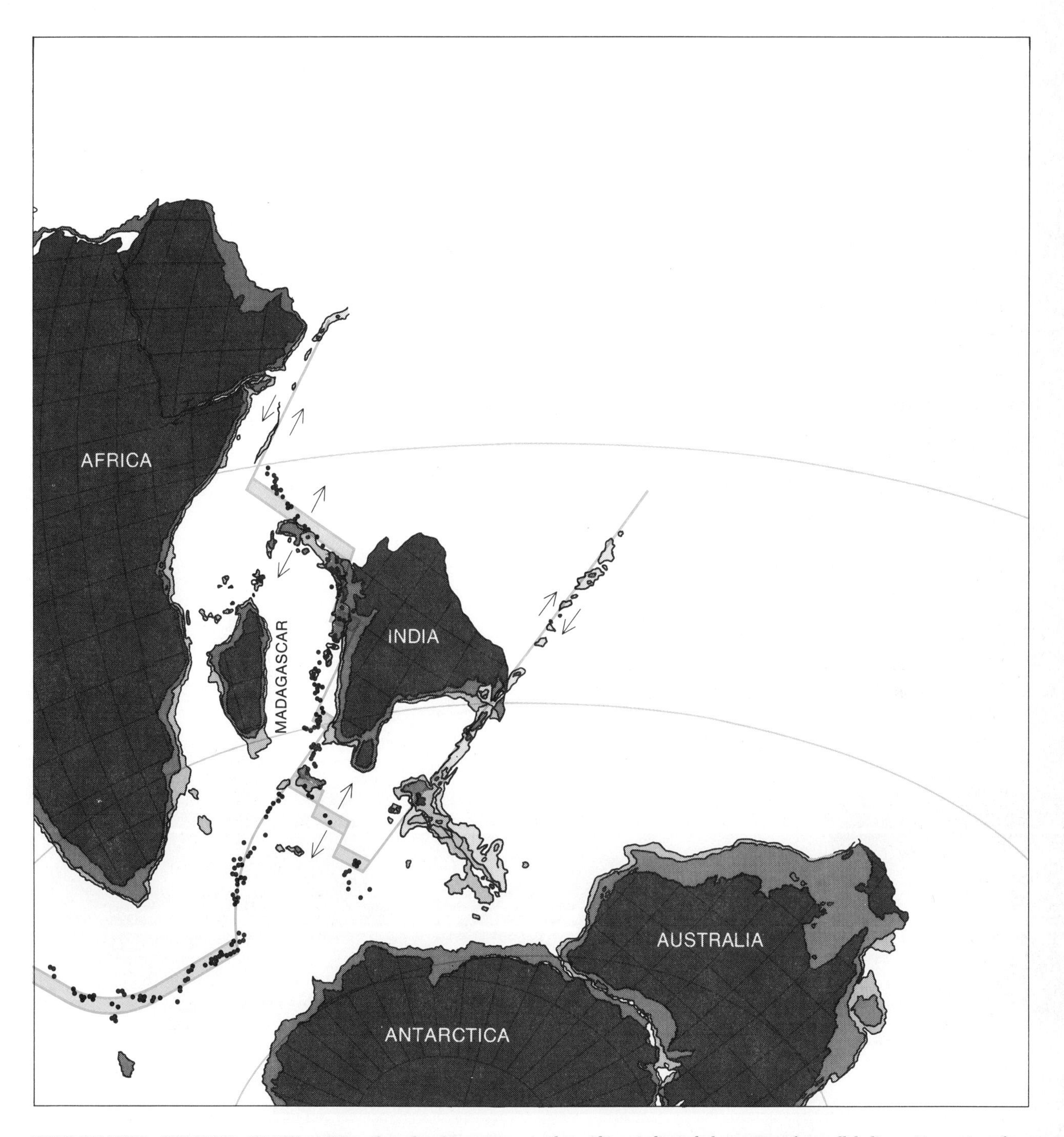

SEVENTY-FIVE MILLION YEARS AGO, after breaking away from the ancient primitive continent named Gondwanaland, the plate carrying India started to move rapidly toward the northeast. The motion was accompanied by the generation of new plate along the ridges indicated by pairs of parallel lines. Large transform faults east and west of the connected series of ridges enabled the Indian plate to move unimpeded. The original fit of India, Africa and Madagascar with Antarctica-Australia is not yet established.

India during the past 75 million years. There are also no other structures like Ninety East Ridge, which was recognized as a transform fault even before the magnetic lineations were mapped. Fortunately the area of sea floor in which the record presumably lies hidden is not great. Last year a series of deep holes were drilled in the floor of the Indian Ocean by the drilling vessel *Glomar Challenger*. The data from these holes have confirmed and amplified our reconstruction of the history of the ocean. They have also added to the evidence needed to reconstruct Gondwanaland.

Meanwhile one can speculate about the original juxtapositions of India, Antarctica and Australia. One guess is that existing reconstructions are wrong because they have attempted to remove practically every piece of sea floor between the continents. That approach has been favored because all the continents believed to have formed Gondwanaland show evidence of having been covered by a huge ice cap some 270 million years ago. We know from the recent glaciation in the Northern Hemisphere, however, that continental ice caps can simultaneously cover landmasses that are separated by oceans. It may be that a small ocean basin, comparable perhaps to the Arctic Ocean, was nestled somewhere among the southern landmasses 270 million years ago. It may be our ignorance of its existence and shape that is preventing the successful reconstruction of Gondwanaland.

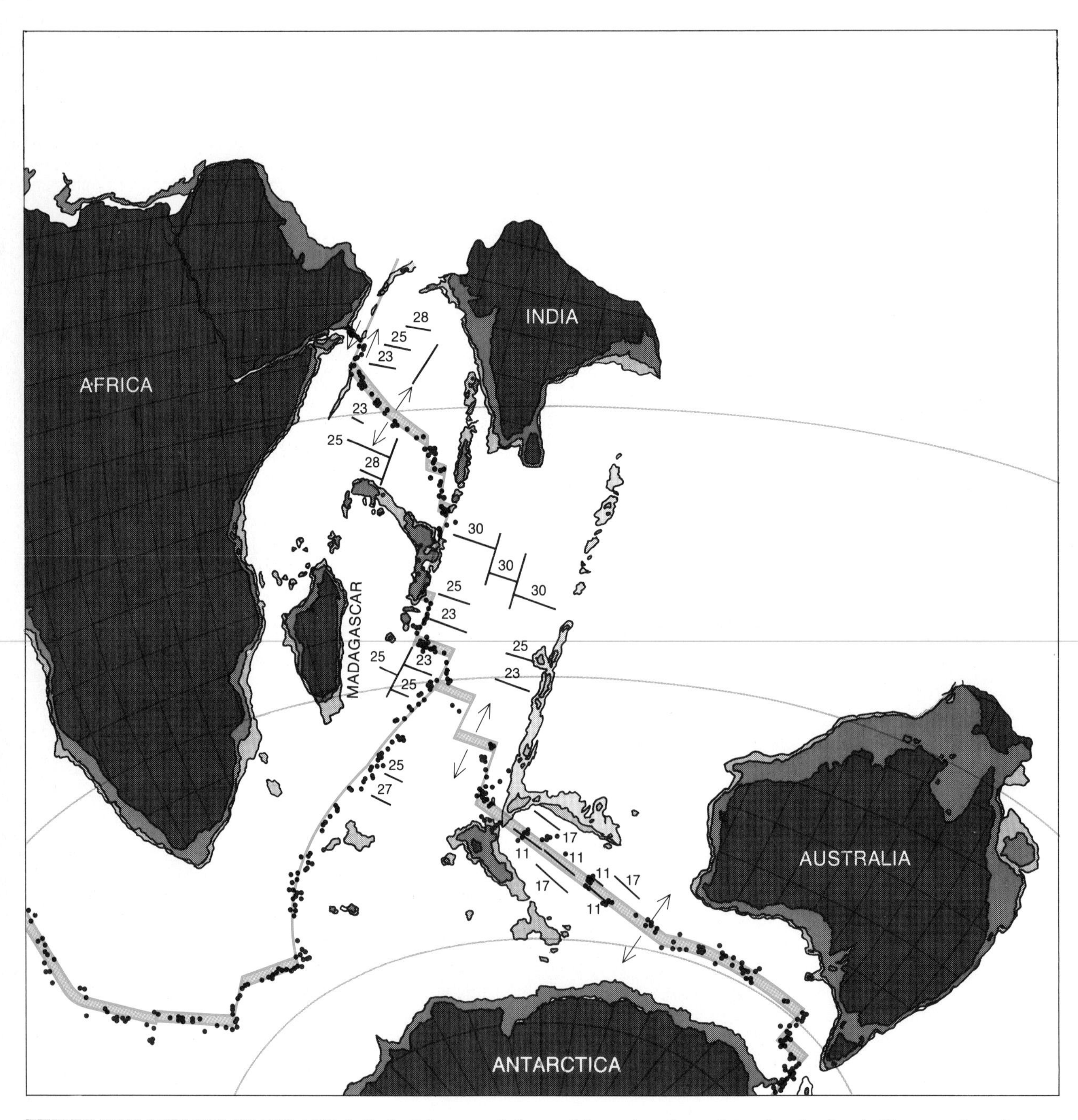

THIRTY-FIVE MILLION YEARS AGO India had been carried some 4,000 kilometers to the north at rates that evidently varied from 16 centimeters per year to seven centimeters or less. It was about this time that the direction of relative motion among several of the major plates changed quite drastically to produce the complicated sea-floor geometry we see today. Australia has been set adrift from Antarctica and has started on its way to its present position. At this point Arabia has still not split off from Africa.

The Evolution of the Pacific

12

Bruce C. Heezen and Ian D. MacGregor
November 1973

Advances in human knowledge often result from the coupling of two separate areas of human activity. An outstanding recent example of this phenomenon is the union of deep-water drilling technology and the scientific theories of continental drift, sea-floor spreading, plate tectonics and the youthfulness of the oceans. In 1968 the National Science Foundation inaugurated the Deep Sea Drilling Project, in which the ship *Glomar Challenger* began to systematically drill holes in the sediments that record the oceans' dynamic history. Among the oceans whose history has begun to emerge from this sampling program, coupled with the theoretical background, is the largest of all: the Pacific.

The *Glomar Challenger* is an unlikely hybrid: a ship with a hole in the middle and a 180-foot drilling rig amidships. With the ship's hull for a platform the men who operate her utilize the techniques of oil-well drilling to bring up the long cylindrical cores of deep-sea sediment. Holding a fixed position over a point on the sea bottom, compensating for the ship's vertical movements and guiding a fresh drill string into a hole only a foot in diameter several thousand feet below the ship's keel are only a few of the more formidable hurdles that the drillers have surmounted. The cores of sediment gathered by the *Glomar Challenger* from around the world constitute one volume after another, so to speak, in the growing library of oceanic history.

The deposition of sediments on the ocean bottom is a variable process. The continual rain of detritus, both organic and inorganic, that slowly sinks to the sea floor is not of the same composition everywhere, nor is it evenly distributed. Consider first the organic component. It is made up of fine fragments of the shells and skeletons and other parts of many different marine animal and plant species. The vast majority of these organisms are small or microscopic: planktonic protozoans and algae such as radiolarians, diatoms, foraminifera and coccoliths that inhabit the upper 400 meters of the ocean. The organisms are cosmopolitan in distribution, but they are found in the greatest numbers only in certain waters. For example, they are abundant in the nutrient-rich zones of deep oceanic upwelling along the western shores of continents. A similar but less well-known zone of abundance exists along the Equator, where the currents of the Northern and Southern hemispheres interact to produce another upwelling of nutrients. One planktonic group, the diatoms, is particularly numerous in two other zones of upwelling: the cold waters of the Arctic and the Antarctic.

Whether or not the calcareous and siliceous remains of these organisms ever actually reach the ocean floor depends on their solubility in seawater. Normally seawater is far from being saturated with calcium carbonate or silica. Because solubility increases with depth, almost all calcareous remains are completely dissolved before they can sink much deeper than 3,700 meters (a little more than two miles). This point is known as the carbonate compensation depth. The point where silica is completely dissolved is somewhat deeper. Compensation depths are, however, kinetic boundaries, and their location depends on both the rate of supply of organic detritus and its rate of dissolution. For example, in areas of high planktonic productivity such as the equatorial zone the carbonate compensation depth may be as much as 5,000 meters below the ocean surface (a little less than three miles). The accumulation of organic detritus on the ocean floor therefore varies from region to region, depending on the ocean depth and the planktonic productivity in the surface waters. As a result in a static model of the Pacific basin the accumulation of organic sediment on the ocean floor would be predictably greatest where the planktonic productivity was highest and the floor itself was shallower than the carbonate compensation level.

What about the sea-floor sediments formed from inorganic detritus? These substances move seaward along a variety of pathways before reaching the bottom. Essentially all inorganic detritus is derived either from continental processes of erosion or from volcanic activity along the margins of the continents. The sediments carried in suspension by streams and rivers make up one class of inorganic detritus. Some of the very smallest particles remain in suspension for a long time and become quite evenly distributed throughout an ocean basin before sinking to the bottom. The same is true of the fine dust that is carried far out to sea by the winds. These two deposition processes are responsible for a thin veneer of red, gray and green abyssal clays that covers much of the floor of all ocean basins. The clays are recognized as distinct sedimentary deposits, however, only where they have not been mingled with other sediments. For example, in the equatorial regions of the Pacific the detritus that elsewhere makes up the abyssal clays is only a trace dilutant, inseparable from the organic oozes that accumulate there in large quantities.

Coarser inorganic sediments reach the ocean in river runoff or as products of shore erosion. They do not usually travel far out to sea before sinking, and so they form a thick wedge of sand, silt and clay adjacent to their continental source. Some may be moved farther offshore by the action of turbidity currents (swift, self-propelled downslope flows of sediment-laden water) but their connection with the mainland remains obvious. Still another characteristic inorganic sediment of the Pacific basin is windblown ash from the active volcanic zones along the margins of the basin. The ash collects in the lee of the volcanoes, forming wedge-shaped deposits that usually thin out to the point of being no longer identifiable 1,000 kilometers or so from their source.

Each sediment, organic or inorganic, accumulates at a different rate. In regions of average planktonic activity the passing of a million years produces a layer of organic detritus 10 meters thick. In zones of greater activity, such as the equatorial region, the rate rises to 15 meters or more per million years. The abyssal clays are laid down at a much lower rate: about one meter per million years. For a few hundred kilometers downwind from an active volcano the average accumulation rate for volcanic ash is about 10 meters per million years. Sediments resulting from river runoff and shore erosion build up the most rapidly: from 50 to 500 meters per million years. (Even this process is not particularly fast; accumulation at the rate of 500 meters per million years means that

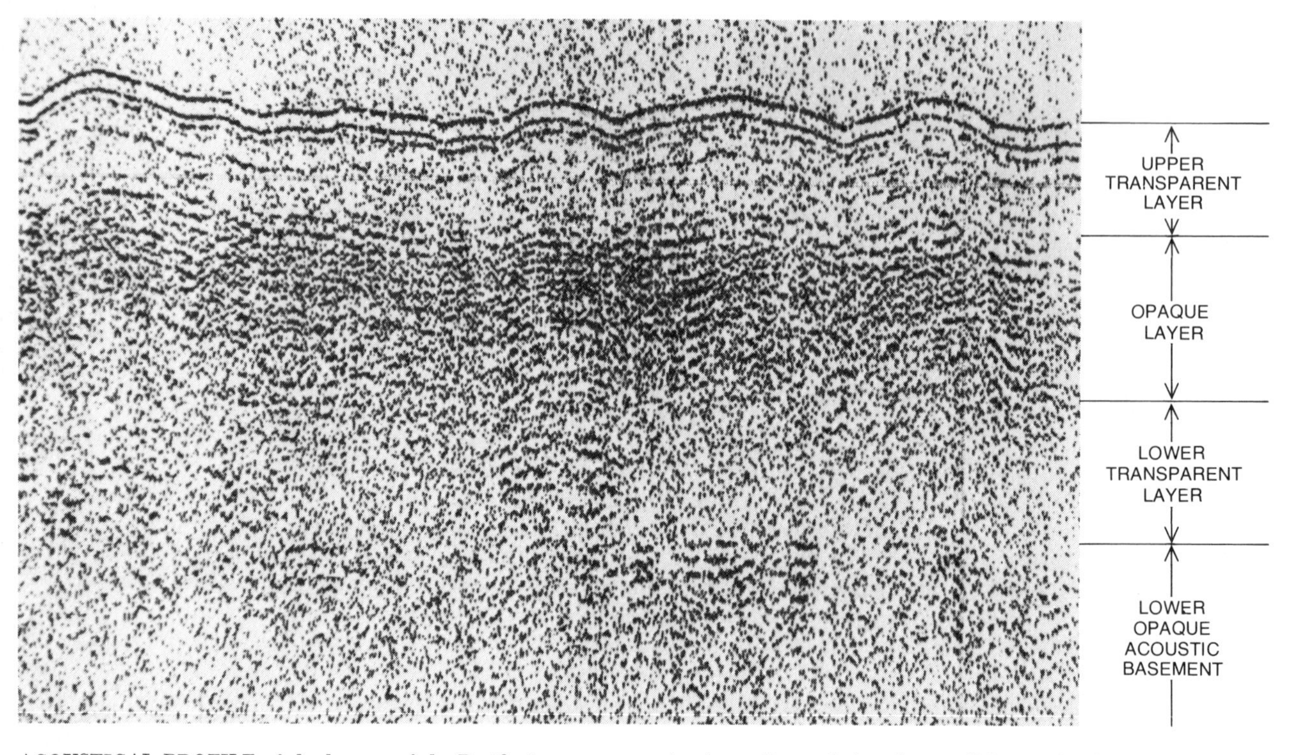

ACOUSTICAL PROFILE of the bottom of the Pacific in an area east of the Japan Trench shows four sedimentary strata, the number characteristic of the western Pacific. A "transparent" layer (*top*) overlies a darker "opaque" layer of sediments. Below the opaque layer lies a less well-defined transparent layer and below that is another opaque layer resting on the basalt oceanic crust.

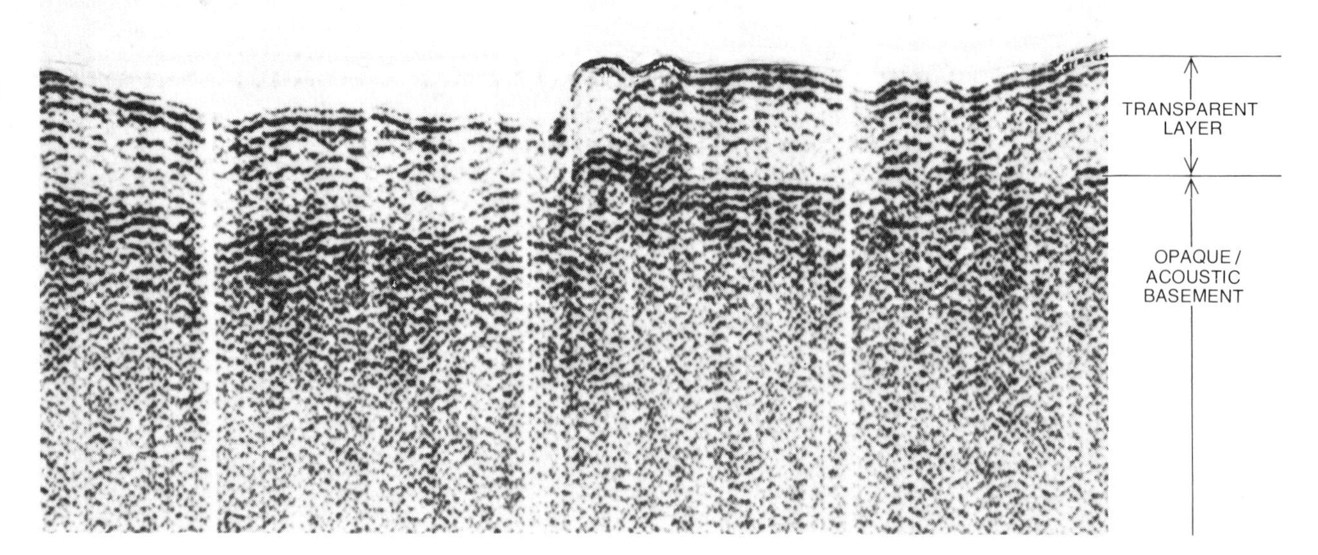

DIFFERENT PROFILE characterizes the eastern Pacific basin. Here only two layers of sediment cover the oceanic crust: a transparent top layer and an underlying opaque one. Deep-sea drill cores confirm the acoustical data. The two profiles are not reproduced at the same scale. In the top one the top three layers are 2,000 feet deep; in the bottom one the top layer is 200 feet deep.

only a twentieth of a centimeter is deposited each year.)

Sediments of these kinds are accumulating today on the floor of the Pacific basin in a clearly patterned fashion [*see illustration below and on page 152*]. In the northern waters and all along the margins of the basin volcanic ash and sediments from runoff and shore erosion build up, mixed in the subpolar region with the tiny skeletons of the diatoms that thrive there. In a broad zone along the Equator the steady rain of planktonic detritus gives rise to a blanket of organic oozes on the sea floor. Between the largely inorganic northern zone and the organic equatorial zone, below waters that are too deep to allow the buildup of organic debris, lies a vast expanse of abyssal clays.

If the present pattern of sedimentation has been constant over the past 100 million years or so, a static model of the Pacific basin would predict that any core of sea-floor sediment taken from one or another of these major zones should consist uniformly of, say, abyssal clays or organic oozes, or sands, silts and ash. In actuality no drill core is that uniform. We are left to conclude either that the pattern of sedimentation has not been constant or that a static ocean-basin model is wrong. Suppose we assume that any changes in the sedimentation pattern over the past 100 million years have been trivial. Can we, by changing the static model into a dynamic one, explain the observed nonuniform nature of the sedimentary deposits in the Pacific basin? We believe the answer is yes.

Current plate-tectonics theory states that the basement rock of the ocean floor is being formed continuously. In the Pacific basin the area of formation is the "rift valley" at the summit of the East Pacific Ridge. The new crust migrates essentially westward, sinking deeper as it creeps along. The generation of new crust at the ridge is presumably balanced by a comparable consumption of old crust as it plunges under the island arcs of the western Pacific. The horizontal and vertical motions of the crust are coupled; the rate of sinking decreases logarithmically with age.

Let us visualize a dynamic model of part of the basin, centered on the East Pacific Ridge. The movement of the newly formed crust will be westward and downward [*see top illustration on page 153*]. From the time that the crust is formed at the top of the ridge, and for a considerable period thereafter, it will be above the carbonate compensation depth. As a result it will acquire a blanket of organic oozes. The sediments will increase in thickness until the outward and downward motion of the crust carries it below the compensation depth. Thereafter, as the crust continues to creep westward into deeper water, a slowly accumulating blanket of inorganic abyssal clays will begin to cover the initial organic deposit. Because the carbonate compensation depth lies 1,000 meters below the ridge crest and because the subsidence rate of the new crust is about 50 meters per million years, organic detritus will accumulate on the new crust over a period of 20 million years. Thus the basal layer of organic sediments, accumulating at a rate of 10 meters per million years, should under

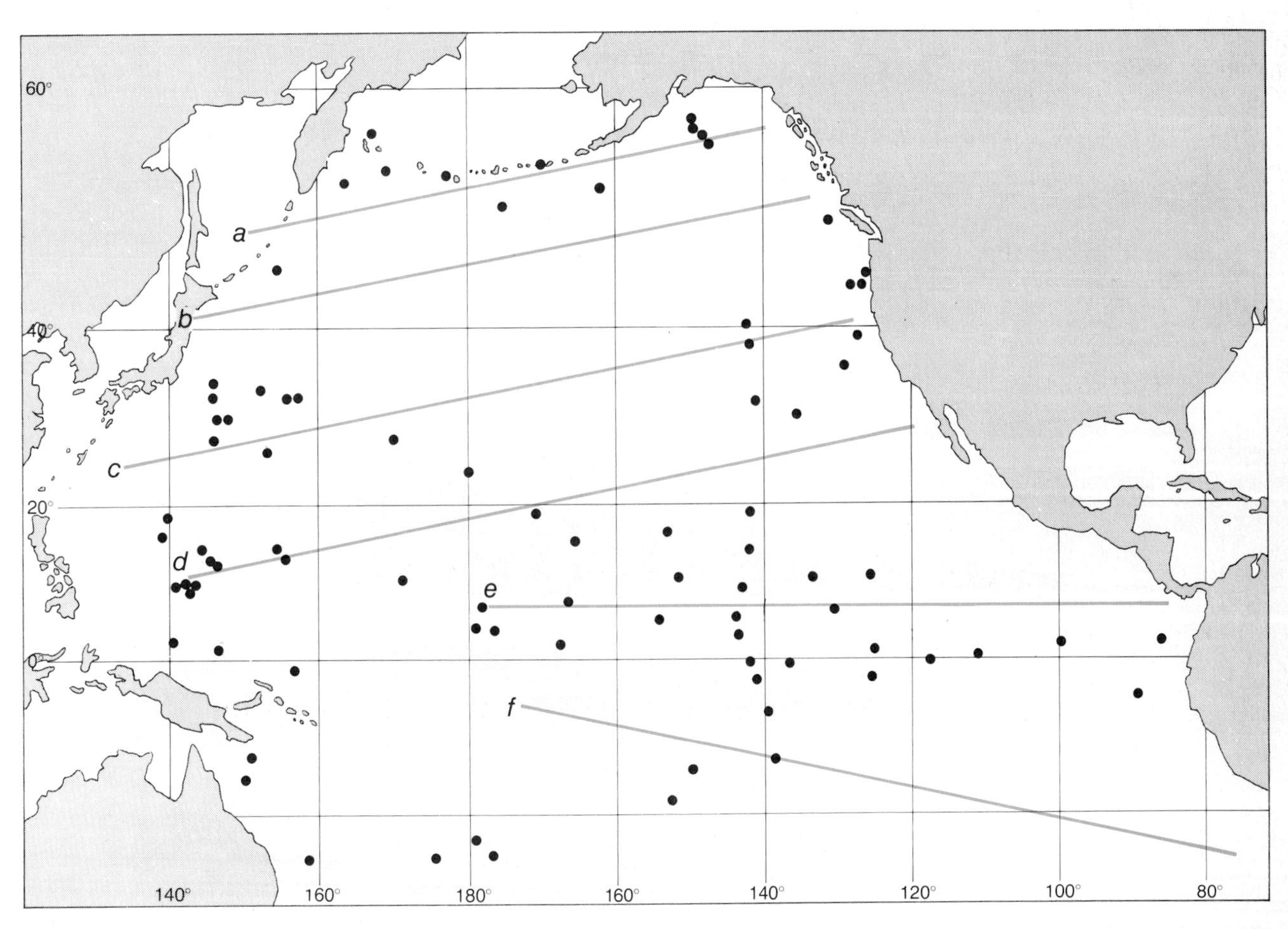

PACIFIC SEDIMENTS WERE SAMPLED by deep-sea cores taken at the sites marked with the black dots on this map. The colored lines labeled *a*, *b*, *c*, *d*, *e* and *f* relate the cores to the cross sections of the sediments shown on the opposite page. The cores were obtained by drilling from the *Glomar Challenger* in the Deep Sea Drilling Project sponsored by the National Science Foundation.

normal circumstances be no more than 200 meters thick.

Now, the sea floor in the western Pacific, the deepest and oldest part of the basin, is at least as old as early Cretaceous times, some 120 million years ago. If during the first 20 million years of its existence this part of the basin accumulated organic sediments at the rate proposed in our model, then 100 million more years were available for the deposition of abyssal clays on top of the organic oozes. At the rate of one meter of clay per million years, therefore, the western Pacific sea floor should have accumulated 100 meters of abyssal strata on top of the 200-meter organic layer. How do the sediments in the western Pacific compare with the predictions of the dynamic model? Before answering the question we must introduce another kind of information.

Drilling is the only way to collect samples of the layers of sediment below the ocean floor but it is by no means the only way to gather information about these strata. One method that has been used increasingly since World War II is acoustical profiling. In this method a strong acoustic signal is returned from the interfaces of layers of contrasty sea-floor sediment and is recorded in a continuous profile. Profiling thus makes it possible to trace the lateral distribution of different sedimentary layers for hundreds and even thousands of miles. In the Pacific profiling indicates that a layer of unconsolidated and relatively "transparent" sediments, less than 100 meters deep, covers the central part of the North Pacific floor [*see lower illustration on page 150*]. Such a region of shallow sediments would correspond to the area where our dynamic model predicts the buildup principally of abyssal clays. Along the northern and northwestern boundaries of the basin the profiles show a much thicker transparent layer; the greater thickness apparently corresponds to the rapid accumulation in these regions of continental debris and volcanic ash.

The acoustical profile along the Equator shows that the sediments there form a massive deposit, triangular in cross section and in places as much as 1,000 meters thick. Such a large accumulation corresponds to the high rate of sedimentation in a zone where upwelling nutrients produce an abundance of plankton. The acoustical profiles in the eastern Pacific are also what would be expected; the transparent layer of surface sediments, presumably abyssal clays, is present and is not unusually thick. In the western Pacific, however, not one or two but four separate sedimentary layers appear in the profiles. Just below the transparent layer on the sea-floor surface is an "opaque" layer. A second transparent layer underlies the opaque layer and a second opaque layer underlies the second transparent layer. This fourth layer rests directly on the oceanic crust [*see upper illustration on page 150*].

The acoustical data are in good agreement with the specific information about stratigraphic sequences provided by deep-sea drilling. The cores show that the bottom layer of sediments, resting directly on the oceanic crust both in the older, deeper western Pacific and in the younger, shallower eastern Pacific, is or-

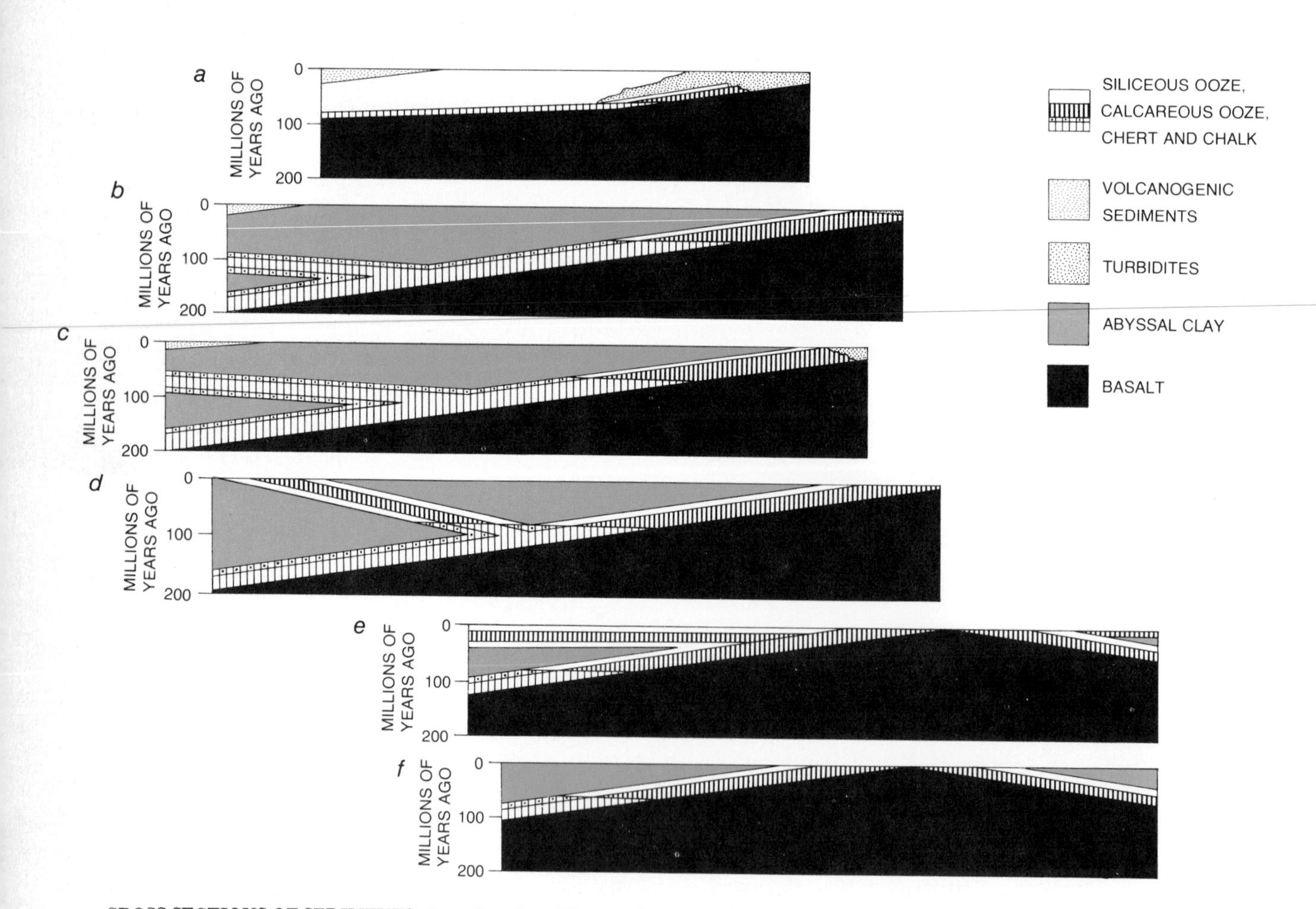

CROSS SECTIONS OF SEDIMENTS along the colored lines in the illustration on the opposite page are shown schematically. The wedge-shaped form of the cross sections results from the fact that, as the sediments were being deposited, the basalt crust below them was moving. The time at which the sediments were being deposited is given by the scale at the left side of each of the cross sections.

ganic in origin. The only difference between west and east is that the older deposits have turned into chalks and cherts whereas the younger ones are still unlithified oozes. In both west and east, the cores show, the basal organic layer is overlain by a layer of abyssal clays.

The clays in the western Pacific are older than the clays in the eastern part of the basin. In addition the western clays are overlain by a second layer of chalks and cherts, and that organic layer is covered by a second layer of inorganic clays. Thus the cores show precisely the same four-layer sequence that is revealed by acoustical profiling, and it is obvious that serious discrepancies exist between the four-layer stratigraphy in the western Pacific and the two-layer prediction of our dynamic model. The model sequence—oceanic crust overlain first by organic oozes and then by inorganic clays—is correct as far as it goes, but it fails to describe the real sequence. There is another difficulty too. We find two depositions of organic sediments in the western Pacific rather than one, and the buildup of this double ration of sediment would have needed more time. The process may have taken 50 million years, or more than twice the 20 million years allotted in the model's timetable.

Can our defective dynamic model be repaired? The answer is yes, if we alter its direction of movement. Let us add a new component—northward drift—to the slow westward creep of the freshly formed oceanic crust. If we do that, the new crust formed north of the Equator will, as before, accumulate only two layers of sediment: first organic oozes and then abyssal clays. The same two layers will also build up on crust newly formed south of the Equator. When the southern crust slowly moves northwestward across the Equator, however, it will receive a further heavy rain of detritus from that nutrient-rich zone, thereby acquiring a second layer of organic oozes [*see middle illustration at right*]. When the crust once again enters waters that contain nothing but inorganic matter, it will start to acquire a second layer of abyssal clays on top of its second layer of organic oozes.

The predictions of our revised dynamic model compare favorably with the results of deep-sea drilling in the Pacific basin, although we should like to see cores taken at many more locations. In the cores from the western Pacific that are available now the oldest chalks and cherts are of lower Cretaceous age (and some are probably even older). This layer evidently represents the organic

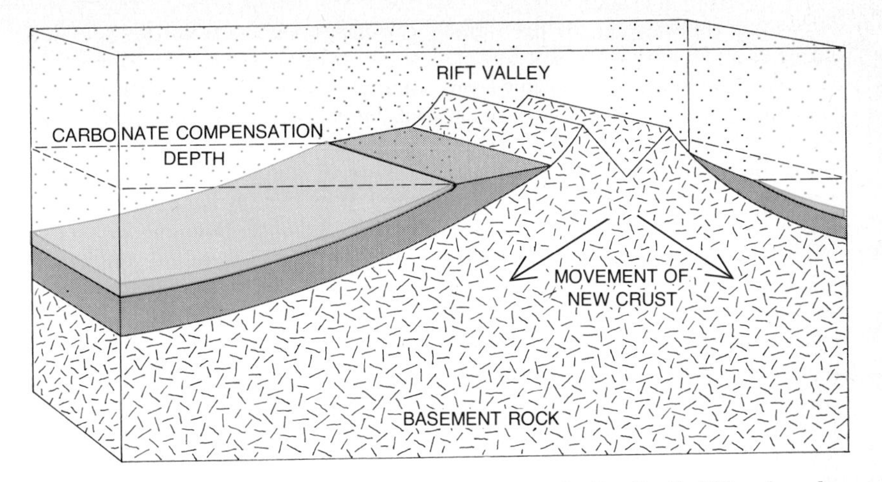

DYNAMIC MODEL of the Pacific basin in the vicinity of the East Pacific Ridge shows how different kinds of bottom sediment accumulate. Oceanic crust, newly generated in the rift valley at the crest of the ridge, slowly moves outward and downward. While the crust remains above the "carbonate compensation depth" most of the detritus that settles on it is organic (*black dots*) and a thick stratum of ooze is formed. When the crust enters deeper water, only inorganic detritus (*colored dots*) reaches it and forms a thin stratum of clay.

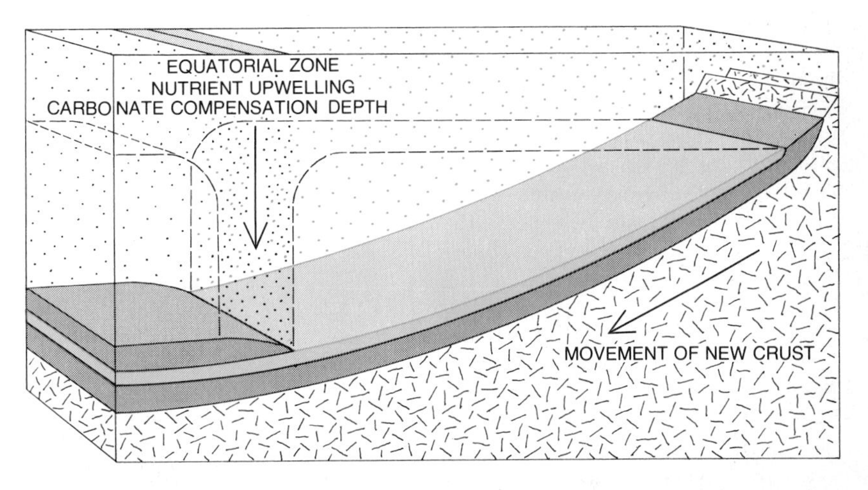

REVISED MODEL makes the oceanic crust travel through the equatorial zone, where abundant planktonic growth brings about an abrupt lowering of the carbonate compensation depth. The heavy "rain" of organic detritus (*black dots*) in this zone deposits a second layer of ooze on the abyssal clay. As a result three separate sedimentary strata overlie crust.

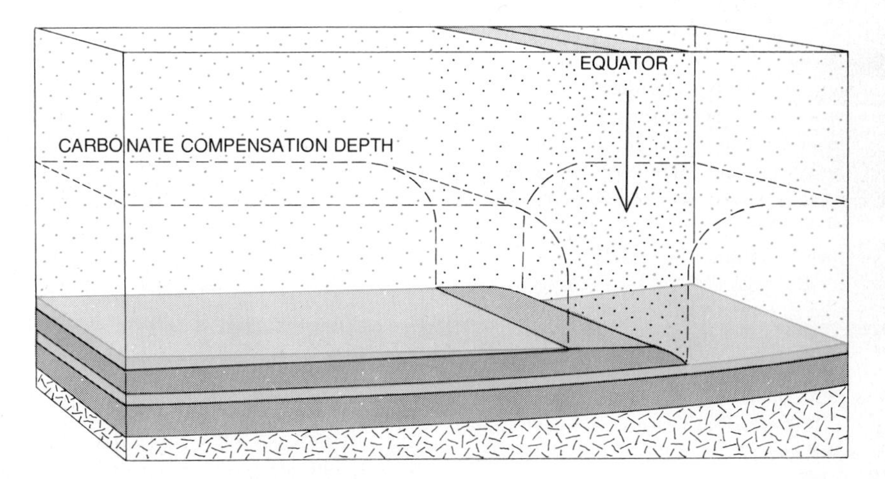

FOURTH LAYER OF SEDIMENTS accumulates above the three earlier layers after the oceanic crust passes beyond the equatorial zone. The crust is once again in water far below the carbonate compensation depth and only inorganic detritus (*colored dots*) reaches the floor of the ocean. As a result a thick second stratum of abyssal clays builds up, covering second stratum of organic ooze. This is the kind of layering characteristic of western Pacific.

detritus accumulated on the new crust during its initial subsidence. The abyssal clays that overlie the basal organic layer similarly represent the deposition of inorganic detritus after the new crust sank below the normal carbonate compensation depth. The chalks and cherts in the layer above the lower abyssal clays are middle Cretaceous in age; they evidently record the passage of this part of the Pacific crustal plate northwestward across the Equator. Finally, the abyssal clays above the second organic layer are late Cretaceous or Cenozoic in age; they must represent the further accumulation of inorganic sediments after the equatorial zone was left behind.

Other kinds of evidence support the concept illustrated by our revised dynamic model. Provocative evidence on the probable directions of motion of the Pacific plate is provided by two long lines of submerged Pacific seamounts: the Hawaiian and the Emperor chains [*see illustration on this page*]. It has been proposed that the seamounts are the truncated remains of former volcanoes that were successively brought into existence by the action of a "hot spot" rooted to the deep mantle below the Pacific crustal plate. In principle the two long seamount chains combined represent a kind of fossil record of the slow movement of the Pacific plate across the hot spot over millions of years.

As a result of drilling and dredging the ages of such features can be estimated. The oldest seamount, at the northern end of the Emperor chain, was evidently submerged some 70 million years ago. Midway Island, near the western end of the Hawaiian chain, is only some 20 million years old. The volcano Kilauea, on the island of Hawaii at the eastern end of the chain, is still erupting. The ages of seamounts located at the "bend" where the two chains meet have not yet been determined, but on the basis of extrapolation they are thought to be from about 30 to 40 million years. Since the ages increase to the north and west it would appear that the Pacific plate has moved west-northwest in a direction parallel to the Hawaiian chain for the past 30 to 40 million years, and that for 30 to 40 million years before that time it had moved in the more northerly direction parallel to the Emperor chain.

It is possible to estimate on the basis of the deep-sea drilling data the average northward component of the plate's motion. The point in any core where the bottom layer of the youngest abyssal clays is in contact with the top layer of the underlying chalks and cherts can be interpreted as a temporal boundary; it indicates approximately when that part of the plate completed its transit of the equatorial zone. In cores taken at a variety of latitudes in the Pacific basin the age of this temporal boundary has been determined from the fossils preserved in the uppermost chalks and cherts. The results suggest that, for a period of some 80 million years beginning in the late Mesozoic and continuing until middle Cenozoic times, the northward movement of the Pacific plate proceeded at a rate of some 4.4 centimeters per year. Since then, for the past 30 million years or so, the northward component of motion has been smaller: some two centimeters per year [*see top illustration on facing page*].

Still further evidence with respect to the motion of the Pacific plate comes from the observation that in oceanic crust generated along an east-west axis the bottom is "striped" with zones of alternating magnetic polarity, the result of past reversals of the earth's magnetic field. The magnetic striping is either nonexistent or poorly developed, however, in crust that has been generated in the equatorial zone; here the crust is magnetically quiet. Now surveys show that a huge area of the North Pacific is also magnetically quiet. The region starts near the present intersection of the Equator and the East Pacific Ridge, in the vicinity of 100 degrees west longitude, and continues generally north by west in a wide swath that terminates off the Kamchatka Peninsula in the northwest Pacific [*see bottom illustration on facing page*].

Up to now the existence of this large quiet area has been explained as evidence that during a considerable part of the Cretaceous there were no reversals of the earth's magnetic field. Although there may have been such a period, our model predicts that a magnetically quiet zone should be present. A better explanation, in our opinion, is that the quiet region represents crust that was formed within the equatorial zone and that has

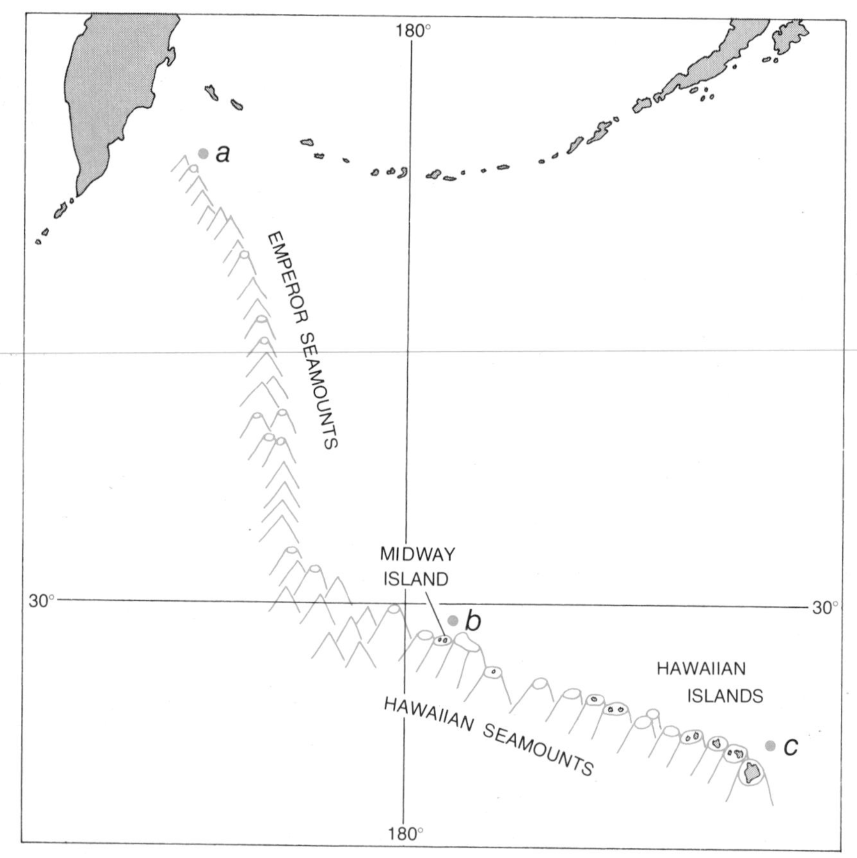

LINE OF SEAMOUNTS, the Emperor chain (*left*) and the Hawaiian chain (*right*), traces the past movement of the Pacific plate over a "hot spot" rooted in the mantle near Hawaii. As the plate moved northwestward over the hot spot, volcanoes appeared; the seamounts are their submerged remnants. A seamount (*a*) at the northern end of the Emperor chain is the oldest known: 70 million years. This contrasts with an age of 20 million years for Midway Island (*b*) and one of less than a million years for the volcano Kilauea on Hawaii (*c*).

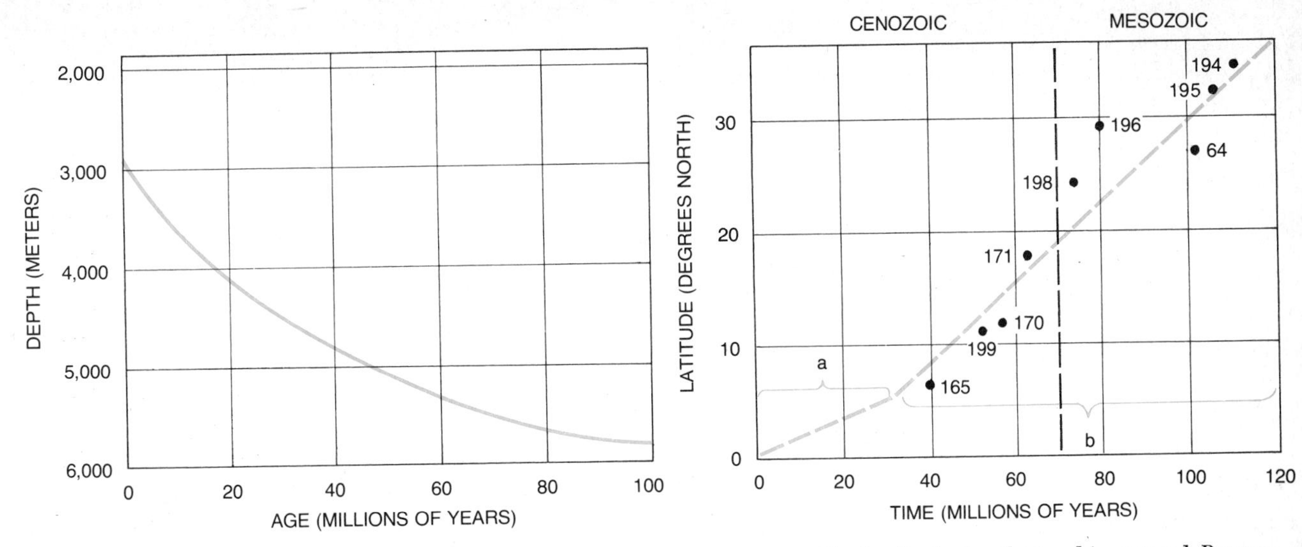

AGE OF FOSSILS in the uppermost organic strata of sea-floor sediments in the Pacific increases with depths (*left*) and distance north of the Equator (*right*). Two rates of plate movement become apparent (*broken line*) when the trend is averaged. Between 120 and 30 million years ago (*b*) the rate was some 4.4 centimeters per year. Since then (*a*) it has diminished to some two centimeters.

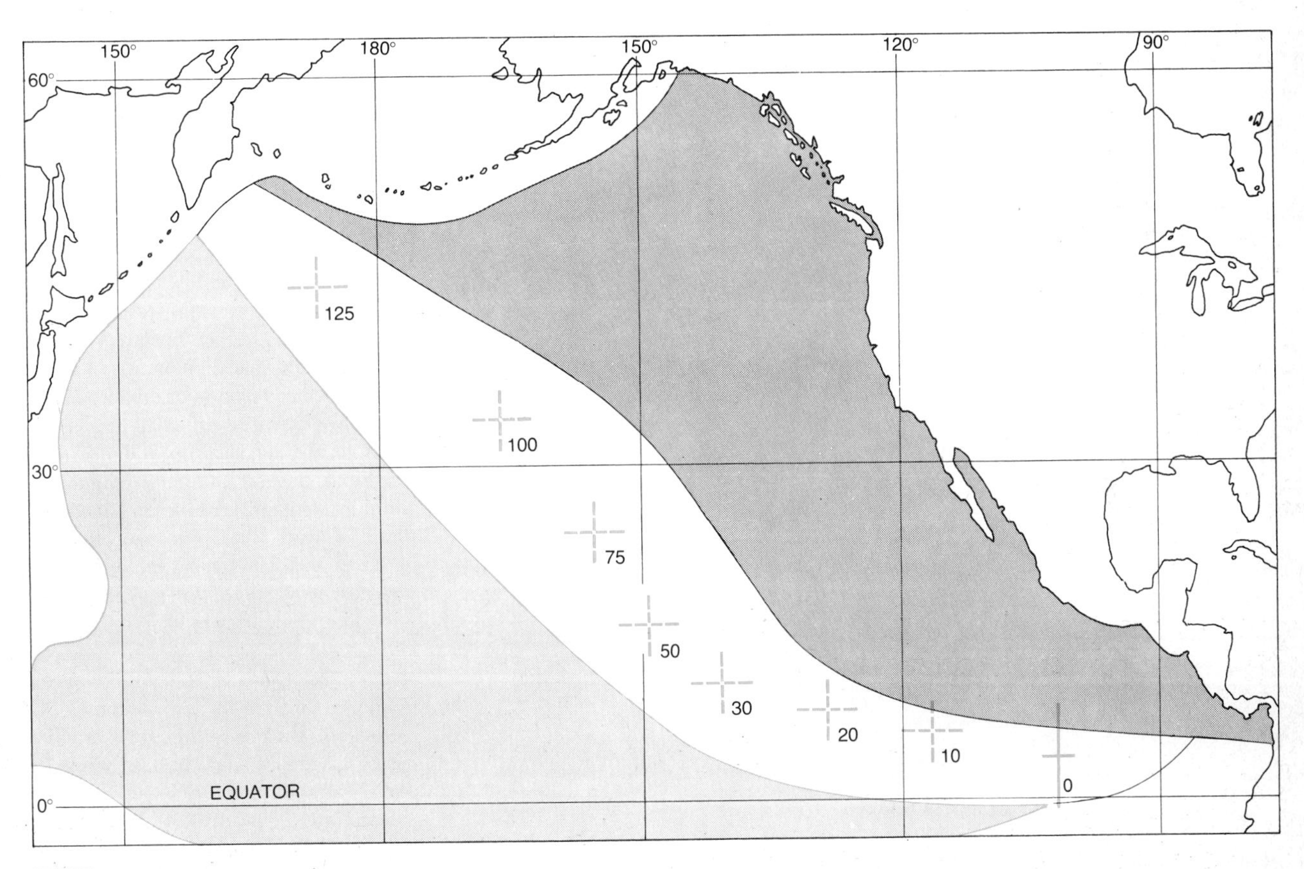

CRUST FORMED WHILE PLATE CROSSED EQUATOR

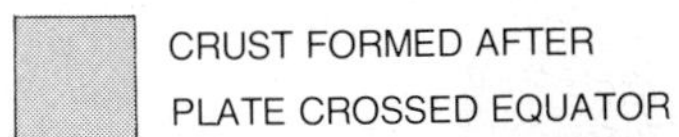

ZONE OF MAGNETIC QUIET, where evidence of reversals of the earth's magnetic field is weak or nonexistent in the crust of the ocean floor, extends from near South America on the Equator to the subpolar waters off the Kamchatka Peninsula. A solid cross (*right*) shows the present intersection of the Equator and the East Pacific Ridge, where new crust is formed continuously. Seven earlier intersections are marked by broken crosses and numbers that suggest how many millions of years ago these paleo-equators began to migrate northwestward; all seven lie within the magnetically quiet zone. The crust north of the quiet zone has never traveled across the Equator. By crossing the Equator the crust south of the zone has collected four layers of sediment rather than two layers collected elsewhere.

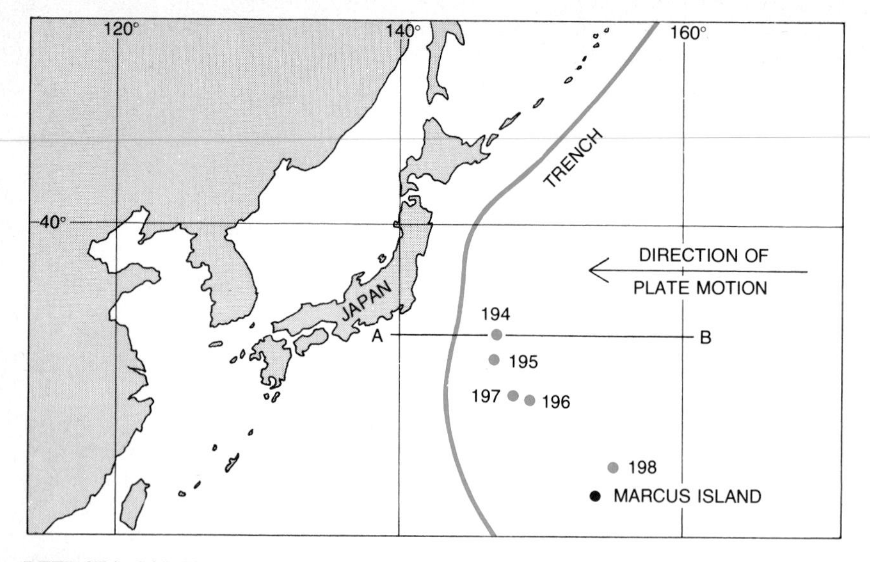

DEEP-SEA CORES from an area in the western Pacific between the Japan Trench and Marcus Island reveal the presence of a wedge of volcanic ash that overlies a part of the four-layer sequence of sediments in this area. In the first illustration below is a cross section.

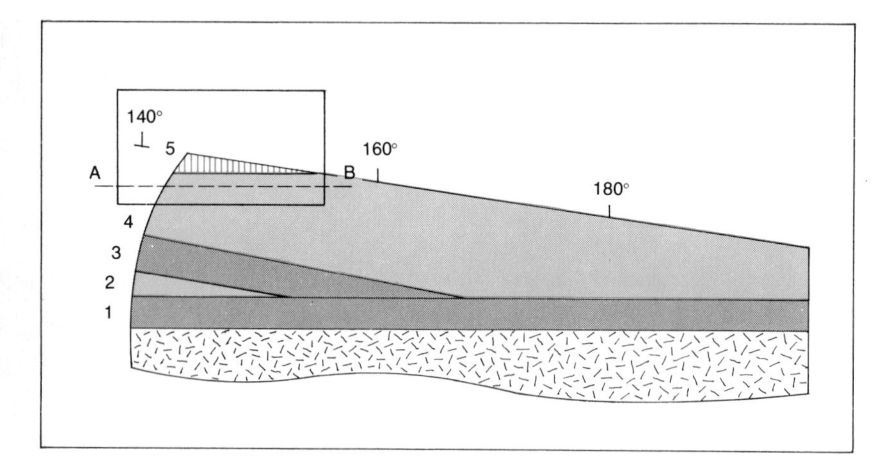

LAYER OF ASH is a fifth sedimentary deposit on the sea-floor crust. The lowest layer is the first deposit of organic detritus, accumulated after the crust was formed. The small wedge above it is the first deposit of inorganic abyssal oozes, which here pinches out. Above that are a second organic layer, which pinches out farther to the east, and a second, very thick inorganic layer that comprises the ocean bottom. A closer view of section is shown below.

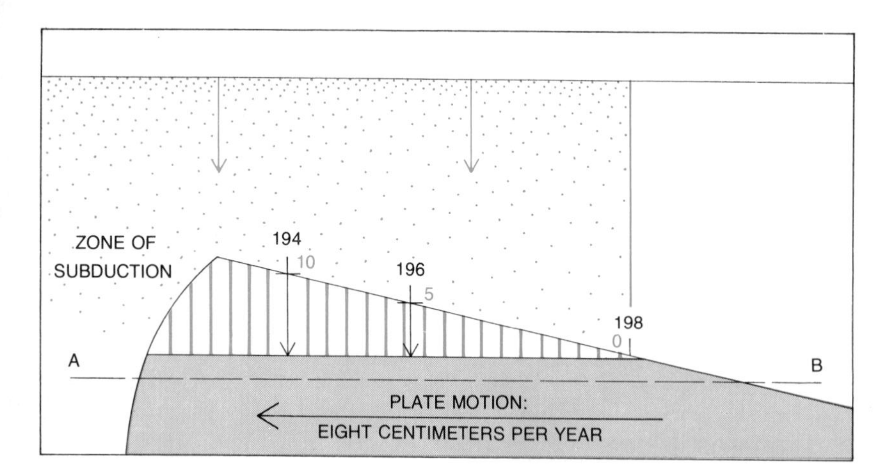

THICKNESS OF LAYER increases from east to west. Assuming a sedimentation rate of 15 meters per million years, the lowest ashes in drill hole No. 196 would be some five million years old and those in hole No. 194 would be some 10 million years old. Ash much older than that has been lost in the zone of subduction at western edge of moving Pacific plate.

reached its present position as a result of the slow northwestward movement of the Pacific plate. An eastward extrapolation of paleo-equator positions determined from deep-sea drilling, together with a westward extrapolation of crustal age based on dated magnetic stripes, enables us to estimate the location and age of a series of points where the East Pacific Ridge and earlier "paleo-equators" once intersected.

In our interpretation the magnetically quiet region represents a boundary between the two principal components of the Pacific crust. North of the quiet region's northern boundary is crust that has never made an equatorial crossing. South of the southern boundary is crust that was formed before a crossing, so that during its subsequent equatorial transit it received a second ration of organic sediments. The quiet belt between these regions would of course be crust formed during the equatorial transit.

Our refined model applies to an oceanographic problem a particular kind of geological reasoning: the facies concept. The concept arises from the observation that sedimentary deposits reflect the circumstances and environment of their origin. In brief, it states that just because similar sets of strata are found in two separate sedimentary deposits this does not necessarily mean that the strata accumulated at the same time. It is just as likely that their similarity is due to the existence of similar environmental conditions at quite different times. For example, if the margin of a continent is subsiding, a shoreline environment will "migrate" inland; the sedimentary deposits of each successive shoreline will become a rock formation essentially identical with the one preceding it, but what may appear to be a uniform lithological unit will not have been deposited at the same time. In the same way our model of the Pacific basin assumes that the ocean's various sedimentary environments have remained constant while the plate comprising the floor of the basin has migrated. The strata that are similar are not contemporaneous; instead they represent the plate's passage through a series of constant environments.

It should be emphasized that the simplicity of our model necessarily gives it a number of shortcomings. For example, it is a depositional model that ignores episodes of erosion, changes in sea level and possible shifts in compensation depths. It scarcely considers such variations in plate motion as rotation or changes in direction and the rate of drift. Neither does it particularly concern itself with the great mass of sediments that comes from continental runoff, shore erosion and volcanic activity.

Concerning these continental sediments, however, one point should be made. The northward and westward motions of the Pacific plate and its subduction under the island arcs of the northern and western Pacific do affect the size and shape of the great sedimentary wedges that extend out from the continents. At the boundary of the subduction zone the dimensions of the wedge will be determined by the combined rates of sedimentation and subduction. A rapid rate of sedimentation will tend to advance the wedge seaward, whereas subduction will counteract this tendency by continuously moving the sediments back toward and under the continent.

Drill cores taken in the vicinity of the Japan Trench, northwest of Marcus Island, provide good examples of the layers of volcanic ash that began to build up on the surface of the abyssal mud in this area late in Miocene times, some 12 million years ago. The ash here has not yet reached the subduction zone, and the volcanic accumulation is some 150 meters thick [*see middle illustration at left*]. Some 800 kilometers farther to the southeast the wedge of ash thins to the vanishing point, leaving the uppermost layer of abyssal clay uncovered. Drillings in between show that the wedge is intermediate in thickness. If we assume that the sedimentation rate for volcanic ash has averaged about 15 meters per million years, then it appears that the Pacific plate is moving into the subduction zone at a rate of eight centimeters per year and that the volcanic ash deposited on the Pacific sea floor in middle Miocene and earlier times has already vanished under the continental plate.

In summary, the fact that our simple model gives a good first-order fit with the data so far available suggests that the evolution of the Pacific basin has been governed by a few equally simple and systematic factors. Among these are the carbonate and silica compensation depths, the gradual sinking of the oceanic crust as it creeps westward from the East Pacific Ridge and the simultaneous northward motion of the crust. Data already exist, of course, that deviate from the model, and future observations will surely produce more such data. Such deviations are welcome. Not only should they reveal local variations in the sedimentary history of the Pacific basin but also they may demonstrate irregularities in the motion of the Pacific plate itself.

The Floor of the Mid-Atlantic Rift

J. R. Heirtzler and W. B. Bryan
August 1975

Some 20 years ago it became apparent that a continuous range of undersea mountains twists and branches for a total length of some 40,000 miles through all the world's oceans. This system of mid-ocean ridges is the site of frequent earthquakes with shallow foci. Geologists had no satisfactory explanation for these earthquakes, and they were even more mystified by the fact that samples of the ocean floor were geologically young. No samples seemed to be more than about 135 million years old, and the closer to the ridge the samples were taken, the younger they were. These findings, together with other discoveries (such as the regular alternation of magnetic polarity in broad bands of the ocean floor running parallel to the ridge), finally led to the hypothesis that the mid-ocean ridge is actually a system of parallel ridges centered on a continuous rift in the ocean floor, and that the floor itself is everywhere spreading outward from the rift. As the rift widens it is filled with lava that wells up from the earth's mantle. The ridges mark the edges of huge plates in the earth's crust that bear the continents. As the sea floor spreads, the continents at the margins of the oceans are either borne apart or compressed into new configurations.

Thus the mid-ocean-rift system reflects processes of the most fundamental importance in the evolution of the earth. Geologists now believe that the processes responsible for the rift system as we see it today have been creating and modifying the crustal plates throughout much of the four to five billion years of the earth's history. Although many details remain to be explained, the general concept of the creation of crust at the mid-ocean ridges has become widely accepted in less than a dozen years and forms the basis for much of present-day geological reasoning.

As has been the case with the spectacular growth of knowledge about the moon and the other planets, the expansion of our knowledge of ocean-bottom geology has been closely connected with technological advances that have made possible the increasingly rapid collection of data. Although much has been learned about the mid-ocean ridges by remote-sensing techniques, by drilling and by random sampling from surface ships, it finally became obvious that direct manned observation, by means of special submersible vessels, would ultimately be essential for any real understanding.

Starting late in 1971 a program was initiated by several interested groups to develop and harness specialized instruments and techniques for a detailed study of the central part of the Mid-Atlantic Ridge. A region of the ridge some 400 miles southwest of the Azores was selected for examination both because it is an area where one can expect good weather and because the port of Ponta Delgada in the Azores offered a convenient base of operations. The U.S. and France took leading roles in the effort because each had had experience with research submersibles of the type that would be needed for conducting manned observations in the later phases of the program. The cooperative venture was dubbed FAMOUS, for French-American Mid-Ocean Undersea Study.

Although the basic technology for deep-sea manned exploration was well developed, further refinements were needed. The French submersible *Archimède* had already gone much deeper than the 8,000-to-10,000-foot depths that would be encountered in the selected study area, but the *Archimède* was bulky, difficult to maneuver and provided only limited visibility. The U.S. submersible *Alvin* had already scored several technical successes in the Atlantic, including the recovery of the hydrogen bomb lost off the coast of Spain. Although the *Alvin* was small, maneuverable and afforded excellent visibility through its three portholes, it was limited to depths of less than 6,000 feet, too shallow for the depths of the Mid-Atlantic Ridge. Accordingly the French undertook to build a new submersible, the *Cyana,* similar in size and capability to the *Alvin,* and the U.S. fitted the *Alvin* with a new titanium pressure hull that would allow dives to at least 10,000 feet.

In further preparation for the venture plans were made to close an "observation gap" that would otherwise exist between the scale of detail that would be recorded from the submersibles, a scale measured in centimeters and meters, and the scale of observations that were then possible from surface vessels, a scale measured in hundreds of meters and kilometers. Strategies for closing the observation gap included echo sounding

TENSION FRACTURE in the floor of the Mid-Atlantic Rift was photographed with an automatic camera mounted on the research submersible *Alvin.* Along the sides of the fissure are numerous "pillow lavas," which are formed when lava is extruded into water and rapidly cooled. At the left is part of the *Alvin*'s external gear; the round object at the bend of the pipe is a compass. Flowerlike objects at top of the fissure are crinoids, invertebrate animals attached to the bottom. The *Alvin* was one of three submersibles that explored the rift as part of Project FAMOUS (French-American Mid-Ocean Undersea Study).

with a high-resolution narrow acoustic beam, photography with new techniques and the use of a variety of deep-towed geophysical instruments. All these methods would require the ability to position ships to an accuracy of a few tens of meters rather than the one kilometer that is usual in deep-sea navigation.

By 1971 there was no longer any doubt that our investigation would be confronted with detailed problems of volcanic geology. It was hard to realize that barely a dozen years earlier serious students of abyssal geology could still speculate that elevated portions of the mid-ocean-ridge system, such as the East Pacific Rise and the Azores Plateau, were remnants of sunken continental land areas. The deep central valley in the Mid-Atlantic Ridge and its transverse fracture zones, so well known today, were just beginning to be defined. The presence of outcrops of volcanic rock on the Mid-Atlantic Ridge was firmly established in 1961 by Earl Hayes and his co-workers on cruise No. 21 of the Woods Hole Oceanographic Institution vessel *Chain,* when they succeeded in photographing and sampling fresh, glassy volcanic rock at 28 degrees 53 minutes north latitude, due west of the Canary Islands. Over the next few years similar studies made by ships from various oceanographic institutions confirmed the presence of essentially similar volcanic rock on the crests of mid-ocean ridges in the South Atlantic, the Pacific and the Indian Ocean. Bizarre as the idea seemed at first, it was becoming evident that the mid-ocean-ridge system was nothing less than a vast unhealed volcanic wound.

Even more remarkable, it was learned that finely divided particles of iron oxide

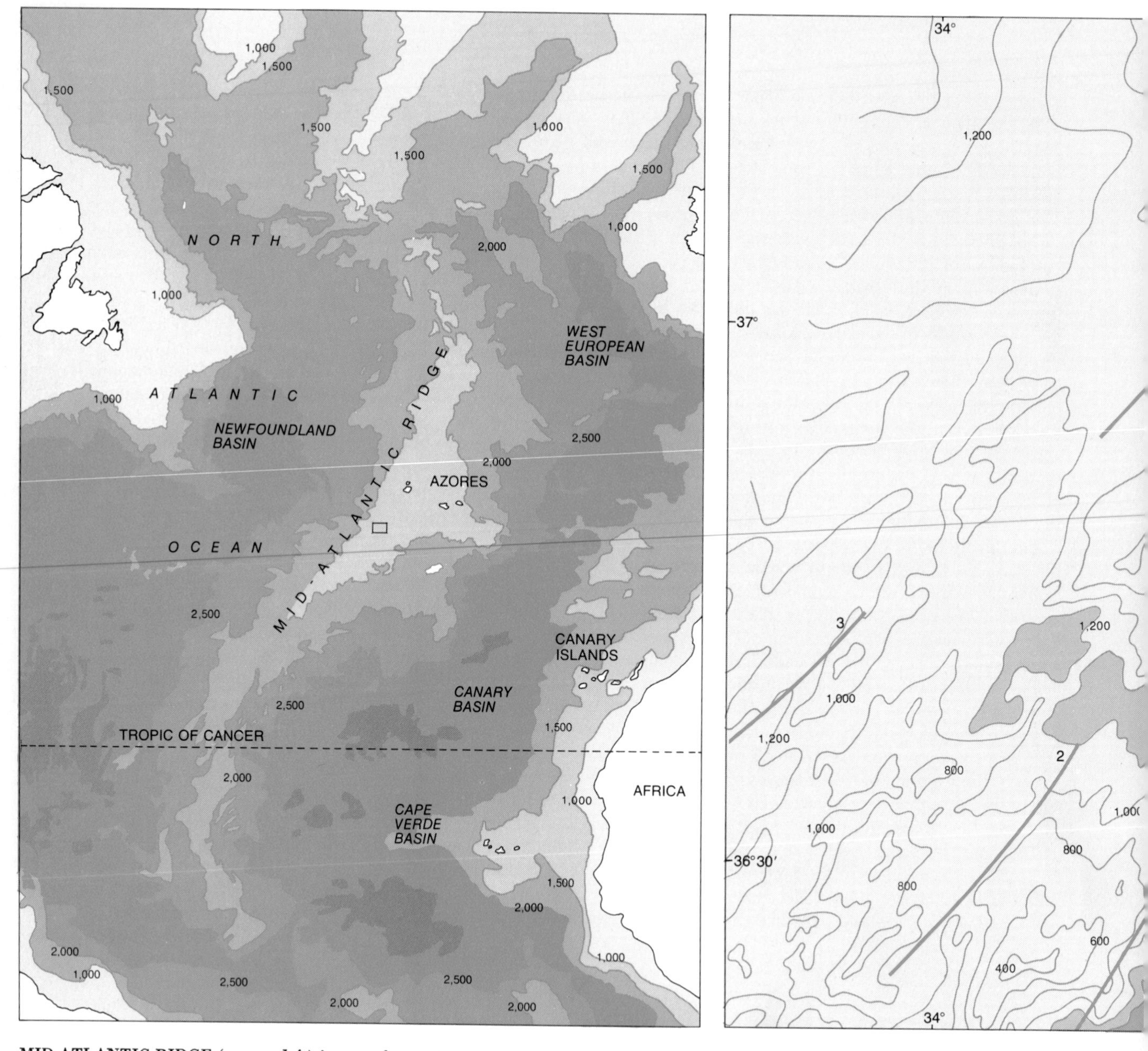

MID-ATLANTIC RIDGE (*map at left*) is part of a continuous system of mid-ocean ridges some 40,000 miles long. Lava extrusions are centered on fissures along the crest of the ridges. These linear zones of extrusion mark the boundaries of huge lithospheric plates that are slowly moving apart, carrying the continents with them. As the lava solidifies, particles of iron oxide in the molten rock become aligned with the earth's magnetic field. Because of periodic reversals in the polarity of the earth's field the solidified and mag-

naturally present in the liquid volcanic rock become aligned with the earth's magnetic field as the molten rock freezes. Records of the fossilized magnetic polarity, obtained by magnetometers towed on the ocean surface, revealed symmetrical bands of periodic reversals of polarity on both sides of the mid-ocean ridges. Concurrent land-based studies of polarity reversals in precisely dated stratigraphic sections that included volcanic rocks provided a time scale for the global reversals in magnetic polarity [see "Sea-Floor Spreading," by J. R. Heirtzler; SCIENTIFIC AMERICAN Offprint 875]. The conclusion seemed incontestable: the volcanic sea-floor basement was spreading away from the mid-ocean-ridge systems, growing by constant additions of fresh volcanic material at the central rift, and in the process was recording the polarity of the earth's magnetic field as a function of time, rather like a giant tape recorder.

Beginning in 1968, deep-sea drilling by the *Glomar Challenger* confirmed the hypothesis of sea-floor spreading by showing that sediments overlying the volcanic basement became progressively older with distance east or west from the Mid-Atlantic Ridge. In mid-1970 the oldest volcanic basement rock yet recovered in the Atlantic was brought up in a core drilled by the *Glomar Challenger* at the base of the continental slope east of Cape Hatteras. At least 150 million years old, the sample exhibits all the mineralogical and chemical characteristics of present-day volcanic extrusions on the Mid-Atlantic Ridge. It appears that the rock was extruded at the Mid-Atlantic Ridge shortly after North America broke away from Africa at the beginning of the current episode of sea-floor spreading.

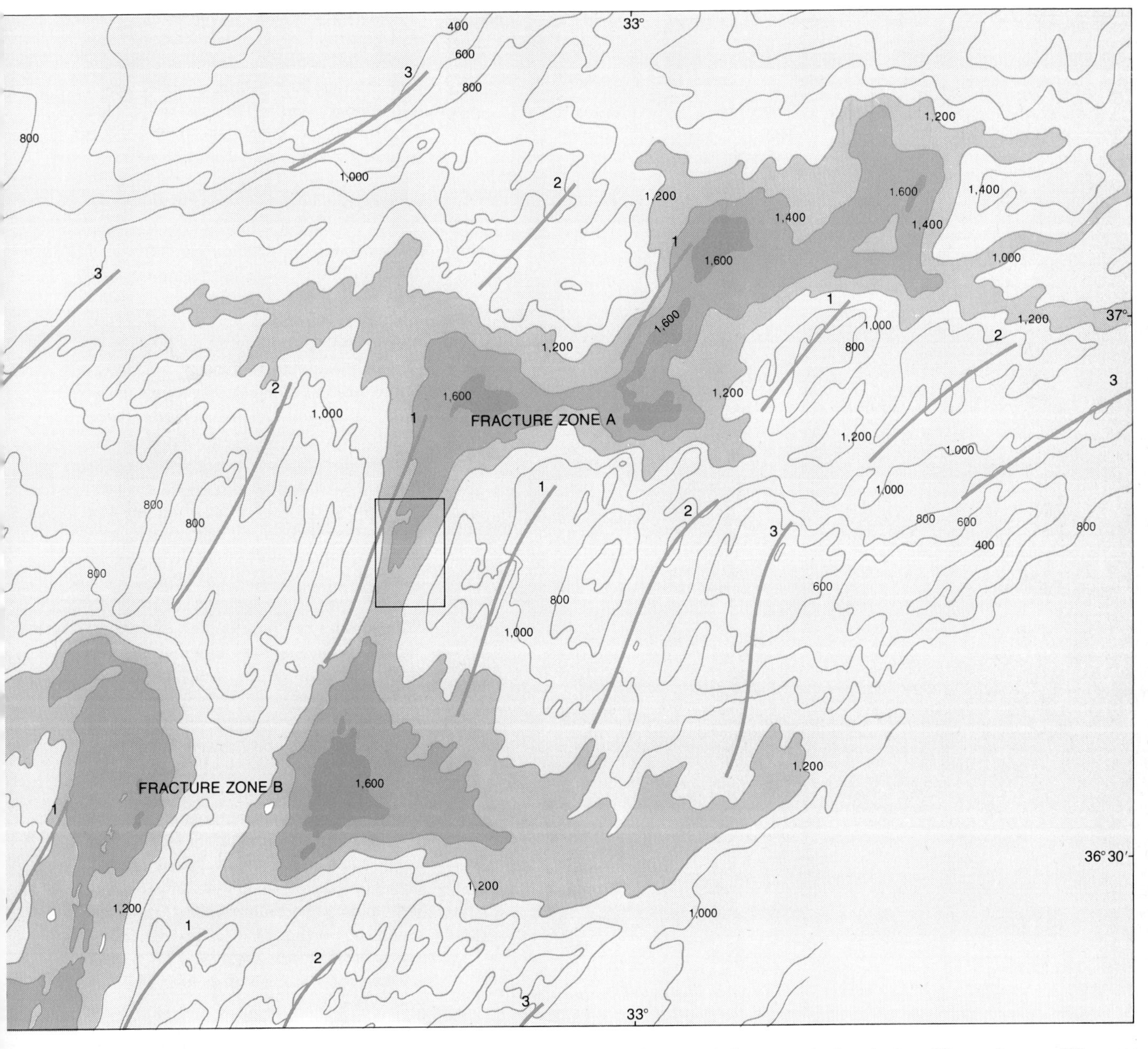

netized rock exhibits symmetrical bands of alternating polarity on both sides of the ridge. In the map at the right, which corresponds to the area within the rectangle in the map at the left, the numbers next to the colored lines that run parallel to the ridge axis represent the ages of the magnetized rocks in millions of years. (The contours are given in fathoms.) The region explored by Project FAMOUS lies within the rectangle in map at the right. A more detailed map of area explored by submersibles appears on next page.

Our primary purpose in Project FAMOUS would be to examine the details of the structure of the median valley in the Mid-Atlantic Ridge and to learn as much as possible about the extrusion and accretion of volcanic rock associated with the spreading process. We planned to begin with a series of broad surveys over the area of interest, gradually focusing more closely on the most promising area as both our knowledge and our technical capability improved. By the time the project was completed we had available for study the data collected by some 25 surface cruises, two aeromagnetic surveys and 47 coordinated submarine explorations of the rift valley.

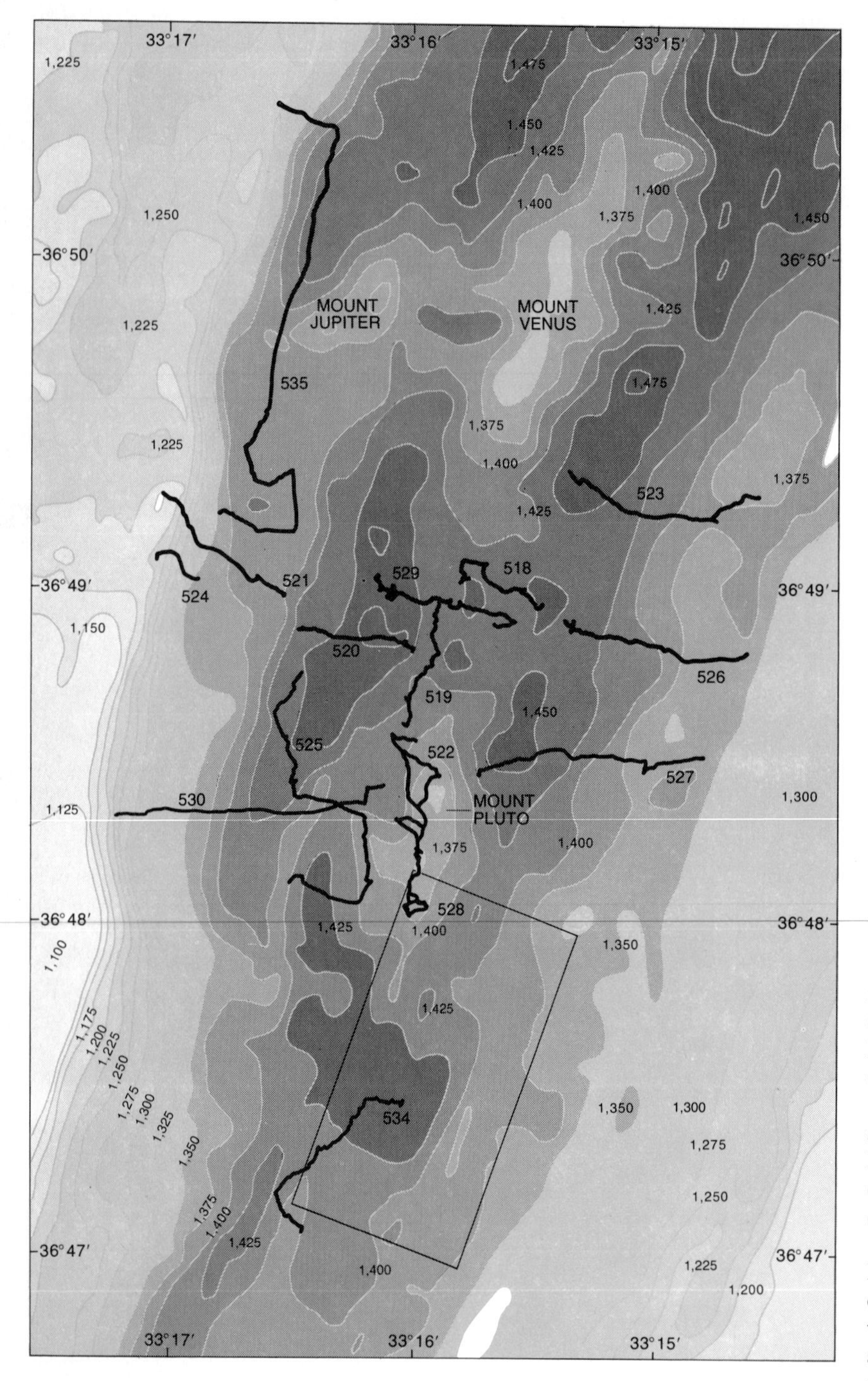

RIFT VALLEY explored in detail embraced an area approximately four kilometers by six kilometers. (Area inside rectangle appears in illustration on opposite page.) The rift valley is bounded on the west by a steep cliff, the West Wall, some 1,000 feet high, and on the east by a series of shallow terraces. The tracks of 15 exploratory dives made by the *Alvin* are depicted and numbered, beginning with dive No. 518. (The *Alvin* had previously made 517 dives elsewhere.) In addition the French submersibles *Cyana* and *Archimède* made dives in the rift-valley-and-fracture zone that lies to the north of Mount Venus in the Mid-Atlantic Rift. The *Cyana* made 12 dives in 1974; the *Archimède* made six in 1973 and 12 in 1974.

In the fall of 1971 a regional aeromagnetic survey established the existence of the typical pattern of magnetic anomalies on both sides of the axis of the mid-ocean ridge and indicated that the African and North American crustal plates were moving away from the axis at the rate of about 1.5 centimeters per year. Later in the year the first precisely navigated wide-beam acoustical survey revealed that the floor of the rift valley was fairly flat. Whereas the entire valley is about 30 kilometers wide, the inner floor is between one kilometer and two kilometers wide. This particular section of the Mid-Atlantic Ridge is offset by two fracture zones about 30 miles apart. The one to the north was designated Fracture Zone A, the one to the south Fracture Zone B [*see illustration on preceding two pages*].

Starting in the spring of 1972 the U.S. Naval Oceanographic Office began a narrow-beam echo-sounding survey of the area between the two fracture zones. The final product of the survey, completed early in 1973, was a set of bathymetric charts with a contour interval of five fathoms (about 10 meters). A different acoustical technique was simultaneously employed by British participants in the project. They used a side-scan instrument, embodied in a seven-ton submerged system, that directs its acoustic radiation at the sea bottom obliquely and thus can record echoes from many topographic features at different distances, all with one outgoing pulse. Since each pulse can irradiate large areas of the sea floor, it does not take long to assemble a regional echo map in which the major linear features of the area are clearly defined. The side-scan technique showed the steep west wall of the valley, the less steep east wall, the "corner hills" near Fracture Zone B and the hills down the center of the rift-valley floor, all in accurate relation to one another.

During the summer of 1972 the French research vessel *Charcot* carried out detailed bathymetry, a magnetic survey and a bottom-sampling program along the northern half of the rift valley south of Fracture Zone A. The survey showed the presence of a central hill on the valley floor (later to be known as Mount Venus) and confirmed that the magnetic-anomaly patterns were broader on the east than on the west, indicating that the spreading to the east is faster than that to the west. Rocks recovered

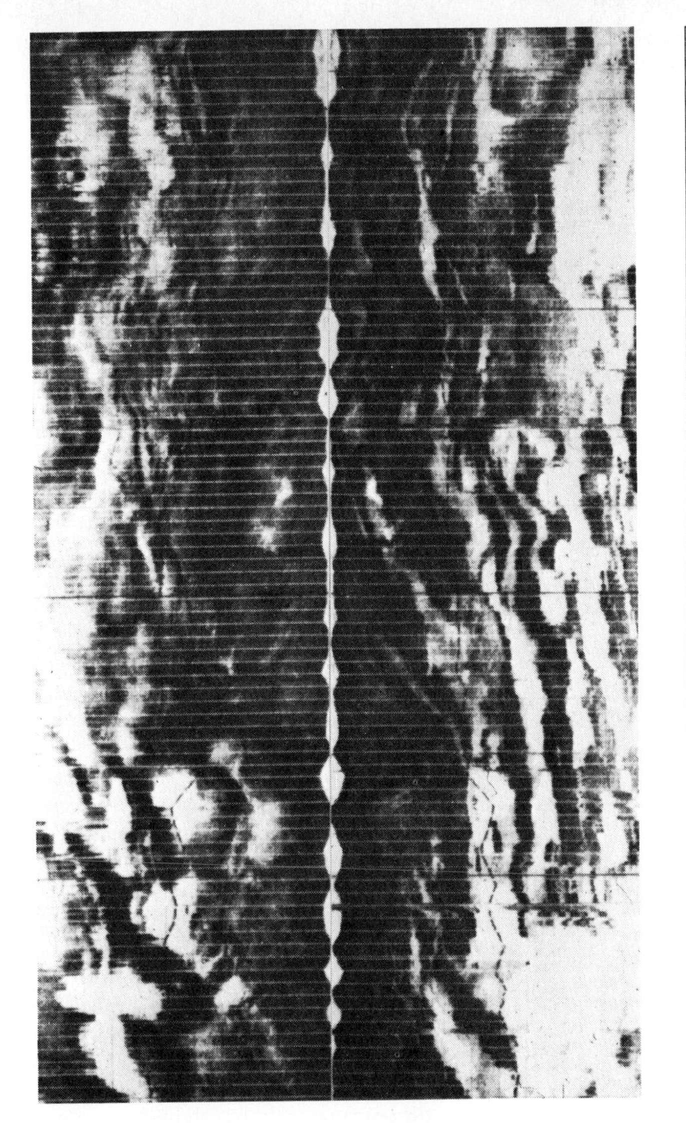

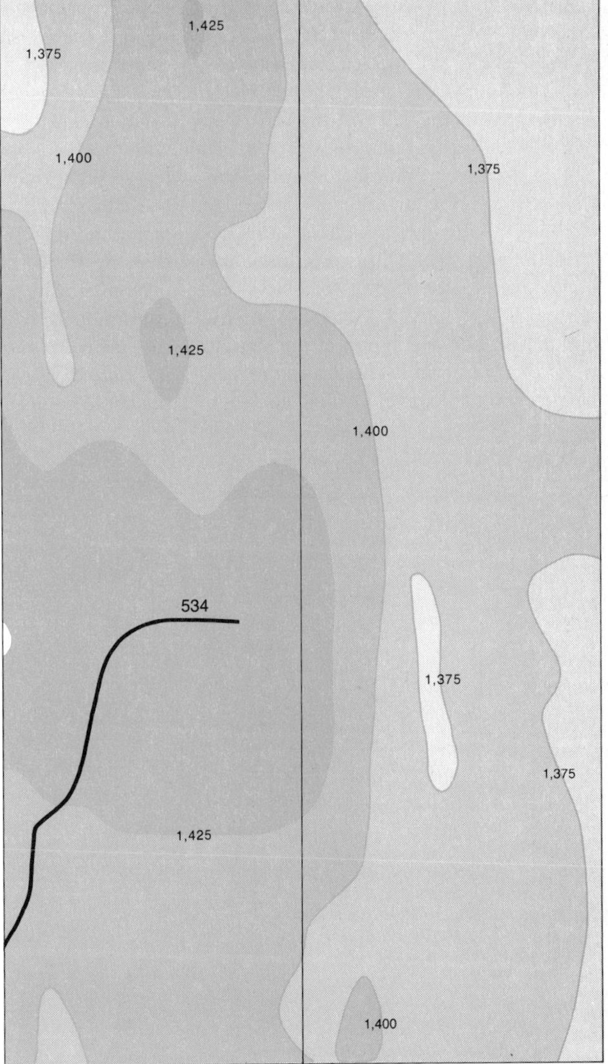

SIDE-SCAN ACOUSTICAL TECHNIQUE produced the images shown at the left. In this technique a "deep-towed fish," designed to operate 100 to 200 meters above the sea floor, directs its acoustic radiation obliquely at the bottom (*see illustration on next page*). Depending on the topography, each outgoing pulse can produce multiple echoes at differing distances from the axis along which the instrument is traveling. The two images that are reproduced here represent the left-hand and right-hand reflections from the sea floor depicted in the contour map of the same area at the right. Elevations in the floor produce reflections that appear bright in the acoustical record. It can be seen that in general the right (east) side of the record contains more linear features than the left (west) side.

from the rift valley popped and exploded on the deck of the ship, apparently as a result of the release of trapped gas. The rocks appeared to be very fresh, suggesting recent volcanic activity. A sled carrying cameras was towed along the bottom and obtained the first photographs of the bulbous and tubular lavas that were to become so familiar to the diving scientists.

The French survey was supplemented and extended south to Fracture Zone B in late 1972 by the *Atlantis II* from Woods Hole. Three radar transponder beacons were anchored on mountain peaks in the rift to serve as navigation reference marks for a series of closely spaced bathymetric and magnetic survey lines. Our data confirmed the asymmetry in the median valley, with the western wall rising very steeply and the eastern wall rising more gradually in a series of steps.

In order to survey and sample this portion of the valley in detail we moored two acoustical transponder beacons on a terrace near the top of the west wall. Since the beacons were supported on floats 100 meters above the sea floor, they would not be affected by storms, unlike the radar buoys, which could swing in a large circle around their mooring. In response to a signal transmitted from the ship, the beacons transmit a characteristic signal, each identifiable by its own wavelength located between 10 and 14 kilohertz. The two-way travel time provides a precise measure of the slant range between the ship and the transponder beacon. The depth of each transponder and the distance between transponders is established by running a survey pattern over them; a shipboard computer is then used to generate a best estimate of the depths and base-line length. With these parameters fixed, the slant ranges can be converted to horizontal distances from points directly above the transponders, thereby uniquely fixing the ship's position in relation to the base line. With a two-transponder base line it is necessary to know, of course, which side of the base line the ship is on, but this ambiguity is easily resolved from the known differences in bathymetry across the base line.

Our dredging and photographic surveys required the lowering of the dredge or the camera frame on a long cable behind the ship. The ship's position could now theoretically be plotted to an accuracy of a few tens of meters. (We now had to ask: For which part of the ship do we want a position?) The towed devices, however, would travel at some considerable but unknown depth and distance behind the ship. Thus a third transponder was attached to the cable close to the dredge or the camera. This transponder transmitted a signal that not only was received back at the ship but also triggered a second set of responses from the base-line transponders. The secondary base-line responses were received by the ship some seconds after the primary responses. Since the ship's position was continuously being fixed on the basis of the primary responses, the depth and position of the "relay" transponder on the cable could be computed. Any depth ambiguity in the solution could be resolved by entering into the calculation an estimated depth (based, for example, on the amount of cable paid out). Essentially the same system would be used later to navigate the manned submersibles, except that an automatic acoustic transmitter would be installed in the submarine to perform the same function as the towed relay transponder [*see top illustration on opposite page*].

In practice all the raw data (acoustic travel times) were stored on magnetic tape for later replay at home, to allow the filtering of spurious positions introduced by acoustic noise or "bottom bounce." Our real-time output on shipboard was a tabletop *X-Y* plotter that drew a small dash to represent the ship's position and a small cross for the position of the towed instrument. On later cruises real-time tracking was provided on a fluorescent screen coupled to a thermal printer. Both systems produced inked or printed records of the intricate maneuvers executed during camera or dredge stations, with the positions of the ship and the instrument being recorded every 20 to 40 seconds. By laying plots of portions of our bathymetric maps on the printer, or on transparent overlays on the screen, we could maneuver the ship and instrument package to photograph or sample any given bathymetric feature. Since the time of each acoustic fix was known, and since the time was recorded on each photograph, the position of any given feature photographed on the sea floor could be established.

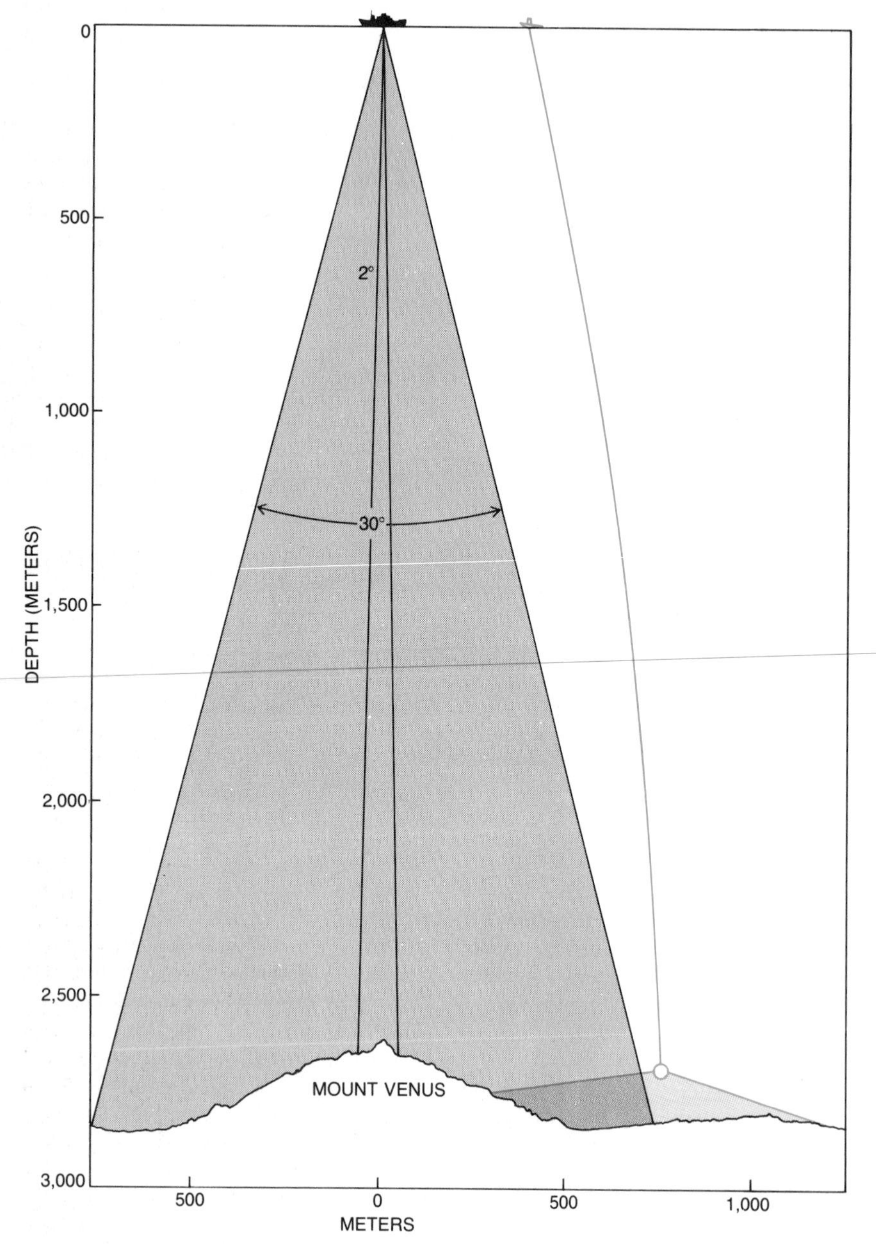

THREE ACOUSTICAL-MAPPING INSTRUMENTS supplement one another in recording the depth and the topography of the ocean floor. The conventional echo sounder, which has a beam angle of 30 degrees, provides quick, rough coverage of large areas. A newer surface instrument with a beam angle of only two degrees yields much finer detail. The deep-towed fish provides even finer detail of certain features, but its images require more interpretation.

We found that when the dredge touched bottom, vibration in the cable destroyed the acoustic signal; gaps in the track of the dredge indicated the positions of bottom contact and hence the location of any samples recovered. We also learned that we could reposition the dredge precisely in areas where rock had been revealed by photographs, and in these places we invariably recovered sizable quantities of rock fragments resembling those observed in the photographs. This procedure soon led to a system we called touchdown dredging; instead of dragging the dredge for half a mile or more across the bottom, as was the common traditional practice, we would maneuver it into position over a specific target, lower it to the bottom, hold it there for a few hundred yards of dragging and then bring it to the surface. In this way it was possible to relate recovered rock samples directly to specific bottom features.

Many of our photographs showed tubular or wormlike lava forms. Dredg-

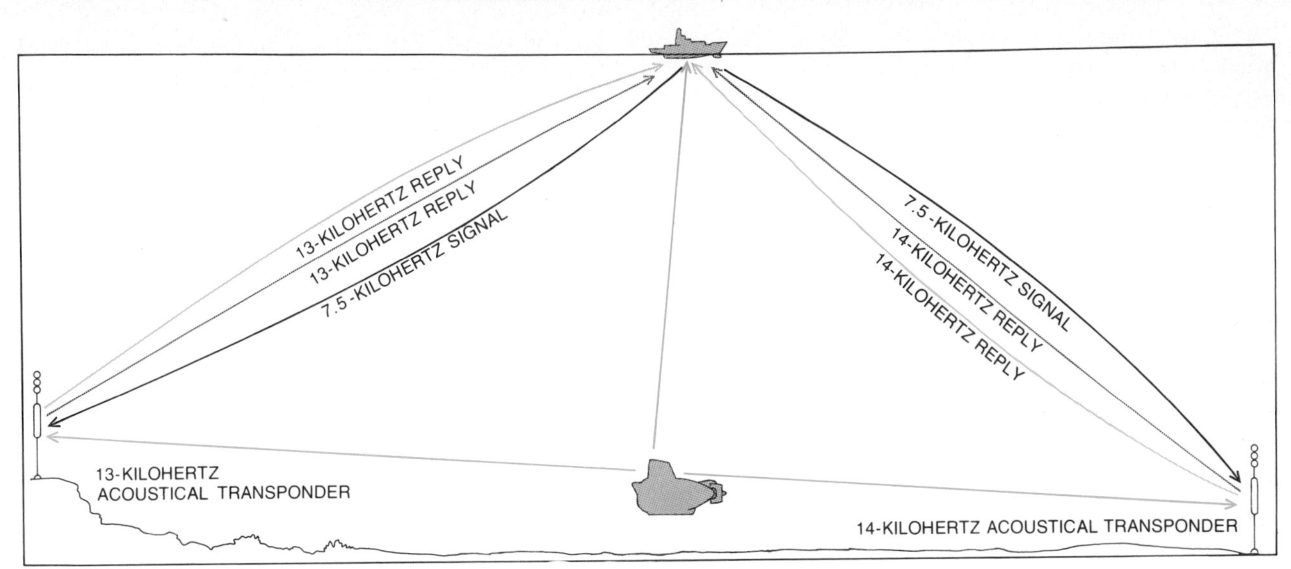

POSITION OF THE ALVIN and the other submersibles during the Project FAMOUS dives was established with the aid of acoustical transponders anchored to the ocean floor, which provided an accurately known 4.5-kilometer base line for computing the travel times of acoustic signals. Three transponders laid out in a shallow triangle were actually used; only two of them are shown here. The surface vessel guiding the submersible established its own position by sending out a 7.5-kilohertz pulse and timing the replies from the anchored transponders at two different frequencies (13 and 14 kilohertz). The submersible also emitted pulses at 7.5 kilohertz that were received directly by the surface tender and that also triggered separate responses from the anchored transponders. The difference in arrival times between the direct signal from the submersible and transponded signals were used to compute submersible's position.

ing in those areas, we recovered samples with a circular or wedge-shaped cross section. Many of the samples could be reassembled on board ship, much as an archaeologist reassembles broken pottery. These reconstructions reproduced exactly the tubular forms observed in our photographs and further confirmed the fact that most of the material had indeed been recovered from a small area, in most cases representing only one or two distinct outcrops. Photographs made in Fracture Zone B showed angular, irregular or slabby rock fragments; again our dredging recovered similar material. Many of these samples had been crushed, sheared and recemented, as one might expect in an active geological fracture zone.

In the summer of 1973 the *Atlantis II* returned to complete the survey of Frac-

LAVA EXTRUSIONS in rift take exotic forms. The large toothpastelike extrusion in this photograph taken from the *Alvin* is typical. A remote manipulator is sampling an adjacent blisterlike extrusion. Photograph was made on dive No. 522 on Mount Pluto.

ture Zone B and of the next rift-valley segment to the south. Data from this survey would help us to make the final decision on the dive site and would extend our regional coverage to help put the detailed dive operations into proper perspective. The acoustical navigation system had been refined on the basis of the previous year's experience. We would also attempt to make detailed studies of heat flow from the bottom in and near the median valley, and we would attempt to monitor and compute the position of earthquake shocks in the median valley and the fracture zones. The navigation system now was programmed to make use of a network consisting of three transponders, which both extended the range of the system and removed the base-line ambiguity.

By working with this system to position the ship over known pockets of sediment in the median valley, we were able to drop sensitive thermocouple probes into the sediment. Then, measuring the temperature differential between two or more points on the probe in the sediment, and knowing the thermal conductivity of the sediment, we were able to compute the rate of heat flow. Such measurements were obtained within a few kilometers of the valley axis and around the active transform section of Fracture Zone B. The results showed low to normal values in the median valley and above-average values on the flanks of the valley and in the fracture zone. In the fracture zone values 10 to 12 times normal were observed. These results, along with veining and cementation in the fracture-zone rock samples, seemed to provide evidence of escaping hydrothermal solutions in or near the fracture zone.

The earthquake survey also made use of a version of the acoustical navigation tracking program, modified to plot the position of three sonobuoys simultaneously. Each sonobuoy is a small floating instrument carrying a radio transmitter and a suspended hydrophone. Three of these sonobuoys were set out in a triangular array that was constantly tracked until the sonobuoys drifted out of range. Earthquake shocks picked up by the hydrophones were transmitted to the ship by radio and were recorded on magnetic tape; the signals were also amplified and broadcast in real time over loudspeakers in the laboratory. Since the positions of the three sonobuoys were known, differences in the arrival times of the earthquake shocks at the three hydrophones could be used to triangulate the position and approximate depth of the shock, just as is done with an array of three or more land-based seismic stations.

ELONGATED AND DRAPING LAVA PILLOWS, which are more characteristic of lava on land than they are of flows under water, appear in this photograph made on the *Alvin*'s dive No. 521 near base of West Wall of the rift. Canisters in foreground contain water samples.

What was to be one of our most exciting stations took place on a calm, starry night. With the sonobuoy array laid out, the *Atlantis II* moved slowly toward the steep west wall of the median valley as we lowered one of our new "Big Bertha" dredges. We planned to tow the dredge up the face of the wall. At this point, although we had begun to refer to the feature as a wall, we were aware of the way echo sounding tends to exaggerate such features, and we therefore assumed that the wall was actually no more than a steep slope, perhaps one of 30 to 40 degrees. As the dredge touched bottom at the base of the slope we began to receive ringing, banging sounds through the sonobuoy loudspeakers. When we slacked off on the cable, the noises stopped; as we took in the cable, they resumed. We watched the tension meter on the dredge cable expectantly as we approached the west wall. The echo sounder showed that the bottom was rising rapidly, when suddenly it disappeared entirely. We changed scales and tried our backup 12-kilohertz echo sounder. There was still no return. The bottom was so steep it could not reflect the signal!

Simultaneously the ringing of the sonobuoys was replaced by an ominous silence. The cable tension built up to 10,000, 12,000, then 15,000 pounds, more than twice its normal working range. Stopping the ship's engines, we drifted slowly backward, taking in cable carefully, trying to keep the tension below the 20,000 to 25,000 pounds that could break it. Finally we were directly back over the dredge, with bottom beginning to show again on the echo sounder. There was nothing more to do now but to take in the cable slowly, waiting for something to give. Suddenly there was a loud report from the loudspeakers, and the ship shuddered and reverberated as the tension meter swung wildly, then dropped back to the normal 5,000 to 6,000 pounds. We began reeling in cable cautiously, wondering when the frayed end would appear. A series of shocks and reverberations continued to roll in over the loudspeakers, although the dredge, if we still had it, should have been well clear of the bottom.

Finally the dredge appeared over the stern of the ship, with an immense freshly fractured hemisphere of pillow lava

BIZARRE SUBMARINE LANDSCAPE of lava extrusions was observed at the intersection of two lava-flow fronts in the Mid-Atlantic Rift. The drawing is based on a sketch of the scene made by one of the authors (Bryan), who served as one of the observers aboard the *Alvin* during its dives into the Mid-Atlantic Rift. The numbers identify some of the lava forms that were observed: (*1*) bulbous pillows with knobby budding, (*2*) a flattened pillow formed by the rapid drainage of lava while the skin was still plastic, (*3*) a hollow blister pillow formed by the drainage of lava after the skin had solidified, (*4*) a hollow layered lava tube formed by temporary halts in a falling lava level, (*5*) a bulbous pillow with a "trapdoor" and toothpaste budding, (*6*) an elongate pillow, typical of a lava extrusion on a steep slope, (*7*) a breccia cascade, formed on very steep slopes where the lower end of an elongate pillow has ruptured, releasing a cascade of fluid lava, and (*8*) an elongate pillow swelling into a bulbous form along a longitudinal spreading crack.

wedged in its jaws and several more large fragments caught in the chain bag. This was the largest single rock recovered during the project. It weighed about 400 pounds and had to be broken into three pieces for handling. The sonobuoys continued to record almost constant small shocks and vibrations for the rest of the night. We speculated that when the dredge broke loose, it triggered a series of rock slides on what must have been a very steep cliff. This speculation was given support the following year, when the submersible scientists observed piles of large talus blocks at the foot of the west wall. They also found many smaller piles of loose rock debris on terraces on the sides of the west wall, which turned out to be a spectacular cliff with a slope of nearly 80 degrees.

For a brief time the *Atlantis II* worked within sight of the French ships *Marcel le Bihan* and *Archimède,* as the first dives were being made near Mount Venus late in 1973. The initial dives showed that it was feasible to work with a submersible in this rugged terrain, to recover rock samples and to do geologic mapping. These preliminary submersible dives, before the major submersible effort planned for 1974, were of immense practical value in developing work and data-handling routines for the following year. For example, the *Archimède* detected bottom currents that were swifter than we had anticipated. A quickly instrumented program for metering ocean-bottom currents disclosed that the currents were nearly all tidal in nature and revealed no evidence of currents strong enough to be dangerous to submersibles.

A deep-towed instrument package, provided by the Scripps Institution of Oceanography, was brought into play at about the same time to study the microtopographic relief and the localized variations in the magnetic anomalies. This near-bottom survey confirmed the central magnetic and bathymetric asymmetry revealed by the surface-ship data, although it suggested that an asymmetry had existed in the opposite direction a few million years ago.

Following closely on the *Atlantis II* cruise the U.S. Naval Research Laboratory introduced a major new bottom photographic technique in the same area. By suspending a strong stroboscopic-flash lamp high above the camera, they were able to illuminate and to photograph an area nearly 100 times as large as the area covered by the usual ocean-bottom camera. In fact, the area photographed with each flash of the system was approximately the same as the area covered by each ping of the narrow-beam echo sounder. Mosaics made from these photographs were later laid out to scale on a gymnasium floor to enable the diving scientists to preview the terrain in which they would work.

Throughout 1972 and 1973 the French submersible *Cyana* and the U.S. submersible *Alvin* were undergoing their refitting for operation at rift-valley depths, and the diving scientists and pilots were engaged in a training program so that they would be able to work effectively over the volcanic terrain of the valley floor. In contrast to the practice of the Apollo lunar program, all the men selected to observe the rift valley from submersibles were professional earth scientists, and several had had previous experience with the submarine and over volcanic terrain. Nevertheless, we would rely heavily on our pilots' understanding of the features we would be searching for. In Hawaii the scientists and pilots spent five very profitable days, accompanied by several French colleagues. There they observed a variety of flow forms and fractured lava flows both above and under water, watched an active volcanic spatter cone and visited an area where divers had recently sampled and photographed flowing lava under water. The French made other trips to Iceland and Africa to observe areas generally supposed to be dry-land extensions of oceanic-rift systems.

In the summer of 1974 the French submersibles *Cyana* and *Archimède*, respectively accompanied by their mother ships *Le Noroit* and *Marcel le Bihan*, and the U.S. submersible *Alvin*, accompanied by its tender *Lulu* and the research vessel *Knorr*, all engaged in a coordinated diving and exploration program in the rift valley and the adjacent fracture zones. During the summer the *Cyana* completed 14 dives, the *Archimède* 13 dives and the *Alvin* 17 dives. These dives, and the preliminary seven made by the *Archimède* the year before, were man's first direct observation of the rift-valley floor.

By good fortune it was possible to schedule the *Glomar Challenger* to drill in the area. The deep-sea drilling vessel was able to penetrate some 600 meters into basement rock 18 miles west of the dive area. The rocks collected by the submersibles and by the *Glomar Challenger*, and dredged by surface ships in the area in between, comprise one of the most remarkable sets of samples ever obtained from the sea floor. The shipboard and submersible samples together will make it possible to examine regional variations in composition over an area measuring some 30 by 60 miles. In addition the drill core provides an opportunity to look at vertical variations in the crust.

The *Archimède* began its diving program on and near Mount Venus, gradually working north toward Fracture Zone A, while the *Alvin* began diving near the next central hill to the south, which we called Mount Pluto. A secondary dive site for the *Alvin* was designated in Fracture Zone B. The *Cyana* was scheduled to begin dives in Fracture Zone A, but rough weather at the start of the program damaged the *Cyana* slightly and delayed the start of the *Alvin* operations for several days. The *Alvin's* first few dives showed that visibility was excellent, that maneuvering was no problem and that sampling was as easy as we had hoped.

As the dives progressed the pilots found that it was possible to maintain almost continuous contact with the bottom. The scientists could recognize lava-flow fronts and conical structures named haystacks, which, because of their size and shape, were difficult to see in bottom photographs. The haystacks appeared to be small centers of lava extrusion. Major lava flows and vent structures were almost always confined to the central hills and were irregularly distributed along the central line of the valley, an observation strongly suggesting episodic volcanic activity. Minor lava flows and haystacks were observed on both sides of the central hills, particularly to the east. The submersibles collected precisely located and oriented rock samples, water samples and sediment cores. Almost continuous photographic coverage was obtained with semiautomatic cameras outside the hull of the vessels and both still and motion-picture cameras inside.

Small cracks in the sea floor near the axis of the mid-ocean ridge had been recognized first in photographs taken during the *Atlantis II* cruises. The extended-area bottom-camera system had clearly shown these features running for at least 100 meters and overlapping in an echelon fashion paralleling both the direction of the ridge and the major walls of the valley. The submersibles followed and crossed numerous fissures, measuring their heading, width, depth, longitudinal tilt, location and cross section. It was found that fissures exist everywhere from the valley's central line to its bounding walls on the east and the west, generally increasing in width with distance from the valley axis. The width varied from a few centimeters near the axis to tens of meters near the walls. Even the narrowest cracks were several meters deep. The wider ones were between 10 and 100 meters deep. In places there were small differences in elevation between one side of a fissure and the other. Across fissures up to a few meters in width it was possible to see matching halves of the same pillow lava on opposed walls.

Little sediment was found on the floor of the rift valley. Near the axis there was not even enough of it to half-bury pillow lavas a few tens of centimeters in diameter. Near the edges of the valley floor the sediment covered many such pillows

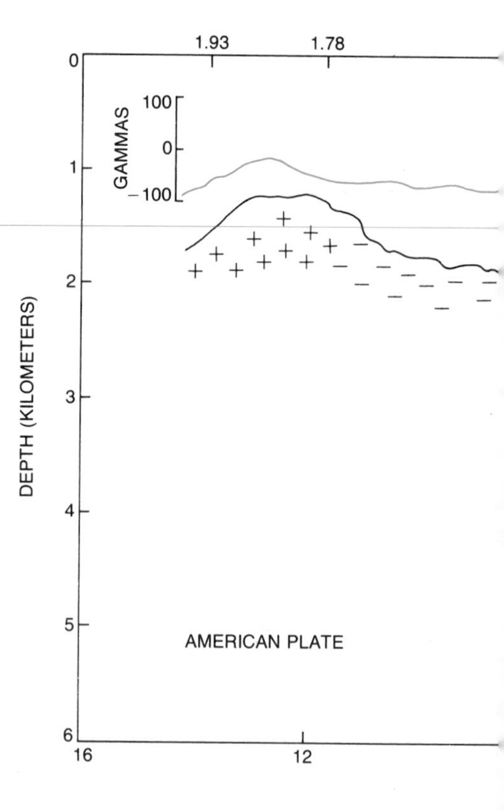

CROSS SECTION of rift is shown in relation to magnetic-field variations. Profile of the bottom is plotted from deep-towed

and possibly reached a depth of a meter or more. Dives in the fracture zones encountered a great deal of semiconsolidated sediment fractured by horizontal shearing. In those zones no fresh volcanic rock was found, even though a high heat flow had been recorded along one narrow fracture line. In Fracture Zone A scientists aboard the *Cyana* observed two manganese-covered areas that appeared to have been created by hot water flowing up through the sea floor. Although sensitive thermocouple probes were carried on the *Alvin* and also were towed near the bottom from surface ships, no significant thermal anomalies were found in the water.

Toward the end of 1974 we began the task of assembling the great mass of data accumulated during the cruises of the surface ships. Although much still remains to be done, some interesting patterns have begun to emerge. For example, by plotting the location and composition of rocks collected by acoustically positioned dredges in and near Fracture Zone B we have found that the sheared and altered rocks, which are typical of fracture zones, are correlated with a band of subdued magnetic anomalies over the fracture zone. The band follows the trend of the fracture zone and is nearly 10 miles wide. The earthquake surveys show that the narrow active transform fault is located near the north side of this band and is connected almost exactly with the limits of the active volcanic rifts in the center of the rift valleys north and south of the fracture zone. Apparently the active fault has shifted over the width of this zone; the volcanic rifts have probably moved north and south with the fault.

The bathymetric and magnetic asymmetry of the rift valley was confirmed in a striking way by the first analyses of the volcanic rocks. The analyses showed a regular variation in mineralogy and chemical composition with distance from the central volcanic hills toward the flanks of the valley. Moreover, the variations are more gradual to the east than they are to the west. In several dives the submersibles climbed the spectacular west wall, which rises some 300 meters in a series of closely spaced steps connected by nearly vertical fault scarps. No comparable feature was observed to the east; there, as the surface-ship data had suggested, the valley rises in a series of much wider steps, separated by short, steep slopes covered by broken fragments of lava.

Low areas flanking the central hills, which were originally thought to be grabens, or collapse depressions, were found to be simply low areas that had not quite been filled by lava flows converging on them from Mount Venus, Mount Pluto and the valley walls. True grabens were found at the base of the west wall and on the largest of the eastern fault scarps. These depressions suggest that the bases of the walls have pulled away from the valley floor, which has collapsed downward and inward toward the center of the valley. Indeed, one general impression is that the valley is near the end of a major period of structural extension and collapse and is just beginning to be inundated by new volcanic extrusions, represented by Mount Venus and Mount Pluto. Such episodes of alternating volcanic activity and quiescence may prove to be typical of slow centers of sea-floor spreading such as the Mid-Atlantic Ridge, where the rate of spreading may not be high enough to keep a rift constantly open to the underlying magma chamber that is the source of the lava flows.

No lava flows in actual progress were

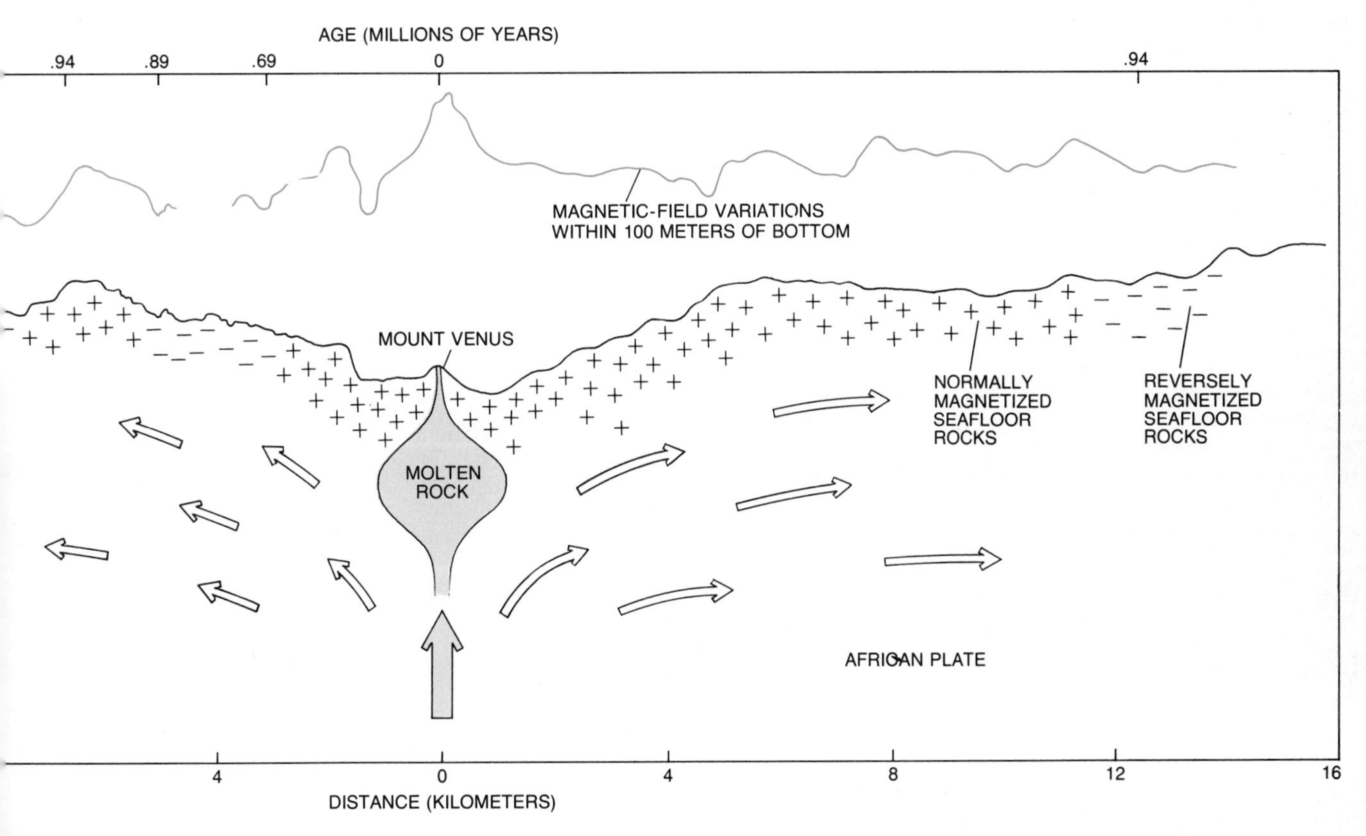

acoustical measurements. Rock of the floor is not symmetrically magnetized as measured from the center line, nor is it magnetized everywhere with equal strength. Ages of rock as identified from magnetic profile indicate that over the past 940,000 years average rate of spreading to the east has been about twice as fast as rate to the west: 1.3 centimeters per year compared with .74 centimeter.

observed by either the American or the French diving groups, although we suspect that the central volcanic hills may be very young; it is perhaps only a few hundred years since the lava last flowed out of them. Preliminary dating suggests that some of the lavas on the flanks of the valley may be much younger than their distance from the valley center would suggest, indicating that they must have emerged on the sides of the valley rather than in its center. The presence of these young flank lavas, together with extensive faulting and mechanical rotation of at least the upper 10 meters of the lava flows on the valley floor, presents difficult problems for the interpretation of the magnetic-anomaly pattern, which seems to persist in spite of such complications. It was formerly supposed that the magnetic patterns had to do with only the uppermost few tens of meters in the lava. That view may now have to be revised. The problem is further complicated by analyses of rock samples in cores taken by the *Glomar Challenger* at its drill site adjacent to the rift valley. The magnetic polarity measured in the rock samples does not seem to agree with the integrated polarity measured by surface-towed instruments. Preliminary results of magnetic-intensity measurements on samples collected by the *Alvin* also do not exhibit any simple relation to the magnetic intensity measured by deep-towed or surface-towed instruments. It now seems possible that the magnetic-anomaly patterns are related to intrusions more deep-seated than the lava flows. These deeper intrusions must be added to the diverging plates in a more regular and consistent way than the surface flows.

Many of the interpretations summarized here will necessarily be modified as the data are further refined and new facts come to light. Moreover, it may be that not all mid-ocean ridges, and probably not all parts of the Mid-Atlantic Ridge, have the same features as those of the area we have studied. Indeed, we expect that fast-spreading ridges, such as the ridge in the eastern Pacific, will turn out to be quite different. Perhaps most important, we have been able to demonstrate that the kind of fine-scale features that eluded detection in the unprecedented concentration of surface-ship studies pursued in the Project FAMOUS area can be examined routinely by manned submersibles. Further investigations of this kind are certain to add new dimensions to our understanding of processes operating on and below the deep-sea floor.

IV

SOME APPLICATIONS OF PLATE TECTONICS

SOME APPLICATIONS OF PLATE TECTONICS

IV

INTRODUCTION

The four articles in this section show some of the applications of plate tectonics to other branches of earth science. Three articles are concerned with evolution, paleontology, and the diversification of living forms, while the fourth shows the important role plate tectonics plays in the search for petroleum and minerals.

In "Continental Drift and Evolution," Björn Kurtén discusses the evolution of reptiles and mammals in light of the theory of continental drift. Kurtén points out that the age of reptiles lasted 200 million years and produced twenty reptilian orders, whereas the age of mammals lasted only about 65 million years—only about a third as long—but gave rise to thirty mammalian orders. He attributes this to the greater number of isolated areas resulting from the separation of continents. It follows that we should expect to find the greatest similarity between the animals of those continents that have separated from one another most recently. Kurtén also shows how examination of fossils demonstrates that the diversification of many types from common stocks can be traced with some certainty, and goes on to show that the variations in types of living and fossil forms are compatible with the history of drift now generally accepted and outlined in previous papers. This not only supports the theory of continental drift, but throws light on the rate and nature of organic evolution itself.

"Continental Drift and the Fossil Record," by A. Hallam, contains two philosophical discussions, illustrated by examples, about the distribution and causes of variation in the fossil record. The first part describes the three kinds of dispersal routes, identified by the paleontologist G. G. Simpson. First are "corridors," which are land connections that allow free migration between continents. Such a corridor formed between North and South America about two million years ago and allowed their distinctive faunas, which had evolved separately, to merge. This led to the extinction of many weaker forms. A second route Simpson called "filter bridges," which are corridors with some additional factor, usually climate, that bars some prospective migrants. Thus cold prevented many species from crossing by the Bering Strait route between Asia and North America. Finally, he called "sweepstakes routes" those chance passages by rafting which at some intervals carried a migrant successfully to another continent. In a converse way, routes between one ocean and another were also available from time to time for marine animals.

The second part of Hallam's article names and describes four different patterns of distribution of fossils observed in pairs of regions. If the resemblance between the fossils of two regions increases with the passage of time, he calls the pattern a "convergence"; if the resemblance decreases, he calls the pattern a "divergence." The pattern of fossils in North and South America since the formation of the Isthmus of Panama offers an example of convergence, while the fossils in the oceans on either side of the Isthmus present a parallel pattern of divergence. The two patterns together Hallam calls a pattern of "complementarity," while the contemporary occurance of similar fossil forms in widely separated parts of the world he refers to as "disjunct endemism." He gives examples of each and indicates the significance of his interpretations for plate motions and other changes in paleogeography.

In "Plate Tectonics and the History of Life in the Oceans," James W. Valentine and Eldridge M. Moores discuss the relationship between present marine life and the present configuration of land and sea, and consider how these conditions changed in the past. They point out that evolutionary processes have been affected by changes in the shape and motion of continents and by such effects due to location as temperature, the nature of seasonal fluctuations, ocean currents, and the distribution of nutrients. Both the abundance and the diversity of living forms depend upon the stability of food supplies. Since the greatest variety and number of marine species live in Indonesia, while other large populations are found in the northwest Indian Ocean and in the Pacific off Panama and Peru, it is clear that continental shelves and the shores of small islands in the tropics provide the most favorable conditions.

In contrast to the present, during the Triassic period, which began 225 million years ago, the number of varieties of fossils was few, but biotas spread over large regions. The change can be accounted for by continental drift or plate tectonics, for during the Triassic all lands were joined into the single supercontinent of Pangaea. It must have had continuous shallow water margins with no physical barriers to the dispersal of marine fauna. Provinciality would have been low and attributable solely to changes in climates. These were kept mild and comparatively uniform by the free circulation of ocean currents. Earlier, during the Paleozoic era, there had been several continents and a correspondingly greater number and variety of fossil types. The interpretation of late Precambrian events and conditions is less certain, but they seem to have resembled those in the Triassic era. Even though the details of interpretation are provisional, it seems that further work will provide a basis for reconstructing the historical sequence of global environmental conditions, which can then be compared with the sequence of organisms revealed in the fossil record.

The interpretation of past littoral conditions and past sedimentation is also important to the petroleum industry for the light it throws on probable areas of petroleum accumulation. Many of the world's greatest petroleum deposits occur in Tertiary and Mesozoic sedimentary rocks underlying coastal plains or areas that were recently along shorelines. These include many of the great deposits of the Middle East, Texas, Louisiana, and California. Closely related to these deposits, and in many cases forming a seaward continuation of them, are the most economically important deposits that have yet been exploited beneath the sea. These are the oil and gas fields in the shallower waters of the continental shelves, especially in the Gulf of Mexico, off California, in the North Sea, and beneath the Gulf of Maracaibo in Venezuela. Many of these have accumulated over salt domes, which are diapirs of rock salt that have risen from buried salt beds because salt is light and able to flow slowly.

Another type of basin, important for both petroleum deposits and the theory of plate tectonics, are the aulacogens, great rifts filled with sedimentary rocks that N. Shatsky first identified and named in the Soviet Union while drilling for oil. The Anadarko Basin in Oklahoma may be a North American example. In "Hot Spots on the Earth's Surface," Kevin Burke and J. Tuzo Wilson describe aulacogens and relate them to continental margins and to great deltas, including those of the Mississippi, Niger, and Mackenzie Rivers, all of which are petroliferous. They also mention that some large basins, such as the Congo or Michigan basins, may have developed on continents at times when they were stationary relative to hot spots. These basins, especially those in the interior plains of North America, contain rich oil and gas deposits.

Finally, as Peter A. Rona points out in "Plate Tectonics and Mineral Resources," the marginal trenches, seas, and volcanic island chains that form along convergent plate boundaries create habitats favorable for the accumulation of petroleum. It is thus clear that the location of most of the petroleum deposits of the world is linked to plate tectonics. This has been realized by the major petroleum companies, many of whose geologists and geophysicists are now engaged in plotting paleogeographic reconstructions of the continental blocks, of sedimentary deposits, of fossils, and of climatic indicators to guide them in their search for new deposits of oil and gas. The recognition by the petroleum industry of the great significance and value of the theory of plate tectonics represents a complete reversal from the view, held as recently as ten years ago, that the continents are forever fixed in place.

Less is known about the relationship between plate tectonics and ore deposits, but evidence is accumulating that this too is important, as Rona's article makes clear. The evidence is harder to obtain because the sedimentary rocks around the margins of continents that contain so much oil and gas hide the basement near shore, so that evidence about ore deposits must be sought in much deeper water or by indirect means. Rona points out that, in spite of all the recent exploration, on the average only three dredge hauls or drill holes have been made in each million square kilometers of the ocean basins, so that their exploration has only begun. The clearest evidence relating ore deposits to plate tectonics comes from the divergent plate boundary along the axis of the Red Sea and from convergent boundaries beneath many island arcs and young mountains. What ore deposits may lie under the bulk of the ocean floors is still largely speculation.

In the deepest parts of the Red Sea, hot brines and the sediments beneath them are so rich in several metals that the deposits would certainly be mined if they did not lie at the bottom of the sea. Debate still rages about whether these metals are rising from deep within the earth or are being carried into the ocean basins from the adjacent coasts. Very much older deposits found on the mountains of Cyprus appear to have a similar geology, and many authorities believe these deposits are a former section of mid-ocean ridge that has been uplifted. It is also held that many of the metal sulphide and gold deposits of island arcs and young mountains appear to have been released from subducted lithospheric plates. Again the geology of many older deposits is similar and suggests that this process of ore formation is of great economic importance. Finally, vast areas of the world's ocean floors are littered with manganese nodules, which in more limited areas contain as much as 2 percent of other, more valuable metals, notably copper and cobalt. These nodules have grown slowly on the top of moving plates by deposition from seawater.

For these reasons many mining geologists are reviewing their ideas about metal deposits in the light of the new theory.

14 Continental Drift and Evolution

Björn Kurtén
March 1969

The history of life on the earth, as it is revealed in the fossil record, is characterized by intervals in which organisms of one type multiplied and diversified with extraordinary exuberance. One such interval is the age of reptiles, which lasted 200 million years and gave rise to some 20 reptilian orders, or major groups of reptiles. The age of reptiles was followed by our own age of mammals, which has lasted for 65 million years and has given rise to some 30 mammalian orders.

The difference between the number of reptilian orders and the number of mammalian ones is intriguing. How is it that the mammals diversified into half again as many orders as the reptiles in a third of the time? The answer may lie in the concept of continental drift, which has recently attracted so much attention from geologists and geophysicists [see "The Confirmation of Continental Drift," by Patrick M. Hurley; Scientific American Offprint 874]. It now seems that for most of the age of reptiles the continents were collected in two supercontinents, one in the Northern Hemisphere and one in the Southern. Early in the age of mammals the two supercontinents had apparently broken up into the continents of today but the present connections between some of the continents had not yet formed. Clearly such events would have had a profound effect on the evolution of living organisms.

The world of living organisms is a world of specialists. Each animal or plant has its special ecological role. Among the mammals of North America, for instance, there are grass-eating prairie animals such as the pronghorn antelope, browsing woodland animals such as the deer, flesh-eating animals specializing in large game, such as the mountain lion, or in small game, such as the fox, and so on. Each order of mammals comprises a number of species related to one another by common descent, sharing the same broad kind of specialization and having a certain physical resemblance to one another. The order Carnivora, for example, consists of a number of related forms (weasels, bears, dogs, cats, hyenas and so on), most of which are flesh-eaters. There are a few exceptions (the aardwolf is an insect-eating hyena and the giant panda lives on bamboo shoots), but these are recognized as late specializations.

Radiation and Convergence

In spite of being highly diverse, all the orders of mammals have a common origin. They arose from a single ancestral species that lived at some unknown time in the Mesozoic era, which is roughly synonymous with the age of reptiles. The American paleontologist Henry Fairfield Osborn named the evolution of such a diversified host from a single ancestral type "adaptive radiation." By adapting to different ways of life—walking, climbing, swimming, flying, plant-eating, flesh-eating and so on—the descendant forms come to diverge more and more from one another. Adaptive radiation is not restricted to mammals; in fact we can trace the process within every major division of the plant and animal kingdoms.

The opposite phenomenon, in which stocks that were originally very different gradually come to resemble one another through adaptation to the same kind of life, is termed convergence. This too seems to be quite common among mammals. There is a tendency to duplication—indeed multiplication—of orders performing the same function. Perhaps the most remarkable instance is found among the mammals that have specialized in large-scale predation on termites and ants in the Tropics. This ecological niche is filled in South America by the ant bear *Myrmecophaga* and its related forms, all belonging to the order Edentata. In Asia and Africa the same role is played by mammals of the order Pholidota: the pangolins, or scaly anteaters. In Africa a third order has established itself in this business: the Tubulidentata, or aardvarks. Finally, in Australia there is the spiny anteater, which is in the order Monotremata. Thus we have members of four different orders living the same kind of life.

One can cite many other examples. There are, for instance, several living and extinct orders of hoofed herbivores. There are two living orders (the Rodentia, or rodents, and the Lagomorpha, or rabbits and hares) whose chisel-like incisor teeth are specialized for gnawing. Some extinct orders specialized in the same way, and an early primate, an ice-age ungulate and a living marsupial have also intruded into the "rodent niche" [*see top illustration on page 181*]. This kind of duplication, or near-duplication, is an essential ingredient in the richness of the mammalian life that unfolded during the Cenozoic era, or the age of mammals. Of the 30 or so orders of land-dwelling mammals that appeared during this period almost two-thirds are still extant.

The Reptiles of the Cretaceous

The 65 million years of the Cenozoic are divided into two periods: the long Tertiary and the brief Quaternary, which includes the present [*see illustration on page 178*]. The 200-million-year age of reptiles embraces the three periods of the Mesozoic era (the Triassic, the Jurassic and the Cretaceous) and the final period (the Permian) of the preceding era. It is instructive to compare the number

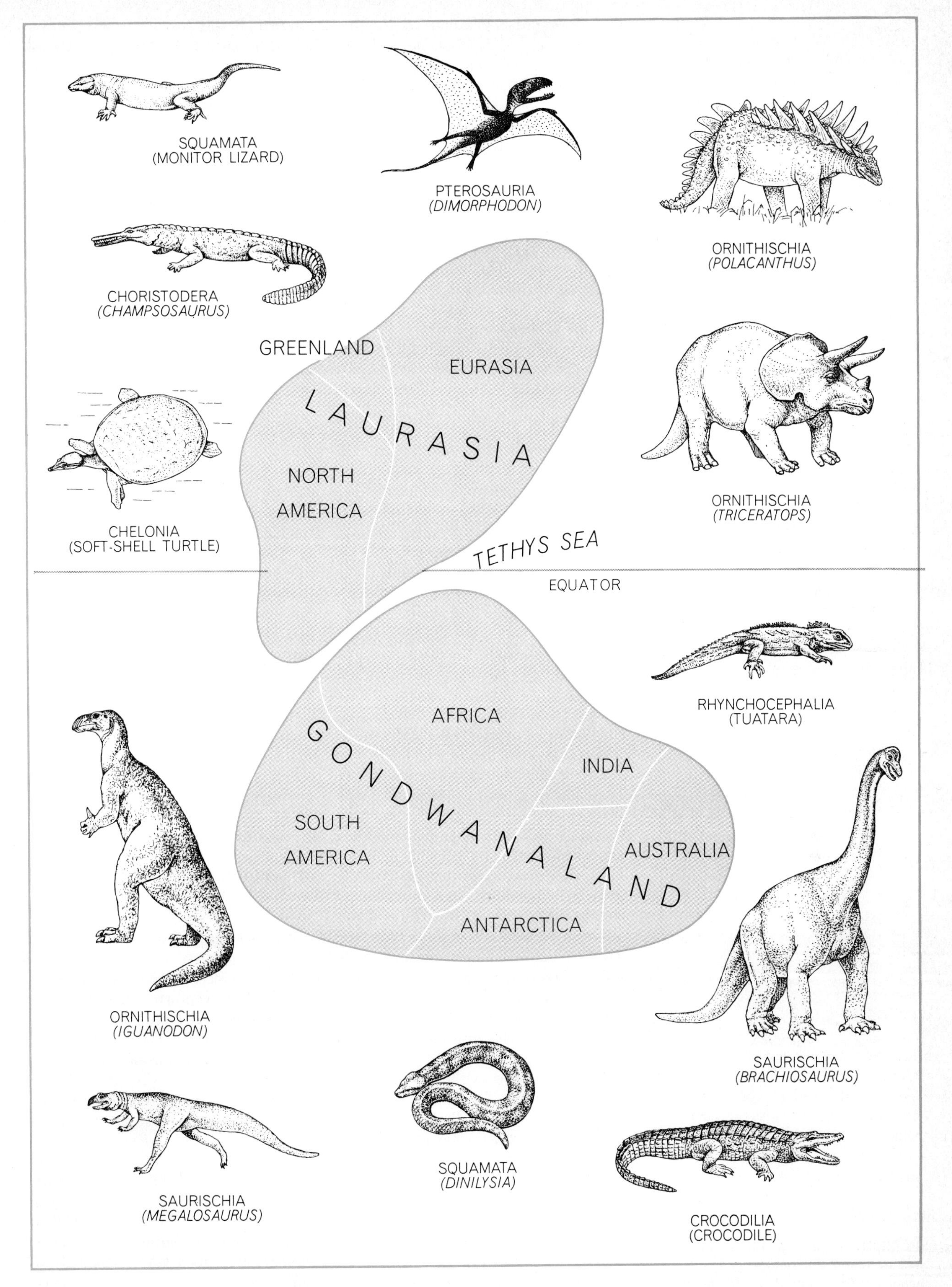

TWO SUPERCONTINENTS of the Mesozoic era were Laurasia in the north and Gondwanaland in the south. The 12 major types of reptiles, represented by typical species, are those whose fossil remains are found in Cretaceous formations. Most of the orders inhabited both supercontinents; migrations were probably by way of a land bridge in the west, where the Tethys Sea was narrowest.

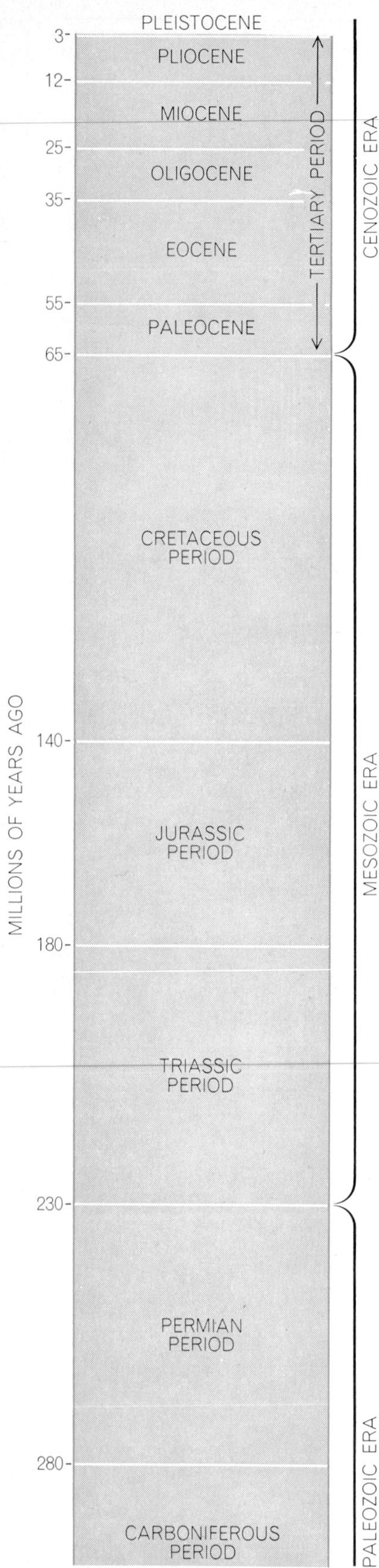

SIX PERIODS of earth history were occupied by the age of reptiles and the age of mammals. The reptiles' rise began 280 million years ago, in the final period of the Paleozoic era. Mammals replaced reptiles as dominant land animals 65 million years ago.

of reptilian orders that flourished during some Mesozoic interval about as long as the Cenozoic era with the number of mammalian orders in the Cenozoic. The Cretaceous period is a good candidate. Some 75 million years in duration, it is only slightly longer than the age of mammals. Moreover, the Cretaceous was the culmination of reptilian life and its fossil record on most continents is good. In the Cretaceous the following orders of land reptiles were extant:

Order Crocodilia: crocodiles, alligators and the like. Their ecological role was amphibious predation; their size, medium to large.

Order Saurischia: saurischian dinosaurs. These were of two basic types: bipedal upland predators (Theropoda) and very large amphibious herbivores (Sauropoda).

Order Ornithischia: ornithischian dinosaurs. Here there were three basic types: bipedal herbivores (Ornithopoda), heavily armored quadrupedal herbivores (Stegosauria and Ankylosauria) and horned herbivores (Ceratopsia).

Order Pterosauria: flying reptiles.

Order Chelonia: turtles and tortoises.

Order Squamata: The two basic types were lizards (Lacertilia) and snakes (Serpentes). Both had the same principal ecological role: small to medium-sized predator.

Order Choristodera (or suborder in the order Eosuchia): champsosaurs. These were amphibious predators.

One or two other reptilian orders may be represented by rare forms. Even if we include them, we get only eight or nine orders of land reptiles in Cretaceous times. One could maintain that an order of reptiles ranks somewhat higher than an order of mammals; some reptilian orders include two or even three basic adaptive types. Even if these types are kept separate, however, the total rises only to 12 or 13. Furthermore, there seems to be only one clear-cut case of ecological duplication: both the crocodilians and the champsosaurs are sizable amphibious predators. (The turtles cannot be considered duplicates of the armored dinosaurs. For one thing, they were very much smaller.) A total of somewhere between seven and 13 orders over a period of 75 million years seems a sluggish record compared with the mammalian achievement of perhaps 30 orders in 65 million years. What light can paleogeography shed on this matter?

The Mesozoic Continents

The two supercontinents of the age of reptiles have been named Laurasia (after Laurentian and Eurasia) and Gondwanaland (after a characteristic geological formation, the Gondwana). Between them lay the Tethys Sea (named for the wife of Oceanus in Greek myth, who was mother of the seas). Laurasia, the northern supercontinent, consisted of what would later be North America, Greenland and Eurasia north of the Alps and the Himalayas. Gondwanaland, the southern one, consisted of the future South America, Africa, India, Australia and Antarctica. The supercontinents may have begun to split up as early as the Triassic period, but the rifts between them did not become effective barriers to the movement of land animals until well into the Cretaceous, when the age of reptiles was nearing its end.

When the mammals began to diversify in the late Cretaceous and early Tertiary, the separation of the continents appears to have been at an extreme. The old ties were sundered and no new ones had formed. The land areas were further fragmented by a high sea level; the waters flooded the continental margins and formed great inland seas, some of which completely partitioned the continents. For example, South America was cut in two by water in the region that later became the Amazon basin, and Eurasia was split by the joining of the Tethys Sea and the Arctic Ocean. In these circumstances each chip of former supercontinent became the nucleus for an adaptive radiation of its own, each fostering a local version of a balanced fauna. There were at least eight such nuclei at the beginning of the age of mammals. Obviously such a situation is quite different from the one in the age of reptiles, when there were only two separate land masses.

Where the Reptiles Originated

The fossil record contains certain clues to some of the reptilian orders' probable areas of origin. The immense distance in time and the utterly different geography, however, make definite inferences hazardous. Let us see what can be said about the orders of Cretaceous reptiles (most of which, of course, arose long before the Cretaceous):

Crocodilia. The earliest fossil crocodilians appear in Middle Triassic formations in a Gondwanaland continent (South America). The first crocodilians in Laurasia are found in Upper Triassic formations. Thus a Gondwanaland origin is suggested.

Saurischia. The first of these dinosaurs appear on both supercontinents in the Middle Triassic, but they are more

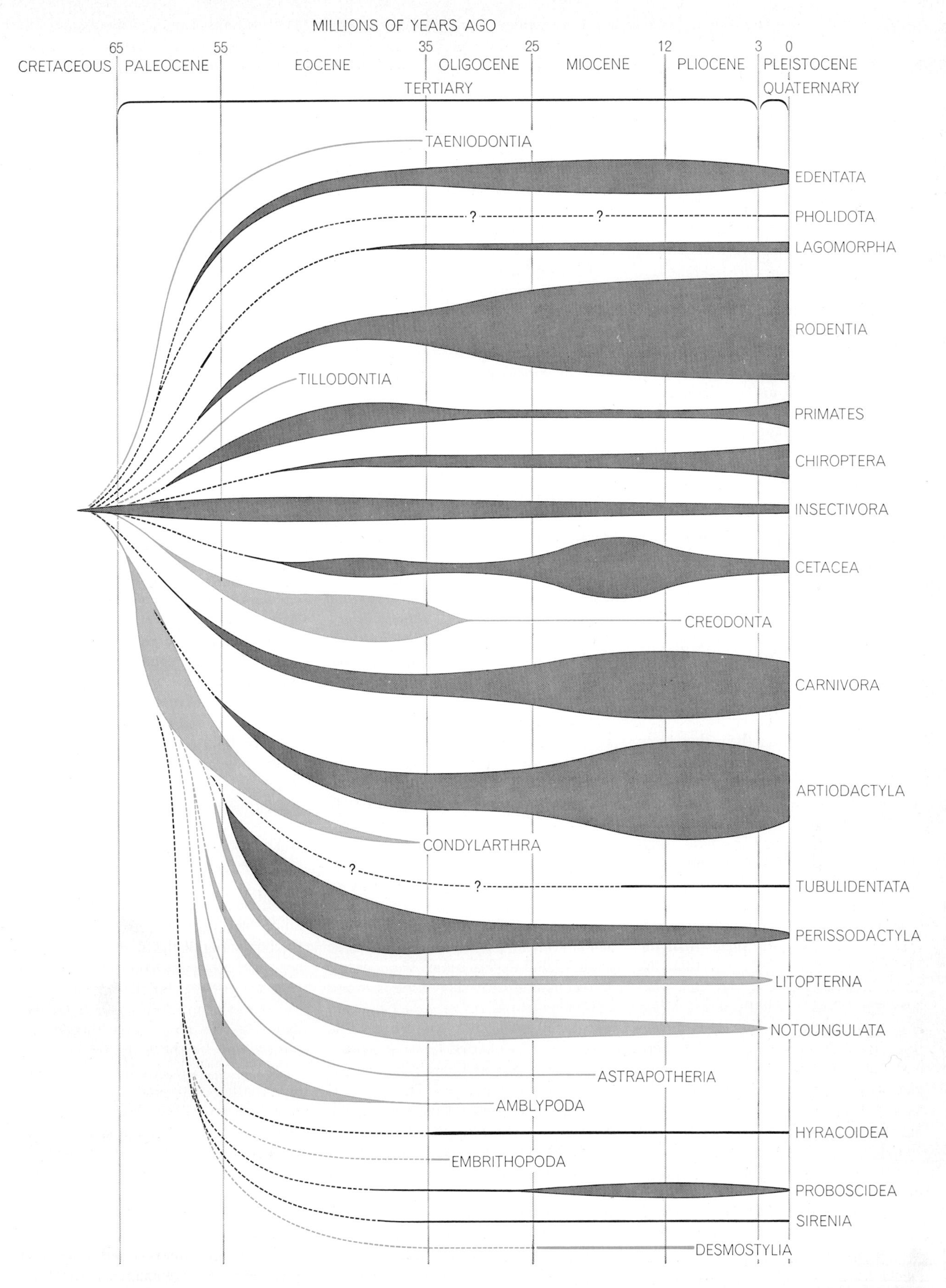

ADAPTIVE RADIATION of the mammals has been traced from its starting point late in the Mesozoic era by Alfred S. Romer of Harvard University. Records for 25 extinct and extant orders of placental mammals are shown here. The lines increase and decrease in width in proportion to the abundance of each order. Extinct orders are shown in color; broken lines mean that no fossil record exists during the indicated interval and question marks imply doubt about the suggested ancestral relation between some orders.

varied in the south. A Gondwanaland origin is very tentatively suggested.

Ornithischia. These dinosaurs appear in the Upper Triassic of South Africa (Gondwanaland) and invade Laurasia somewhat later. A Gondwanaland origin is indicated.

Pterosauria. The oldest fossils of flying reptiles come from the early Jurassic of Europe. They represent highly specialized forms, however, and their antecedents are unknown. No conclusion seems possible.

Chelonia. Turtles are found in Triassic formations in Laurasia. None are found in Gondwanaland before Cretaceous times. This suggests a Laurasian origin. On the other hand, a possible forerunner of turtles appears in the Permian of South Africa. If the Permian form was in fact ancestral, a Gondwanaland origin would be indicated. In any case, the order's main center of evolution certainly lay in the northern supercontinent.

Squamata. Early lizards are found in the late Triassic of the north, which may suggest a Laurasian origin. Unfortunately the lizards in question are aberrant gliding animals. They must have had a long history, of which we know nothing at present.

Choristodera. The crocodile-like champsosaurs are found only in North America and Europe, and so presumably originated in Laurasia.

The indications are, then, that three orders of reptiles—the crocodilians and the two orders of dinosaurs—may have originated in Gondwanaland. Three others—the turtles, the lizards and snakes and the champsosaurs—may have originated in Laurasia. The total number of basic adaptive types in the Gondwanaland group is six; the Laurasia group has four. The Gondwanaland radiation may well have been slightly richer than the Laurasian because it seems that the southern supercontinent was somewhat larger and had a slightly more varied climate. Laurasian climates seem to have been tropical to temperate. Southern parts of Gondwanaland were heavily glaciated late in the era preceding the Mesozoic, and its northern shores (facing the Tethys Sea) had a fully tropical climate.

Although some groups of reptiles, such as the champsosaurs, were confined to one or another of the supercontinents, most of the reptilian orders sooner or later spread into both of them. This means that there must have been ways for land animals to cross the Tethys Sea. The Tethys was narrow in the west and wide to the east. Presumably whatever land connection there was—a true land bridge or island stepping-stones—was located in the western part of the sea. In any case, migration along such routes meant that there was little local differentiation among the reptiles of the Mesozoic era. It was over an essentially uniform reptilian world that the sun finally set at the end of the age of reptiles.

Early Mammals of Laurasia

The conditions of mammalian evolution were radically different. In early and middle Cretaceous times the connections between continents were evidently close enough for primitive mammals to spread into all corners of the habitable world. As the continents drifted farther apart, however, populations of these primitive forms were gradually isolated from one another. This was particularly the case, as we shall see, with the mammals that inhabited the daughter continents of Gondwanaland. Among the Laurasian continents North America was drifting away from Europe, but at the beginning of the age of mammals the distance was not great and there is good evidence that some land connection remained well into the early Tertiary. North American and European mammals were practically identical as late as early Eocene times. Furthermore, throughout the Cenozoic era there was a connection between Alaska and Siberia, at least intermittently, across the Bering Strait. On the other hand, the inland sea extending from the Tethys to the Arctic Ocean formed a complete barrier to direct migration between Europe and Asia in the early Tertiary. Migrations could take place only by way of North America.

In this way the three daughter continents of ancient Laurasia formed three semi-isolated nuclear areas. Many orders of mammals arose in these Laurasian nuclei, among them seven orders that are now extinct but that covered a wide spectrum of specialized types, including primitive hoofed herbivores, carnivores, insectivores and gnawers. The orders of mammals that seem to have arisen in the northern daughter continents and that are extant today are:

Insectivora: moles, hedgehogs, shrews and the like. The earliest fossil insectivores are found in the late Cretaceous of North America and Asia.

Chiroptera: bats. The earliest-known bat comes from the early Eocene of North America. At a slightly later date bats were also common in Europe.

Primates: prosimians (for example, tarsiers and lemurs), monkeys, apes, man. Early primates have recently been found in the late Cretaceous of North America. In the early Tertiary they are common in Europe as well.

Carnivora: cats, dogs, bears, weasels and the like. The first true carnivores appear in the Paleocene of North America.

Perissodactyla: horses, tapirs and other odd-toed ungulates. The earliest forms appear at the beginning of the Eocene in the Northern Hemisphere.

Artiodactyla: cattle, deer, pigs and other even-toed ungulates. Like the odd-toed ungulates, they appear in the early Eocene of the Northern Hemisphere.

Rodentia: rats, mice, squirrels, beavers and the like. The first rodents appear in the Paleocene of North America.

Lagomorpha: hares and rabbits. This order makes its first appearance in the Eocene of the Northern Hemisphere.

Pholidota: pangolins. The earliest come from Europe in the middle Tertiary.

The fact that a given order of mammals is found in older fossil deposits in North America than in Europe or Asia does not necessarily mean that the order arose in the New World. It may simply reflect the fact that we know much more about the early mammals of North America than we do about those of Eurasia. All we can really say is that a total of 16 extant or extinct orders of mammals probably arose in the Northern Hemisphere.

Early Mammals of South America

The fragmentation of Gondwanaland seems to have started earlier than that of Laurasia. The rifting certainly had a much more radical effect. Looking at South America first, we note that at the beginning of the Tertiary this continent was tenuously connected to North America but that for the rest of the period it was completely isolated. The evidence for the tenuous early linkage is the presence in the early Tertiary beds of North America of mammalian fossils representing two predominantly South American orders: the Edentata (the order that includes today's ant bears, sloths and armadillos) and the Notoungulata (an order of extinct hoofed herbivores).

Four other orders of mammals are exclusively South American: the Paucituberculata (opossum rats and other small South American marsupials), the Pyrotheria (extinct elephant-like animals), the Litopterna (extinct hoofed herbivores, including some forms resembling

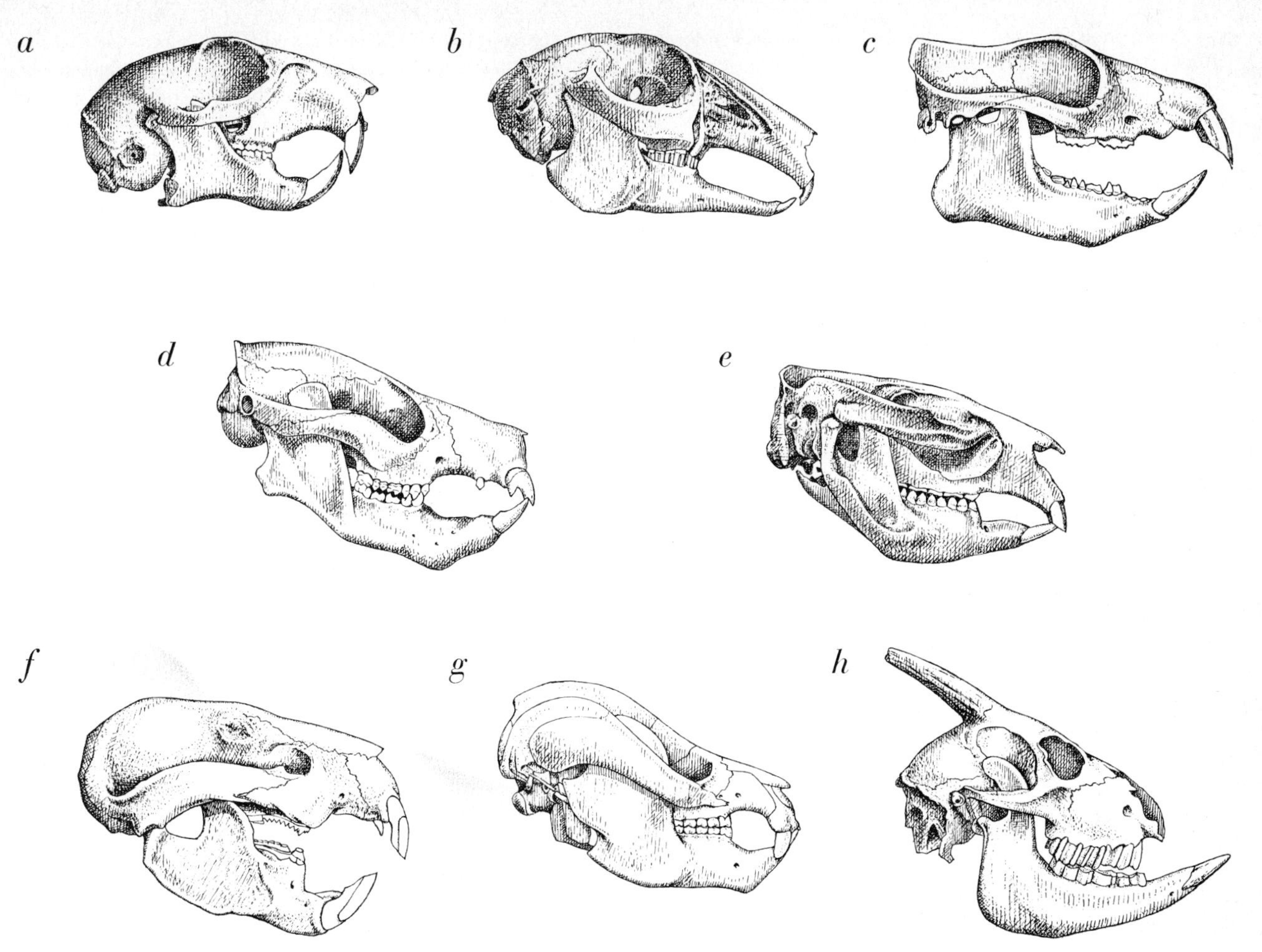

CHISEL-LIKE INCISORS, specialized for gnawing, appear in animals belonging to several extinct and extant orders in addition to the rodents, represented by a squirrel (*a*), and the lagomorphs, represented by a hare (*b*), which are today's main specialists in this ecological role. Representatives of other orders with chisel-like incisor teeth are an early tillodont, *Trogosus* (*c*), an early primate, *Plesiadapis* (*d*), a living marsupial, the wombat (*e*), one of the extinct multituberculate mammals, *Taeniolabis* (*f*), a mammal-like reptile of the Triassic, *Bienotherium* (*g*), and a Pleistocene cave goat, *Myotragus* (*h*), whose incisor teeth are in the lower jaw only.

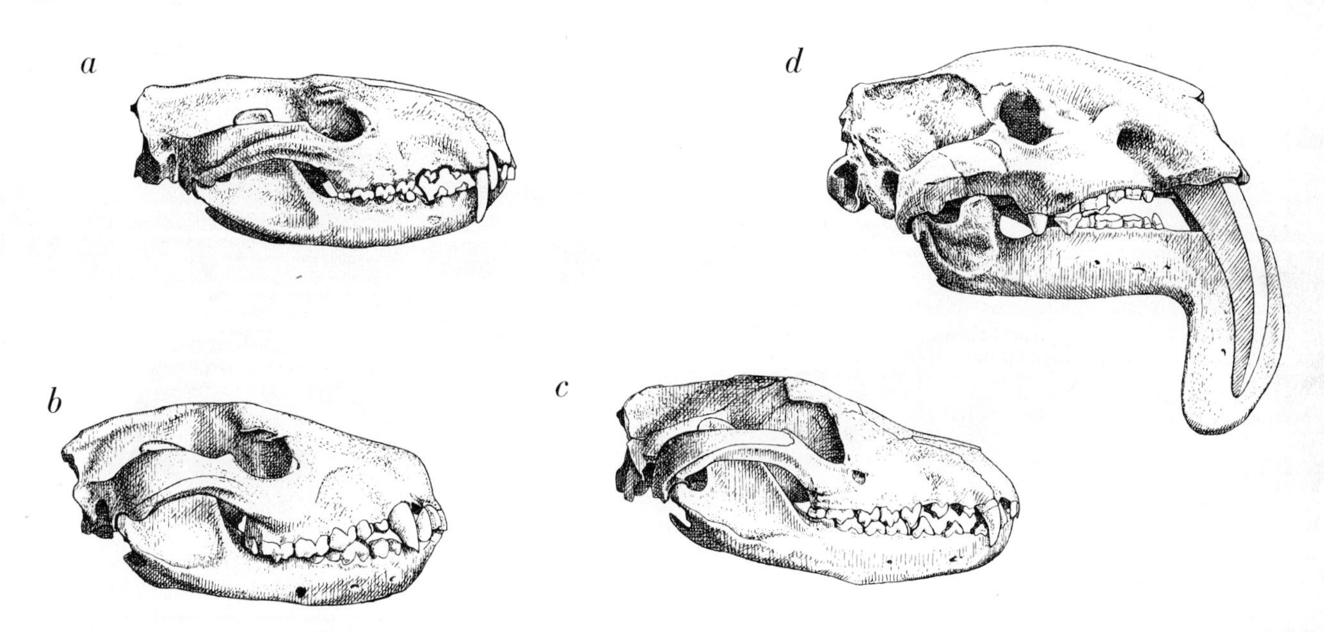

CARNIVOROUS MARSUPIALS, living and extinct, fill an ecological niche more commonly occupied by the placental carnivores today. Illustrated are the skulls of two living forms, the Australian "cat," *Dasyurus* (*a*), and the Tasmanian devil, *Sarcophilus* (*b*). The Tasmanian "wolf," *Thylacinus* (*c*), has not been seen for many years and may be extinct. A tiger-sized predator of South America, *Thylacosmilus* (*d*) became extinct in Pliocene times, long before the placental sabertooth of the Pleistocene, *Smilodon*, appeared.

horses and camels) and the Astrapotheria (extinct large hoofed herbivores of very peculiar appearance). Thus a total of six orders, extinct or extant, probably originated in South America. Still another order, perhaps of even more ancient origin, is the Marsupicarnivora. The order is so widely distributed, with species found in South America, North America, Europe and Australia, that its place of origin is quite uncertain. It includes, in addition to the extinct marsupial carnivores of South America, the opossums of the New World and the native "cats" and "wolves" of the Australian area.

The most important barrier isolating South America from North America in the Tertiary period was the Bolívar Trench. This arm of the sea cut across the extreme northwest corner of the continent. In the late Tertiary the bottom of the Bolívar Trench was lifted above sea level and became a mountainous land area. A similar arm of sea, to which I have already referred, extended across the continent in the region that is now the Amazon basin. This further enhanced the isolation of the southern part of South America.

Africa's role as a center of adaptive radiation is problematical because practically nothing is known of its native mammals before the end of the Eocene. We do know, however, that much of the continent was flooded by marginal seas, and that in the early Tertiary, Africa was cut up into two or three large islands. Still, there must have been a land route to Eurasia even in the Eocene; some of the African mammals of the following epoch (the Oligocene) are clearly immigrants from the north or northeast. Nonetheless, the majority of African mammals are of local origin. They include the following orders:

Proboscidea: the mastodons and elephants.

Hyracoidea: the conies and their extinct relatives.

Embrithopoda: an extinct order of very large mammals.

Tubulidentata: the aardvarks.

In addition the order Sirenia, consisting of the aquatic dugongs and manatees, is evidently related to the Proboscidea and hence presumably also originated in Africa. The same may be true of another order of aquatic mammals, the extinct Desmostylia, which also seems to be related to the elephants. The one snag in this interpretation is that desmostylian fossils are found only in the North Pacific, which seems rather a long way from Africa. Nonetheless, once they were waterborne, early desmostylians might have crossed the Atlantic, which was then only a narrow sea, navigated the Bolívar Trench and, rather like Cortes (but stouter), found themselves in the Pacific.

Early Mammals of Africa

Thus there are certainly four, and possibly six, mammalian orders for which an African origin can be postulated. Here it should be noted that Africa had an impressive array of primates in the Oligocene. This suggests that the order Primates had a comparatively long history in Africa before that time. Even though the order as such does not have its roots in Africa, it is possible that the higher primates—the Old World monkeys, the apes and the ancestors of man—may have originated there. Most of the fossil primates found in the Oligocene

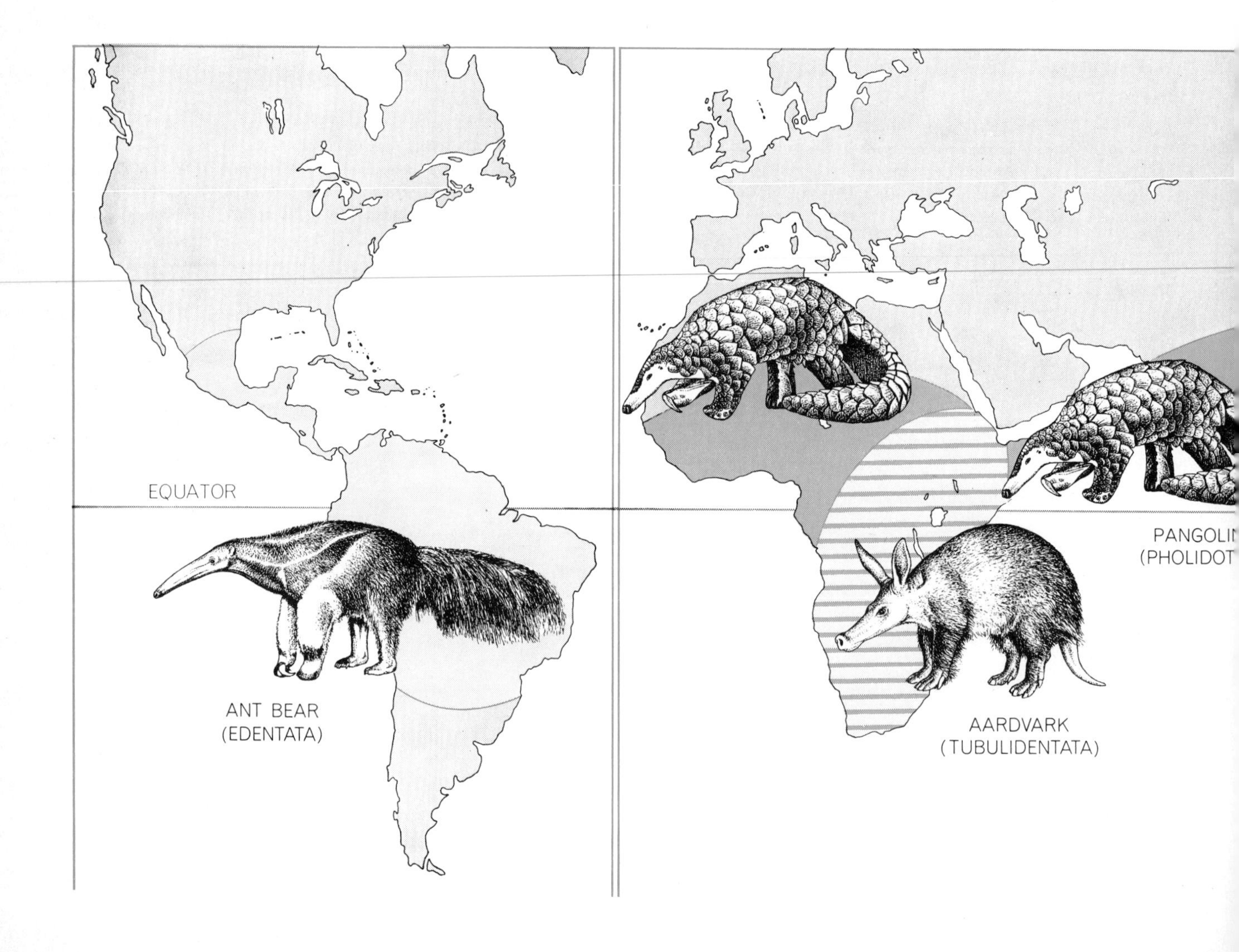

formations of Africa are primitive apes or monkeys, but there is at least one form (*Propliopithecus*) whose dentition looks like a miniature blueprint of a set of human teeth.

The Rest of Gondwanaland

We know little or nothing of the zoogeographic roles played by India and Antarctica in the early Tertiary. Mammalian fossils from the early Tertiary are also absent in Australia. It may be assumed, however, that the orders of mammals now limited to Australia probably originated there. These include two orders of marsupials: the Peramelina, comprised of several bandicoot genera, and the Diprotodonta, in which are found the kangaroos, wombats, phalangers and a number of extinct forms. In addition the order Monotremata, a very primitive group of mammals that includes the spiny anteater and the platypus, is likely to be of Australian origin. This gives us a total of three orders probably founded in Australia.

Summing up, we find that the three Laurasian continents produced a total of 16 orders of mammals, an average of five or six orders per continent. As for Gondwanaland, South America produced six orders, Africa four to six and Australia three. The fact that Australia is a small continent probably accounts for the lower number of orders founded there. Otherwise the distribution—the average of five or six orders per subdivision—is remarkably uniform for both the Laurasian and Gondwanaland supercontinents. The mammalian record should be compared with the data on Cretaceous reptiles, which show that the two supercontinents produced a total of 12 or 13 orders (or adaptively distinct suborders). A regularity is suggested, as if a single nucleus of radiation would tend in a given time to produce and support a given amount of basic zoological variation.

As the Tertiary period continued new land connections were gradually formed, replacing those sundered when the old supercontinents broke up. Africa made its landfall with Eurasia in the Oligocene and Miocene epochs. Laurasian orders of mammals spread into Africa and crowded out some of the local forms, but at the same time some African mammals (notably the mastodons and elephants) went forth to conquer almost the entire world. In the Western Hemisphere the draining and uplifting of the Bolívar Trench was followed by intense intermigration and competition among the mammals of the two Americas. In the process much of the typical South American mammal population was exterminated, but a few forms pressed successfully into North America to become part of the continent's spectacular ice-age wildlife.

India, a fragment of Gondwanaland that finally became part of Asia, must have made a contribution to the land fauna of that continent but just what it was cannot be said at present. Of all the drifting Noah's arks of mammalian evolution only two—Antarctica and Australia—persist in isolation to this day. The unknown mammals of Antarctica have long been extinct, killed by the ice that engulfed their world. Australia is therefore the only island continent that still retains much of its pristine mammalian fauna. [*see illustration on page 98*].

If the fragmentation of the continents at the beginning of the age of mammals promoted variety, the amalgamation in the latter half of the age of mammals has promoted efficiency by means of a large-scale test of the survival of the fittest. There is a concomitant loss of variety; 13 orders of land mammals have become extinct in the course of the Cenozoic. Most of the extinct orders are island-continent productions, which suggests that a system of semi-isolated provinces, such as the daughter continents of Laurasia, tends to produce a more efficient brood than the completely isolated nuclei of the Southern Hemisphere. Not all the Gondwanaland orders were inferior, however; the edentates were moderately successful and the proboscidians spectacularly so.

As far as land mammals are concerned, the world's major zoogeographic provinces are at present four in number: the Holarctic-Indian, which consists of North America and Eurasia and also northern Africa; the Neotropical, made up of Central America and South America; the Ethiopian, consisting of Africa south of the Sahara, and the Australian. This represents a reduction from seven provinces with about 30 orders of mammals to four provinces with about 18 orders. The reduction in variety is proportional to the reduction in the number of provinces.

In conclusion it is interesting to note that we ourselves, as a subgroup within the order Primates, probably owe our origin to a radiation within one of Gondwanaland's island continents. I have noted that an Oligocene primate of Africa may have been close to the line of human evolution. By Miocene times there were definite hominids in Africa, identified by various authorities as members of the genus *Ramapithecus* or the genus *Kenyapithecus*. Apparently these early hominids spread into Asia and Europe toward the end of the Miocene. The cycle of continental fragmentation and amalgamation thus seems to have played an important part in the origin of man as well as of the other land mammals.

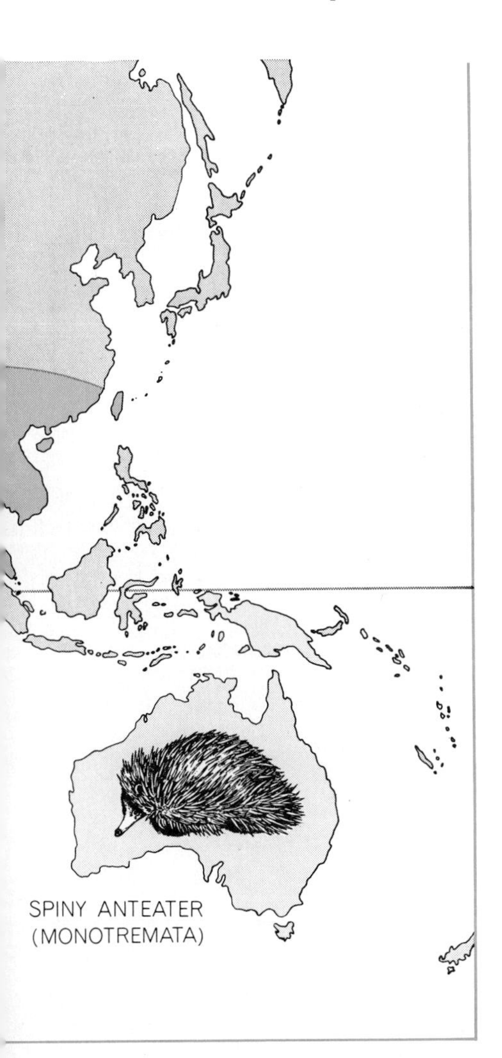

FOUR ANT-EATING MAMMALS have become adapted to the same kind of life although each is a member of a different mammalian order. Their similar appearance provides an example of an evolutionary process known as convergence. The ant bears of the New World Tropics are in the order Edentata. The aardvark of Africa is the only species in the order Tubulidentata. Pangolins, found both in Asia and in Africa, are members of the order Pholidota. The spiny anteater of Australia, a very primitive mammal, is in the order Monotremata.

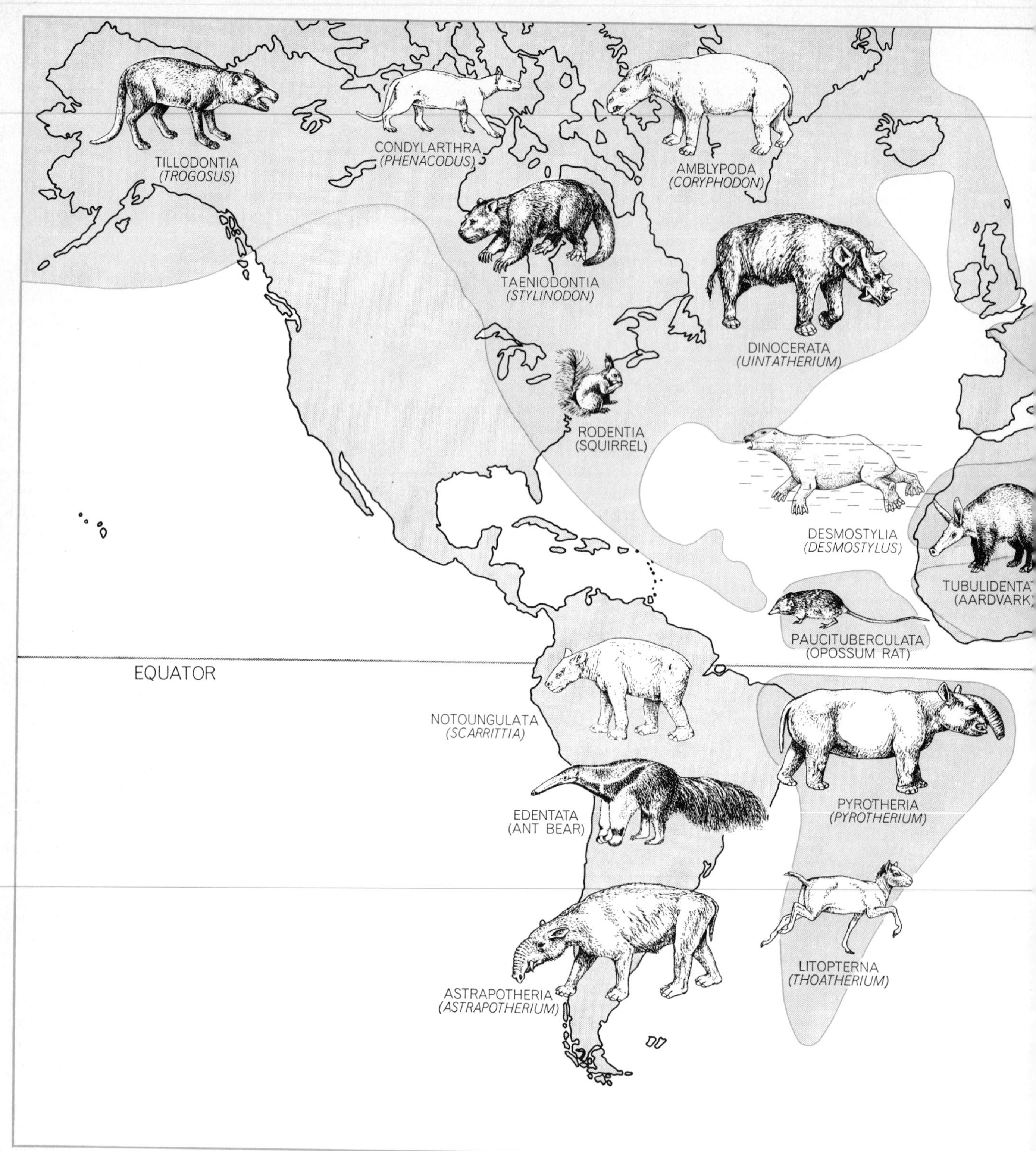

CONTINENTAL DRIFT affected the evolution of the mammals by fragmenting the two supercontinents early in the Cenozoic era. In the north, Europe and Asia, although separated by a sea, remained connected with North America during part of the era. The

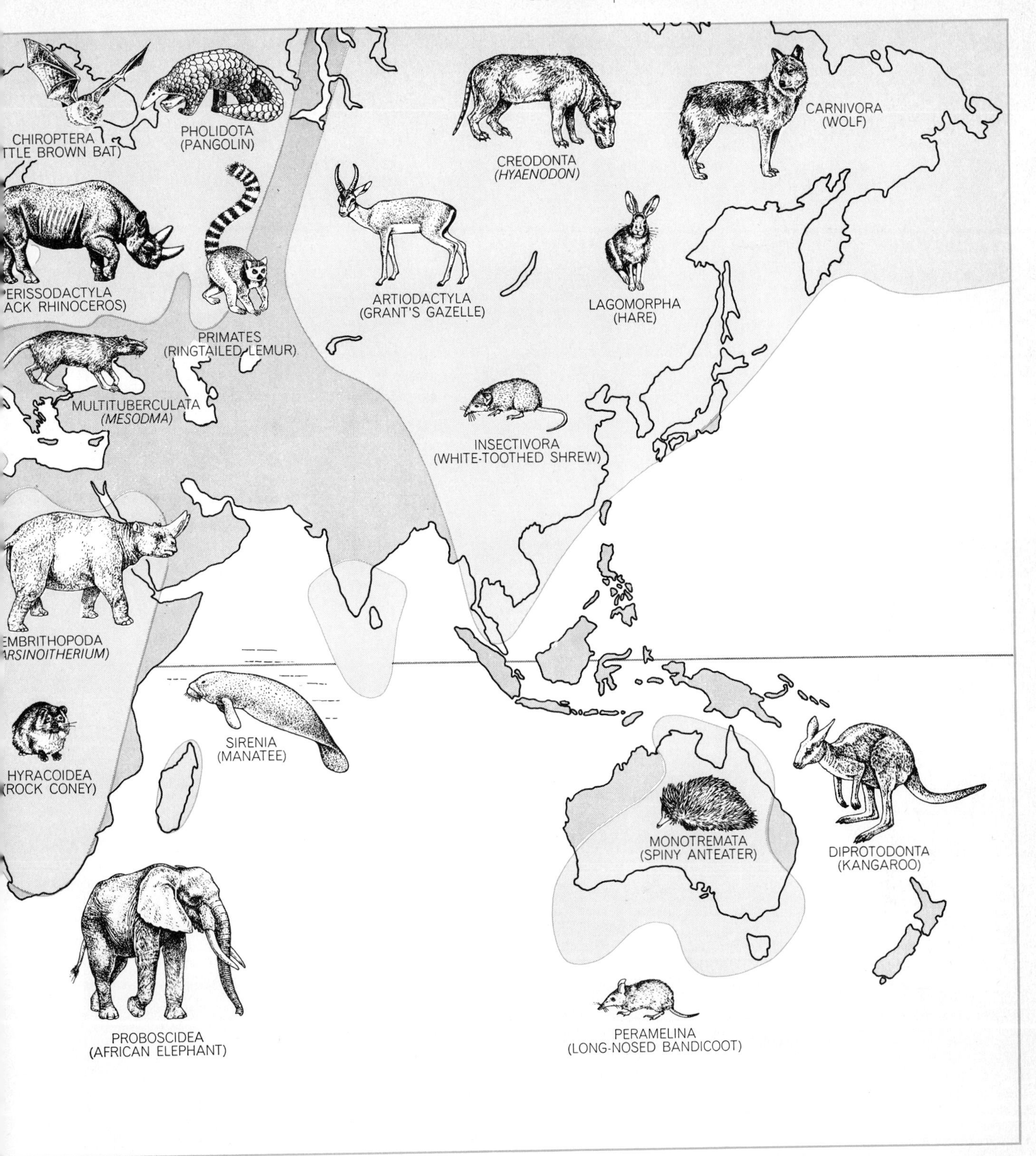

free migration that resulted prevents certainty regarding the place of origin of many orders of mammals that evolved in the north. The far wider rifting of Gondwanaland allowed the evolution of unique groups of mammals in South America, Africa and Australia.

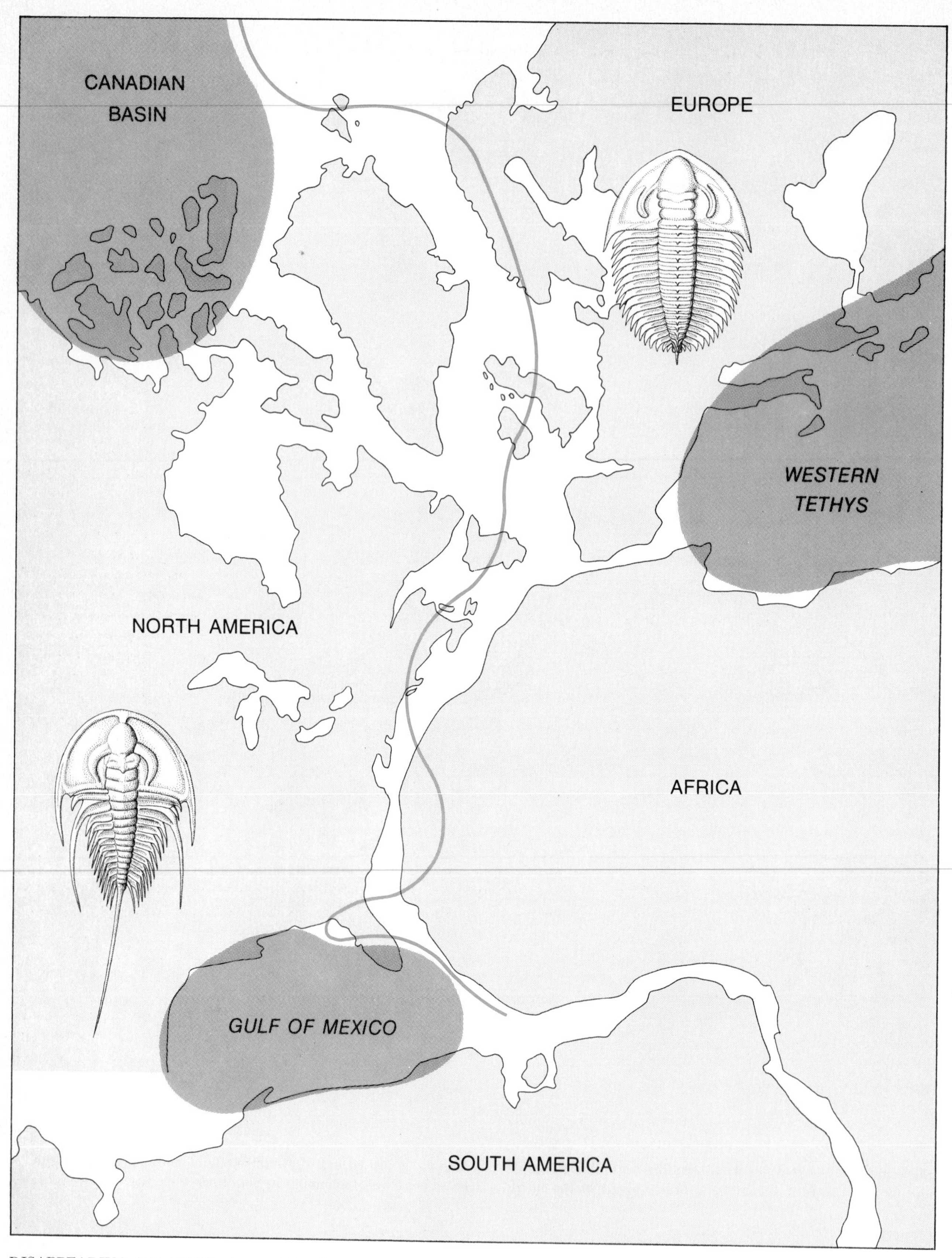

DISAPPEARING ATLANTIC of some 400 million years ago may have been reduced to two comparatively minor bodies of water: the Canadian basin and the Gulf of Mexico. The broad colored line divides regions that are connected today, such as the Scandinavian peninsula, the British Isles and parts of eastern North America. It marks the boundary between faunal realms that were distinctly different in Cambrian and early Ordovician times. A fossil typical of each realm is illustrated. The American form is the trilobite *Paedeumias transitans;* the European form is the trilobite *Holmia kjerulfi.* The difference between the two realms grew progressively less, and by the middle of the Paleozoic era it had vanished. Convergence of the two faunas suggests that what had once been a wide, deep proto-Atlantic was swallowed up along zones of subduction as American and European continental plates came together.

Continental Drift and the Fossil Record

A. Hallam
November 1972

In the past 10 years geologists have been widely converted to plate tectonics, a concept that implies the lateral migration of the continents. If Alfred Wegener, who put forward the hypothesis of continental drift around the turn of the century, were alive today, he might be wryly amused. The workers who revitalized his conception paid comparatively little attention to the evidence in its favor provided by fossils. Yet by Wegener's own account he began to take the idea of drifting continents seriously only after learning of the fossil evidence for a former land connection between Brazil and Africa. The fossil record, rather than the much noted physical "fit" between the opposing coastlines, was what inspired him.

The fossil record is no less important to students of continental drift today. The similarities and differences between fossils in various parts of the world from Cambrian times onward are now helping paleontologists both to support the drift concept and to provide a reasonably precise timetable for a number of the key events before and after the breakup of the ancestral continent of Pangaea.

In Wegener's day there was nothing particularly novel about the idea that the various continents had been connected in various ways off and on in the distant past. Biologists and paleontologists in the 19th century and in the early 20th readily invoked such land connections to explain the strong resemblances between plants and animals on different continents. It was generally agreed that links had existed between Australia and other regions bordering the Indian Ocean until early in the Jurassic period, some 180 million years ago. The same was held to be true of a link between Africa and Brazil until early in the Cretaceous period, some 140 million years ago, and a link between Madagascar and India up to the start of the Cenozoic era, only 65 million years ago.

At the same time the orthodox explanation of these connections was that the position of the continents was fixed but that "land bridges" had spanned the considerable distances of open ocean between them. In the orthodox view these extensive bridges had later sunk without a trace [*see upper illustration on page 189*]. Wegener dismissed such explanations in trenchant terms. The earth's crust, he pointed out, is composed of rocks that are far less dense than the material of the earth's interior. If the floors of the oceans were paved with vast sunken bridges composed of the same thickness of light crustal material as the continental areas that lie above sea level, then gravity measurements made at sea should reveal that fact. The gravity measurements indicate the exact opposite: the underlying rock of the ocean floor is much denser than the crustal material of the continents.

The essential improbability of sunken land bridges can also be stated in terms of isostatic balance. If the low-density crustal rocks of the vanished bridges had indeed been somehow forced downward into the denser sea bottom, the bridges would tend to rise again. None of the hypothetical land bridges, however, has reemerged. This makes it necessary to assume the existence of some colossal unspecified force that continues to hold the bridges submerged. The existence of such a force seems improbable in the extreme. Unless one chose to dismiss the fossil evidence out of hand, Wegener concluded, the only feasible means of explaining intercontinental plant and animal resemblances was by the drifting of the continents themselves.

It is odd that neither paleontologists nor geophysicists paid much heed to Wegener's cogent argument. The paleontologists were almost unanimous in rejecting the notion, perhaps because they did not fully appreciate the force of Wegener's geophysical proposals. The main effect of his hypothesis on this group was that quite narrow land bridges became more popular than the embarrassingly broad avenues that had been favored at the turn of the century [*see lower illustration on page 189*]. Of course even when the bridges were pared down in this fashion, serious isostatic problems remained. As for the geophysicists, they largely ignored the considerable body of fossil evidence for continental drift that Wegener had assembled. Perhaps they failed to appreciate its significance or perhaps they mistrusted data of a merely qualitative kind and were suspicious of the seemingly subjective character of taxonomic assessments.

The Zoogeography of the Past

In recent years there has been a major resurgence of interest in the zoogeography of the past. This is no doubt due in part to the prospect that fossil evidence will enable paleontologists to test independently the conclusions that have been reached on purely geological and geophysical grounds about plate tectonics. Here I shall describe how some of this evidence, in particular the remains of higher animals on land and of simpler bottom-dwelling marine organisms offshore, can shed light on the making and breaking of connections between continents in the past and can help to determine how long some of these continental linkages and separations endured.

Two principal factors control the geo-

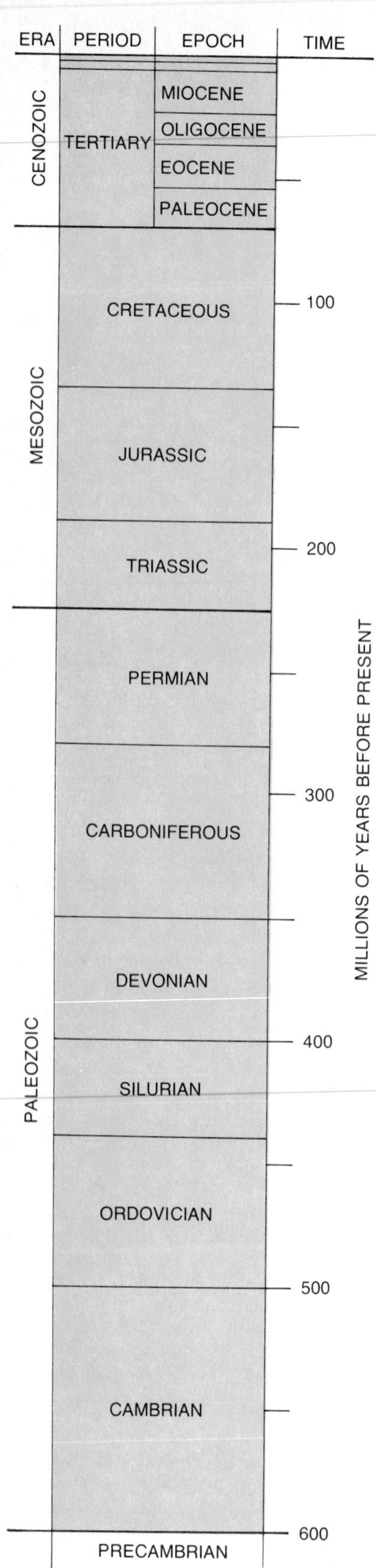

THREE SUCCESSIVE ERAS occupy the 600-million-year span seen in this geological time scale. Patterns of distribution found in the fossil record of each era throw light on plate-tectonic theory of continental drift.

graphical distribution of land animals. They are on the one hand the climate and on the other various water barriers and in particular wide stretches of sea. The effectiveness of climate as a barrier is nicely illustrated by the fact that the animals that live near the poles are far less diverse than those that live in the Tropics. Polar ice, however, is not a permanent feature of the planet. In times of a more equable world climate such as prevailed during the Mesozoic era far and away the most significant deterrents to movement must have been ocean barriers.

An animal need not be exclusively terrestrial to be confined by an ocean barrier. For example, a large number of fishes cannot survive except in fresh water. Amphibians are also severely circumscribed by the sea, although frogs are better able to colonize islands by swimming across the sea than newts or salamanders. As for the probability that reptiles or mammals might successfully move across the sea by accidental rafting, the chances are obviously best for the small and the rapidly reproducing. By the same token, however, the small animals are the very ones that would first die of starvation if the rafting were prolonged. Hence even fairly narrow marine straits can be highly effective barriers to terrestrial animals.

The paleontologist George Gaylord Simpson has made a useful distinction between three kinds of dispersal route. The first, which he calls "corridors," are land connections that allow free migration of animals in both directions. The second, called "filter bridges," combine a land connection with some additional factor, such as climate, in a way that bars some prospective migrants. As an example, it seems doubtful that warm-weather-loving animals crossed the Bering Strait bridge between Asia and North America during the Pleistocene. The passage was open only when the sea level was low during the colder phases of that glacial epoch.

Simpson's third category, "sweepstakes routes," takes its name from the small proportion of winners compared with losers. The rare winners are those that survive chance rafting and succeed in colonizing isolated areas. Unlike corridors (or even filter bridges), which favor the eventual homogeneity of the faunas at both ends of the passage, sweepstakes routes lead to the development of populations that are low in diversity and ecologically unbalanced. The reason is that, in addition to the high mortality rate, chance rafting can only be possible for a very small fraction of any continental fauna. One result of this double selectivity is that islands are often a refuge for a comparatively primitive group of animals. The tuataras of New Zealand, the sole surviving representatives of one major order of reptiles, are one example; the lemurs of Madagascar are another.

The three routes Simpson defines are the ones that are available to land animals in general and to the higher vertebrates in particular. It is obvious, however, that the same kinds of connection would also influence the dispersal of marine organisms such as bottom-dwelling invertebrates. Of course, the effects would be exactly reversed. For example, the establishment of a corridor between two landmasses would simultaneously raise a barrier between two segments of a previously homogeneous marine fauna. The disappearance of a land connection, in turn, would result in the establishment of a corridor as far as marine organisms were concerned. Analogous to the filter bridge for land animals would be the marine "filter barrier." Here a bottom-dweller on one side of an ocean basin might migrate freely to the other side as long as it could drift in tropical waters, but it would be unable to survive the rigors of such a journey in the cooler waters of the temperate zones.

Isolation and Homogenization

Bearing these considerations in mind, what does evolutionary theory predict with respect to continental drift? Clearly when a formerly unified landmass splits up, the result is genetic isolation (and hence morphological divergence) among the separated segments of a formerly homogeneous land fauna. Conversely, the suturing of two continental areas is followed by the homogenization of the corresponding faunas as there is cross-migration. The process will quite probably be accompanied by the extinction of any less well-adapted groups that may now face stronger competition.

In land areas that are unconnected two factors, parallel evolution and convergence, may produce animal species that develop a similar morphology because they occupy identical ecological niches. A well-known instance is provided by the ant bear of South America, the aardvark of Africa and the cosmopolitan pangolin. Forces of this kind are unlikely, however, to affect entire faunas.

So much for the land animals. What are the effects of continental drift on the invertebrates that inhabit the shallow ocean floor? It is obvious that a stretch of land that separates two oceans acts as a barrier to marine migration. It is less well appreciated that a wide stretch of deep ocean may be almost as effective as a land barrier in preserving the genetic isolation of the marine shelf dwellers on opposite shores. This isolation is due to the fact that such animals disperse at only one time in their life cycle: when they are newly hatched larvae that join the plankton community at the ocean surface or close to it. A bottom dweller's larval stage is normally not long enough to enable the animal to survive a slow ocean crossing.

This matter was quantified some time ago by the Danish biologist Gunnar Thorson. He studied the larval stages of no fewer than 200 species of marine invertebrates, and he concluded that only 5 percent could survive in the plankton for more than three months. That length of time is too short to allow transoceanic colonization except by an occasional "transport miracle." More recently Thorson has been criticized for confining his attention to organisms that inhabit cool temperate waters. New data gathered by towing fine nets through the plankton indicate that a significant number of larvae of tropical species can survive an Atlantic crossing in either direction; they drift with the equatorial surface current one way and with the subsurface countercurrent the other way. It is also true

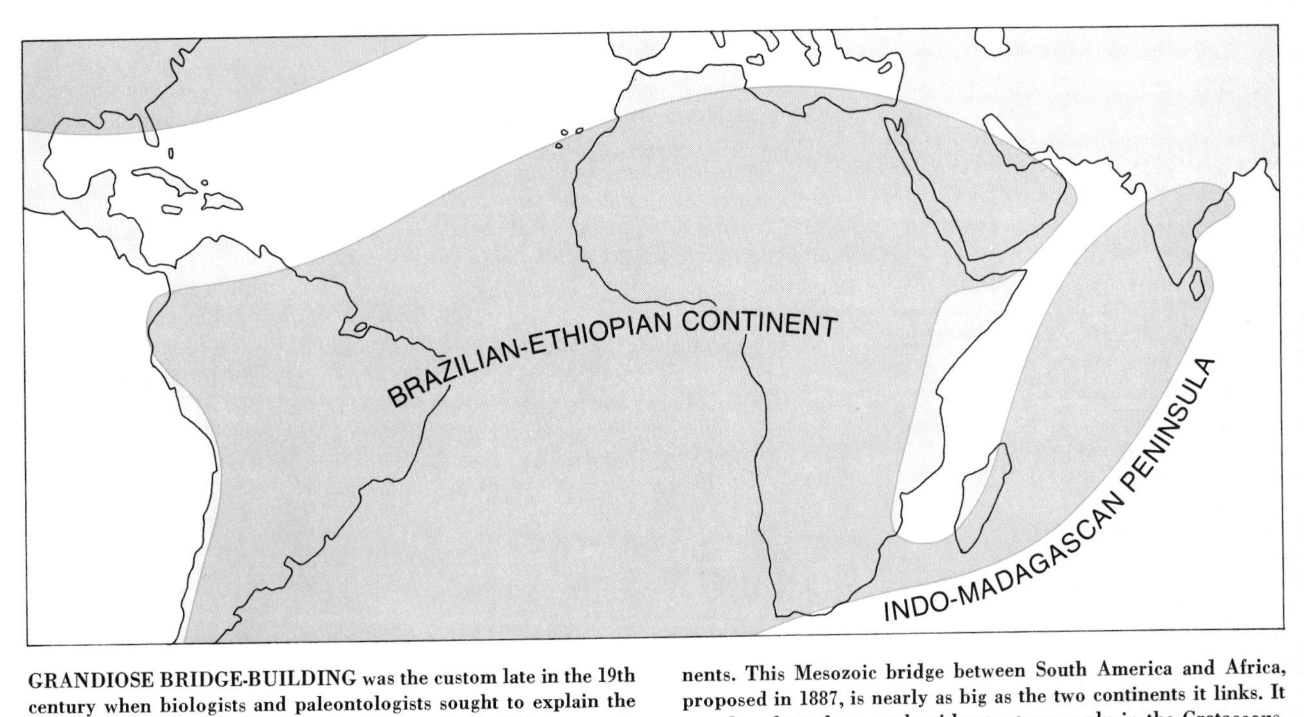

GRANDIOSE BRIDGE-BUILDING was the custom late in the 19th century when biologists and paleontologists sought to explain the strong similarities between the fossil records of different continents. This Mesozoic bridge between South America and Africa, proposed in 1887, is nearly as big as the two continents it links. It was thought to have sunk without a trace early in the Cretaceous.

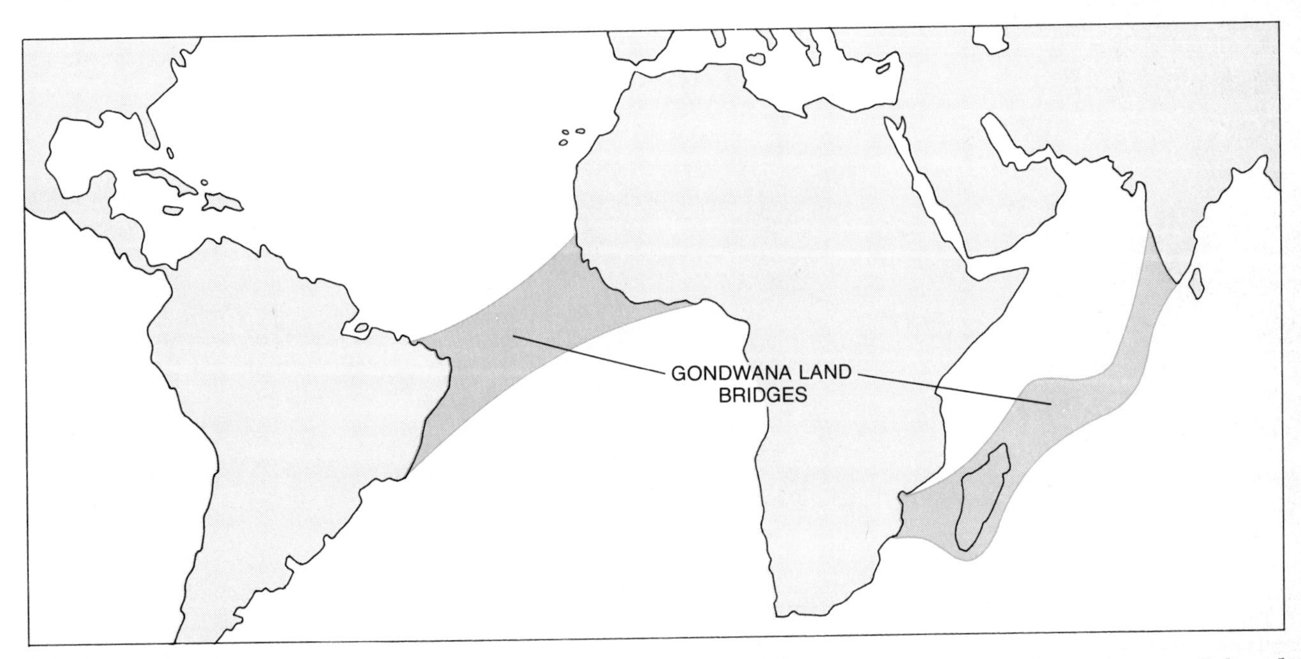

MORE MODEST BRIDGES were proposed in the 20th century as geophysical studies found no evidence in support of vast drowned intercontinental connections. This pared-down link between Brazil and West Africa and another (supposedly in existence until the end of Cretaceous times) between East Africa and India represent an effort to satisfy those who were critical of the broader bridges.

that some mollusks and some corals are found everywhere in the Tropics, a fact that gives added support to the notion that transoceanic migration is possible for long-lived larvae. At the same time it is possible that organisms with prolonged larval stages represent a late evolutionary development that has taken place only since the present Atlantic and Indian oceans began to open up.

Be this as it may, it is evident that, although Thorson's conclusions must be somewhat amended, an ocean barrier is effective in restricting the migration of a majority of the shelf-dwelling marine invertebrates. Moreover, the capacity for migration being in direct proportion to the length of the larval stage, it follows that the wider the deep-ocean barrier is, the more the faunas of the opposing shelves should differ. That is true today. There are fewer species in common among the faunas on opposite sides of the Pacific than there are on opposite sides of the Atlantic. Therefore among marine fossil faunas the degrees of similarity or difference between two coastal assemblages should allow an estimate of the amount of deep-ocean separation between the two coasts.

Given the present state of knowledge, it would be futile to seek any absolute measure of similarity between fossil faunas. If, however, one adopts a dynamic approach and deals with changes in the degree of resemblance during successive intervals of geologic time, a fair amount of progress can be made. Such an approach has several advantages. For one, animals belonging to widely separated phyla can be grouped together. For another, the specific factors that control a particular species' migration potential need not be precisely understood. Further, one can use the work of many different taxonomists, without regard to whether they are "lumpers" or "splitters" in their method of classification, as long as the work is self-consistent. Best of all, the approach takes fully into account the factor that is all-important to the geologist: the factor of time.

The Distribution of Fossils

There are four principal patterns revealed by the distribution of fossils. They are closely interrelated and I shall define them in order, accompanying each definition with one or more examples. All the examples will concern the making or breaking of land or sea connections, but not all will be the result of continental drift.

The first pattern is "convergence." The term describes an increase, as time passes, in the degree of resemblance between faunas of different regions. An example unrelated to continental drift is provided by the history of South America during the Cenozoic era, beginning some 65 million years ago. Throughout most of the era South America had a highly distinctive land fauna [*see illustration below*]. The fossil mammals from Cenozoic formations in Argentina were among the most spectacular finds of Darwin's voyage aboard H.M.S. *Beagle*. The strongly endemic nature of this fauna, comparable in this respect to the fauna of Australia today, is clear evidence that the continent was isolated for many millions of years.

At about the end of the Pliocene epoch, two million or so years ago, a drastic change took place. A land connection—the Isthmus of Panama—was established between North and South America and many New World animals

DIVERGENCE OF MAMMALS in the New World occurred during a period of millions of years when North and South America were unconnected. Mammals then unknown in the south included mastodons, tapirs, primitive camels and various carnivores. The mammals of the south included many that are now extinct. Illustrated here are (*a*) *Mylodon*, a giant sloth, (*b*) *Paedotherium*, a notoungulate, (*c*) *Prodolichotis*, a rodent, (*d*) *Macrauchenia*, an ungulate, and (*e*) *Plaina*, an early relative of the more successful armadillo.

that had been unable to move south now crossed the bridge. Among them were the mastodon, the tapir, primitive camels and a number of carnivores. Simultaneously many of the indigenous South American faunas became extinct; the losers included all but two genera of the many primitive marsupials that had been sheltered there and were evidently incapable of competing with the better-adapted migrants from North America. The traffic was not, however, entirely one-way. Armadillos soon extended their range northward throughout Central America and into the southern U.S. [*see illustration below*].

Simpson has estimated that before North America and South America were united perhaps 29 families of mammals lived in the area south of the Isthmus of Panama and perhaps 27 entirely different families of mammals lived to the north. After the union the faunas of both continents had 22 families of mammals in common. This is a particularly dramatic example of convergence, even though the establishment of the Panama bridge seems to have owed nothing to continental drift.

The mammalian fauna of Africa during the Cenozoic era provides another example of convergence. Up to some 25 million years ago the African fauna had a strong endemic element. The animals ancestral to the living elephants, the manatees and the hyrax were found only there. This suggests that the continent had been isolated for a substantial period.

Early in the Miocene epoch a number of mammals from Eurasia entered Africa by means of one or more land connections. The migration led to the reduction and even the extinction of some of the indigenous African fauna. At the same time the ancestral elephants crossed over to Eurasia and had soon spread around the world. Less than 25 million years later mastodons were waiting at the emerging Isthmus of Panama to enter the last major continental area in the world still barred to the proboscoids. The Miocene bridge-building that allowed a substantial degree of convergence between the faunas of Africa and Eurasia can be attributed to continental drift, specifically a northward movement of the Africa-Arabia plate.

A third example of convergence takes us all the way back to the Cambrian and early Ordovician periods, some 500 to 600 million years ago. The most spectacular marine organisms of the Cambrian, the early anthropods known as trilobites, turn out to be sharply separated into two distinct faunas; the line that divides the two faunal provinces runs through eastern North America and through the British Isles and Scandinavia [*see illustration on page 186*]. Over the next 75 million years or so, during late Ordovician and Silurian times, the two trilobite faunas tend to lose their regional distinctiveness. So do a number of other early marine invertebrates: corals, brachiopods, graptolites and conodonts. So do two groups of early freshwater fishes: primitive jawless ostracoderms belonging to the orders Anaspida and Thelodonti. Freshwater fishes are, of course, prisoners of the continental streams they inhabit. By late Silurian or early Devonian times, some 400 million years ago, only one faunal province existed in the North Atlantic region.

The paleontologist A. W. Grabau noted the difference between the trilobites of adjacent areas many years ago, and he suggested that the sharp delineation might be attributable to the former

CONVERGENCE OF MAMMALS began about the end of the Pliocene epoch, after a land bridge was established between North and South America. One mammal native to South America, the armadillo, migrated northward. So many mammals formerly unknown south of Panama moved to South America, however, that the two continents soon came to have 22 families of mammals in common. At the same time many South American mammals were unable to compete with the immigrants from the north and became extinct.

existence of a deep-ocean barrier. More recently J. Tuzo Wilson of the University of Toronto followed up Grabau's suggestion and proposed that the border between the two faunal provinces marked the closure of a proto-Atlantic ocean that was eliminated by continental drift in Paleozoic times. Since then John F. Dewey of the State University of New York at Albany has developed this concept with considerable success in terms of plate tectonics and subduction: the process in which the leading edge of a drifting plate is destroyed by plunging under another plate [see "Plate Tectonics," by John F. Dewey, beginning on page 34].

Dewey's view, which is based primarily on geological grounds, envisions the loss of an ancient segment of ocean down one or more zones of subduction as an American and a European plate drifted together. Compression and subsequent uplift in the region of subduction formed the Caledonian mountain belt in northwestern Europe and the "old" Appalachian Mountains in eastern North America. The record of faunal convergence in the interval between Cambrian and Devonian times evidently reflects the steady narrowing of the proto-Atlantic, a process that continued for tens of millions of years until most of the ancient ocean was swallowed up. Here we have an example of the fossil record supporting a reconstruction of a drift episode that has been independently inferred from geological data.

The Case of the Urals

The Caledonian belt is not the only ancient mountain range that hints of a collision between two drifting plates. The Urals, the mountains that separate the European and Asiatic parts of the U.S.S.R., have also been interpreted as a collision feature on geological grounds. Let us see whether the fossil record supports this interpretation.

In Devonian times an order of jawless freshwater fishes, cousins to the orders that once flourished on opposite sides of the proto-Atlantic, inhabited the streams of the region that is now the European and Asiatic flanks of the Urals. These fishes are the Heterostraci. Specimens of the order from fossil formations on the European side of the Urals clearly belong to a faunal province that is distinct from the province on the Asiatic slope. It follows that, at least during the 50 million years of Devonian times, a marine barrier separated the two landmasses that now meet in the Urals.

There is negative evidence to suggest that the same barrier persisted during the succeeding interval: the Carboniferous period. Fossil deposits of Carboniferous age in the European U.S.S.R. contain the remains of amphibians and reptiles. In spite of more than a century of prospecting, however, no Carboniferous amphibians or reptiles have been found in the Asiatic U.S.S.R. Yet 50 million years later, at the beginning of the Mesozoic era, the amphibians and reptiles of Asia closely resemble those found elsewhere in the world. Evidently by that time land connections between the two regions were firmly established.

Does this array of fossil evidence, which seems very much like the evidence for a proto-Atlantic, show that the Urals are indeed the products of continental drift? Not necessarily. A further example will show why. Early in the Cenozoic era the mammals of Europe comprised a fauna that was different in many respects from the mammalian fauna of Asia. This was not, however, because two continental plates had drifted apart. Instead the regions were separated at that time by the intrusion of a long arm of shallow sea. As far as land animals are concerned, a shallow sea is quite as effective a barrier as a true deep-ocean basin. How do we know that the separation of the two regions in Paleozoic times was not a similar shallow sea?

Examination of another fossil group, the bottom-dwelling marine invertebrates, should enable us to resolve this question. If a deep-ocean basin separated the European and Asiatic parts of the U.S.S.R. in Paleozoic times, then the wider that ocean was, the more divergent the invertebrate fossils on opposite sides of the Urals should be. Unfortunately, when we apply this test, the existing data prove to be somewhat indecisive. Still, the weight of evidence seems to suggest that the Paleozoic marine gap between Europe and Asia was not very large. This fossil finding suggests that, if continental drift formed the

THREE PAIRS OF MOLLUSKS, each pair belonging to the same genus but to different species, exemplify the divergence that marine animals undergo when a barrier divides a once uniform fauna. The cowries (*top*) are of the genus *Cypraea;* the Caribbean species, *C. zebra* (*left*), has diverged from the Pacific species, *C. cervinetta* (*right*). The same is true of a second gastropod pair (*middle*), both of the genus *Strombus,* and the bivalve pair (*bottom*), both of the genus *Arca.* Divergence began when the Isthmus of Panama arose.

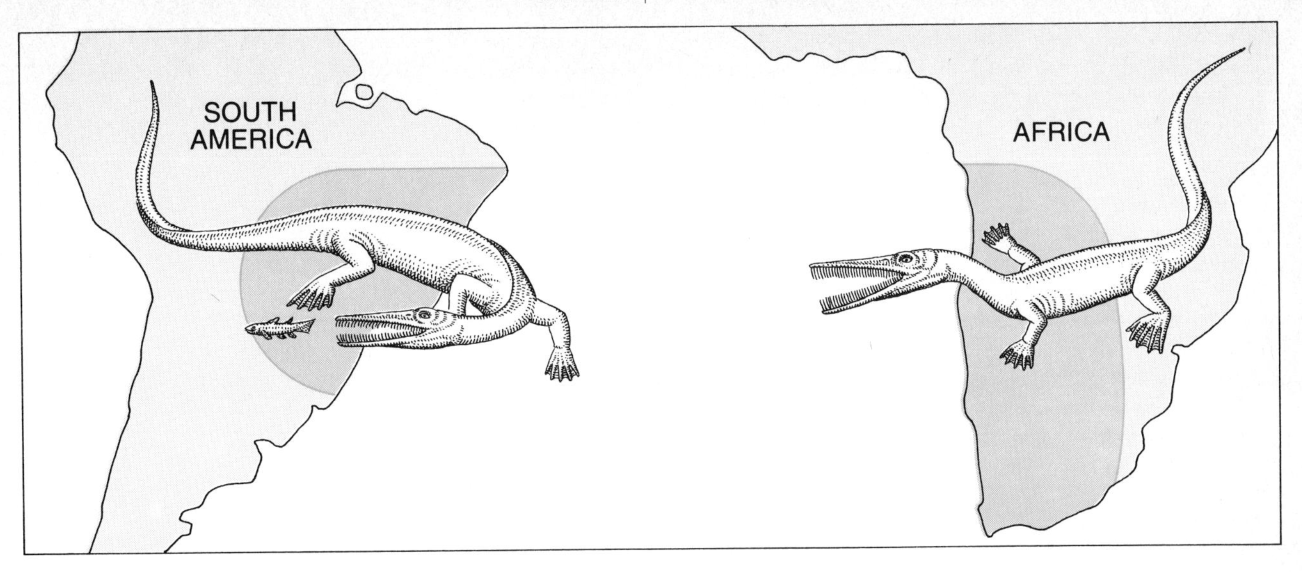

FOSSILS OF MESOSAURUS, a late Paleozoic reptile seen in a restoration here, are found on both sides of the South Atlantic and nowhere else in the world. If *Mesosaurus* was able to swim well enough to cross the ocean, it should have diffused far more widely. Since it did not, this example of "disjunct endemism" suggests that South America and Africa must have been joined at that time.

Urals, the ocean that disappeared as a consequence was a narrow one.

The Pattern of Divergence

We can now consider the second principal pattern of fossil distribution. This is "divergence," which is simply the reverse of convergence. To illustrate the pattern I shall use three examples drawn from the Cenozoic fossil record and one from the Mesozoic. The first is the late Cenozoic rise of the Isthmus of Panama that, as we have seen, allowed the convergence of the land faunas of North America and South America. Simultaneously the rise cut in two a marine region that until then had been inhabited by a homogeneous population of bottom-dwelling invertebrates. The consequence of this genetic isolation was divergence; during the Pleistocene epoch a number of "twin" species, the descendants of identical but isolated genera of marine invertebrates, have evolved independently on opposite sides of the isthmus.

The second example concerns the invertebrate faunas of the Tethys seaway, an ancient span of ocean that in early Cenozoic times reached all the way from the Caribbean, by way of the Mediterranean basin, to the western shores of Indonesia. Throughout this vast region in early Cenozoic times the invertebrate faunas were markedly homogeneous. Beginning about 25 million years ago, however, during the Miocene period, the homogeneity of the Tethys faunas was abruptly disturbed. Thereafter the marine invertebrates of the Indian Ocean differed sharply from those of the Mediterranean. And whereas the faunas of the Mediterranean continued in general to resemble the faunas of the Atlantic, there are indications that some groups, in particular bottom-dwelling foraminifera, also began to show divergence.

The land animals of the Cenozoic era provide the third example. During that era the fossil faunas of the various continents, the mammals in particular, differ in many respects. The most obvious differences are evident in Australia, South America and Africa—the continents of the Southern Hemisphere. During the preceding Mesozoic era, in contrast to this Cenozoic pattern of divergence, the land-dwelling animals (most of them reptiles) were quite homogeneous irrespective of their continued residence. The Cenozoic pattern of divergence is the inevitable consequence of the breakup of Pangaea in late Mesozoic times.

Two groups of late Mesozoic marine invertebrates yield a fourth example of divergence. They flourished during the Cretaceous period, when the breakup of Pangaea was well under way. When one compares certain species of bivalves and foraminifera from fossil strata of the lower Cretaceous in the Caribbean with similar organisms in the Mediterranean region, the two faunas prove to be very much alike. By late Cretaceous times, however, new genera of bivalves and foraminifera have appeared that are unique to one or the other of the formerly homogeneous regions.

This late Cretaceous divergence conforms with the inference, drawn from geological and geophysical data, that the final period of the Mesozoic era witnessed a progressive enlargement of the deep-ocean separation between the Mediterranean and the Caribbean. To the north of the Tropics in the newborn Atlantic a sea-shelf connection between the Old World and the New persisted, but there is little doubt that the diverging marine faunas (including the rudists, a peculiar group of reef-building bivalves) were unable to migrate in any but warm waters and were thus inhibited from crossing from one side of the Atlantic to the other through the cool shelf sea.

The Pattern of Complementarity

I term the third of the patterns of fossil distribution "complementarity" because the faunas in adjacent areas of shore and ocean shelf react to alterations of the environment in a complementary way. For example, when a land connection forms, the newly united land faunas tend to converge and the newly divided marine faunas tend to diverge, whereas the breaking of a land connection gives rise to the opposite effect. The pattern of complementarity is significant because it provides a cross-check on interpretations of the fossil record that depend exclusively on either the marine faunas or the terrestrial ones.

One additional example will suffice to define complementarity. As we have already noted, there is evidence from the Miocene of divergence among the

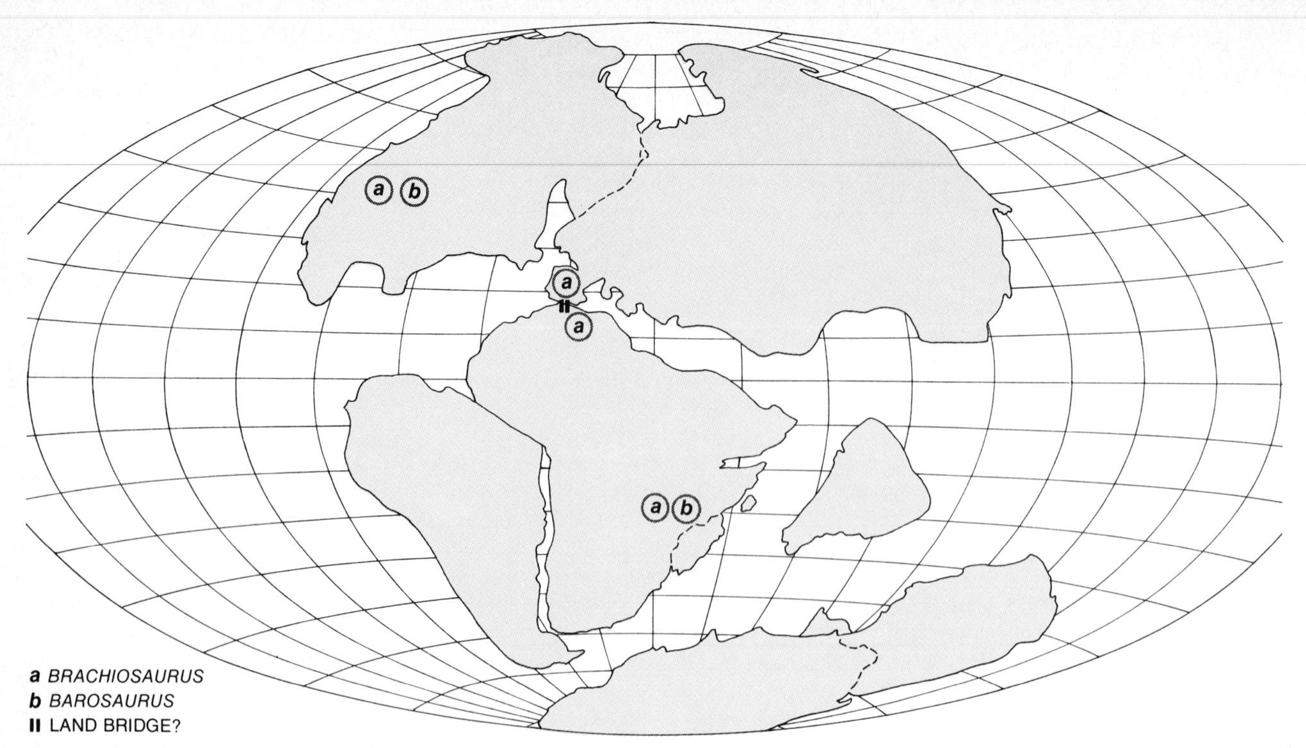

FOSSIL-RICH STRATA that contain the remains of two genera of late Jurassic dinosaurs are located in western North America and in East Africa respectively. Fossils of one genus (*a*) are also found in Portugal and Algeria. At that time the ancient world continent, Pangaea, had broken apart; its components are shown here at the positions calculated by Robert S. Dietz and John C. Holden of the National Oceanic and Atmospheric Administration. Unless a land bridge existed in the late Jurassic where North Africa and Spain nearly touch, the presence of identical fossils on separate continents is another example of disjunct endemism. But where the bridge should be there are oceanic rocks of Jurassic age instead, making the existence of a land connection at that time improbable.

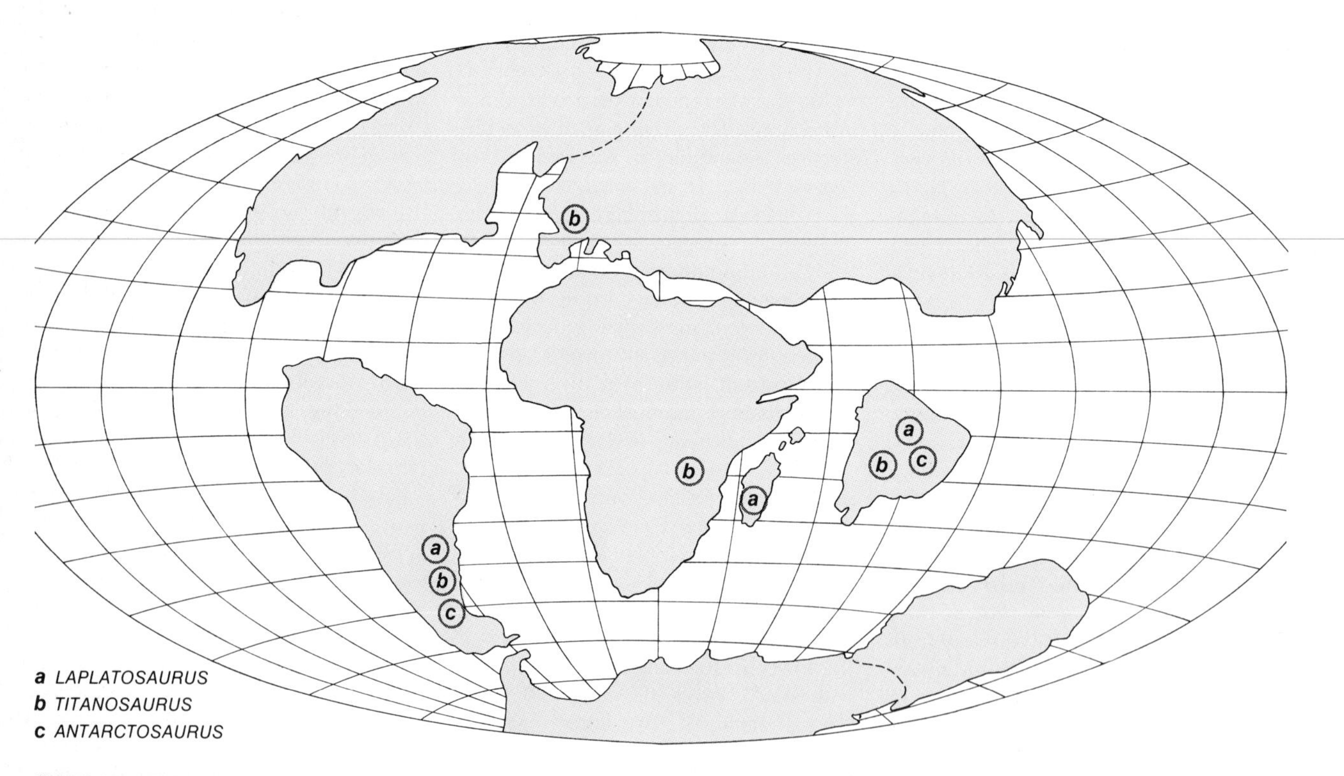

SIMILAR DILEMMA is posed by the disjunct endemism of three genera of late Cretaceous dinosaurs. All three genera are present in fossil formations in South America and in India. One genus (*b*) is also present in both Europe and Africa and another (*a*) is also found in Madagascar. The map shows the various continents in the positions calculated for the late Cretaceous by Dietz and Holden. By that time, students of continental drift generally agree, India had moved well away from the Africa-Arabia plate, continuing a trend that supposedly first isolated the drifting subcontinent some 100 million years earlier. Unless the dinosaur fossil identifications are mistaken, however, the land connection between India and South America could scarcely have been severed at so early a date.

marine invertebrates along the length of the Tethys seaway on the one hand and of convergence among the land mammals of Africa and Eurasia on the other. It appears that this pattern of complementarity in the Miocene fossil record signals the withdrawal of the Tethys seaway from the region of the Near East and Middle East. From the viewpoint of plate tectonics the withdrawal of the Tethys must have been a consequence of the Africa-Arabia plate impinging on Eurasia. The fossil evidence is thus in accord with the geological evidence of compressive, generally north-south earth movements and mountain uplift in southern Spain, Turkey and Iran in Miocene times.

There is also geological evidence of compressive tectonic activity in the Near East and Middle East in late Cretaceous times, some 50 million years earlier. This episode could not, however, have eliminated the Tethys seaway; the fossil record shows no matching record of complementarity in the late Cretaceous. When the story of the faunal migrations between Europe and Africa is finally known in detail, it will surely prove to be a complex narrative. It will probably not call, however, for any fundamental modification of the general picture I have presented here.

Disjunct Endemism

We come now to the fourth and last of the fossil-distribution patterns, which I call "disjunct endemism." The term describes the following situation. A group of fossil organisms is limited in its geographical distribution but nonetheless appears in two or more parts of the world that are now separated by major geographical barriers such as zones of deep ocean. The classic case in point is *Mesosaurus*, a small, snaggle-toothed reptile that lived in late Paleozoic times, some 270 million years ago. Strata that contain fossils of *Mesosaurus* are found only in Brazil and in South Africa [*see illustration on page 193*]. This animal, measuring some 18 inches from snout to tip of tail, was evidently aquatic. It could hardly have been able to swim very far, however, without having diffused into many parts of the world other than Brazil and South Africa. The application of Occam's razor suggests that in late Paleozoic times Brazil and South Africa were contiguous; this bit of fossil evidence in favor of continental drift was first noted many years ago.

In at least two instances that involve the ruling reptiles of the Mesozoic era, the dinosaurs, the fossil evidence seems to require sharp revisions of the drift timetable. The dinosaurs involved are five genera of sauropods, the line that gave rise to such museum favorites as *Brontosaurus* and *Diplodocus*. Two of the genera flourished in late Jurassic times and three in the late Cretaceous, some 70 million years afterward.

The Jurassic dinosaur genera are *Brachiosaurus*, the biggest of all the sauropods, and *Barosaurus*. The remains of dinosaurs of both genera are found in the Morrison Formation of the western U.S. and in the Tendaguru fossil beds of Tanzania; *Brachiosaurus* fossils have also been found in Portugal and in Algeria. These huge herbivores seem to have occupied an ecological niche comparable to an elephant's. Although they might possibly have negotiated swamps, they would have had trouble swimming across a wide river, let alone an ocean. Their presence in both Africa and North America thus points to the existence of a land connection between these areas in late Jurassic times. The necessity for this connection in turn imposes a constraint on the timing of the oceanic separation of the northern and southern halves of Pangaea.

Now, one popular reconstruction of the continental array in Jurassic times, prepared by Robert S. Dietz and John C. Holden of the National Oceanic and Atmospheric Administration, shows a possible place of crossing between Eurasia (which was then linked to North America) and Africa. That is where Spain and North Africa nearly touch; a Jurassic land bridge here would solve the problem of the seemingly disjunct endemism of the two sauropods [*see top illustration, facing page*]. The existence of marine deposits of Jurassic age in the parts of Spain and North Africa that might have been joined, however, seems to rule out any such land bridge.

Another way to be rid of this supposed example of disjunct endemism is to attack the biological classifications involved. One could assert that the African species of these two dinosaur genera are actually quite different from the North American species and attribute this divergence to a break in the land connection between the two regions. By and large, however, the dinosaurs are a group that has been excessively split by taxonomists. An admitted resemblance even at the genus level is likely to signify a close genetic affinity if not an actual ability to interbreed. On balance, then, close similarities between the late Jurassic dinosaur faunas of the Tendaguru beds and of the Morrison Formation pose a problem for the continental-drift timetable that has not yet been satisfactorily resolved.

The disjunct endemism of the three late Cretaceous sauropods—*Titanosaurus*, *Laplatosaurus* and *Antarctosaurus*—presents an even more clear-cut contradiction between the fossil record and the accepted timetable for continental drift. These three dinosaurs are known from fossil formations both in South America and in India. Yet the Indian subcontinent had supposedly become isolated by surrounding ocean some 100 million years before late Cretaceous times, toward the close of the Triassic [*see bottom illustration, facing page*]. Unless the fossil identifications are in error, such an early date for the severance of India from the rest of Gondwanaland is clearly inadmissible. Several similar arguments from the fossil record, which I need not give in detail, can be used to support the persistence of land connections (if only intermittent ones) between the Africa–South America landmass and the Australia-Antarctic landmass until quite late in the Mesozoic period.

I have tried to show here how the fossil record can contribute to plate-tectonic theory by helping to establish a more refined timetable for continental drift. Critics may say that the fossil data are often imprecise and are in addition subject to ambiguous interpretation. That is undoubtedly true in some instances. Moreover, we must beware of oversimplification and of overinterpretation of patchy evidence. We should also acknowledge that endemism cannot be explained in every instance on the basis of continental movements. Nonetheless, the general level of agreement between paleontology and the other earth sciences is sufficiently high to warrant some confidence in our fossil-based conclusions.

As time passes and the reconstruction of past plate movements becomes more precise, I believe interest in several biological questions that arise from the new view of earth history will increase. For one thing, we shall be in a better position to learn more about the rates of evolution among isolated organisms. For another, we shall be able to make findings about the relative ease of migration and colonization under different geographical circumstances. The disjunct distribution of many living animals—for example the lungfishes, the marsupials and the giant flightless birds—will be better understood. Perhaps most intriguing of all, we may acquire new insights into why many groups of plants and animals have become extinct.

16

Plate Tectonics and the History of Life in the Oceans

James W. Valentine and Eldridge M. Moores
April 1974

During the 1960's a conceptual revolution swept the earth sciences. The new world view fundamentally altered long established notions about the permanency of the continents and the ocean basins and provided fresh perceptions of the underlying causes and significance of many major features of the earth's mantle and crust. As a consequence of this revolution it is now generally accepted that the continents have greatly altered their geographic position, their pattern of dispersal and even their size and number. These processes of continental drift, fragmentation and assembly have been going on for at least 700 million years and perhaps for more than two billion years.

Changes of such magnitude in the relative configuration of the continents and the oceans must have had far-reaching effects on the environment, repatterning the world's climate and influencing the composition and distribution of life in the biosphere. These more or less continual changes in the environment must also have had profound effects on the course of evolution and accordingly on the history of life.

Natural selection, the chief mechanism by which evolution proceeds, is a very complex process. Although it is constrained by the machinery of inheritance, natural selection is chiefly an ecological process based on the relation between organisms and their environment. For any species certain heritable variations are favored because they are particularly well suited to survive and to reproduce in their prevailing environment. To answer the question of why any given group of organisms has evolved, then, one needs to understand two main factors. First, it is necessary to know what the ancestral organisms were that formed the "raw material" on which selection worked. And second, one must have some idea of the sequence of environmental conditions that led the ancestral stock to evolve along a particular pathway to a descendant group. Given these factors, one can then infer the organism-environment interactions that gave rise to the evolutionary events. The study of the relations between ancient organisms and their environment is called paleoecology.

The new ideas of continental drift that came into prominence in the 1960's revolve around the theory of plate tectonics. According to this theory, new sea floor and underlying mantle are currently being added to the crust of the earth at spreading centers under deep-sea ridges and in small ocean basins at rates of up to 10 centimeters per year. The sea floor spreads laterally away from these centers and eventually sinks into the earth's interior at subduction zones, which are marked by deep-sea trenches. Volcanoes are created by the consumption process and flank the trenches. The lithosphere, or rocky outer shell of the earth, therefore comprises several major plates that are generated at spreading centers and consumed at subduction zones. Most lithospheric plates bear one continent or more, which passively move with the plate on which they rest. Because the continents are too light to sink into the trenches they remain on the surface. Continents can fragment at new ridges, however, and hence oceans may appear across them. Conversely, continents can be welded together when they collide at the site of a trench. Thus continents may be assembled into supercontinents, fragmented into small continents and generally moved about the earth's surface as passive riders on plates. In tens or hundreds of millions of years entire oceans may be created or destroyed, and the number, size and dispersal pattern of continents may be vastly altered.

The record of such continental fragmentation and reassembly is evident as deformed regions in the earth's mountain belts, particularly those mountain belts that contain the rock formations known as ophiolites. These formations are characterized by a certain sequence of rocks consisting (from bottom to top) of ultramafic rock (a magnesium-rich rock composed mostly of olivine), gabbro (a coarse-grained basaltic rock), volcanic rocks and sedimentary rocks. The major ophiolite belts of the earth are believed to represent preserved fragments of vanished ocean basins [*see illustration on pages 198 and 199*]. The existence of such a belt within a continent (for example the Uralian belt in the U.S.S.R.) is evidence for the former presence there of an ocean basin separating two continental fragments that at some time in the past collided with each other and were welded into the single larger continent. The timing of such events as the opening of ocean basins, the dispersal of continents and the closing of oceans by continental collisions can accordingly be "read" from the geology of a given mountain system.

Of course, the biological environment is constantly being altered as well. For example, the changes in continental configuration will greatly affect the ocean currents, the temperature, the nature of seasonal fluctuations, the distribution of nutrients, the patterns of productivity and many other factors of fundamental importance to living organisms. Therefore evolutionary trends in marine animals must have varied through geologic time in response to the major environmental changes, as natural selection acted to adapt organisms to the new conditions.

It should in principle be possible to detect these changes in the fossil record. Indeed, paleontologists have long recognized that vast changes in the composition, distribution and diversity of marine life are well documented by the fossil record. Now for the first time, however, it is possible to reconstruct the sequence of environmental changes based on the theory of plate tectonics, to determine their environmental consequences and to attempt to correlate them with the sequence of faunal changes that is seen in the fossil record. Such a thorough reconstruction ultimately may explain many of the enigmatic faunal changes known for many years. Even at this early stage paleontologists have succeeded in shedding much new light on a number of major extinctions and diversifications of the past.

As a first step toward understanding the relation between plate tectonics and the history of life it is helpful to investigate the relations that exist today between marine life, the present pattern of continental drift and plate-tectonic theory. The vast majority of marine species (about 90 percent) live on the continental shelves or on shallow-water portions of islands or subsurface "rises" at depths of less than about 200 meters

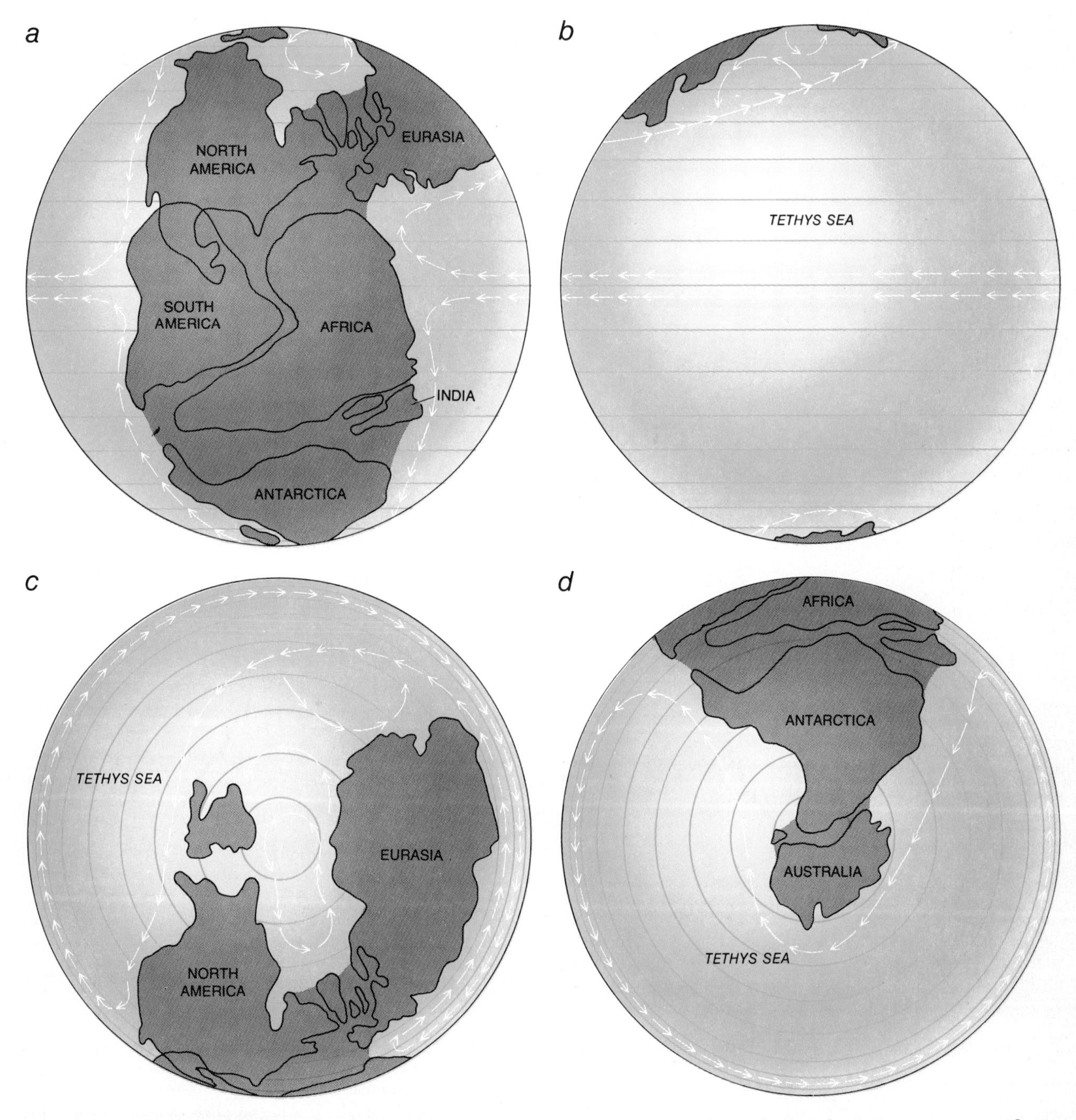

ANCIENT OCEAN CURRENTS in the vicinity of Pangaea, the single "supercontinent" that is believed to have existed near the beginning of the Triassic period some 225 million years ago, are indicated here in two equatorial views (*a*, *b*) and two polar views (*c*, *d*). Owing to a combination of geographic and environmental factors, including the predominantly warm-water currents shown, one would expect the continuous shallow-water margin that surrounded Pangaea to have been populated by comparatively few but widespread species. Such low species diversity combined with low provinciality is precisely what the fossil record indicates.

(660 feet); most of the fossil record also consists of these faunas. Therefore it is the pattern of shallow-water sea-floor animal life that is of particular interest here.

The richest shallow-water faunas are found today at low latitudes in the Tropics, where communities are packed with vast numbers of highly specialized species. Proceeding to higher latitudes, diversity gradually falls; in the Arctic or Antarctic regions less than a tenth as many animals are living as in the Tropics, when comparable regions are considered [*see illustration on pages 200 and 201*]. The diversity gradient correlates well with a gradient in the stability of food supplies; as the seasons become more pronounced, fluctuations in primary productivity become greater. Although this strong latitudinal gradient dominates the earth's overall diversity pattern, there are important longitudinal diversity trends as well. In regions of similar latitude, for example, diversity is lower where there are sharp seasonal changes (such as variations in the surface-current pattern or in the upwelling of cold water) that affect the nutrient supply by causing large fluctuations in productivity.

At any given latitude, therefore, diversity is highest off the shores of small islands or small continents in large oceans, where fluctuations in nutrient supplies are least affected by the seasonal effects of landmasses, whereas diversity is lowest off large continents, particularly when they face small oceans,

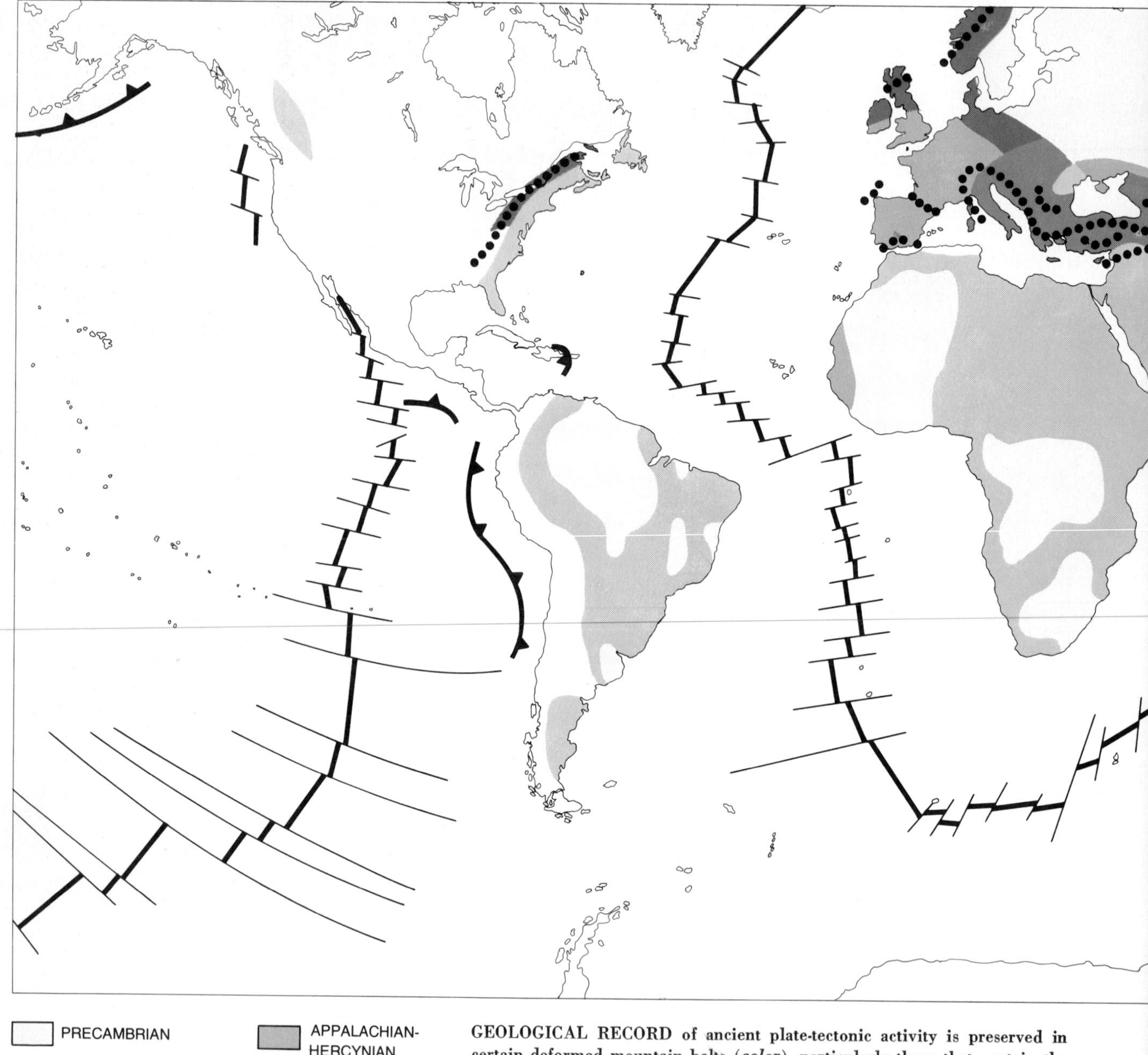

GEOLOGICAL RECORD of ancient plate-tectonic activity is preserved in certain deformed mountain belts (*color*), particularly those that contain the characteristic rock sequences known as ophiolites (*black dots*). The Pan-African–Baikalian belt, for example, is made up of rocks dating from 873 to 450 million years ago and may represent the assembly of all or nearly all the landmasses near the beginning of Phanerozoic time. This supercontinent may then have fragmented into four or more smaller continents, sometime just before and during the Cambrian period. The Caledonian mountain system may represent the collision of two continents at about late Silurian or early De-

where shallow-water seasonal variations are greatest. In short, whereas latitudinal diversity increases generally from high latitudes to low, longitudinal diversity increases generally with distance from large continental landmasses. In both of these trends the increase in diversity is correlated with increasing stability of food resources. The resource-stability pattern depends largely on the shape of the continents and should also be sensitive to the extent of inland seas and to the presence of coastal mountains. Seas lying on continental platforms are particularly important: not only do extensive shallow seas provide much habitat area for shallow-water faunas but also such seas tend to damp seasonal climatic changes and to have an ameliorating influence on the local environment.

Today shallow marine faunas are highly provincial, that is, the species living in different oceans or on opposite sides of the same ocean tend to be quite different. Even along continuous coastlines there are major changes in species composition from place to place that generally correspond to climatic changes. The deep-sea floor, generated at oceanic ridges, forms a significant barrier to the dispersal of shallow-water organisms, and latitudinal climatic changes clearly form other barriers. The present dominantly north-south series of ridges forms a pattern of longitudinally alternating oceans and continents, thereby creating a series of barriers to shallow-water marine organisms. The steep latitudinal climatic gradient, on the other hand, creates chains of provinces along north-south coastlines. As a result the marine faunas today are partitioned into more than 30 provinces, among which there is in general only a low percentage of common species [*see illustration on pages 202 and 203*]. It is estimated that the shallow-water marine fauna represents more than 10 times as many species today as would be present in a world with only a single province, even a highly diverse one.

The volcanic arcs that appear over subduction zones form fairly continuous island chains and provide excellent dispersal routes. When long island chains are arranged in an east-west pattern so as to lie within the same climatic zone, they are inhabited by wide-ranging faunas that are highly diverse for their latitude. Indeed, the widest ranging marine province, and also by far the most diverse, is the Indo-Pacific province, which is based on island arcs in its central regions. The faunal life of this province spills from these arcs onto tropical continental shelves in the west (India and East Africa) and also onto tropical intraplate volcanoes (the Polynesian and Micronesian islands) that are reasonably close to them. This vast tropical biota is cut off from the western American mainland by the East Pacific Barrier, a zoogeographic obstruction formed by a spreading ridge.

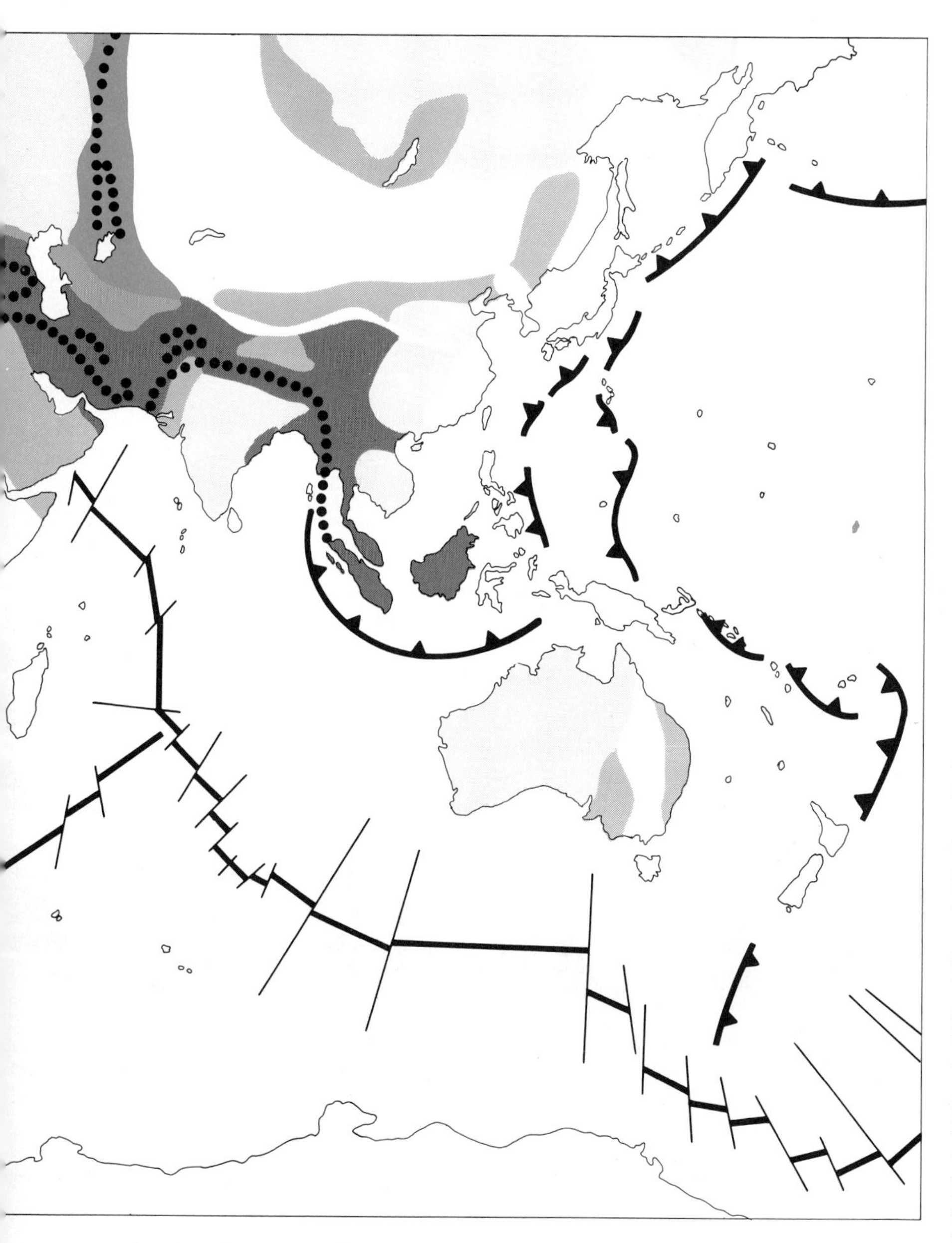

vonian time (approximately 400 million years ago). The Appalachian-Hercynian system may represent a two-continent collision during the late Carboniferous period (300 million years ago). The Uralian mountains may represent a similar collision at about Permo-Triassic time (220 million years ago). The Cordilleran-Tethyan system represents regions of Mesozoic mountain-building and includes the continental collisions that resulted in the Alpine-Himalayan mountain system. The ophiolite belts shown are the preserved remnants of ocean floor exposed in the mountain systems in question. Spreading ridges such as the Mid-Atlantic Ridge are indicated by heavy lines cut by lighter lines, which correspond to transform faults. Subduction zones are marked by heavy black curved lines with triangles.

Since current patterns of marine provinciality and diversity fit closely with the present oceanic and continental geography and the resulting environmental patterns, one would expect ancient provinces and ancient diversity patterns also to fit past geographies. One of the best-established of ancient geographies is the one that existed near the beginning of the Triassic period, about 225 million years ago. The continents were then assembled into a single supercontinent named Pangaea, which must have had a continuous shallow-water margin running all the way around it, with no major physical barriers to the dispersal of shallow-water marine animals [*see illustration on page 197*]. Therefore provinciality must have been low compared with today, and it must have been attributable entirely to climatic effects. It is likely that the marine climate was quite mild and that even in high latitudes water temperatures were much warmer than they are today. As a result climatic provinciality must have been greatly reduced also. Furthermore, the seas at that time were largely confined to the ocean basins and did not extend significantly over the continental shelves. Thus the habitat area for shallow-water marine organisms was greatly reduced, first by the diminution of coastline that accompanies the creation of a supercontinent from smaller continents, and second by the general withdrawal of seas from continental platforms. The reduced habitat area would make for low species

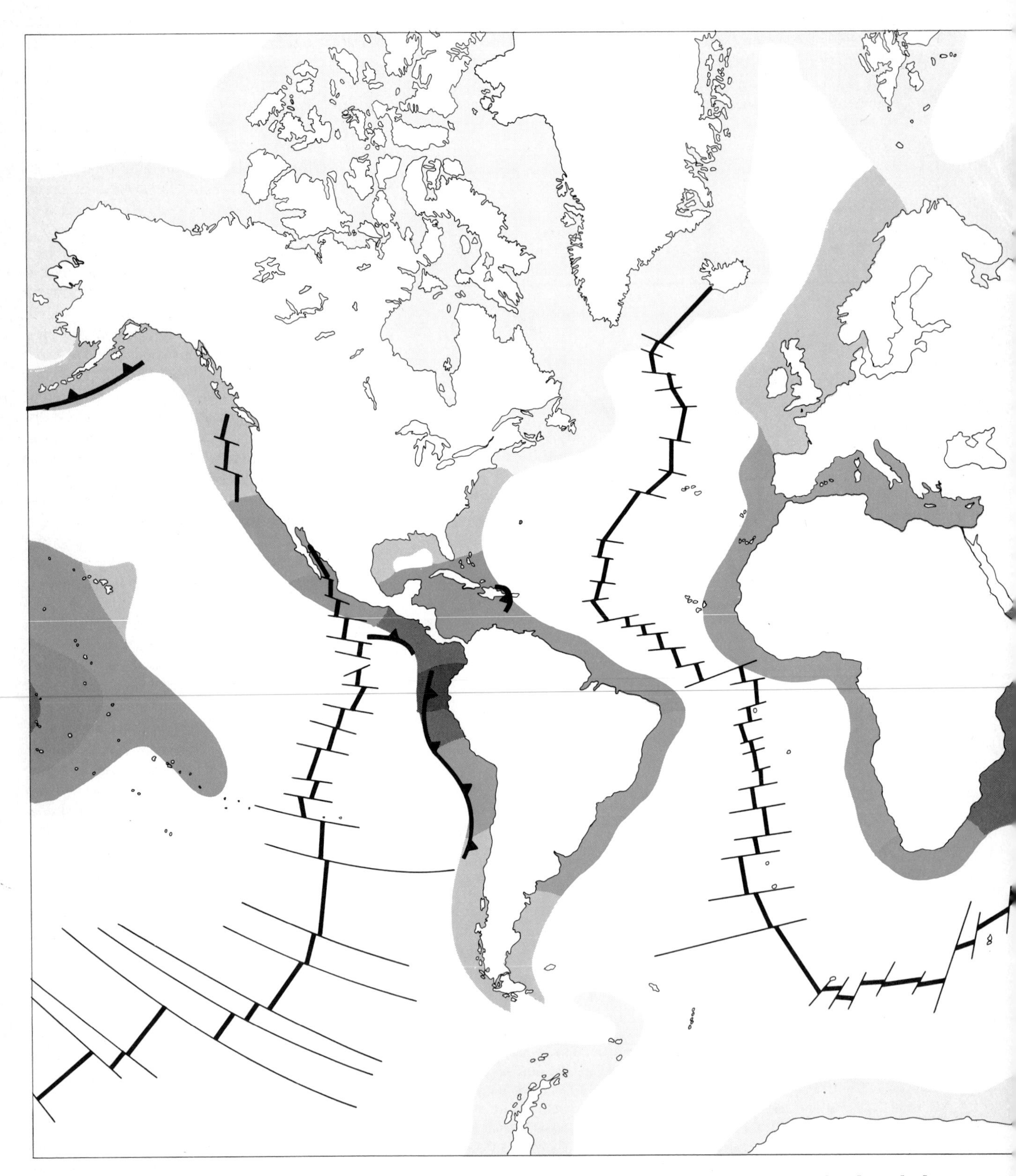

RELATIVE DIVERSITY of shallow-water, bottom-dwelling species in the present oceans is suggested by the colored patterns in this world map. The diversity classes are not based on absolute counts but are inferred from the diversity patterns of the best-

diversity. Finally, the extreme emergence of such a supercontinent would provide unstable nearshore conditions, with the result that food resources would have been very unstable compared with those of today. All these factors tend to reduce species diversity; hence one would expect to find that Triassic biotas were widespread and were made up of comparatively few species. That is precisely what the fossil record indicates.

Prior to the Triassic period, during the late Paleozoic, diversity appears to have been much higher [*see top illustration on page 205*]. It was sharply reduced again near the close of the Permian period during a vast wave of extinction that on balance is the most severe known to have been suffered by the marine fauna. The late Paleozoic species that were the more elaborately adapted specialists became virtually extinct, whereas the surviving descendants tended to have simple skeletons. A high proportion of these survivors appear to have been detritus feeders or suspension feeders that harvested the water layers just above the sea floor. These successful types seem to be ecologically similar to the populations found today in unstable environments, for instance in high latitudes; the unsuccessful specialists, on the other hand, seem ecologically similar to the populations found in stable environments, for instance in the Tropics. Thus the extinctions appear to have been caused by the reduced potential for diversity of the shallow seas, a trend associated with less provinciality, less habitat area and less stable environmental conditions.

In the period following the great extinction, as Pangaea broke up and the resulting continents themselves gradually fragmented and migrated to their present positions, provinciality increased, communities in stabilized regions became filled with numerous specialized animals and the overall diversity of species in the world ocean rose to unprecedented heights, even though occasional waves of extinctions interrupted this long-range trend.

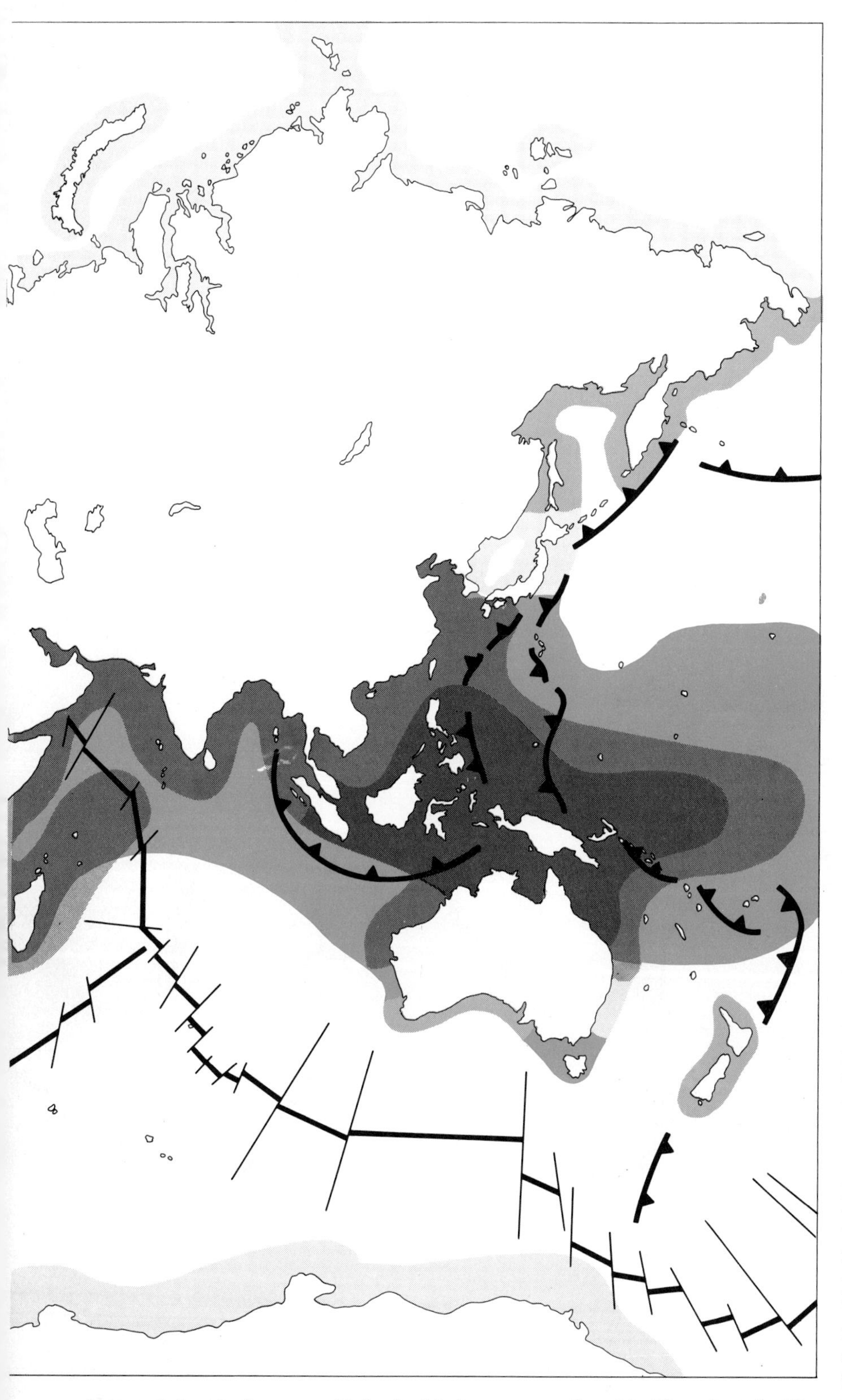

known skeletonized groups, chiefly the bivalves, gastropods, echinoids and corals. The highest class (*darkest color*) is about 20 times as diverse as the lowest (*lightest color*).

There is another time in the past besides the early Triassic period when low provinciality and low diversity were coupled with the presence of a high proportion of detritus feeders and near-bottom suspension feeders. That is in the late Precambrian and Cambrian periods, when a widespread, soft-bodied fauna of low diversity gave way to a slightly provincialized, skeletonized fauna of somewhat higher diversity. It seems likely that the late Precambrian environment was quite unstable and that there may well have been a supercontinent in existence, or at least that the continents then were collected into a more compact assemblage than at present. In the late Precambrian period one finds the first unequivocal records of invertebrate life, including burrowing forms that were probably coelomic, or hollow-bodied, worms. In the Cambrian four continents may have existed although they were not arranged in the present pattern. During the Cambrian a skeletonized fauna appears that is at first almost entirely surface-dwelling and that includes chiefly detritus-feeding and suspension-feeding forms, probably with some browsers.

It seems possible, therefore, that the late Precambrian species were adapted to highly unstable conditions and became diversified chiefly as a bottom-living, detritus-feeding assemblage. The coelomic body cavity, evidently a primitive adaptation for burrowing, was developed and diversified into a variety of forms, perhaps as many as five basic ones: highly segmented worms that lived under the ocean floor and were detritus feeders; slightly segmented worms that lived attached to the ocean floor and were suspension feeders; slightly segmented worms that lived attached to the ocean floor and were detritus feeders; "pseudosegmented" worms that lived on the ocean floor and were detritus feeders or browsers, and nonsegmented worms

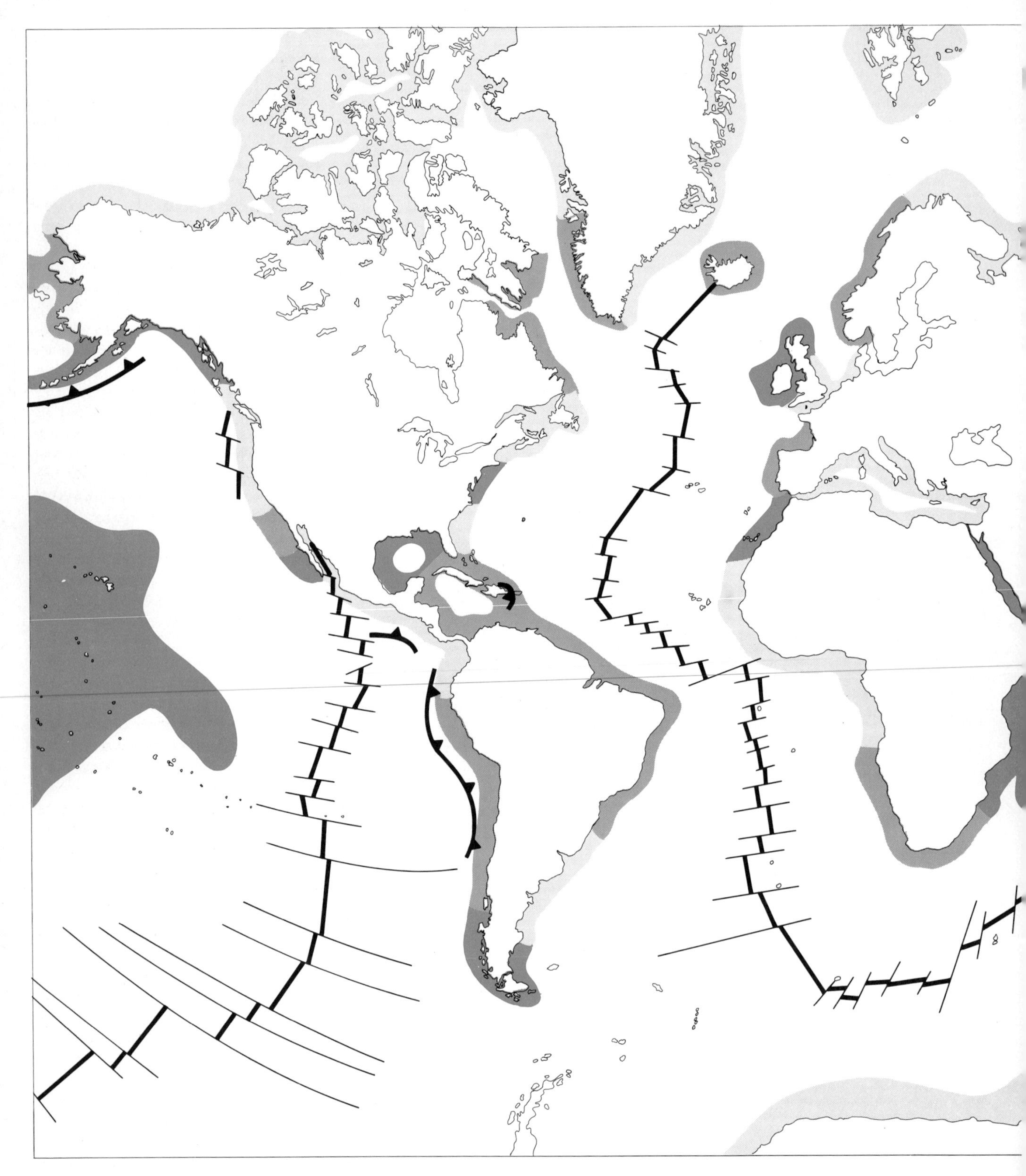

PRINCIPAL SHALLOW-WATER MARINE PROVINCES at present are indicated by the colored areas. The dominant north-south chains of provinces along the continental coastlines are created by the present high latitudinal gradient in ocean temperature

that lived under the ocean floor and fed by means of an "introvert." In addition to these coelomates there were a number of coelenterate stocks (such as corals, sea anemones and jellyfishes) and probably also flatworms and other noncoelomate worms.

From the chiefly wormlike coelomate stocks higher forms of animal life have originated; many of them appear in the Cambrian period, when they evidently first became organized into the groups that characterize them today. Animals with skeletons appeared in the fossil record at that time. Presumably the invasion of the sea-floor surface by coelomates and the origin of numerous skeletonized species accompanied a general amelioration of environmental conditions as the continents became dispersed; the skeletons themselves can be viewed as adaptations required for worms to lead various modes of life on the surface of the sea floor rather than under it. The sudden appearance of skeletons in the fossil record therefore is associated with a generalized elaboration of the bottom-dwelling members of the marine ecosystem. Later, free-swimming and underground lineages developed from the skeletonized ocean-floor dwellers, with the result that skeletons became general in all marine environments.

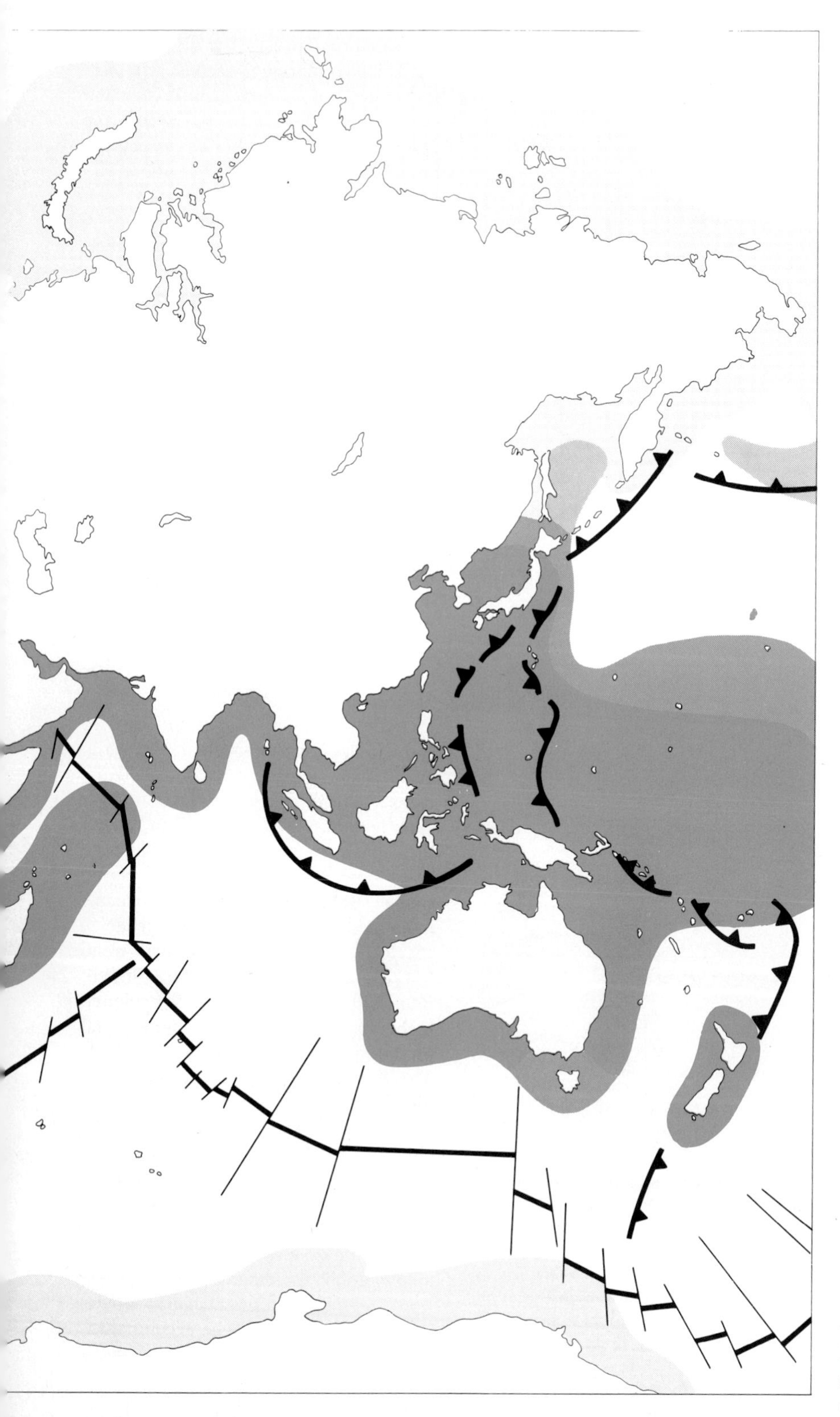

and by the undersea barriers formed by spreading ridges. The vast Indo-Pacific province (*darkest color*) spills out onto scattered islands as indicated. There are 31 provinces shown.

The correlation of major events in the history of life with major environmental changes inferred from plate-tectonic processes is certainly striking. Even though details of the interpretation are still provisional, it seems certain that further work on this relation will prove fruitful. Indeed, the ability of geologists to determine past continental geographies should provide the basis for reconstructing the historical sequence of global environmental conditions for the first time. That sequence can then be compared with the sequence of organisms revealed in the fossil record. The following tentative account of such a comparison, on the broadest scale and without detail, will indicate the kind of history that is emerging; it is based on the examples reviewed above and on similar considerations.

Before about 700 million years ago bottom-dwelling, multicellular animals had developed that somewhat resembled flatworms. As yet no fossil evidence for their evolutionary pathways exists, but evidence from embryology and comparative anatomy suggests that they arose from swimming forms, possibly larval jellyfish, which in turn evolved from primitive single-celled animals.

Approximately 700 million years ago, perhaps in response to the onset of fluctuating environmental conditions brought about by continental clustering, a true coelomic body cavity was evolved to act as a hydrostatic skeleton in roundworms; this adaptation allowed burrowing in soft sea floors and led to the diversification of a host of worm architectures as that mode of life was explored. Burrows of this type are still preserved in some late Precambrian rocks. As the environment later became more stable, several of the worm lineages evolved more varied modes of life. The changes

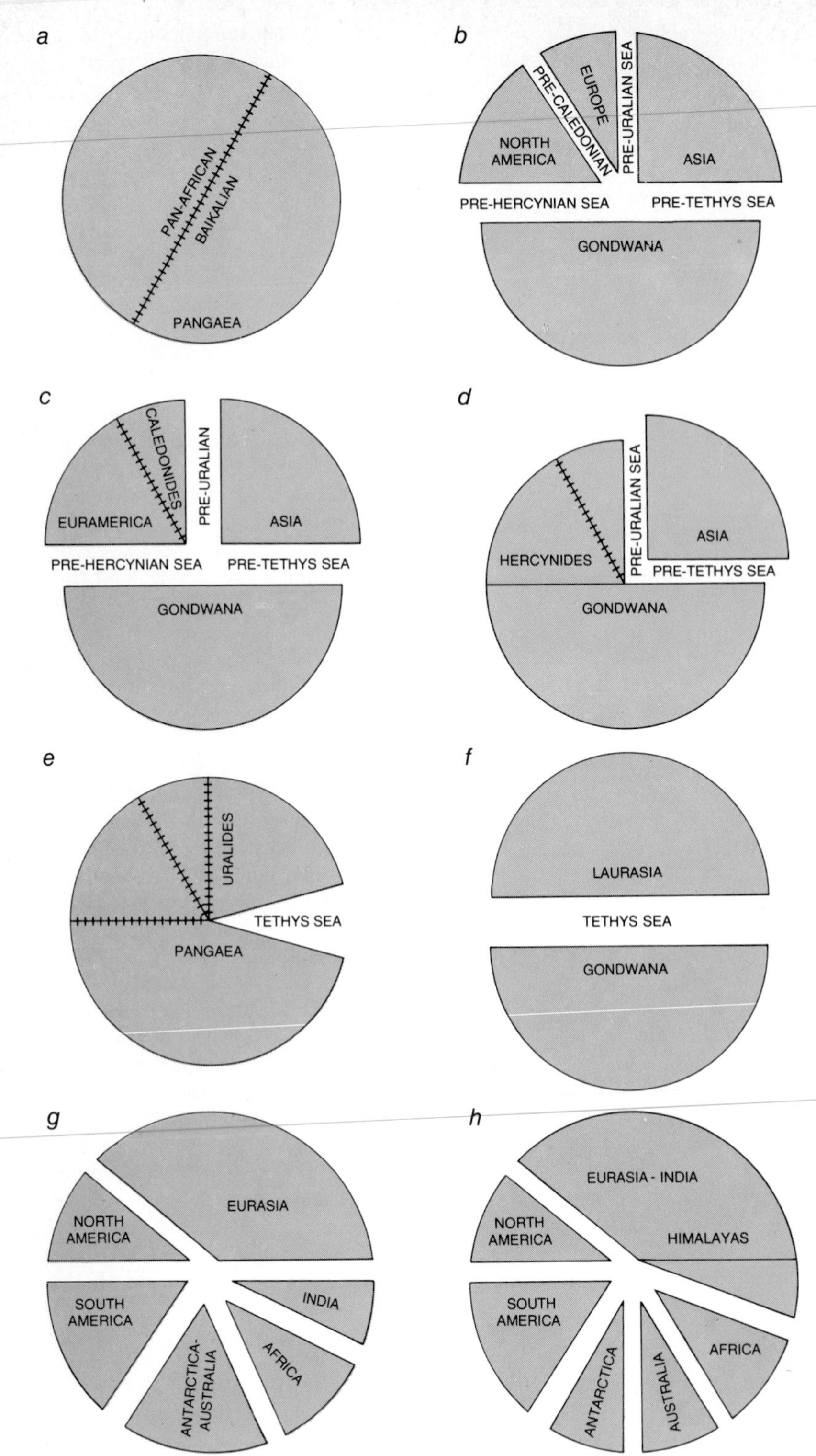

SIMPLIFIED DIAGRAMS are employed to suggest the relative configuration of the continents and the oceans during the past 700 million years. The late Precambrian supercontinent (*a*), which probably existed some 700 million years ago, may have been formed from previously separate continents. The Cambrian world (*b*) of about 570 million years ago consisted of four continents. The Devonian period (*c*) of about 390 million years ago was distinguished by three continents following the collapse of the pre-Caledonian Ocean and the collision of ancient Europe and North America. In the late Carboniferous period (*d*), about 300 million years ago, Euramerica became welded to Gondwana along the Hercynian belt. In the late Permian period (*e*), about 225 million years ago, Asia was welded to the remaining continents along the Uralian belt to form Pangaea. In early Mesozoic time (*f*), about 190 million years ago, Laurasia and Gondwana were more or less separate. In the late Cretaceous period (*g*), about 70 million years ago, Gondwana was highly fragmented and Laurasia partially so. The present continental pattern (*h*) shows India welded to Eurasia.

in body plan necessary to adapt to such a life commonly involved the development of a skeleton. There were evidently three or four main types of worms that are represented by skeletonized descendants today. One type was highly segmented like earthworms, and presumably burrowed incessantly for detrital food; these were represented in the Cambrian period by the trilobites and related species. A second type was segmented into two or three coelomic compartments and burrowed weakly for domicile, afterward filtering suspended food from the seawater just above the ocean floor; these evolved into such forms as brachiopods and bryozoans. A third type consisted of long-bodied creepers with a series of internal organs but without true segmentation; from these the classes of mollusks (such as snails, clams and cephalopods) have descended. Probably a fourth type consisted of unsegmented burrowers that fed on surface detritus and gave rise to the modern sipunculid worms. These may also have given rise to the echinoderms (which include the sea cucumber and the spiny sea urchin), and eventually to the chordates and to man. Although the lines of descent are still uncertain among these primitive and poorly known groups, the adaptive steps are becoming clearer.

The major Cambrian radiation of the underground species into sea-floor surface habitats established the basic evolutionary lineages and occupied the major marine environments. Further evolutionary episodes tended to modify these basic animals into more elaborate structures. After the Cambrian period shallow-water marine animals became more highly specialized and richer in species, suggesting a continued trend toward resource stabilization. Suspension feeders proliferated and exploited higher parts of the water column, and predators also became more diversified. This trend seems to have reached a peak (or perhaps a plateau) in the Devonian period, some 375 million years ago. The characteristic Paleozoic fauna was finally swept away during the reduction in diversity that accompanied the great Permian-Triassic extinctions. Thus the rise of the Paleozoic fauna accompanied an amelioration in environmental conditions and increased provinciality, whereas the decline of the fauna accompanied a reestablishment of severe, unstable conditions and decreased provinciality. The subsequent breakup and dispersal of the continents has led to the present biosphere.

Today we live in a highly diverse world, probably harboring as many species as have ever lived at any time, associated in a rich variety of communities and a large number of provinces, probably the richest and largest ever to have existed at one time. We have been furnished with an enviably diverse and interesting biosphere; it would be a tragedy if we were to so perturb the environment as to return the biosphere to a low-diversity state, with the concomitant extinction of vast arrays of species. Of course, natural processes might eventually recoup the lost diversity, if we waited patiently for perhaps a few tens of millions of years. Alternatively we can work to preserve the environment in its present state and therefore to preserve the richness and variety of nature.

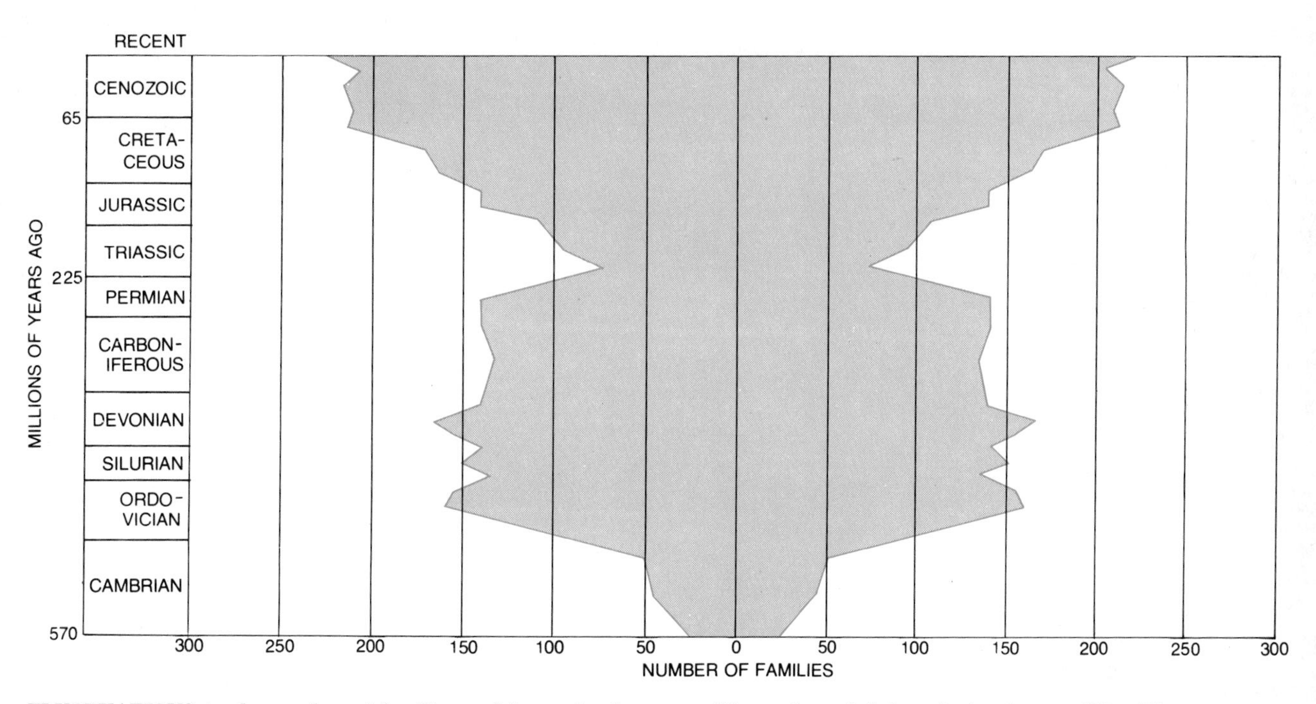

FLUCTUATIONS in the number of families, and hence in the level of diversity, of well-skeletonized invertebrates living on the world's continental shelves during the past 570 million years are plotted by geologic epoch in this graph. Time proceeds upward.

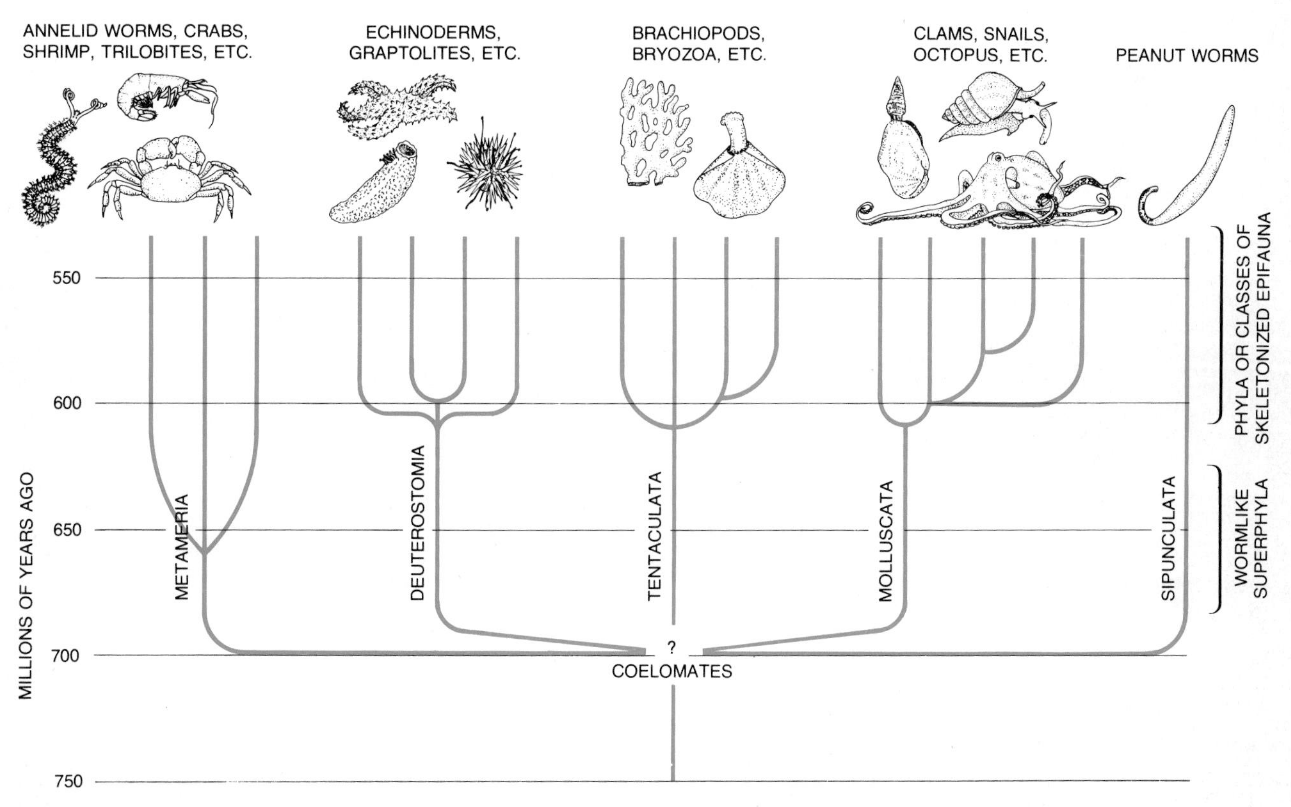

PHYLOGENETIC MODEL of the evolution of coelomate, or hollow-bodied, marine organisms is based on inferred adaptive pathways. The late Precambrian lineages were chiefly worms, which gave rise to epifaunal (bottom-dwelling), skeletonized phyla during the Cambrian period. The organisms depicted in the drawings at top are modern descendants of the major Cambrian lineages.

Plate Tectonics and Mineral Resources

Peter A. Rona
July 1973

A scientific revolution is in progress that over the past five years has already changed our understanding of the earth as profoundly as the Copernican revolution changed medieval man's understanding of the solar system. The Copernican revolution entailed a fundamental change in man's world view from an earth-centered planetary system to a sun-centered one and led to the development of modern astronomy and the exploration of space. The current scientific revolution entails a fundamental change in man's world view from a static earth to a dynamic one and presages comparable benefits. Some of the benefits may even be economic. The implications of the new global tectonics for mineral resources, particularly the mineral resources of the ocean floor, are only now beginning to emerge.

At present the only undersea mineral resources that certainly have economic value are the vast oil and gas reserves found under many continental shelves and continental slopes, gravel, sand, shells and placer deposits on the continental shelves, various other minerals buried under the continental shelves in specific relation to adjacent continental deposits, and fields of manganese nodules that blanket large areas of the deep-sea floor. Even this limited knowledge is remarkable in the light of the difficulty that was encountered in obtaining it. Consider how much we would know about the mineral deposits of the continents if our sampling procedure were limited to flying in a balloon at an altitude of up to six miles and suspending a bucket at the end of a cable to scrape up loose rocks from the surface of the land. What are the chances that we would find the major known ore bodies, which generally underlie areas of less than a square mile?

Yet this farfetched analogy accurately describes man's present capacity for sampling the sediments and rocks of the ocean bottom, utilizing a variety of coring, drilling and dredging devices lowered from ships through the water column over an area twice as large as that of the continents. Averaged over the world's oceans, the distribution of ocean-floor rocks that have been sampled to date is only about three dredge hauls per million square kilometers!

In recent years every major discovery of a hidden mineral resource has been anticipated by a theoretical vision. For example, once field geologists realized that there was a definite association between the type of sedimentary structure termed an anticline and accumulations of oil, they knew where to drill and the rate of discovery of oil deposits accelerated accordingly. In the same way the right conceptual framework can be used to extend man's limited direct knowledge of resources of the ocean basin toward a realistic appraisal of their potential. The test of the value of such a conceptual framework is how well it explains what one sees and predicts what one does not see.

The old conceptual framework of a static earth held that the continents and ocean basins were permanent features that had existed in their present form since early in the 4.5-billion-year history of the earth. Only the most accessible continental mineral deposits were discovered, largely by trial and error, with little understanding of why or where they existed. The recent change to a conceptual framework based on a dynamic-earth model, in which continents are constantly moving and ocean basins are opening and closing, is leading toward a better understanding of the global distribution of mineral deposits in both space and time.

The basis of the new conceptual framework is the theory of plate tectonics, the essentials of which have already been reported in SCIENTIFIC AMERICAN [see "Plate Tectonics," by John F. Dewey, page 34]. "Tectonics" is a geological term pertaining to earth movements. The movements in question involve the lithosphere, the rigid outer shell of the earth, which is of the order of 60 miles thick. The lithosphere, which behaves as if it were floating on an underlying plastic layer, the asthenosphere, is segmented into about six primary slabs, or plates, each of which may encompass a continent and part of an adjacent ocean basin [*see top illustration on page 211*].

The boundaries of the lithospheric plates are delineated by narrow earthquake zones where the plates are moving with respect to each other. Three types of boundary are recognized. One type, called a convergent plate boundary, is where two adjacent plates move together and collide or where one plate plunges downward under the other plate and is absorbed into the interior of the earth.

The second type of boundary, called a divergent plate boundary, is where two adjacent plates move apart because new lithosphere is added to each plate by the process of sea-floor spreading. The new lithosphere, which moves more or less symmetrically to each side of the divergent plate boundary, acts like a conveyor belt, carrying the continents apart in the motion that has become known as continental drift. The dual existence of convergent boundaries where lithosphere is destroyed and divergent boundaries where lithosphere is created implies that the diameter of the earth is not changing radically.

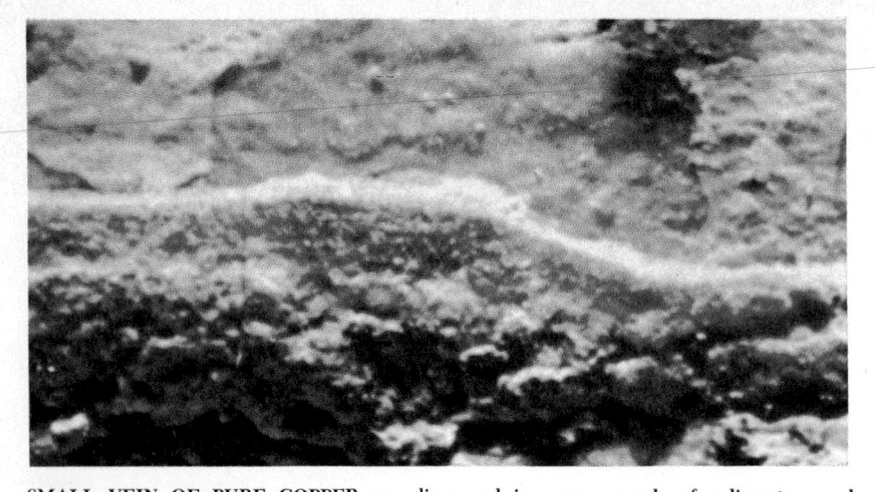

SMALL VEIN OF PURE COPPER was discovered in a core sample of sedimentary rock obtained by the Deep Sea Drilling Project some 350 miles southeast of New York City. The copper vein, the horizontal reddish structure in this longitudinal section of a piece of the original core, is about half an inch long. It was found in sediment about 65 feet above the volcanic basement rocks under the lower continental rise at a water depth of 17,000 feet.

METAL-RICH CORE, collected from the Atlantis II Deep, one of the hot-brine pools located along the axial valley of the Red Sea at a depth of about 6,600 feet below sea level, represents the most concentrated submarine metallic sulfide deposits known. The muddy sediments containing the sulfide minerals fill the Red Sea basins to a thickness estimated at between 65 and 330 feet. The deposits are saturated with (and overlain by) salty brines considered to be the hydrothermal solutions from which the sulfide minerals were precipitated. The photograph was made by David A. Ross of the Woods Hole Oceanographic Institution.

The third type of tectonic-plate boundary is the parallel plate boundary, where two adjacent plates move edge to edge along their common interface.

Hydrothermal mineral deposits, that is, mineral deposits formed by precipitation from solutions, constitute a major part of our useful metallic ores on the continents. Economically the most important types of hydrothermal deposit are the sulfides, in which various metals combine with sulfur to precipitate from the hydrothermal solution. About a year ago Frederick Sawkins, a geologist at the University of Minnesota, pointed out that most of the sulfide deposits of the world are located along present or former convergent plate boundaries where an oceanic lithospheric plate plunges under the margin of a continent (including the continental shelf) or under a chain of volcanic islands. The processes that concentrate the sulfide deposits along convergent plate boundaries, which are at present only partly understood, involve mineralizing solutions that emanate from the plunging lithospheric plate, which melts as it is absorbed into the interior of the earth.

Metallic sulfide deposits along convergent plate boundaries include the Kuroko deposits of Japan, the sulfide ore bodies of the Philippines and the deposits extending along the mountain belts of western North America and South America (the Coast Ranges, the Rockies and the Andes) and from the eastern Mediterranean region to Pakistan. Gold-bearing deposits are not sulfides but often accompany sulfide minerals. The majority of gold deposits in Alaska, Canada, the southeastern U.S., California, Venezuela, Brazil, West Africa, Rhodesia, southern India and

TROODOS MASSIF on the island of Cyprus, the site of economically important mineral deposits that originated at a divergent tectonic-plate boundary, stands out clearly as the dark-colored mountainous region in the middle of the satellite photograph on the opposite page. The photograph was made recently from an altitude of nearly 600 miles by a multispectral camera system on board the first Earth Resources Technology Satellite (ERTS I). Region is believed to be a slice of oceanic lithosphere that was formed by the process of sea-floor spreading from a submerged mid-oceanic ridge and was subsequently thrust upward.

southeastern and western Australia occur in rocks that can be associated with former convergent plate boundaries.

Divergent plate boundaries are formed by the spreading of lithospheric plates in the central portions of ocean basins. The Red Sea and the island of Cyprus in the Mediterranean Sea provide important clues to the potential of metallic sulfide deposits at divergent plate boundaries.

The Red Sea, the product of a divergent plate boundary developing between the African plate and the Eurasian plate, provides an accessible natural laboratory for the study of mineral processes associated with divergent plate boundaries. About five years ago the richest submarine metallic sulfide deposits known were found in three rather small basins along the center of the Red Sea at a depth of about 6,600 feet below sea level. The sulfide minerals are disseminated in sediments that fill the basins to a thickness estimated at between 65 and 330 feet. The top 30 feet or so of sediment, which has been explored by coring the largest of the basins, has a total dry weight of about 80 million tons, with average metal contents of 29 percent iron, 3.4 percent zinc, 1.3 percent copper, .1 percent lead, .005 percent silver and .00005 percent gold. The deposits are saturated with (and overlain by) salty brines carrying the same metals in solution as those present in the sulfide deposits. The salty brines are considered to be the hydrothermal solutions from which the sulfide minerals are precipitated. It remains controversial whether the brines are being charged with minerals from volcanic sources under the Red Sea or from sediments with high copper, vanadium and zinc contents adjacent to the basins where the metallic sulfide deposits are found [see "The Red Sea Hot Brines," by Egon T. Degens and David A. Ross; SCIENTIFIC AMERICAN, April, 1970].

The Red Sea represents the earliest stage in the growth of an ocean basin: the stage where a divergent plate boundary rifts a continent in two. The most advanced growth stage of a divergent plate boundary is the mid-oceanic-ridge system, a 47,000-mile undersea mountain chain that extends through all the major ocean basins and girdles the globe. The mid-oceanic-ridge system has not been adequately sampled to determine whether or not concentrations of metallic sulfides comparable to the Red Sea deposits are present at sites along its crest or in basins in its flanks. Measurements of the distribution of heat emanating from mid-oceanic ridges and of the chemical alteration of ridge rocks indicate that seawater forms a hydrothermal solution by penetrating fissures, dissolving minerals from rocks underlying the ridges and precipitating those minerals in concentrated deposits.

A limited amount of sampling indicates that hydrothermal processes are actively concentrating metals from volcanic sources underlying mid-oceanic ridges. Sediments on active mid-oceanic ridges are generally enriched in iron, manganese, copper, nickel, lead, chromium, cobalt, uranium and mercury, with trace amounts of vanadium, cadmium and bismuth. The concentrations typical of sediments covering widespread areas on mid-oceanic ridges are not economic, but much higher concentrations exist locally.

Metallic sulfides are found in rocks dredged from the Indian Ocean Ridge. In addition small veins of pure copper have been recovered by the Deep Sea Drilling Project at several sites. At the crest of the Ninety East Ridge near the Equator in the Indian Ocean, for example, veins of copper are found in volcanic rocks overlain by 1,440 feet of sediment at a water depth of 7,380 feet. Some 350 miles southeast of New York City a small vein of pure copper and clusters of copper crystals have been discovered in sediment about 65 feet above the volcanic basement rocks under the lower continental rise at a water depth of 17,000 feet [*see top illustration on page 208*].

A specimen of manganese 1.7 inches thick recently dredged from a water depth of about 12,000 feet in the median valley of the Mid-Atlantic Ridge by the Trans-Atlantic Geotraverse of the National Oceanic and Atmospheric Administration has particular significance. The composition, form and thickness of this manganese sample, which accumulated at a rate about 100 times faster than the manganese in nodules, indicates a hydrothermal origin and demonstrates that hydrothermal mineral deposits are actively accumulating at certain divergent plate boundaries in ocean basins. Because the sea floor is supposed to originate by spreading from mid-oceanic ridges, a mineral deposit on a mid-oceanic ridge would be expected to extend in a linear zone from the ridge across the ocean basin to the adjacent continental margin if the depositional process is a

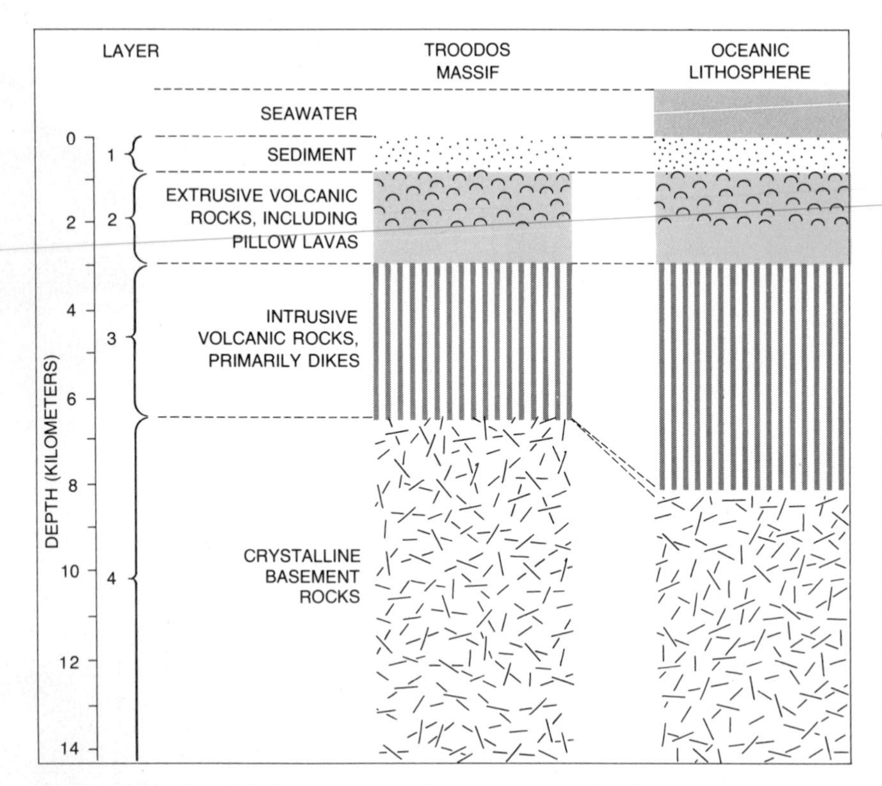

CLOSE CORRESPONDENCE between the layered sequence of rocks in the Troodos Massif (*left*) and that of the oceanic lithosphere (*right*) is evident from this comparison. The geological structure of the Troodos Massif was determined directly from rock outcrops; the structure of the oceanic lithosphere was determined indirectly by seismic-refraction techniques. The sulfide ore bodies of the Troodos Massif are in the upper portion of layer made up of extrusive volcanic rocks. Pillow shapes form when volcanic lava cools on the sea floor.

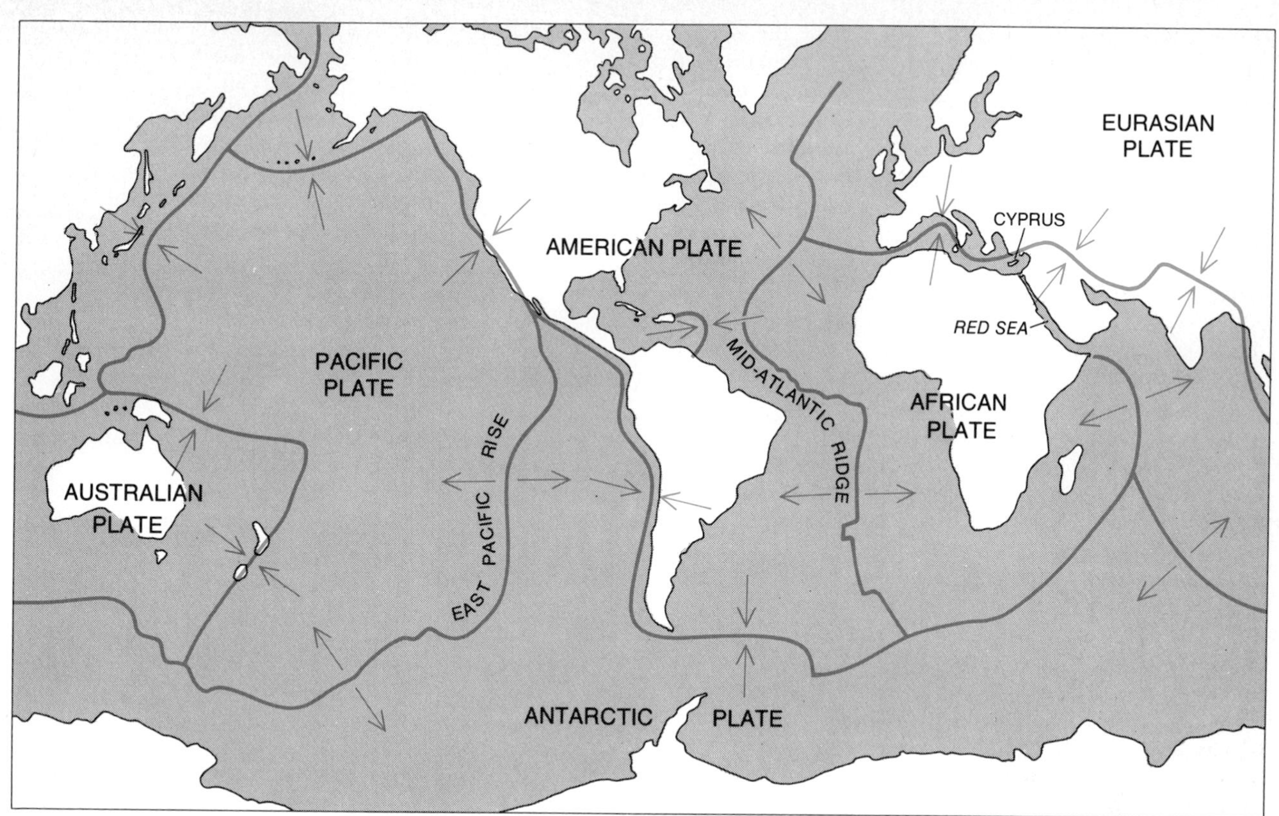

SIX PRINCIPAL TECTONIC PLATES of the lithosphere, the rigid outer shell of the earth, are delineated by the heavy color lines on this world map. The paired arrows indicate whether a plate boundary is convergent or divergent (*see illustration below*).

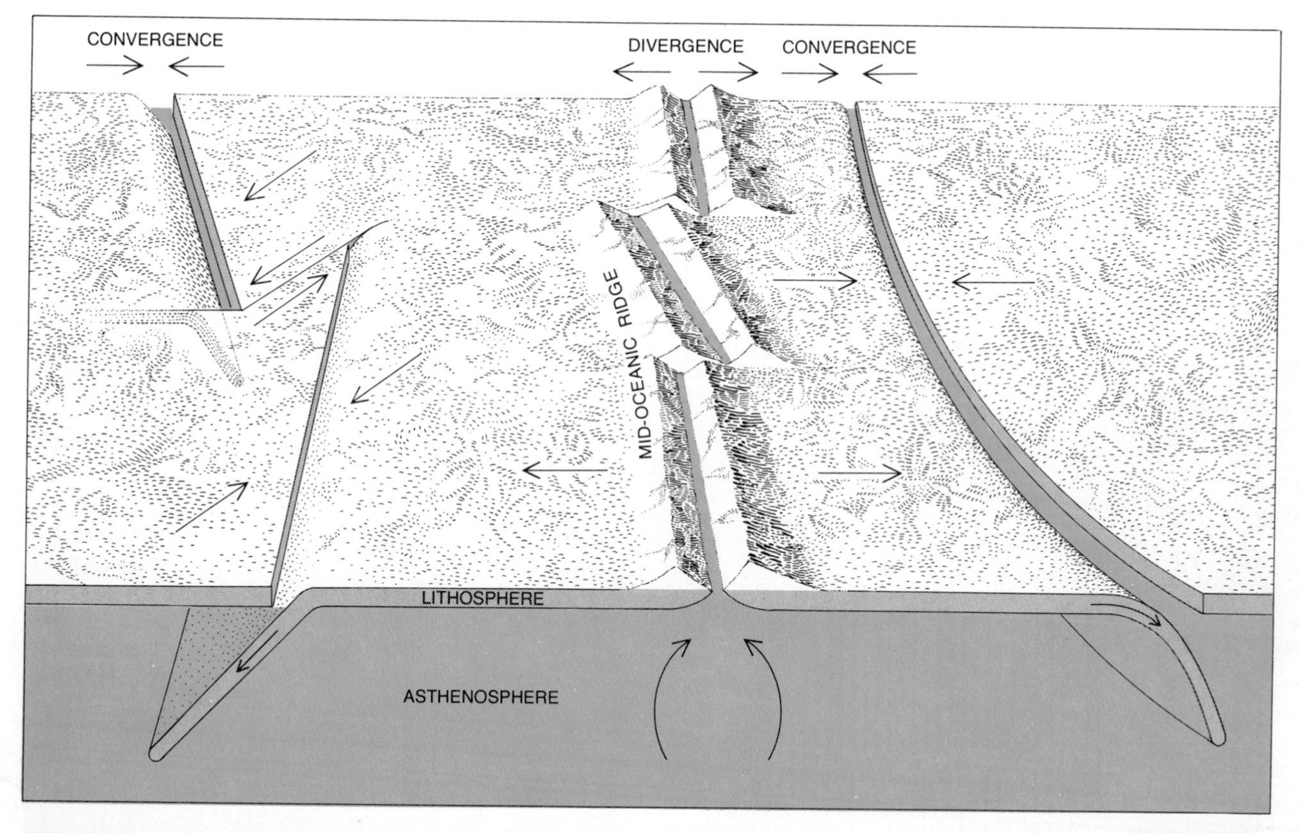

TWO TYPES OF PLATE BOUNDARY are illustrated schematically in this block diagram. The 60-mile-thick lithospheric plates move outward like conveyor belts from the mid-oceanic ridges (divergent plate boundaries) and plunge downward under the deep-sea trenches (convergent plate boundaries). The third major type of plate boundary, not shown here, is the parallel plate boundary.

continuous one [*see illustration on facing page*].

At this point in man's exploration of the oceans it would seem to be too much to expect that it would be possible to make detailed observations on an economically important metallic sulfide deposit that originated at a divergent plate boundary on a submerged mid-oceanic ridge. Yet such a deposit is known and has been extensively studied. The Troodos Massif on the island of Cyprus is interpreted as being a slice of oceanic lithosphere that was formed by the process of sea-floor spreading from a mid-oceanic ridge and was subsequently thrust upward to its present position [*see illustration on page 209*]. The composition and layered sequence of rocks that constitute the Troodos Massif are the

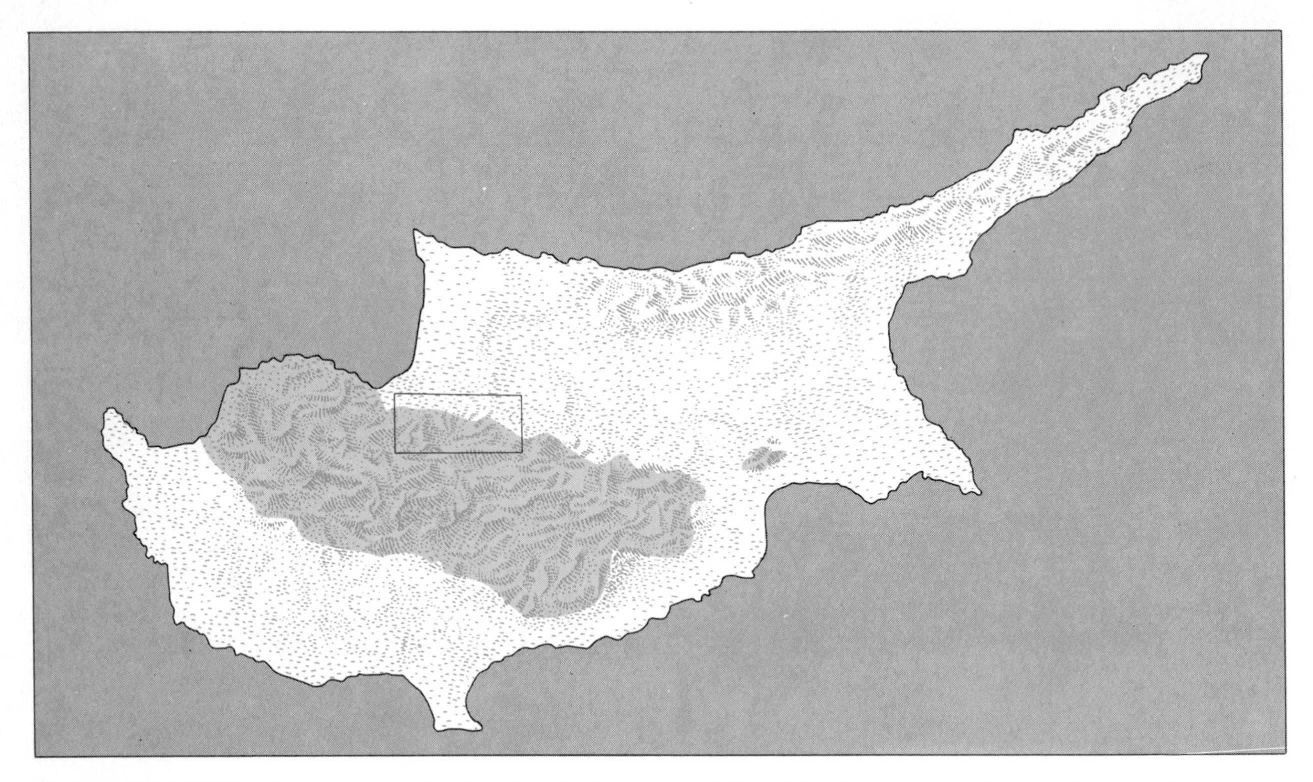

ISLAND OF CYPRUS has been famous for its mineral wealth since Phoenician times. The principal ore bodies are in the uppermost volcanic layers of the Troodos Massif, the total extent of which is indicated by the dark-colored area. The hatched area represents sediments, including alluvium. A geological map of a portion of the Troodos igneous complex (*small rectangle*) is shown below.

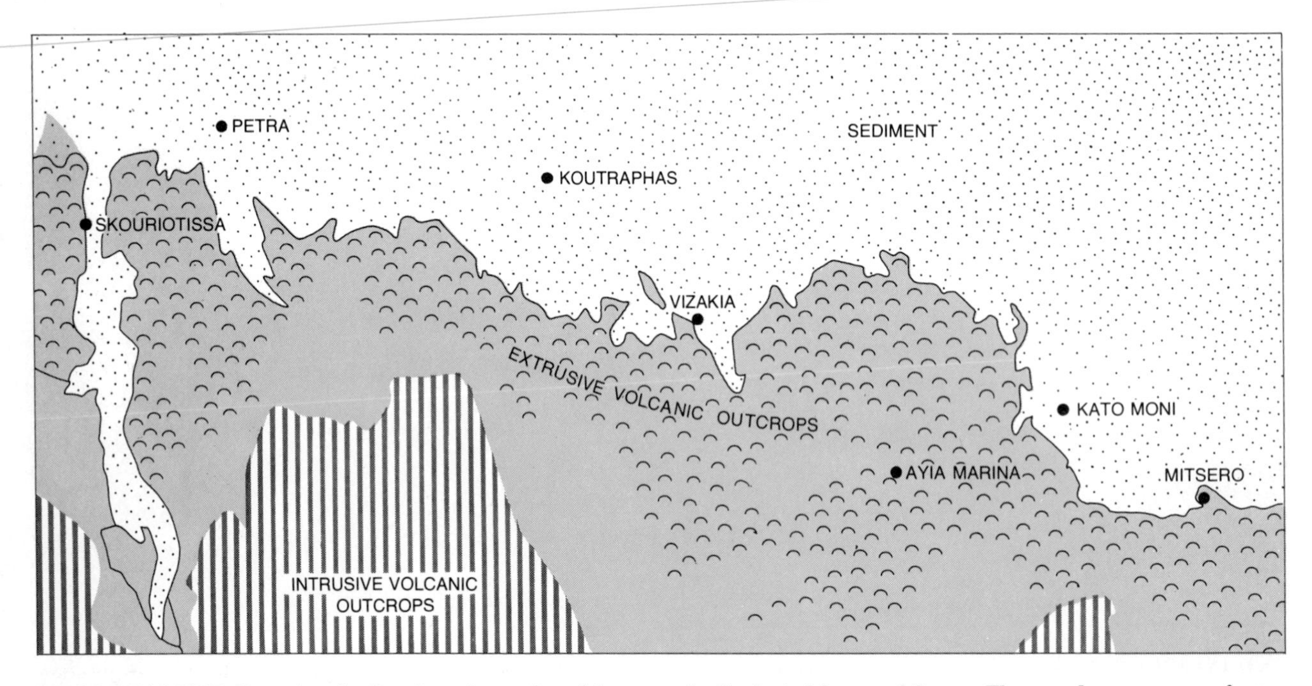

GEOLOGICAL MAP of a region that lies along the northern fringe of the Troodos Massif is based on studies that were undertaken by the Geological Survey of Cyprus. The map shows outcrops of extrusive volcanic rock that incorporate bodies of metallic sulfide ore.

same as those known to underlie the seabed.

Cyprus has long been famous for its mineral wealth. The mining of copper (for which the island is named) was an important industry in Roman and even in Phoenician times. The brilliant green stains of copper sulfides on ancient mine tailings have attracted modern prospectors. Between 1965 and 1970 the average annual exports amounted to about a million tons each of iron pyrites, chromite and gypsum, about 150,000 tons of copper pyrites and 100,000 tons of copper concentrates. The estimated value of the mineral products exported from Cyprus in 1970 amounted to $30 million.

The principal ore bodies are in the uppermost volcanic layers of the Troodos Massif. It has been uncertain whether the Troodos sulfide-ore bodies originated before the upthrust of the Troodos Massif or afterward. In the first instance the ore bodies would be representative of the seabed. In the second the ore bodies would be attributed to special conditions unrelated to the seabed. The sulfide deposits are clearly related to the volcanic rocks in which they occur. Recent studies reveal that iron-rich and manganese-rich sediments interlayered with the volcanic rocks and associated with the ore bodies of the Troodos Massif are chemically identical with those metal-enriched sediments found on active mid-oceanic ridges, indicating that both the sediments and the ore bodies were formed on the sea floor by hydrothermal processes.

The Troodos ore bodies may provide the first firm evidence on the nature of metallic sulfide deposits in ocean basins. The Skouriotissa ore body, for example, is roughly elliptical in plan view, measures approximately 2,000 feet long by 600 feet wide and is lens-shaped in cross section. Its estimated mass is six million tons. The average composition of the ore is 2.25 percent copper (ranging to greater than 5 percent), 48 percent sulfur and 43 percent iron.

The Mavrovouni ore body is also roughly elliptical in plan view, measures approximately 1,000 feet long by 600 feet wide and forms a lens that attains a thickness of 800 feet in cross section. Its estimated mass is greater than 15 million tons. The average composition of the ore is 4.2 percent copper, 48 percent sulfur, 43 percent iron, .4 percent zinc, .25 ounce per ton gold and .25 ounce per ton silver.

Sediments underlying the Skouriotissa ore body, presumably a disintegration product of the pyrite in the ore body, contain 2.12 ounces of gold per ton and 12.96 ounces of silver per ton. Exposed patches of metallic oxides indicate the presence of the ore bodies under the mountainous surface of the Troodos Massif. The Skouriotissa ore body is exploited by underground shafts and the Mavrovouni ore body by strip-mining.

What kind of target for exploration would a Troodos ore body make if it were submerged under thousands of feet of water on the crest or flank of a mid-oceanic ridge? It is unlikely that any of the present exploration methods would be capable of detecting the ore body. The resolution of present geophysical exploration methods will have to be improved in order to detect such an ore body under the sea. Both the exploration methods and the engineering development involved will be costly.

The prerequisites for the accumulation of petroleum consist of a source of organic matter to generate the petroleum, a natural reservoir to contain it

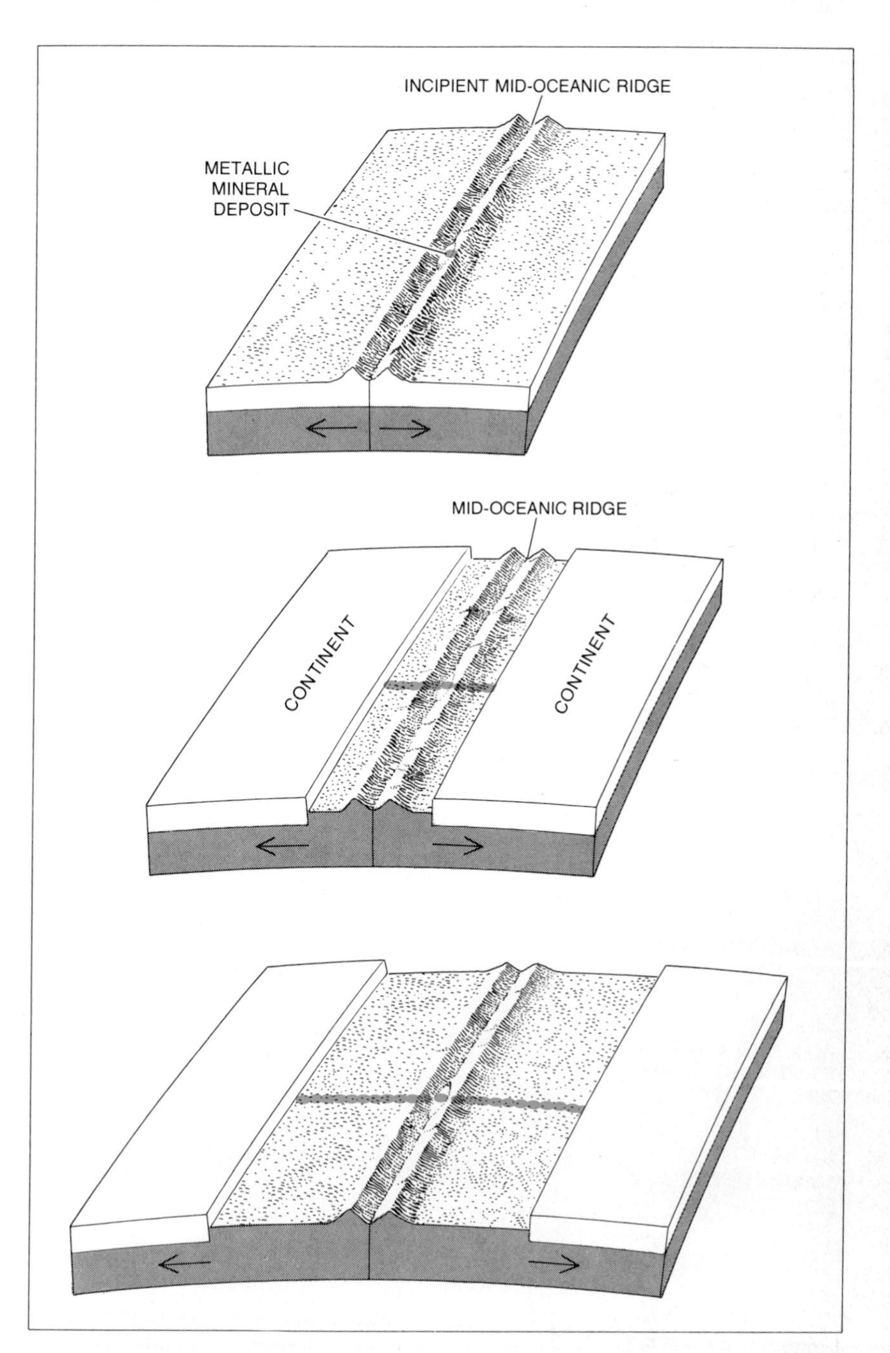

HYDROTHERMAL MINERAL DEPOSIT (*color*) **formed in a hot-brine pool on the axis of a mid-oceanic ridge would be expected to extend in a linear zone from the ridge across the ocean basin to the adjacent continental margins as the ocean basin progressively widens** (*from top to bottom*) **as a consequence of sea-floor spreading from the mid-oceanic ridge.**

and a trap to concentrate its fluid and gas constituents. Petroleum is hydrocarbons derived from the remains of plants and animals. As the progenitor of the petroleum, the organic matter must accumulate in an environment where it is preserved. The preservation of organic matter is favored by an environment that is toxic to life (so that the organic matter is not consumed as food) and deficient in oxygen (so that the hydrocarbon is not decomposed). How do conditions favorable to the accumulation of petroleum relate to convergent and divergent plate boundaries?

Convergent plate boundaries where the oceanic portion of a lithospheric plate plunges under the margin of a continent are characterized by the presence of a deep-sea trench running along their length. A system of deep-sea trenches runs along the entire western margin of North America and South America where the Pacific lithosphere is plunging under the continents. In addition to a deep-sea trench, chains of volcanic islands are present along some convergent plate boundaries; they are located between the trench and the continent. There are many such chains of volcanic islands at the western margin of the Pacific, including the Aleutians, the Kuriles, Japan, the Ryukyus, the Philippines and Indonesia. Other such chains are the Marianas, the South Sandwich Islands and the West Indies. The island chains divide an ocean basin into smaller basins partially enclosed between the islands and the adjacent continent; such basins include the Bering Sea, the Sea of Okhotsk, the Sea of Japan, the Yellow Sea, the East China Sea and the South China Sea.

Both the marginal trenches and the volcanic-island chains create a habitat that is favorable for the accumulation of petroleum in several respects. First, the trenches and island chains act as barriers that catch sediment and organic matter from the continent and the ocean basin. Second, the shape of the trenches and the small ocean basins acts to restrict the circulation of the ocean, so that oxygen is not replenished in the seawater and organic matter is preserved. Third, the accumulation of sediments and the geological structures that develop as a result of the deformation of the sediments by tectonic forces provide reservoirs and traps for the accumulation of petroleum. According to Hollis D. Hedberg of Princeton University, "these marginal semienclosed basins constitute some of the most promising areas in the world for petroleum accumulation."

The development of divergent plate boundaries may also create a habitat favorable for the accumulation of oil, a finding that would open immense possibilities for petroleum resources in the deep ocean basin. When a divergent plate boundary develops under a continent, the continent is rifted in two and the continental fragments are carried apart on a conveyor belt of new lithosphere generated at the divergent plate boundary. As the two continental fragments move apart, a sea forms between them. The surrounding continents act as barriers to restrict the circulation of the sea. As a result organic matter is preserved and, if the evaporation of the seawater exceeds its replenishment, layers of rock salt are deposited along with the organic matter. As the continental fragments continue to move apart and to subside along with the adjacent sea floor,

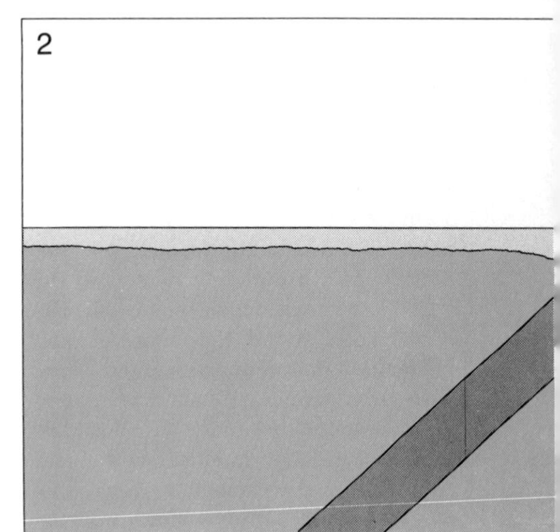

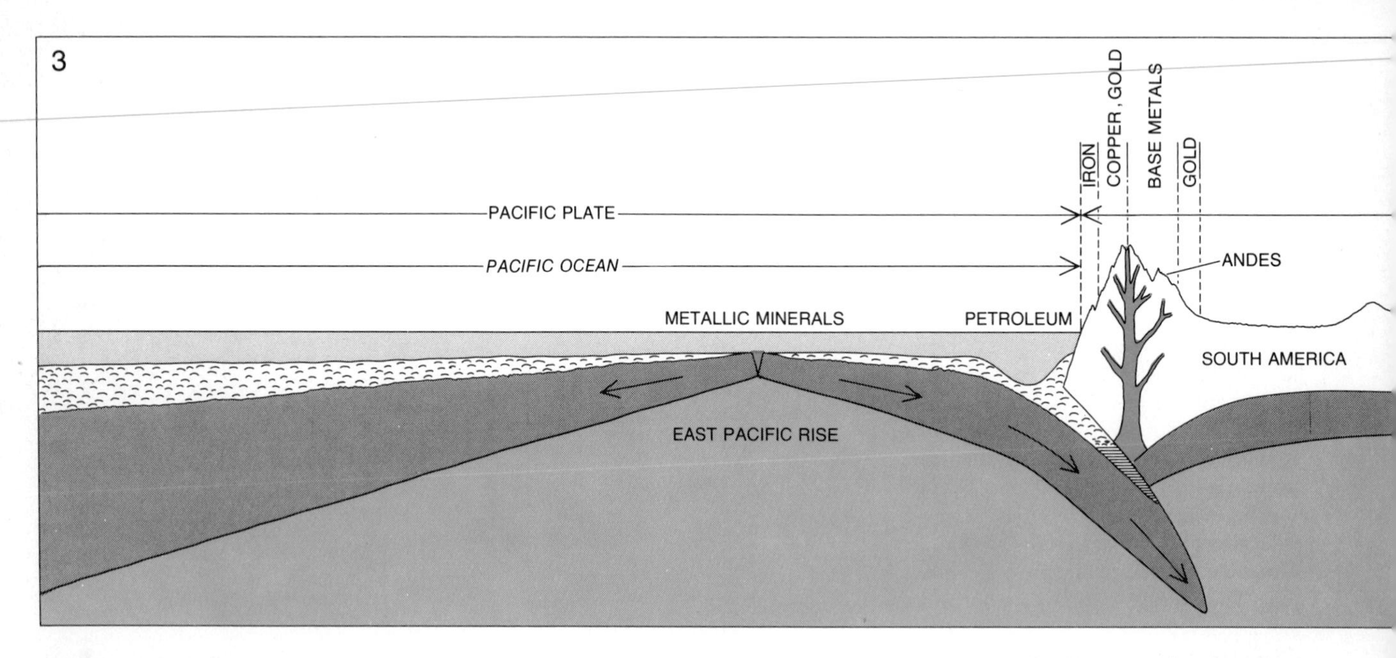

ROLE OF PLATE BOUNDARIES in the accumulation of mineral deposits is exemplified in this sequence of cross-sectional views of the development of the South Atlantic Ocean. The position of Africa is assumed to be stationary throughout the sequence of cross sections. In stage *1* a single ancestral continent, called Pangaea, is rifted into two continents (South America and Africa) about a divergent plate boundary. In stage *2* the oceanic crust created by the process of sea-floor spreading from the divergent plate boundary (a precursor of the Mid-Atlantic Ridge) rafts South America westward and is compensated for by the consumption of oceanic crust at a trench (a convergent plate boundary) that develops to the west of South America. Thick layers of rock salt, organic matter and metallic minerals accumulate in the Atlantic Sea during this early stage of continental drift. In stage *3* continued sea-floor spreading

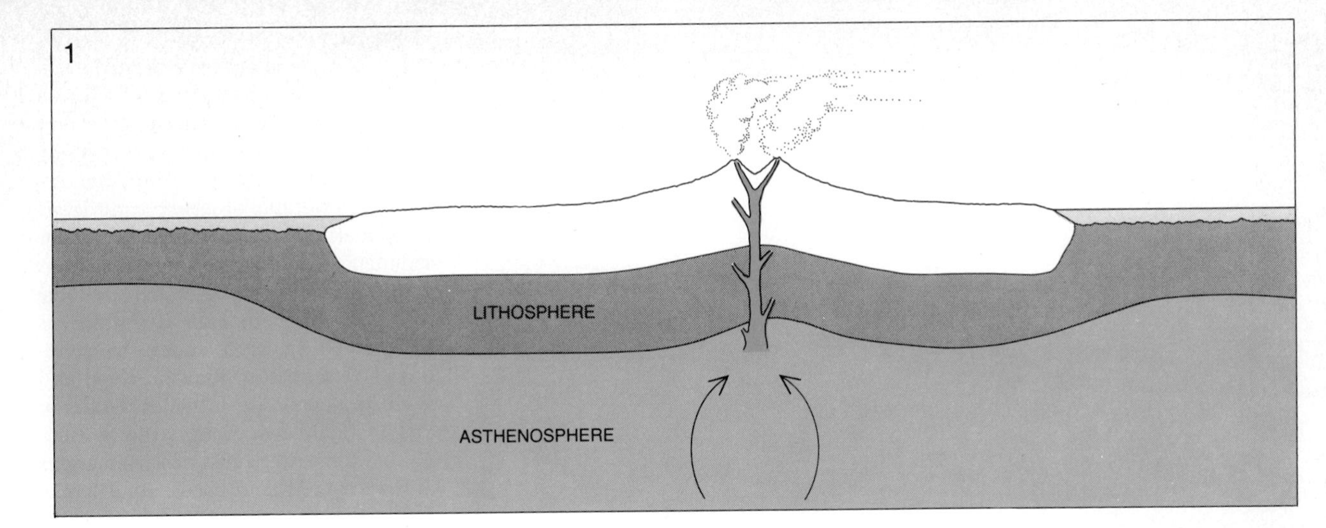

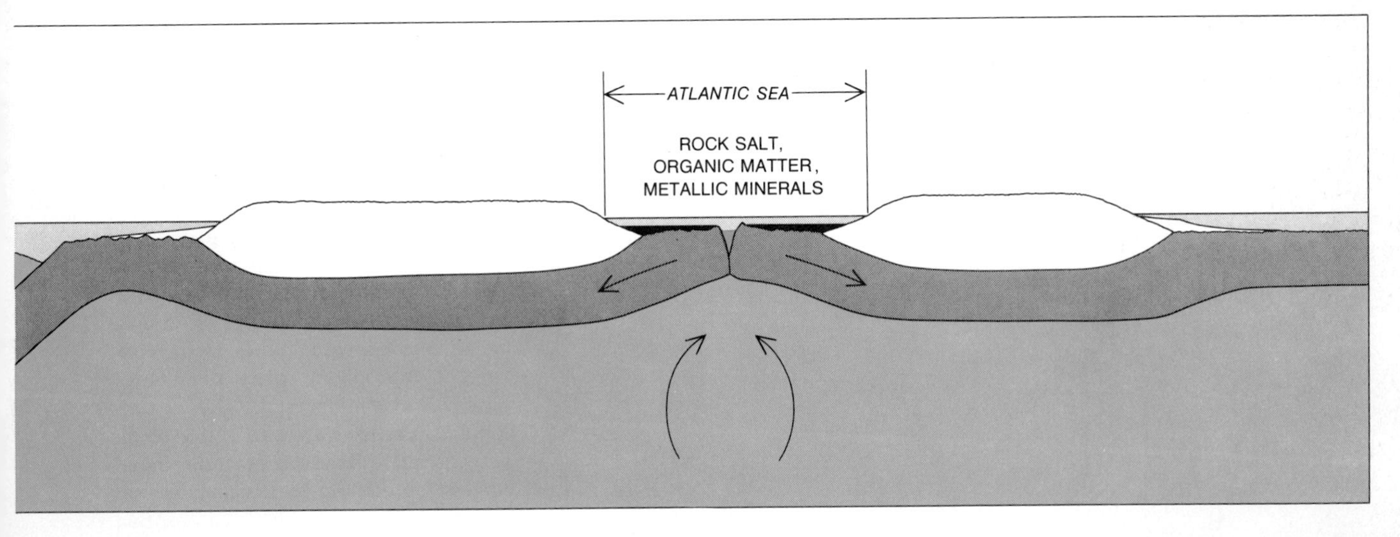

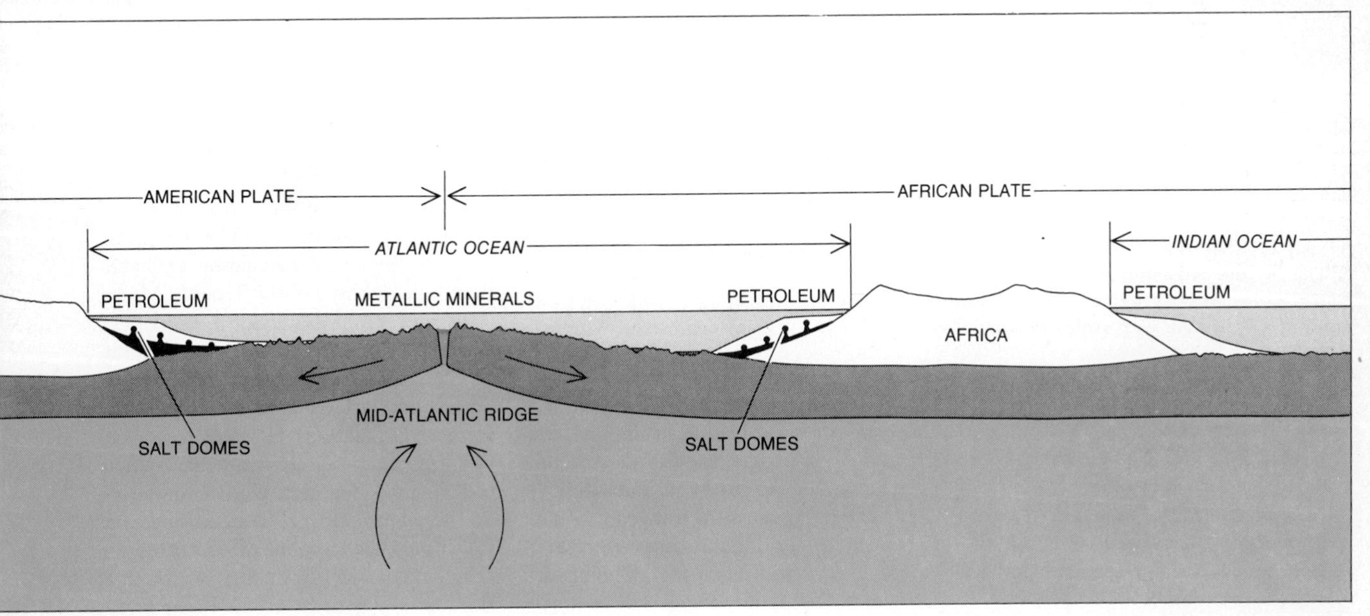

from the Mid-Atlantic Ridge widens the Atlantic into an ocean, rafts South America westward over the trench, reversing the inclination of the trench and producing the Andes mountain chain as a consequence of the deformation that develops at the convergent plate boundary along the western margin of South America. Metallic minerals that are melted from the Pacific plate as it plunges under South America ascend through the overlying crustal layers and are deposited in them to form the metal-bearing provinces of the Andes. Meanwhile in the Atlantic Ocean metallic minerals continue to accumulate about the Mid-Atlantic Ridge. Salt originating in the thick layers of rock salt that have been buried under the sediments of the continental margins rises in large, dome-shaped masses that act to trap the oil and gas that are generated from the organic matter that was preserved in the former Atlantic Sea.

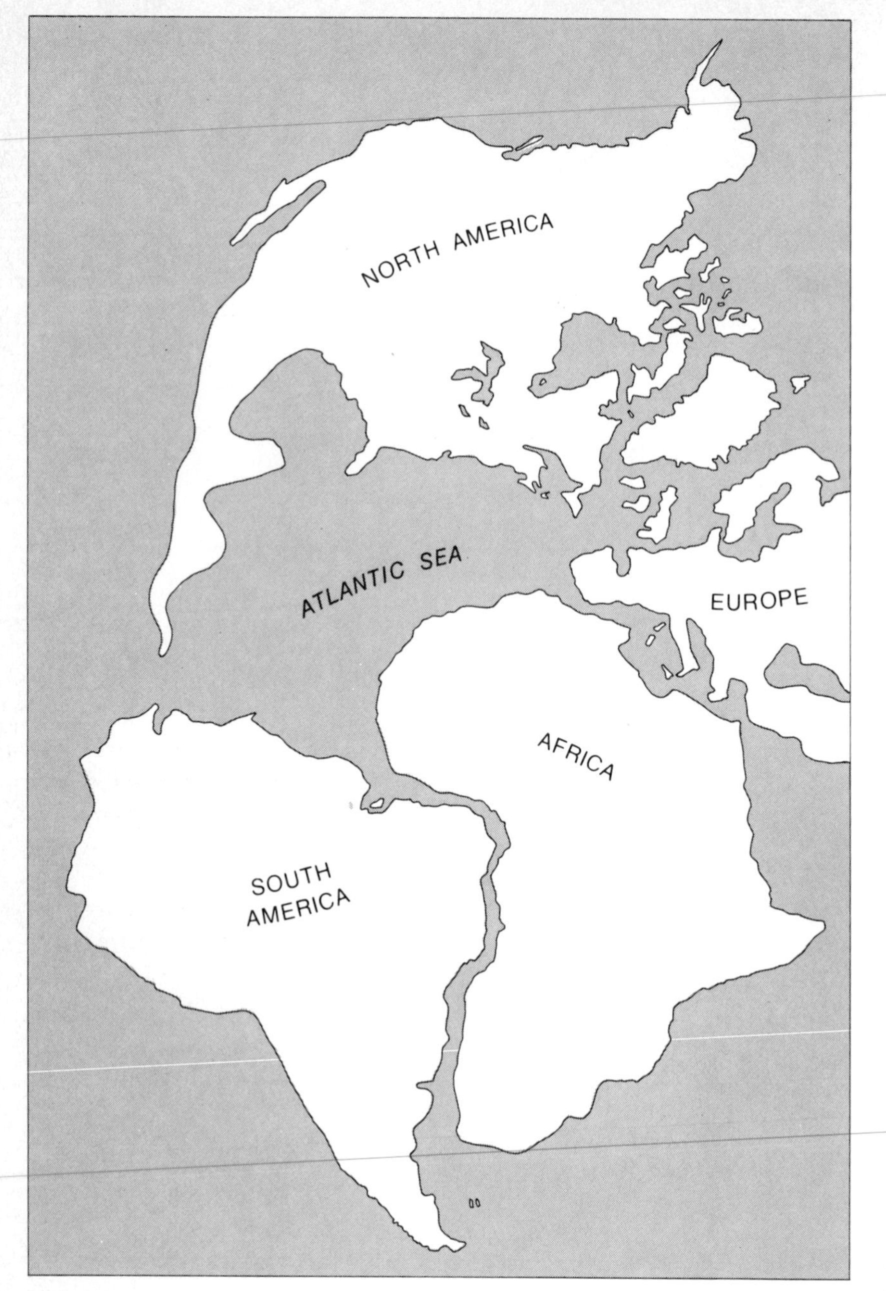

AT AN EARLY STAGE of continental drift the Atlantic was a sea with its circulation restricted by the surrounding continents. As in the present Red Sea, conditions in the Atlantic Sea favored the preservation of organic matter and the deposition of rock salt, leading to the formation of petroleum accumulations under the present continental margins.

the restricted sea becomes an open ocean. The layers of organic matter and salt are buried under sediments. The organic matter subsequently develops into petroleum (by processes that are only partly understood) and the salt forms into dome-shaped masses that act to trap the petroleum.

The Red Sea is an example of a restricted sea formed at an early stage of development of the divergent plate boundary along which Arabia is rifting from Africa. Layers of rock salt up to 17,000 feet thick and organic muds have been found under it. Along both the eastern and western margins of the North Atlantic and the South Atlantic apparent salt domes have been discovered extending seaward from continental shelves to continental rises in water depths of up to 16,500 feet. The occurrence of these salt domes in the deep Atlantic indicates that at an early stage of continental drift the Atlantic was a sea with its circulation restricted by the surrounding continents in their positions at that time [*see illustration above*].

Like the present Red Sea, conditions in the Atlantic Sea favored the preservation of organic matter and the deposition of rock salt. As the Atlantic widened in response to the symmetric creation of new lithosphere by sea-floor spreading from the Mid-Atlantic Ridge, the Atlantic Sea became an ocean and the organic matter and salt were buried under sediments, forming the present margins of the Atlantic Ocean. It is reasonable to expect that petroleum accumulations will extend seaward under the continental shelf, the continental slope and the continental rise to water depths of about 18,000 feet along large portions of both the eastern and western margins of the North Atlantic and South Atlantic. Petroleum may likewise be found in other ocean basins that have grown through the stage of a restricted sea by sea-floor spreading.

In short, the patterns of mineral distribution that are emerging from the conceptual framework provided by the new global tectonics will clearly help to guide man's search for new mineral deposits. Hydrothermal processes have concentrated the majority of known metallic sulfide ore bodies along convergent lithospheric plate boundaries originally at continental margins. Hydrothermal processes are also active at divergent plate boundaries from initial stages (represented by the metallic sulfide deposits accumulating in the Red Sea) to advanced stages (represented by the metal concentration in sediments on mid-oceanic ridges and by possible metallic sulfide deposits of the Troodos Massif type). The Troodos Massif metallic sulfide ore bodies provide an actual example of the type of deposits that can be expected in sea-floor rock generated by mid-oceanic ridges. The confirmation and economic evaluation of metallic sulfide deposits of the Troodos Massif type in ocean basins await technological advances in marine exploration methods.

With regard to petroleum, convergent plate boundaries create conditions that form accumulations in small ocean basins and deep-sea trenches marginal to continents. Divergent plate boundaries, on the other hand, create conditions that favor the development of oil accumulations extending from the continental shelf into the deep ocean basin under the continental rise.

The global patterns of mineral distribution that are emerging from such models can be expected to accelerate the discovery of resources not only on the seabed but also on the continents.

CONCLUSION

The seventeen articles in this anthology have described the recent revolutionary growth in our knowledge of the interior of the earth, its ocean floors, and its behavior. Thirteen years ago, when the first of the articles was published, the concept of plate tectonics had not been formulated and most earth scientists still opposed the theory of continental drift. The identification by Ewing and Heezen of the continuity of the mid-ocean ridge system had not achieved its full impact; Hess's statement on sea-floor spreading had only just been published; and the epoch-making explanation by Vine and Matthews of the striped pattern of magnetic anomalies on the ocean floor had not been written. Many doubted the value of paleomagnetic observations. So recent have been the advances in our understanding of the earth that two-thirds of the articles presented here have been written in the past five years. They mark a complete scientific revolution in our concept of the earth.

Within a short time, the world's oceanographers have mapped the ocean floors and have discovered that their geology is relatively simple and quite unlike that of the continents. Geophysicists, aided by many new seismograph stations, have investigated the interior of the earth and have shown how it enables the surface to break into the moving plates that carry the continents about. The great variety and freedom thus introduced into the study of the earth has resolved many hitherto baffling problems, including the causes of earthquakes and the nature of mountains. Many more problems will be tackled and many details—indeed, most details—await interpretation.

One very large question remains unanswered: What is the nature of the forces that move the plates about? Wegener believed that continents raised above the sea floor on a spinning spheroidal earth should be propelled towards the equator and should drift west. Although such forces exist, they are practically negligible, and most authorities, including the authors of several of the preceding articles, have favored convection currents driven by the earth's radioactive heat. The nature of these currents has proved elusive; if they exist, they must meet many rather strict criteria. Heirtzler has shown, for example, that the rate of opening of the South Atlantic has not changed during the past 80 million years, and that therefore the currents must be extraordinarily steady and persistent. This is difficult to understand when we consider that much of the crest of the mid-ocean ridge must have moved: in Mesozoic time other continents were clustered around Africa and Antarctica, so when breakup began, the mid-ocean ridge system must have lain around their shores—but it does no longer.

The plates are vast and they move steadily from mid-ocean ridges towards trenches many thousands of miles away, which suggests a system of convection currents of broad dimensions. However, as Peter J. Wyllie points out in "The Earth's Mantle," the thickness of the layer in which flow can most readily take place is at most only a few hundred kilometers, and may be much less. It is difficult to conceive of convection in so thin a shell of the earth.

An alternative mechanism that is finding some favor arises from the concept of hot spots—that is, deep-seated upwellings in the mantle—which Burke and Wilson, and Dietz and Holden, mentioned in their articles. Recently, W. J. Morgan proposed that these upwellings of light hot rock lift the lithosphere over them into domes a couple of miles high. Such upwellings may exist under islands along the Mid-Atlantic Ridge, including Jan Mayen, Iceland, the Azores, St. Peter and St. Paul Rocks, Ascension, Tristan da Cunha, and Bouvet. If this is so, is it not possible that the crest of this mid-ocean ridge represents a crack that has formed to join the upwellings, and that the Eurasian, African, and American plates are very slowly sliding off the domes? The same could be true in the other oceans, and all the plates could be sliding together to overlap beneath subduction zones and mountains.

As fantastic as the idea may seem, Anton L. Hales and Wolfgang R. Jacoby have shown that it only requires that the asthenosphere have no permanent strength—gravity will do the rest. If the asthenosphere is indeed weak, such a mechanism could satisfy the requirements for the motion of continents—that is, long-lived, steady motion and moving plates of great lateral and small vertical extent. This mechanism could explain why, in contrast to the uniformity of the South Atlantic and the Pacific, other oceans—for example, the Arctic basins, the Norwegian Sea, Baffin Bay, the Tasman Sea, and the Indian Ocean—have complex patterns and many ridges. If only the East Pacific Rise and the southern part of the Mid-Atlantic Ridge remained steady over their dominant uplifts, the crests of the other ridges would be forced to migrate off their domes. At intervals, the underlying upwellings might break through to create new lengths of active crest and abandon the old ridges; this could account for the complexity of these regions.

A third type of motion, favored by D. Forsyth and S. Uyeda, is that the plates are dragged down by their own greater density and sink into the asthenosphere, causing them to move horizontally. This view gains support from the experiments of D. W. Oldenburg and J. N. Brune. They have shown that if the thin film of frozen wax on the surface of a tank of cooling molten wax is pulled apart, the break will assume the same orthogonal pattern of spreading ridges and transform faults as is seen on mid-ocean ridges. This shows that oceanic plates in tension could form mid-ocean ridges, so that these could, perhaps, be formed because plates are pulled apart as they sink at subduction zones.

E. Kanasewich has just pointed out a hitherto unnoticed symmetry in the pattern of plates. On theoretical grounds it would be neat to observe an axial symmetry in the arrangement of the plates that is parallel to the spin axis of the earth, but this does not exist. Kanasewich points out that the two largest plates, the African and the Pacific, are roughly circular and that their centers lie at opposite ends of a diameter through the equator. The other plates form a roughly symmetrical ring between them. He therefore suggests that the plates are arranged in a pattern with axial symmetry, but that its axis is normal to the earth's spin axis.

Still others believe that there may be two layers of convection cells, one deeper in the mantle than the other, or that the convection takes the form of a hundred or more small plumes rising beneath hot spots, with the rest of the mantle slowly sinking everywhere else to form the return current.

Of course, if D. L. Anderson's suggestion that the oceanic parts of plates increase in density as they age and thicken is true, it follows that in time their behavior would change. This leads to the possibility that different rules apply in different regions.

One can only conclude that the difficult problem of what mechanism (or mechanisms) drives the plates is far from settled. On the other hand, there is now widespread agreement that the plates do move. The pattern of their motions in Phanerozoic time is becoming clear. It is also evident that plates moved during much of Precambrian time, although only fragments of the patterns of motion are yet known.

This acceptance of plate tectonics as the dominant theory of the solid earth's behavior is leading to a complete revision of interpretation (but not of the basic observations) in many branches of geology: in tectonics, stratigraphy, sedimentary geology, petrology, paleontology, and even geomorphology. The profound importance of the new theory for prospecting and for economic geology is now widely accepted. However, the number of unanswered questions shows that much remains to be done. A few years ago, most geological work was rather routine and many scientists in other fields regarded geology with disdain; the acceptance of the theory of plate tectonics has opened many new avenues of research and restored the good name of the earth sciences.

In other scientific revolutions, the periods following the unfolding of a new theory have always proved fruitful and rewarding. In the same way the immediate future of the earth sciences promises to be a period of excitement and discovery.

BIBLIOGRAPHIES

I MOBILITY IN THE EARTH

1. Alfred Wegener and the Hypothesis of Continental Drift

Alfred Wegener. Else Wegener. F. A. Brockhaus, Wiesbaden, 1960.

Memories of Alfred Wegener. J. Georgi in *Continental Drift,* edited by S. K. Runcorn. Academic Press, 1962.

Continental Drift: The Evolution of a Concept. Ursula B. Marvin. Smithsonian Institution Press, 1973.

A Revolution in the Earth Sciences: From Continental Drift to Plate Tectonics. A. Hallam. Oxford University Press, 1973.

2. Continental Drift

Did the Atlantic close and then reopen? J. Tuzo Wilson in *Nature,* Vol. 211, No. 5050, pages 676–681; August 13, 1966.

Our Wandering Continents: An Hypothesis of Continental Drifting. Alex. L. Du Toit. Hafner Publishing Company, Inc., 1937.

Submarine Fracture Zones, Aseismic Ridges and the International Council of Scientific Unions Line: Proposed Western Margin of the East Pacific Ridge. J. Tuzo Wilson in *Nature,* Vol. 207, No. 5000, pages 907–911; August 28, 1965.

Evolution of the Earth. R. H. Dott, Jr., and R. L. Batten. McGraw-Hill Book Company, Inc., 1971.

Convection Plumes in the Lower Mantle. W. J. Morgan in *Nature,* Vol. 230, No. 5288, pages 42–43; March 5, 1971.

The Structure of Scientific Revolutions, 2nd ed. The University of Chicago Press, 1970.

The Origin of the Oceanic Ridges. E. Orowan in *Scientific American,* Vol. 221, No. 5, pages 102–118; November, 1969.

3. Plate Tectonics

Speculations on the Consequences and Causes of Plate Motion. D. P. McKenzie in *Geophysical Journal of the Royal Astronomical Society,* Vol. 18, No. 1, pages 1–32; September, 1969.

Mountain Belts and the New Global Tectonics. John F. Dewey and John M. Bird in *Journal of Geophysical Research,* Vol. 75, No. 14, pages 2625–2647; May 10, 1970.

4. The Earth's Mantle

Ultramafic and Related Rocks. Edited by Peter J. Wyllie. John Wiley & Sons, Inc., 1967.

The Dynamic Earth: Textbook in Geosciences. Peter J. Wyllie. John Wiley & Sons, Inc., 1971.

Continents Adrift: Readings from *Scientific American.* Introductions by J. Tuzo Wilson. W. H. Freeman and Company, 1972.

Earth. Frank Press and Raymond Siever. W. H. Freeman and Company, 1974.

Earth, Mechanical Properties of; Earth, Structure and Composition of; Earth as a Planet; Earthquakes in *The New Encyclopedia Britannica: Macropaedia, Vol. 6.* Encyclopaedia Britannica, Inc., 1974.

Planet Earth: Readings from *Scientific American.* Edited by Frank Press and Raymond Siever. W. H. Freeman and Company, 1974.

5. Hot Spots on the Earth's Surface

PHYSICS AND GEOLOGY. J. A. Jacobs, R. D. Russell and J. Tuzo Wilson. McGraw-Hill Book Company, Inc., 1959.

IS THE AFRICAN PLATE STATIONARY? K. Burke and J. T. Wilson in *Nature*, Vol. 239, No. 5372, pages 387–390; October 13, 1972.

TWO TYPES OF MOUNTAIN. J. Tuzo Wilson and Kevin Burke in *Nature*, Vol. 239, No. 5373, pages 448–449; October 20, 1972.

PLUME GENERATED TRIPLE JUNCTIONS: KEY INDICATORS IN APPLYING PLATE TECTONICS TO OLD ROCKS. Kevin Burke and J. F. Dewey in *The Journal of Geology*, Vol. 81, No. 4, pages 406–433; July, 1973.

RELATIVE AND LATITUDINAL MOTION OF ATLANTIC HOT SPOTS. Kevin Burke, W. S. F. Kidd and J. Tuzo Wilson in *Nature*, Vol. 245, No. 5421, pages 133–137; September 21, 1973.

II SEA-FLOOR SPREADING, TRANSFORM FAULTS, AND SUBDUCTION

6. Sea-Floor Spreading

DEBATE ABOUT THE EARTH: APPROACH TO GEOPHYSICS THROUGH ANALYSIS OF CONTINENTAL DRIFT, 2nd ed. H. Takeuchi, S. Uyeda and H. Kanamori, translated by Keiko Kanamori. Freeman, Cooper & Company, 1967.

SPREADING OF THE OCEAN FLOOR: NEW EVIDENCE. F. J. Vine in *Science*, Vol. 154, No. 3755, pages 1405–1415; December 16, 1966.

SEISMOLOGY AND THE NEW GLOBAL TECTONICS, Bryan Isacks, Jack Oliver and Lynn R. Sykes in *Journal of Geophysical Research*, Vol. 73, No. 18, pages 5855–5899; September 15, 1968.

THE DEEP SEA DRILLING IN THE SOUTH ATLANTIC. A. F. Maxwell, R. P. Von Heezen, K. J. Hsu, J. E. Andrews, T. Saito, S. R. Percival, Jr., E. D. Milow and R. E. Boyle in *Science*, Vol. 168, No. 3935, pages 1047–1059; May 29, 1970.

AGE OF THE NORTH ATLANTIC OCEAN FROM MAGNETIC ANOMALIES. W. C. Pitman, III, M. Talwani and J. R. Heirtzler in *Earth & Planetary Science Letters*, Vol. 11, No. 3, pages 195–200; June, 1971.

DISCONTINUITIES IN SEA-FLOOR SPREADING. P. R. Vogt, O. E. Avery, E. D. Schneider, C. N. Anderson and D. R. Bracey in *Tectonophysics*, Vol. 8, Nos. 4–6, pages 285–317; November, 1969.

MAGNETIZED BASEMENT OUTCROPS ON THE SOUTHEAST GREENLAND CONTINENTAL SHELF. P. R. Vogt in *Nature*, Vol. 226, No. 5247, pages 743–744; May 23, 1970.

7. The San Andreas Fault

RELATIONSHIP BETWEEN SEISMICITY AND GEOLOGIC STRUCTURE IN THE SOUTHERN CALIFORNIA REGION. C. R. Allen, P. St. Amand, C. F. Richter and J. M. Nordquist in *Bulletin of the Seismological Society of America*, Vol. 55, No. 4, pages 753–797; August, 1965.

PROCEEDINGS OF CONFERENCE ON GEOLOGIC PROBLEMS OF SAN ANDREAS FAULT SYSTEM. Edited by William R. Dickinson and Arthur Grantz in *Stanford University Publication: Geological Sciences, Vol. XI.* School of Earth Sciences, 1968.

IMPLICATIONS OF PLATE TECTONICS FOR THE CENOZOIC TECTONIC EVOLUTION OF WESTERN NORTH AMERICA. Tanya Atwater in *Geological Society of America Bulletin*, Vol. 81, No. 12, pages 3513–3535; December, 1970.

EARTHQUAKE PREDICTION AND CONTROL. L. C. Pakiser, J. P. Eaton, J. H. Healy and C. B. Raleigh in *Science*, Vol. 166, No. 3912, pages 1467–1474; December 19, 1969.

EARTHQUAKE PREDICTION AND CONTROL. Allen L. Hammond in *Science*, Vol. 173, No. 3993, page 316; July 23, 1971.

8. Geosynclines, Mountains, and Continent-Building

NORTH AMERICAN GEOSYNCLINES: THE GEOLOGICAL SOCIETY OF AMERICA, MEMOIR 48. Marshall Kay. Geological Society of America, 1951.

COLLAPSING CONTINENTAL RISES: AN ACTUALISTIC CONCEPT OF GEOSYNCLINES AND MOUNTAIN BUILDING. Robert S. Dietz in *The Journal of Geology*, Vol. 71, No. 3, pages 314–333; May, 1963.

MIOGEOCLINES (MIOGEOSYNCLINES) IN SPACE AND TIME. Robert S. Dietz and John C. Holden in *The Journal of Geology*, Vol. 74, No. 5, Part 1, pages 566–583; September, 1966.

CONTINENTAL MARGINS, GEOSYNCLINES, AND OCEAN FLOOR SPREADING. Andrew H. Mitchell and Harold G. Reading in *The Journal of Geology*, Vol. 77, No. 6, pages 629–646; November, 1969.

MOUNTAIN BELTS AND THE NEW GLOBAL TECTONICS. John F. Dewey and John M. Bird in *Journal of Geophysical Research*, Vol. 75, No. 14, pages 2625–2647; May 10, 1970.

THEORIES OF BUILDING OF CONTINENTS. J. Tuzo Wilson in *The Earth's Mantle*, edited by T. F. Gaskell. Academic Press, 1967.

9. The Subduction of the Lithosphere

SEISMOLOGY AND THE NEW GLOBAL TECTONICS. Bryan Isacks, Jack Oliver and Lynn R. Sykes in *Journal of Geophysical Research,* Vol. 73, No. 18, pages 5855–5899; September 15, 1968.

MOUNTAIN BELTS AND THE NEW GLOBAL TECTONICS. John F. Dewey and John M. Bird in *Journal of Geophysical Research,* Vol. 75, No. 14, pages 2625–2647; May 10, 1970.

TEMPERATURE FIELD AND GEOPHYSICAL EFFECTS OF A DOWNGOING SLAB. M. Nafi Toksöz, John W. Minear and Bruce R. Julian in *Journal of Geophysical Research,* Vol. 76, No. 5, pages 1113–1138; February 10, 1971.

EVOLUTION OF THE DOWNGOING LITHOSPHERE AND THE MECHANISMS OF DEEP FOCUS EARTHQUAKES. M. Nafi Toksöz, Norman H. Sleep and Albert T. Smith in *The Geophysical Journal of the Royal Astronomical Society,* Vol. 35, Nos. 1–3, pages 285–310; December, 1973.

III EXAMPLES OF PLATE MOTION

10. The Breakup of Pangaea

CONTINENT AND OCEAN BASIN EVOLUTION BY SPREADING OF THE SEA FLOOR. Robert S. Dietz in *Nature,* Vol. 190, No. 4779, pages 854–857; June 3, 1961.

GEOTECTONIC EVOLUTION AND SUBSIDENCE OF BAHAMA PLATFORM. Robert S. Dietz, John C. Holden, and Walter P. Sproll in *Geological Society of America Bulletin,* Vol. 81, No. 7, pages 1915–1927; July, 1970.

THE FIT OF THE CONTINENTS AROUND THE ATLANTIC. Sir Edward Bullard, J. E. Everett and A. Gilbert Smith in *A Symposium on Continental Drift,* edited by P. M. S. Blackett, Sir Edward Bullard and S. K. Runcorn in *Philosophical Transactions of the Royal Society of London, Series A,* Vol. 258, No. 1088, pages 41–51; October 28, 1965.

PALEOMAGNETISM. E. Irving. John Wiley & Sons, Inc., 1964.

THE FIT OF THE SOUTHERN CONTINENTS. A. Smith and A. Hallam in *Nature,* Vol. 225, No. 5228, pages 139–144; January 10, 1970.

SEA-FLOOR SPREADING. F. J. Vine and H. H. Hess in *The Sea* (Arthur E. Maxwell, General Editor), Vol. 4, Part III, pages 587–622. Wiley-Interscience, Inc., 1970.

THE INDIAN OCEAN AND THE HIMALAYAS—A GEOLOGICAL INTERPRETATION. A. Gansser in *Eclogae Geologicae Helvetiae,* Vol. 59, No. 2, pages 831–848; December, 1966.

11. The Evolution of the Indian Ocean

THE EARTH AS A DYNAMO. Walter M. Elsasser in *Scientific American,* Vol. 198, No. 5, pages 44–48; May, 1958.

THE EAST PACIFIC RISE. Henry W. Menard in *Scientific American,* Vol. 205, No. 6, pages 52–61; December, 1961.

THE PRINCIPLES OF PHYSICAL GEOLOGY. Arthur Holmes. The Ronald Press Company, 1965.

REVERSALS OF THE EARTH'S MAGNETIC FIELD. Allan Cox, G. Brent Dalrymple and Richard R. Doell, in *Scientific American,* Vol. 216, No. 2, pages 44–54; February, 1967.

SEA-FLOOR SPREADING. J. R. Heirtzler in *Scientific American,* Vol. 219, No. 6, pages 60–70; December, 1968.

THE ORIGIN OF THE OCEANIC RIDGES. Egon Orowan in *Scientific American,* Vol. 221, No. 5, pages 102–119; November, 1969.

12. The Evolution of the Pacific

200,000,000 YEARS BENEATH THE SEA: THE STORY OF THE GLOMAR CHALLENGER. Peter Briggs. Holt, Rinehart and Winston, Inc., 1968.

DEEP OCEAN DRILLING WITH GLOMAR CHALLENGER. M. N. A. Peterson and N. T. Edgar in *Oceans,* Vol. 1, No. 5, pages 17–32; June, 1969.

DIACHRONOUS DEPOSITS: A KINEMATIC INTERPRETATION OF POST JURASSIC SEDIMENTARY SEQUENCE ON THE PACIFIC PLATE. B. C. Heezen, I. D. MacGregor, H. P. Foreman, G. Forristal, H. Hekel, R. Hesse, R. H. Hoskins, E. J. W. Jones, A. Kaneps, V. A. Krasheninnikov, H. Okada and M. H. Ruef in *Nature,* Vol. 241, No. 5384, pages 25–32; January 5, 1973.

13. The Floor of the Mid-Atlantic Rift

UNDERSTANDING THE MID-ATLANTIC RIDGE: A COMPREHENSIVE PROGRAM. Edited by J. R. Heirtzler. National Academy of Sciences, 1972.

FLOW OF LAVA INTO THE SEA. James G. Moore, R. L. Phillips, Richard W. Grigg, Donald W. Peterson and Donald A. Swanson in *Geological Society of America Bulletin,* Vol. 84, No. 2, pages 537–546; February, 1973.

FAMOUS: A PLATE TECTONIC STUDY OF THE GENESIS OF THE LITHOSPHERE. J. R. Heirtzler and X. Le Pichon in *Geology*, Vol. 2, No. 6, pages 273–274; 1974.

ATLANTIC OCEAN FLOOR: GEOCHEMISTRY AND PETROLOGY OF BASALTS FROM LEGS 2 AND 3 OF THE DEEP SEA DRILLING PROJECT. Fred A. Frey, Wilfred B. Bryan and Geoffrey Thompson in *Journal of Geophysical Research*, Vol. 79, No. 35, pages 5507–5527; December 10, 1974.

IV SOME APPLICATIONS OF PLATE TECTONICS

14. Continental Drift and Evolution

VERTEBRATE PALEONTOLOGY. Alfred Sherwood Romer. The University of Chicago Press, 1966.

THE AGE OF THE DINOSAURS. Björn Kurtén. World University Library, 1968.

THE AGE OF MAMMALS. Björn Kurtén. Columbia University Press, 1972.

GONDWANALAND REVISITED: NEW EVIDENCE FOR CONTINENTAL DRIFT. Edited by Gerard Piel in *Proceedings of the American Philosophical Society*, Vol. 112, No. 5, pages 307–353; October, 1968.

TRIASSIC TETRAPODS FROM ANTARCTICA: EVIDENCE FOR CONTINENTAL DRIFT. D. H. Elliot and others in *Science*, Vol. 169, No. 3950, pages 1197–1201; September 18, 1970.

15. Continental Drift and the Fossil Record

THE ORIGIN OF CONTINENTS AND OCEANS. Alfred L. Wegener. Methuen, 1966.

THE BEARING OF CERTAIN PALAEOZOOGEOGRAPHIC DATA OF CONTINENTAL DRIFT. Anthony Hallam in *Palaeogeography Palaeoclimatology Palaeoecology*, Vol. 3, pages 201–241; 1967.

CONTINENTAL DRIFT AND THE EVOLUTION OF THE BIOTA ON SOUTHERN CONTINENTS. Allen Keast in *The Quarterly Review of Biology*, Vol. 46, No. 4, pages 335–378; December, 1971.

ATLAS OF PALAEOBIOGEOGRAPHY. Edited by Anthony Hallam. American Elsevier Publishing Co., 1974.

16. Plate Tectonics and the History of Life in the Oceans

DYNAMICS IN METAZOAN EVOLUTION. R. B. Clark. Oxford University Press, 1964.

GLOBAL TECTONICS AND THE FOSSIL RECORD. James W. Valentine and Eldridge M. Moores in *The Journal of Geology*, Vol. 80, No. 2, pages 167–184; March, 1972.

EVOLUTIONARY PALEOECOLOGY OF THE MARINE BIOSPHERE. J. W. Valentine. Prentice-Hall, Inc., 1973.

A REVOLUTION IN THE EARTH SCIENCES: FROM CONTINENTAL DRIFT TO PLATE TECTONICS. A. A. Hallam. Oxford University Press, 1973.

17. Plate Tectonics and Mineral Resources

EXPLORATION METHODS FOR THE CONTINENTAL SHELF: GEOLOGY, GEOPHYSICS, GEOCHEMISTRY. P. A. Rona. National Oceanic and Atmospheric Administration Technological Report ERL 238-AOML 8, U.S. Government Printing Office, 1972.

SULFIDE ORE DEPOSITS IN RELATION TO PLATE TECTONICS. F. Sawkins in *Journal of Geology*, Vol. 80, pages 377–397; 1972.

HYDROTHERMAL MANGANESE IN THE MEDIAN VALLEY OF THE MID-ATLANTIC RIDGE. Martha R. Scott, Robert B. Scott, Andrew J. Nalwalk, P. A. Rona and Louis W. Butler in *EOS: American Geophysical Union Transactions*, Vol. 54, No. 4, page 244; April, 1973.

INDEX